Astrophysics and Space Science Library

Volume 460

The Astrophysics and Space Science Library is a series of high-level monographs and edited volumes covering a broad range of subjects in Astrophysics, Astronomy, Cosmology, and Space Science. The authors are distinguished specialists with international reputations in their fields of expertise. Each title is carefully supervised and aims to provide an in-depth understanding by offering detailed background and the results of state-of-the-art research. The subjects are placed in the broader context of related disciplines such as Engineering, Computer Science, Environmental Science, and Nuclear and Particle Physics. The ASSL series offers a reliable resource for scientific professional researchers and advanced graduate students.

More information about this series at http://www.springer.com/series/5664

Roberto Mignani • Andrew Shearer • Agnieszka Słowikowska • Silvia Zane
Editors

Astronomical Polarisation from the Infrared to Gamma Rays

Springer

Editors
Roberto Mignani
INAF/IASF - Milan
Milan, Italy

Andrew Shearer
School of Physics
National University of Ireland Galway
Galway, Ireland

Agnieszka Słowikowska
Faculty of Physics, Astronomy
and Informatics
Center for Astronomy
Nicolaus Copernicus University
Toruń, Poland

Silvia Zane
Mullard Space Science Laboratory
University College London
Dorking, UK

ISSN 0067-0057 ISSN 2214-7985 (electronic)
Astrophysics and Space Science Library
ISBN 978-3-030-19714-8 ISBN 978-3-030-19715-5 (eBook)
https://doi.org/10.1007/978-3-030-19715-5

Cover illustration: An artist's concept of NASA's Imaging X-ray Polarimetry Explorer observatory. Credit: Ball Aerospace.

This Springer imprint is published by the registered company Springer Nature Switzerland AG.
The registered company address is: Gewerbestrasse 11, 6330 Cham, Switzerland

Preface

Electromagnetic radiation from celestial sources contains a wealth of fundamental information about its properties, e.g. the physical processes that produced the radiation, the properties of the source environment, and the mechanisms influencing the radiation propagation in the interstellar medium (ISM). In particular, photons carry information on their energy and arrival direction, on the orientation of the electric and magnetic field vectors, and, ultimately, on the geometry of the space-time in which they propagate. This information can be decoded using a variety of techniques and deploying different instrument setups.

Polarimetry, long regarded as a powerful technique, is at the same level as photometry and spectroscopy, leading to an understanding of astrophysical processes. In particular, the role of polarimetry is crucial to investigate the effects of light propagation in the vacuum or the ISM and, in combination with spectroscopy, to obtain information on the physical processes that produce electromagnetic radiation and their dependence on, for example, the magnetic field strength and the source emission geometry.

Spectroscopy, photometry, and timing are techniques traditionally applied at all wavelengths; however, polarimetry has been mainly restricted to the radio and in the optical domains. Whilst all radio telescopes are built as polarimeters by default, in all other ranges of the electromagnetic spectrum, dedicated instruments are needed to measure the polarisation state of the radiation, thus making polarimetry less widely available. In spite of that, the optical polarimetry community has always been at the forefront, and it is now well established worldwide and extremely productive, with over 4000 refereed publications issued since 2000, covering an amazingly wide variety of subjects. This breakthrough result has only been possible, thanks to a continuous and restless effort pursuing new research projects in optical polarimetry. Instrumentation for optical/infrared polarimetry is now becoming a standard resource of many telescopes worldwide, including the ESO Very Large Telescope, the 10.4 m Gran Telescopio Canarias (GTC) in La Palma, or the 4 m Canadian French Hawai Telescope (CFHT) on Mauna Kea.

Optical polarimetry has been pivotal in studying a variety of incoherent non-thermal emission mechanisms from several types of sources: compact objects (neutron stars, white dwarfs, black holes), interacting binaries, active galactic nuclei, gamma-ray bursts, etc. Some of these studies investigate the properties of the extreme magnetic fields around neutron stars and map the magnetic and electric fields in diffuse emission environments, such as supernova remnants (SNR) or pulsar wind nebulae (PWN). On a broader scale, the study of optical polarisation has been key to study the thermal emission and physical properties of basically all astronomical sources, from stars across the whole Hertzsprung-Russell diagram, to solar system bodies, such as planets and asteroids, to the Sun itself. Last, but not least, optical polarimetry has been crucial in many experiments in fundamental physics, such as testing for the first time the effects of vacuum birefringence in extreme magnetic fields and verifying the predictions of quantum electrodynamics theories, a result which strengthened the connection between astrophysics and fundamental physics. This heritage makes the astronomical community well prepared for pursuing further advances in the field of optical polarimetry, providing a very fertile ground to implement new research ideas and observing techniques.

The importance of polarimetry in the multidimensional approach to modern astrophysics has been now recognised in other energy domains, but X- and γ-ray polarimetry is yet to be fully established. Polarimetry X-ray measurements of the Crab Nebula and pulsar were carried for the first time 40 years ago, and very little has been done since then. Thanks to the breakthrough improvement in technology, X-ray polarimetry missions have been studied and proposed and are foreseen to launch in the immediate future. The NASA's *Imaging X-ray Polarimetry Explorer* (*IXPE*) is now on the way to be launched, and the Chinese *enhanced X-ray Timing and Polarimetry* mission (*eXTP*) is expected to continue this exploratory effort with many powerful instruments. In the soft γ-rays, the *e-ASTROGAM* mission had been proposed to ESA, with a downsized version, *All-Sky-ASTROGAM*, being preselected. In parallel, a similar mission concept, *AMEGO*, has been proposed to NASA. At least a mission with soft γ-ray polarimetry capability is likely to be launched in the next decade. A new era is coming, where polarimetry will be a cornerstone of modern multiwavelength astronomy. Efforts also continue around the world to develop technology to enable polarimetric missions capable of operating in the soft and hard X-ray band. We are then on the way to be capable of performing unique polarimetric observations of astrophysical sources at multi-wavelength, a possibility that will open a whole new window on our understanding of cosmic sources. In parallel, there is a substantial theoretical effort in the scientific community to develop models and predictive tools to exploit the future multiwavelength data.

The massive revolution that this will bring to our understanding of the Universe and of fundamental physics has motivated the idea to publish this book. The idea originated from a meeting that we organised as part of the European Week of Astronomy and Space Science that took place in Liverpool in Spring 2018 under

Special Session 20 entitled "Multi-wavelength Polarimetry."[1] Based on the success of this session, we developed this into a book fully dedicated to the subject.

This book covers a vast variety of topics, starting from the fundamentals of polarimetry (including the definition of the Stokes parameters, Mueller matrices, the Poincaré sphere, and Jones calculus), moving to the description of the various instruments and techniques used in the broad range of the electromagnetic spectrum with an accent on calibration methods, and proceeding to the application of polarimetry to the study of a multitude of astronomical sources, from the Sun to small bodies of our Solar System, stars, compact objects, interstellar medium, AGNs, and GRBs. As indicated by its title, this book focuses on polarimetry measurements at wavelengths other than radio, spanning from the consolidated field of optical/infrared polarimetry to X- and γ-ray polarimetry with future facilities. Whilst the importance of radio polarimetry in astronomy can hardly be over-estimated, the implementation of this technique on different radio telescope is a topic of its own, whereas infrared-to-γ-ray polarimetry is covered here for the first time. This book is aimed at the whole scientific community, including students and early and as experienced researchers who will find here an updated view of the multiwavelength polarimetry approach to the study of our Universe.

We hope that the readers will adopt this book as a state-of-the-art reference on astronomical polarimetry, both in terms of science and observational techniques, and will appreciate the potentials of polarimetry across the electromagnetic spectrum, which will be fully exploited over the next decade. Hopefully, those that have never been involved in astronomical polarimetry will find here their inspiration for new scientific projects and start their adventure on the Q-U plane.

The editors warmly thank all the contributing authors for their work and dedication in writing the book chapters and for keeping the text up to date by including a discussion of the most recent achievements in the field of astronomical polarimetry.

Milan, Italy
Galway, Ireland
Toruń, Poland
London, UK
March 2019

Roberto Mignani
Andrew Shearer
Agnieszka Słowikowska
Silvia Zane

[1] https://eas.unige.ch//EWASS2018/session.jsp?id=SS20.

Acknowledgement

Agnieszka Słowikowska acknowledges the support through the Polish National Science Centre grant 2017/25/B/ST9/02805.

Contents

Contributors

Editors

Roberto Mignani INAF – Istituto di Astrofisica Spaziale e Fisica Cosmica Milano, Milano, Italy

Janusz Gil Institute of Astronomy, University of Zielona Góra, Zielona Góra, Poland

Andrew Shearer Centre for Astronomy, School of Physics, National University of Ireland, Galway, Ireland

Agnieszka Słowikowska Faculty of Physics, Astronomy and Informatics, Centre for Astronomy, Nicolaus Copernicus University, Toruń, Poland

Silvia Zane Mullard Space Science Laboratory, UCL, Holbury St Mary, Dorking, Surrey, UK

Authors

Irina Belskaya Institute of Astronomy, V.N. Karazin Kharkiv National University, Kharkiv, Ukraine

Andrei Berdyugin Department of Physics and Astronomy, University of Turku, Turku, Finland

Stefano Bagnulo Armagh Observatory and Planetarium, College Hill, Armagh, UK

Ilaria Caiazzo Department of Physics and Astronomy, University of British Columbia, Vancouver, BC, Canada

Alberto Cellino INAF – Osservatorio Astrofisico di Torino, Pino Torinese, Italy

Enrico Costa IAPS/INAF, Roma, Italy

Alice Harding NASA Goddard Space Flight Center, Greenbelt, MD,USA

Jeremy Heyl Department of Physics and Astronomy, University of British Columbia, Vancouver, BC, Canada

Swetlana Hubrig Leibniz-Institut für Astrophysik Potsdam (AIP), Potsdam, Germany

Fiona Kenny School of Physics, National University of Ireland, Galway, Ireland

Shiho Kobayashi Astrophysics Research Institute, Liverpool John Moores University, Liverpool, UK

Philippe Laurent CEA/DRF/IRFU/DAp, CEA Saclay, Gif sur Yvette,France
Laboratoire APC, Paris Cedex 13,France

Anny-Chantal Levasseur-Regourd LATMOS, Sorbonne Université, CNRS, UVSQ, Paris, France

Frédéric Marin Université de Strasbourg, CNRS, Observatoire Astronomique de Strasbourg, UMR 7550, Strasbourg, France

Mark L. McConnell Space Science Center, University of New Hampshire, Durham, NH, USA
Department of Earth, Oceans, and Space, Southwest Research Institute, Durham, NH, USA

Christian Peest Universität Potsdam, Potsdam, Germany

Vilppu Piirola Department of Physics and Astronomy, University of Turku, Turku, Finland

Juri Poutanen Department of Physics and Astronomy, University of Turku, Turku, Finland
Nordita, KTH Royal Institute of Technology and Stockholm University, Stockholm, Sweden
Space Research Institute of the Russian Academy of Sciences, Moscow, Russia

Markus Schöller European Southern Observatory, Garching, Germany

Tariq Shahbaz Instituto de Astrofísica de Canarias (IAC), La Laguna, Tenerife, Spain
Departamento de Astrofísica, Universidad de La Laguna (ULL), La Laguna, Tenerife, Spain

Ralf Siebenmorgen European Southern Observatory, München, Germany

Paolo Soffitta IAPS/INAF, Roma, Italy

David Orozco Suárez Instituto de Astrofísica de Andalucía, Glorieta de la Astronomía, Granada, Spain

Vincent Tatischeff Centre de Sciences Nucléaires et de Sciences de la Matière (CSNSM), CNRS/IN2P3, Univ. Paris-Sud, Université Paris-Saclay, Orsay, France

Sascha Trippe Seoul National University, Seoul, Korea

Roberto Turolla Department of Physics, University of Padova, Padova, Italy

Mullard Space Science Laboratory, UCL, Holbury St. Mary, Dorking, Surrey, UK

Chapter 1
When Polarimetry Made History

Sascha Trippe

Abstract Polarimetry is a well-known and well-established method of astronomy, and has been so for more than a century. Over the last four decades, the astronomical literature has featured about one polarimetry-related work for every eight spectroscopy-related works. Important mile stones of astronomical polarimetry have been the observations of the solar corona since the 1840s; the theory and observation of polarised light reflected from solar system bodies since the 1900s; the discovery of astrophysical synchrotron radiation in the 1950s; the measurements of galactic magnetic fields since the 1950s; space telescope observations of polarised X-radiation in the 1970s; the viewing-angle unification scheme of active galactic nuclei since the 1980s; and the hunt for B-mode polarisation in the cosmic microwave background.

1.1 Introduction

Polarimetry is a well-known and well-established method of astronomy, and has been so for more than a century. Quantifying its impact on astronomy is not straightforward, but we may assess the importance of polarisation studies by tracing the evolution of polarimetry-related publications with time (while keeping the many caveats of bibliometrics in mind). Figure 1.1 provides a comparison of the number of polarimetry-related works per decade to the number of spectroscopy-related works, with the latter ones also being used as a proxy for the overall activity in observational astronomy. Both numbers have increased drastically over time, with a sharp jump in the 1960s and 1970s, and a slower but still exponential growth since then (left panel of Fig. 1.1). The ratio of polarimetry and spectroscopy works (right panel of Fig. 1.1) seems to suggest the presence of three distinct phases in the history of astronomical polarisation studies: an "early era" until about the end of the 1950s,

---

S. Trippe (✉)
Seoul National University, Gwanak-gu, Seoul, Korea
e-mail: trippe@astro.snu.ac.kr

R. Mignani et al. (eds.), *Astronomical Polarisation from the Infrared to Gamma Rays*, Astrophysics and Space Science Library 460,
https://doi.org/10.1007/978-3-030-19715-5_1

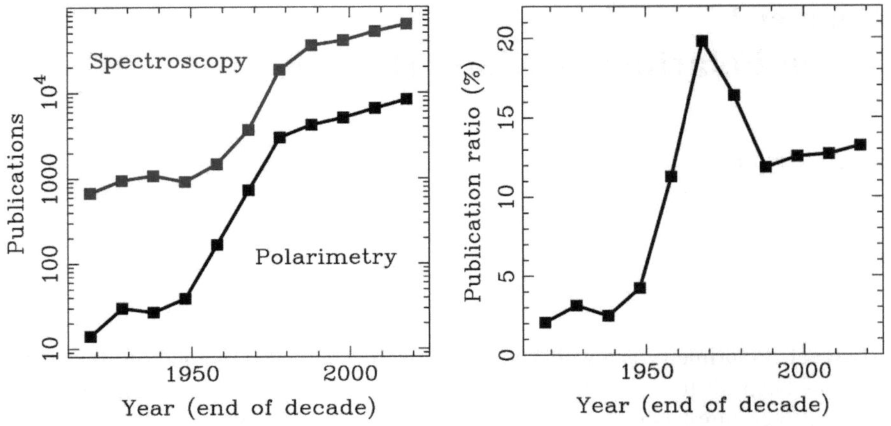

Fig. 1.1 Polarimetry in the astronomical literature. *Left:* Number of refereed works per decade containing the key words "polarisation" or "polarimetry" in their abstracts (black data), compared to the number of works with the key words "spectrum" or "spectroscopy" (grey data). Decades are 1909–1918, 1919–1928, ..., 2009–2018. Bibliometric data are from the SAO/NASA Astrophysics Data System. *Right:* Number of polarimetry-related works divided by number of spectroscopy-related works, in units of per cents, using the data from the left panel

when the ratio was around 3% initially; a "golden era" in the 1960s and 1970s, when the ratio peaked at about 20%; and a "modern era" with the ratio being around 13% throughout the last four decades. Evidently, the importance of polarimetry has grown with time but is still well below the prominence of spectroscopy.

Ever since the first polarimetric observations of celestial bodies in the mid of the nineteenth century, polarimetry has provided important astrophysical insights (Trippe 2014). At various points in the history of astronomy, observations of polarised light led to insights that were crucial for the understanding of astrophysical phenomena. A selection of such key events, from the nineteenth to the twenty-first century, is presented in the following.

1.2 1840s: The Solar Corona

In 1860, the Swedish physicist Erik Edlund published a report titled *On the polarisation of the light of the corona during total solar eclipses* in the journal *Astronomische Nachrichten* (Edlund 1860). This paper was itself a recap of a report Edlund had submitted to the Swedish Royal Academy of Sciences in 1851; therein, he described how he observed the solar corona during a solar eclipse that occurred in Sweden in the same year. Edlund was not the first to do so: the French astronomer François Arago performed polarimetric observations of the solar corona in 1842, making use of a total solar eclipse in that year.

For their observations, Arago, Edlund, and others used a device they called a *polariscope* (today, we would call it a polarimeter) that used two optical elements: a calcite analyser and a quartz plate; both minerals are birefringent. In 1842, Arago found the light from the solar corona to be linearly polarised. Even though, it seems that Edlund was the first to determine the orientation of polarisation. During the 1851 solar eclipse, he pointed his instrument at four different locations of the solar corona: to the top, bottom, left, and right side of the Moon. His observations told him that the light was polarised everywhere (which was not a surprise anymore) and that the orientation of the polarisation was radial—the plane of polarisation pointed away from the centre of the Moon (and thus the Sun) everywhere.

The observations by Arago, Edlund, and others were initially unexplained. Only in the 1870s, Rayleigh and Schuster realised that such a linear polarisation pattern can result from scattering of light at small particles within the corona (Inhester 2015). At that time, it was unclear what the nature of the scattering particles was; nevertheless, quantitative calculations were possible by assuming Rayleigh scattering at particles much smaller than the wavelength of the scattered light. Only in 1896, Thomson postulated the existence of electrons which were, eventually, identified as the scattering particles responsible for the polarisation of the solar corona. From a modern point of view, the observations by Arago, Edlund, and others were groundbreaking—indeed, they indicated the need for "new physics". As it turned out, this new physics was the existence of electrons and the structure of the atom.

1.3 1900s: The Surfaces of Solar System Objects

Whenever sunlight is reflected at a solid surface, it becomes partially linearly polarised due to scattering polarisation. By geometry, the observed degree of polarisation depends on the relative orientation of Sun, reflecting body, and observer; the maximum polarisation will be observed when sunlight is scattered by 90°. The maximum degree of polarisation depends on the properties of the surface material.

In 1905, the Russian physicist Nikolay Umov discovered (Umov 1905) what is today referred to as *Umov effect* or *Umov's law*. This law states: "The maximum degree of polarisation of the reflected light is the inverse of the albedo of the reflecting material" (for degree of polarisation and albedo both being normalised to the interval [0, 1]). The Umov effect originates from different levels of depolarisation at surfaces of different reflectance. At surfaces composed of dark material, only photons that do not interact with the surface particles multiple times are reflected; those that undergo multiple scatterings are absorbed eventually. At surfaces composed of bright material, even photons undergoing multiple scatterings escape eventually. The brighter the material, the larger the number of scatterings a photon can experience without being absorbed. The larger the number of scatterings, the more the orientation of the linear polarisation of an individual photon becomes randomised, meaning that the linear polarisation averages out partially when taking the average over many photons.

The Umov effect can be exploited to study the surface composition of solar system bodies (Zubko et al. 2011). In the beginnings of space flight in the 1950s, the first target was the Moon. Using both polarisation and albedo measurements, it was possible to estimate the composition of the surface regolith. For bodies whose size is not known, as is the case for many asteroids, the Umov effect is the best, and sometimes only, way to estimate the albedo of the body. Measuring the degree of linear polarisation at several wavelengths provides the albedo as function of wavelength, which makes it possible to constrain the chemical composition of the surface material (Cellino et al. 2005). Variations of the albedo with time provide information on the rotation of the body.

The Sun is not the only source of electromagnetic radiation we use to study solar system objects. Another option is the use of planetary radar (Ostro 1993). In this case, however, we cannot make use of the Umov effect because the angle between inbound and outbound light rays is close to zero from the point of view of the target. Umov's law breaks down because the degree of scattering polarisation becomes marginal regardless of albedo. We can, however, exploit the fact that we can control the polarisation of the emitted radio light. Usually, radar uses radiation with frequencies of a few GHz that is fully circularly polarised. When reflected at a "perfect mirror", i.e., an ideal dielectric surface, the sense of polarisation (i.e., the sign of Stokes V) is inverted. Thus the light emitted and received by the radio antenna has opposite sense of polarisation—if the target surface is indeed an ideal mirror. If the light experiences multiple scatterings at the target surface, the sense of polarisation may be flipped multiple times. This will lead to some of the reflected light being in the *same* polarisation state as the emitted light. Accordingly, the ratio of the fluxes of light in the two possible polarisation states (in radar astronomy nomenclature: the ratio of the two radar cross sections) provides information on the structure of the surface of the target. For most solar system objects, less than 25% of the reflected light is flipped to the same polarisation state as the emitted light; for the icy moons of Jupiter, this number is as high as about 50%.

1.4 1950s: Synchrotron Radiation

The name of Jan Hendrik Oort is a household name for astronomers, although different astronomers might give different answers when asked for the reason. To planetary scientists, his postulate of the *Oort cloud*, a spherical shell of minor bodies around the solar system that supplies the inner solar system with comets, was groundbreaking. Galactic astronomers admire Oort for his insights on the structure and dynamics of the Milky Way: he showed that the Galaxy is rotating, that the Sun is located at least 5.9 kpc (modern value: 8 kpc) away from the galactic centre, that the mass of the Milky Way is on the order of 100 billion solar masses, and that the galactic disc is surrounded by a spherical stellar halo. For those interested in astronomical polarimetry, Oort is remembered for his studies of the Crab nebula.

In the 1950s, radio observations found multiple sources of radiation that could only be explained by a non-thermal origin. One of these sources was the Crab nebula, the remnant of the supernova SN 1054. In 1956, Oort and his colleague Theodore Fjeda Walraven published a long (26 pages in two-column format) paper where they outlined their findings from polarimetric observations of this nebula (Oort and Walraven 1956). For their observations, they used a photometer equipped with a high-quality photomultiplier tube that was to be mounted at the 13$''$ telescope of the Leiden Observatory. Initially, they planned to use the instrument for a measurement of the integrated brightness of the Crab nebula; however, bad weather had prevented a full commissioning of the photometer. Then, in 1954, Soviet astronomers reported that they had found indication that the light from the Crab nebula was linearly polarised. After becoming aware of this Oort and Walraven turned to their, as yet untested, photometer. They replaced several of the colour filters in the instrument's filter wheel with pieces of polaroid, thus turning the photometer into a polarimeter. (Polaroids are sheets of organic polymers with long-chain molecules which are aligned by stretching and are very efficient dichroic polarisers.) When turning their new polarimeter to the Crab nebula, they indeed found a strong linear polarisation of about 19%.

Oort and Walraven realised immediately that a non-standard emission mechanism was required in order to understand the high degree of polarisation; especially, bremsstrahlung, the mechanism favoured by astronomers until then, seemed to be ruled out because it would not produce linear polarisation. They concluded that the best explanation was synchrotron radiation—the radiation emitted by electrons gyrating around magnetic field lines at relativistic speeds. In synchrotron radiation, the magnetic field lines introduce a preferred orientation of the planes of oscillation of electromagnetic waves, leading to substantial (as high as 75% in idealised cases) linear polarisation. From this insight, Oort and Walraven went on to calculate several physical parameters. They found the (spatially averaged) magnetic field strength in the nebula to be about 1 mG and concluded that the gyrating electrons radiate their energy away too fast to be powered by the original supernova explosion. As we know today, there is indeed another, persistent, source of energy in the nebula: the Crab pulsar, a magnetised neutron star with a spin period of 33 ms, that was discovered with the Arecibo radio observatory in 1968.

The observation of Oort and Walraven established synchrotron radiation in astronomy. The various observations of the 1950s and the resulting discussion led to a detailed understanding of synchrotron radiation from astronomical sources (Ginzburg and Syrovatskii 1965). We now understand that synchrotron radiation is the mechanism behind radiation from multiple types of emitters: large-scale radio radiation from galaxies, neutron stars, γ-ray bursts, and active galactic nuclei (see Fig. 1.2 for an example).

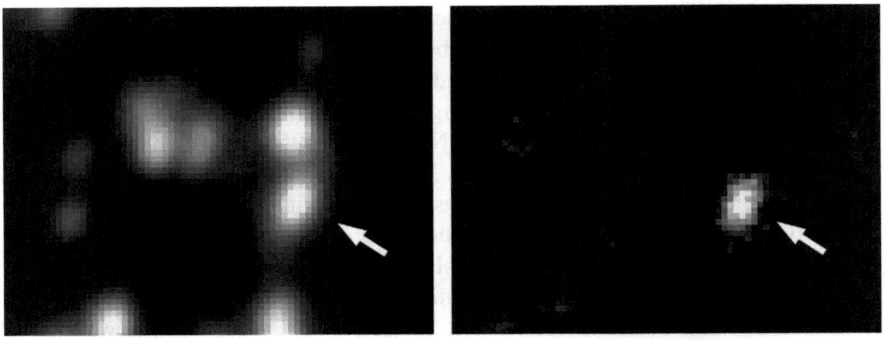

Stokes *I* Stokes Q

Fig. 1.2 Near-infrared images illustrating the linear polarisation of synchrotron radiation. Both panels show the central arcsecond (about 8000 astronomical units) of the Milky Way. The total intensity (Stokes I) image shows multiple stars plus Sagittarius A* (Sgr A*, marked by an arrow), the central supermassive black hole which appears starlike. The Stokes Q image leaves only Sgr A* whose radiation is linearly polarised; the unpolarised stars have been subtracted. (Images: S. Trippe/MPE Garching)

1.5 1950s: Galactic Magnetic Fields

In 1953, the *Astrophysical Journal* published a paper that was remarkable for several reasons. It was unusually short, barely covering three pages. Its authors were a past and a future Nobel laureate, Enrico Fermi and Subrahmanyan Chandrasekhar. Fermi was well-known for his work on nuclear and particle physics, Chandrasekhar for his work on stellar, statistical, and mathematical physics. In this paper, they joined forces to tackle a problem rather different from their usual lines of work: the galactic magnetic field. The result was the aforementioned article (Chandrasekhar and Fermi 1953), which may well be one of the most efficient scientific papers ever written.

Chandrasekhar and Fermi noted that the presence of magnetic fields in the Milky Way had recently been deduced from the linear polarisation of star light. As stars themselves are unpolarised, the best explanation for this polarisation was the presence of an interstellar magnetic field that followed the galactic spiral arms (or, at least, the local arm wherein the Sun is located). The magnetic field can force a preferred orientation of elongated interstellar dust grains. As a result, the absorption of star light passing through the dust would be different parallel and perpendicular to the magnetic field lines, the light becomes linearly polarised, with the direction of polarisation being parallel to the local arm, or almost parallel. Chandrasekhar and Fermi observed that the polarisation measured from multiple stars follows the orientation of the local arm globally but shows significant scatter, of approximately 10°, about the mean orientation. They arrived at the insight that this scatter in the polarisation angles can be used to measure the strength of the galactic magnetic field. Assume a magnetic field permeating the interstellar medium. The interstellar gas is turbulent. If the magnetic field is strong, so is the corresponding magnetic force, the field will be "stiff", and the field lines will be straight regardless of the turbulent motions of the gas. If the magnetic field is weak, the field lines will be dragged around by the turbulent gas and become rather wavy. With this being said, Chandrasekhar and Fermi required a few lines of math to conclude that the magnetic field strength is proportional to the square root of the gas density, proportional to the root-mean-squared gas speed due to turbulence, and inversely proportional to the root-mean-square of the polarisation angles. This formula quantifies what is known nowadays as the *Chandrasekhar–Fermi effect*. From the data available to them, they found a magnetic field strength in the local arm of about 7.2 μG.

Remarkably, Chandrasekhar and Fermi were not finished yet. They went on to provide a second, independent estimate of the magnetic field strength. This estimate was based on the need for an equilibrium between gravitational pressure in the gas on the one hand and kinetic (from the turbulent motion) and magnetic pressure on the other hand. This time, they arrived at a magnetic field strength of about 6 μG, which agreed with their first estimate to within 20%.

Today, the presence of galactic magnetic fields is well established, although their origin is still a matter of debate. The Chandrasekhar–Fermi effect continues to be an important tool for observations of galactic magnetic fields, along with Faraday rotation of light from pulsars or extragalactic background sources, weak Zeeman effect line splitting especially of H I lines, synchrotron radiation, and dust scattering polarisation.

1.6 1970s: Polarised X-Ray Light

Measurements of polarisation at optical wavelengths are (relatively) simple because it is straightforward to manipulate the light falling into the telescope with standard optical elements. Once the wavelengths are shorter than the typical distances between atomic layers of crystals, i.e., a few nanometers (corresponding to photon energies higher than a few hundred eV), optical elements like lenses, mirrors, or prisms become fully transparent; we have to look out for other analysis methods. X-ray telescopes therefore employ Bragg reflection at small angles to collimate and focus light. In addition, X-ray light is absorbed by the atmosphere (a fact that is unfortunate for astronomers but very fortunate for life on Earth); thus the rise of X-ray astronomy had to await the beginning of the space age.

X-ray polarimetry remains challenging until the present day but has been attempted occasionally. In 1975, NASA launched the 8th *Orbiting Solar Observatory* (*OSO-8*). Else than its name suggests, the satellite carried several instruments dedicated to celestial X-ray sources other than the Sun. Among them were two X-ray polarimeters. Details aside, each polarimeter featured a mosaic graphite crystal that was tilted by 45° relative to the line of sight to the target source. Via Bragg diffraction, the incident light was reflected onto a proportional counter located at a right angle relative to the line of sight. By geometry, only the component of the light polarised parallel to the surface of the crystal was reflected to the counter. In order to obtain the full linear polarisation information for a target, the satellite was rotated around an axis parallel to the line of sight to the target. If the light from the target is linearly polarised, the intensity measured as function of rotation angle θ should show a characteristic $\cos^2\theta$ pattern, with the amplitude of the variation encoding the degree of polarisation and the phase corresponding to the polarisation angle.

In March 1976, *OSO-8* was pointed at the Crab nebula for six days. The polarimeters were able to measure the intensity as function of rotation angle at 2.6 and 5.2 keV from the first and second order Bragg diffraction, respectively. The intensity indeed showed a clear $\cos^2\theta$ pattern: the Crab nebula was linearly polarised, with a degree of polarisation of about 17% and a polarisation angle of about 160° at both energies—values that are very similar to those found at radio and optical wavelengths. The conclusion was inevitable: the X-ray radiation from the Crab nebula is synchrotron radiation, as it is at lower energies (Weisskopf et al. 1976).

OSO-8 was the first and as yet only X-ray satellite equipped with a polarimeter. (In addition, there were a few experiments with ballistic rockets.) The next one is supposed to be the *Imaging X-ray Polarimetry Explorer* (*IXPE*) which is scheduled to launch in the year 2021. *IXPE*'s polarimeters are composed of gas pixel detectors which exploit the fact that the direction of the emission of photoelectrons from the gas atoms depends on the orientation of the linear polarisation of the light (see Fig. 1.3). *IXPE* will be the first polarimetric X-ray telescope in 45 years. Just in time: the principal investigator of *IXPE* is Martin Weisskopf—the very same person who analysed the *OSO-8* data.

1.7 1980s: What Are Active Galactic Nuclei?

In 1918, the American astronomer Heber Curtis studied images of the elliptical galaxy M 87 when he noted a "curious straight ray ... apparently connected with the nucleus by a thin line of matter" in the outskirts of the galaxy. As we know today, Curtis' report marks the discovery of jets from active galactic nuclei (AGN). In 1918, neither Curtis nor anyone else could explain his observation; indeed, it would take four more years until Edwin Hubble recognised that M 87 is a galaxy outside the Milky Way. Already in 1908, Edward Fath and Vesto Slipher had found that the spectrum of the galaxy—which they called a "spiral nebula"—NGC 1068 shows several emission lines, at odds with the absorption line spectra of galaxies usually observed. In 1943, Carl Seyfert reported (Seyfert 1943) the observation of such emission lines from the nuclear regions of a total of six spiral galaxies; nowadays, such galaxies are aptly referred to as "Seyfert galaxies". From the 1950s on, radio observations revealed a large number of extragalactic radio sources that were dubbed "quasi-stellar radio sources", or quasars, initially. In the 1980s, a growing number of observations had led to the assembly of a whole zoo of AGN types: Seyfert galaxies, quasars, radio galaxies, quasi-stellar objects (QSOs), blazars, type

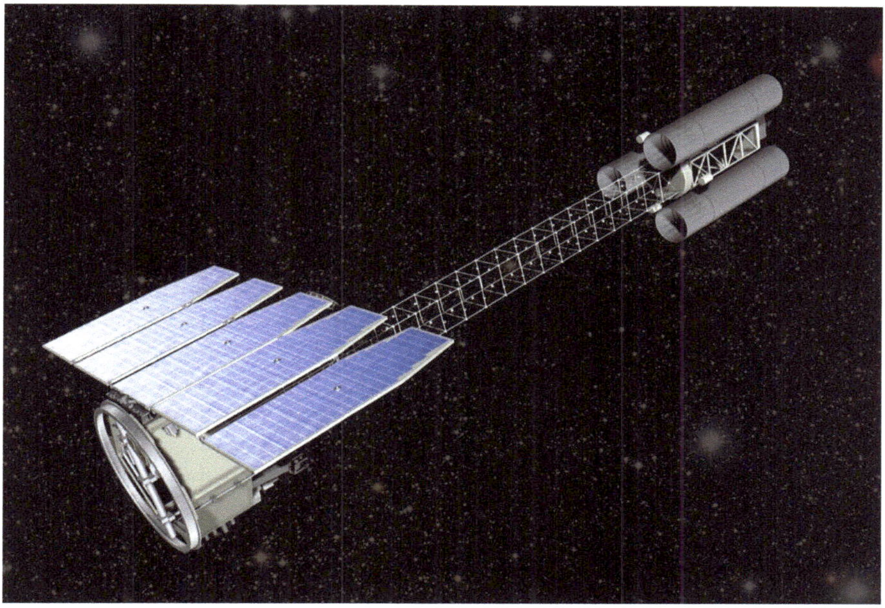

Fig. 1.3 Artist's view of the *Imaging X-ray Polarimetry Explorer* (*IXPE*). Three identical X-ray telescopes are placed at the right end of the carrier structure. They focus the light onto three detector units located in the structure below the solar panels. (Image: NASA [https://ixpe.msfc.nasa.gov/about/index.html])

I and type II AGN, flat-spectrum radio quasars, and others. What was lacking was an understanding how all these phenomena were related.

In the early 1980s, Robert Antonucci and Joseph Miller turned their attention to NGC 1068. This object was known as a Seyfert 2 galaxy, i.e., a Seyfert galaxy showing (relatively) narrow emission lines in the spectrum of its centre. In addition, the spectral continuum from the nucleus was linearly polarised, with degrees of polarisation around 16%. Using the 3-m Lick Shane Telescope, Antonucci and Miller combined both spectroscopy and polarimetry into a full spectropolarimetric study of NGC 1068 (Antonucci and Miller 1985). The key idea behind their study was to compare two spectra: one spectrum of the total light and one spectrum of the polarised component of the light. To their surprise, the two spectra were notably different. In their own words: "Plots of polarised flux [...] eliminate the effects of unpolarised starlight, and suppress low-polarisation line emission. [Our figures] show the polarised flux, and the appearance of these figures is surprising. They look like the spectra of a Seyfert Type 1 object! Broad symmetric Balmer lines and Fe II emission are clearly seen". In other words: NGC 1068 was a Seyfert 2 galaxy in total flux but a Seyfert 1 galaxy (with broader emission lines) in polarised flux. It was as if the observations by Antonucci and Miller had revealed two galaxies in one.

In a way, they had. Antonucci and Miller realised that the polarisation of the emission lines was caused by scattering. They proposed "a mysterious central continuum source surrounded by a very thick absorbing disk". The broad-line emission region is located very close to the central continuum source, which is, as we know today, the accretion disc around a black hole. The "thick absorbing disk", rather a donut-shaped torus of gas and dust, surrounds the broad-line region and obscures it from direct view. The gas emitting the narrow emission lines is located further outward outside the torus and directly visible. Therefore, spectroscopy of the total flux reveals a narrow-line Seyfert 2 spectrum. The broad-line emission is not lost however: a fraction of the radiation can be scattered off the surrounding torus into the line of sight. Scattered light becomes partially polarised. When taking the spectrum of the polarised flux, the result is a broad-line Seyfert 1 spectrum. Antonucci and Miller thus realised that Seyfert 1 and Seyfert 2 galaxies were intrinsically identical; which spectrum dominates the total flux depends only on the viewing angle.

The observations of NGC 1068 were a crucial step to the "viewing angle unification scheme" of AGN (Urry and Padovani 1995). As we know today, AGN are powered by the accretion of gas into supermassive black holes and surrounded by broad-line and narrow-line emission regions plus a dusty torus and, in some cases, produce plasma jets perpendicular to the accretion disc. Almost the entire zoo of AGN can be explained with just three parameters: the viewing angle, the accretion rate, and the luminosity of jets.

1.8 2010s: Cosmic Inflation

Since Penzias and Wilson discovered the cosmic microwave background (CMB) in 1965, the conclusion has become inevitable that the universe originated in some sort of singular event: the "big bang". More and more detailed observations of the CMB revealed that it corresponds to blackbody radiation with an average temperature of 2.73 K (and thus a spectrum peaking at wavelength of 1.1 mm) and spatial fluctuations with amplitudes on the order of tens of μK. The angular power spectrum of these temperature fluctuations is a key observation for constraining the "standard" ΛCDM cosmology (although one may ask how sustainable a cosmological model can be whose main ingredients, the cosmological constant Λ and cold dark matter [CDM], we do not understand). In addition, the CMB is linearly polarised, with different polarisation patterns arising from different mechanisms. "Electric field mode" ("E-mode") polarisation is caused by Thomson scattering of CMB light toward the observer; the electric field vectors form ring-like patterns around local temperature (and thus flux) maxima. "Magnetic field mode" ("B-mode") polarisation is caused by perturbations of spacetime by gravitational fields; the electric field vectors form curly patterns centred at local temperature maxima. B-mode polarisation can be caused by gravitational lensing of CMB light by foreground masses. Another possible source are gravitational waves propagating through the CMB. A possible source of such "primordial" gravitational waves is cosmic inflation—a cornerstone of modern cosmology. Measuring the B-mode polarisation of the CMB therefore corresponds to a quest for the origins of the universe.

When speaking of measuring the CMB polarisation, this is easier said than done. CMB polarisation is weak, with temperature amplitudes on the order of μK for all polarisation components combined. The B-mode polarisation from primordial gravitational waves is about ten times weaker. Fortunately, the different polarisation components dominate at different angular scales: E-mode polarisation and B-mode polarisation from gravitational lensing both occur on scales of about 0.2°. B-mode polarisation from gravitational waves is most prominent on angular scales of around 2°. The inflationary B-mode polarisation can be separated from the other effects by means of spatial filtering: when observing only polarisation on angular scales of about 2°, there is a good chance that one has found the B-mode polarisation caused by primordial gravitational waves. One class of instruments that provides spatial filtering is interferometers: the minimum angular scale observed is given by the ratio of observing wavelength and maximum extension of the instrument (maximum distance between a pair of telescopes). The maximum angular scale follows from the ratio of wavelength and the minimum distance of two telescopes in the interferometer array. Indeed, several CMB polarimeters were designed as interferometers sensitive to angular scales below 1°.

In 2010, an international team of scientists began observations of the CMB using a telescope located at the south pole. This instrument, BICEP2, combined techniques from optical and radio astronomy. The telescope itself was a 26-cm aperture refracting telescope using polyethylene lenses to collect radio light. The light was focused onto an array of 16 × 16 dipoles, each measuring about half a millimetre in length, located in the focal plane. The dipoles were oriented horizontally and vertically (relative to a pre-selected reference direction), with neighbouring dipoles being orthogonal. Because dipoles are sensitive only to light parallel to their orientation, they are polarimeters by construction. With its dipole array, BICEP2 was sensitive to two orthogonal linear polarisations at once. By turning the telescope by 45° around its longitudinal axis, the diagonal linear polarisations could be covered. Combining measurements obtained at rotation angles of 0° and 45°, the Stokes parameters I, Q, and U could be measured. The telescope scanned the sky at a rate of about 1.5° per second. By means of *temporal* filtering of the output signals from the dipole array, the instrument could be made sensitive to a specific angular scale: keeping only signals varying with frequencies between 0.2 Hz and 1.2 Hz made BICEP2 sensitive to angular scales between 0.5° and 3°.

BICEP2 monitored its target field from early 2010 till late 2012, collecting 590 days of data. In March 2014, the BICEP2 collaboration held a press conference to announce the discovery of a strong polarisation signal at exactly the angular scale expected for B-mode polarisation from primordial gravitational waves (Ade et al. 2014). This signal provided the first observational support for cosmic inflation. It was a spectacular discovery, a triumph of observational astrophysics and theoretical cosmology.

It was not within days from the first announcement of the results, several experts raised doubts, noting, among other things, that the BICEP2 data were inconsistent with the data collected by other instruments at smaller angular scales and/or different frequencies, including data from the PLANCK satellite. Eventually, the BICEP2 collaboration had to face a hard truth: the signal they had been monitoring for almost three years did not originate from the CMB, but from interstellar dust (Flauger et al. 2014). The quest for CMB polarisation from gravitational waves is still open—until the present day.

As the examples mentioned above show, observing and understanding polarised light from various cosmic sources has been crucial for progress in astrophysics. Importantly, some discoveries were entirely unexpected. Therefore, polarisation will keep playing an important role for sure; new windows like the polarised γ-ray sky (e.g., with *AMEGO*) or the polarised sub-millimetre radio sky (with ALMA and other facilities) promise new insights.

References

Ade PAR et al (2014) Phys Rev Lett 112:241101
Antonucci RRJ, Miller JS (1985) Astrophys J 297:621
Cellino A et al (2005) Icarus 179:297
Chandrasekhar S, Fermi E (1953) Astrophys J 118:113
Edlund E (1860) AN 52:305
Flauger R, Hill JC, Spergel DN (2014) J Cosmol Astropart Phys 8:39
Ginzburg VL, Syrovatskii SI (1965) Annu Rev Astron Astrophys 3:297
Inhester B (2015) arXiv:1512.00651
Oort JH, Walraven TF (1956) Bull Astron Inst Neth 12:285
Ostro SJ (1993) Rev Mod Phys 65:1235
Seyfert CK (1943) Astrophys J 97:28
Trippe S (2014) J Kor Astron Soc 47:15
Umov NA (1905) Phys Z 6:674
Urry CM, Padovani P (1995) Publ Astron Soc Pac 107:803
Weisskopf MC et al (1976) Astrophys J 208:L125
Zubko E et al (2011) Icarus 212:403

Chapter 2
Introduction to Polarisation and Polarimetry

Fiona Kenny

Abstract The polarisation state of light can often be described using the Stokes parameters or Jones vector. Changes that can occur to the polarisation state as it propagates through a system can be modelled mathematically using Mueller or Jones algebra. The Poincaré sphere is another useful tool to describe the change in polarisation state due to the effects of a component. A polarimeter is an instrument that can be used to measure the polarisation state of a beam of light or the polarisation effects of a specimen. This can be accomplished using a number of methods, among them division-of-amplitude or division-of-time polarimetry. Thorough calibration of a polarimeter should be carried out so that reliable and accurate measurements are obtained. In this chapter we describe a number of robust calibration methods that do not require any assumptions to be made about the system.

2.1 Polarised Light

The polarisation state of an electromagnetic wave is equivalent to the orientation of vibration of its electric field (Shurcliff and Ballard 1964). This orientation may be constant, as is the case for linearly polarised light, or it can vary periodically with time, which is the case for circularly or elliptically polarised light. Circularly polarised light carves out a circular pattern in space as it propagates. Elliptically polarised light can be described as any combination of circularly and linearly polarised light; its electric field vector will form an elliptical pattern as it propagates towards an observer. If the polarisation state of the beam is varying continually and randomly with time, then the light is said to be unpolarised. In this case there will be no fixed phase relationship between any two orthogonal components of the field, and the phases will be uncorrelated in time (Azzam and Bashara 2003).

F. Kenny (✉)
School of Physics, National University of Ireland, Galway, Galway, Ireland
e-mail: fiona.kenny@nuigalway.ie

R. Mignani et al. (eds.), *Astronomical Polarisation from the Infrared to Gamma Rays*, Astrophysics and Space Science Library 460,
https://doi.org/10.1007/978-3-030-19715-5_2

Circularly polarised light can be resolved into two orthogonally polarised components with equal amplitudes. These two components will have a phase difference with respect to each other of one quarter of a wave, or $\pi/2$. It is this relative phase that gives rise to the circular pattern made by the electric field. The two types of circular polarisation are left- and right-circular; the difference between them is in the direction of rotation of the electric field as they propagate. Left-circularly polarised light rotates anticlockwise as it propagates towards an observer, while right-circular rotates clockwise when viewed from the same position. These two differ in the sign of their relative phases: $\pi/2$ will give rise to left-circular, while $-\pi/2$ results in right-circular.

2.1.1 Stokes Parameters

The Stokes parameters, or Stokes vector, is a useful mathematical tool for the representation of the state of polarisation (SOP) of an electromagnetic wave. Measurements of the SOP usually involve the calculation of the Stokes parameters. These parameters are equivalent to the time-averaged measurement of the orientation of the electric field (Shurcliff and Ballard 1964; Born and Wolf 1999; Goldstein 2011). The Stokes parameters, **s**, can be defined as

$$\mathbf{s} = \begin{pmatrix} s_0 \\ s_1 \\ s_2 \\ s_3 \end{pmatrix} = \begin{pmatrix} I_{\text{total}} \\ I_H - I_V \\ I_{+45^\circ} - I_{-45^\circ} \\ I_R - I_L \end{pmatrix} = \begin{pmatrix} I \\ Q \\ U \\ V \end{pmatrix}, \tag{2.1}$$

where I_{total} is the total intensity of the light being described, I_H and I_V are the intensities of the light resolved into its horizontal and vertical components, respectively. I_{+45° and I_{-45° correspond to radiation linearly polarised at $+45^\circ$ and -45°, and I_R and I_L are right-circularly and left-circularly polarised. These quantities are often written and referred to as I, Q, U, and V.

The polarisation degree (PD) of the light under consideration depends on the ratio of polarised light to the total intensity, and is given by

$$\text{PD} = \frac{\sqrt{(s_1)^2 + (s_2)^2 + (s_3)^2}}{s_0}. \tag{2.2}$$

Additionally, the degree of linear polarisation (PD_L) and degree of circular polarisation (PD_C) can also be calculated from the Stokes parameters:

$$\text{PD}_\text{L} = \frac{\sqrt{(s_1)^2 + (s_2)^2}}{s_0}, \tag{2.3}$$

$$\mathrm{PD_C} = \frac{|s_3|}{s_0}. \tag{2.4}$$

2.1.2 Mueller Matrices

In polarimetry, it is often useful to consider the effect of an optical element on the polarisation state of light. If the four-element Stokes vector is used to represent the polarisation state of the light, then matrices with dimensions 4×4 can be used to model the effects of an optical component on any polarisation state. These matrices are known as Mueller matrices and they can represent any physically realisable component of a system (Goldstein 2011).

The resulting Stokes vector, $\mathbf{s}_{\mathrm{out}}$, after experiencing the polarisation effects of the elements of an optical system, is calculated by pre-multiplying the Mueller matrices of the individual elements by the Stokes vector of the incident beam, $\mathbf{s}_{\mathrm{in}}$:

$$\mathbf{s}_{\mathrm{out}} = \mathbf{M}_N \cdot \mathbf{M}_{N-1} \cdot \ldots \cdot \mathbf{M}_1 \cdot \mathbf{s}_{\mathrm{in}}, \tag{2.5}$$

where and $\mathbf{M}_i$ are the Mueller matrices representing each of the components affecting the polarisation state; $\mathbf{M}_1$ is the first element encountered by the beam, $\mathbf{M}_2$ is the second, and so on until $\mathbf{M}_N$, which represents the final element of the system.

We will now examine a simple example of the use of Mueller algebra to model the effect of a polariser on an incident beam that is horizontally polarised. In this example we take the incident beam to be vertically polarised and the polariser to be horizontally transmitting. The Mueller matrix of a horizontal polariser is

$$\mathbf{M}_{\mathrm{pol}}(0^\circ) = \tau \begin{pmatrix} 1 & 1 & 0 & 0 \\ 1 & 1 & 0 & 0 \\ 0 & 0 & 0 & 0 \\ 0 & 0 & 0 & 0 \end{pmatrix}, \tag{2.6}$$

where τ is the transmittance of the polariser, and is equal to 0.5 for a perfect polariser. The Stokes vector representing light vertically polarised is $\mathbf{s}_{\mathrm{in}}$:

$$\mathbf{s}_{\mathrm{in}} = \begin{pmatrix} 1 \\ -1 \\ 0 \\ 0 \end{pmatrix}. \tag{2.7}$$

The Stokes vector of the beam after the horizontal polariser is calculated as follows:

$$\mathbf{s}_{\mathrm{out}} = \mathbf{M}_{\mathrm{pol}}(0^\circ) \cdot \mathbf{s}_{\mathrm{in}}. \tag{2.8}$$

Computation of the above equation results in

$$\mathbf{s}_{\mathrm{out}} = \begin{pmatrix} 0 \\ 0 \\ 0 \\ 0 \end{pmatrix}, \tag{2.9}$$

which implies that, for a perfect polariser, all of the vertically polarised light will be absorbed by the horizontal polariser. This agrees with the expected observation for such a system in the laboratory. The polarisation effects of any physically realisable polarimetric system can be modelled using Mueller matrices and Stokes parameters.

A diattenuating material is one that preferentially absorbs a particular state of polarisation, such as a linear polariser, while a retarding material is one that creates or changes the relative phase between perpendicular polarisation states. The polarisation effects of many optical elements amount to a combination of these two properties, and another useful Mueller matrix is one that describes any linear-diattenuator and/or linear-retarder. The Mueller matrix of this type of element is

$$\mathbf{M}_{gen}(\tau, \Psi, \delta) = \tau \cdot \begin{pmatrix} 1 & -\cos 2\Psi & 0 & 0 \\ -\cos 2\Psi & 1 & 0 & 0 \\ 0 & 0 & \sin 2\Psi \cos\delta & \sin 2\Psi \cos\delta \\ 0 & 0 & -\sin 2\Psi \cos\delta & \sin 2\Psi \cos\delta \end{pmatrix}, \tag{2.10}$$

where τ is the total intensity transmittance, δ is the retardance, and Ψ is related to the relative transmission of parallel ($\tau_{\|}$) and perpendicular ($\tau_{\perp}$) states as follows:

$$\tan\Psi = \sqrt{\frac{\tau_{\|}}{\tau_{\perp}}}. \tag{2.11}$$

A rotation of an optical component can be represented in Mueller algebra by using the Mueller rotation matrix, $\mathbf{M}_{\mathrm{rot}}$:

$$\mathbf{M}_{\mathrm{rot}}(\theta) = \begin{pmatrix} 1 & 0 & 0 & 0 \\ 0 & \cos 2\theta & -\sin 2\theta & 0 \\ 0 & \sin 2\theta & \cos 2\theta & 0 \\ 0 & 0 & 0 & 1 \end{pmatrix}, \tag{2.12}$$

where θ is the angle by which the element is rotated. Rotation of an optical element by an azimuthal angle, θ, is represented by the following product:

$$\mathbf{M}(\theta) = \mathbf{M}_{\mathrm{rot}}(\theta) \cdot \mathbf{M}(0^{\circ}) \cdot \mathbf{M}_{\mathrm{rot}}(-\theta), \tag{2.13}$$

where $\mathbf{M}$ is the Mueller matrix of the optical element, and $\mathbf{M}_{\mathrm{rot}}$ is the Mueller rotation matrix from Eq. (2.12).

The rotation of the Mueller matrix in Eqs. (2.10) and using Eq. (2.13) can be used as a general matrix to describe most elements, as the polarisation properties of most components can be considered to be a combination of retardance and diattenuation rotated at some angle.

2.1.3 The Poincaré Sphere

The Poincaré sphere is a visual tool which is useful for the understanding of polarisation states and changes to the polarisation state due to any physically realisable component (Collett 2005). It can be understood as a three-dimensional plot of the Stokes parameters; a single point on the sphere is representative of a set of Stokes parameters, i.e. a state of polarisation. The six poles of the sphere correspond to the six polarisation states referred to in the definition of the Stokes parameters; one axis spans the states from vertically to horizontally polarised light, a second will represent ±45° polarised light, while the vertical axis is conventionally reserved for left- and right-circularly polarisation states. All fully polarised states correspond to points on the surface of the sphere, while depolarised states are represented by the interior of the sphere. A plot of the Poincaré sphere is shown in Fig. 2.1.

All linear polarisation states are represented along the equator of the sphere; this is the region of the sphere where s_3 is equal to zero and no ellipticity is present. These states are illustrated by the dotted line in Fig. 2.1. Purely circularly polarised polarisation states correspond to the north and south poles of the sphere, with right-circular polarisation at "north" and left-circular polarisation at "south". Any point that lies between the equator and either of the poles represents an elliptically polarised state. The angle of ellipticity depends on its angle along the equator. The azimuthal angle with respect to the equator indicates the degree of ellipticity present. In astronomical polarisation the position angle of linear or elliptically polarised light is defined relative to North, and the magnitude of the position angle increases in the counter-clockwise direction (di Serego Alighieri 2017).

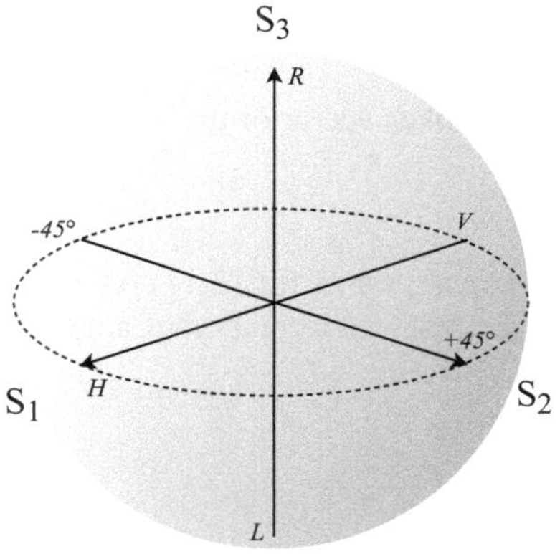

Fig. 2.1 The Poincaré sphere can be thought of as a three-dimensional plot of the Stokes parameters. Linearly polarised states are positioned on the equator of the sphere, marked by the dotted line in this figure, while the two pure circular polarisation states are situated at the extrema of the vertical axis

Looking at Eq. (2.2), which represents the DOP of a polarisation state, it is clear that states with DOP $= 1$ will be located on the surface of the Poincaré sphere. This implies that purely polarised states are represented on the outer surface or outer shell of the sphere. Calculation of the DOP is equivalent to calculating the displacement of any point within the volume of the sphere from the origin. The DOP will be less than 1 when the light is partially polarised, implying that depolarised states will therefore be located somewhere in the interior of the sphere. Totally unpolarised light will be represented by a point at the origin.

The Poincaré sphere is also a useful tool for visualising the effect of an optical element on any polarisation state. For example, the effect of a linear retarder on any polarisation state can be thought of as a rotation on the Poincaré sphere; the axis of this rotation is defined by the orientation of the fast axis of the retarder. Mueller analysis is used below to see this algebraically. Using Eq. (2.10), and assuming no diattenuation, the Mueller matrix of a retarder of retardance δ and with its fast axis at 0° is

$$\mathbf{M}_{\text{ret}}(\delta, 0^\circ) = \begin{pmatrix} 1 & 0 & 0 & 0 \\ 0 & 1 & 0 & 0 \\ 0 & 0 & \cos\delta & \sin\delta \\ 0 & 0 & -\sin\delta & \cos\delta \end{pmatrix}. \tag{2.14}$$

In order to convert this into the Mueller matrix of a retarder with fast axis at an angle of 45°, the following matrix product is computed:

$$\mathbf{M}_{\text{ret}}(\delta, 45^\circ) = \mathbf{M}_{\text{rot}}(45^\circ) \cdot \mathbf{M}_{\text{ret}}(\delta, 0^\circ) \cdot \mathbf{M}_{\text{rot}}(-45^\circ), \tag{2.15}$$

where $\mathbf{M}_{\text{rot}}(\theta)$ is the Mueller matrix representation for a rotation around the optical axis of an angle θ, previously given in Eq. (2.12). We now consider the effect of this retarder on an incident beam which is linearly polarised in the horizontal direction. Its Stokes vector is

$$\mathbf{s}_{\text{in}} = \begin{pmatrix} 1 \\ 1 \\ 0 \\ 0 \end{pmatrix}. \tag{2.16}$$

The Stokes vector of the beam after the retarder at 45° is

$$\mathbf{s}_{\text{out}} = \mathbf{M}_{\text{ret}}(\delta, 45^\circ) \cdot \mathbf{s}_{\text{in}}. \tag{2.17}$$

Computation of Eq. (2.17) reveals that the state of polarisation of the output beam is the result of a rotation of angle δ on the surface of the sphere:

$$\mathbf{s}_{\text{out}} = \begin{pmatrix} 1 \\ \cos\delta \\ 0 \\ \sin\delta \end{pmatrix}. \tag{2.18}$$

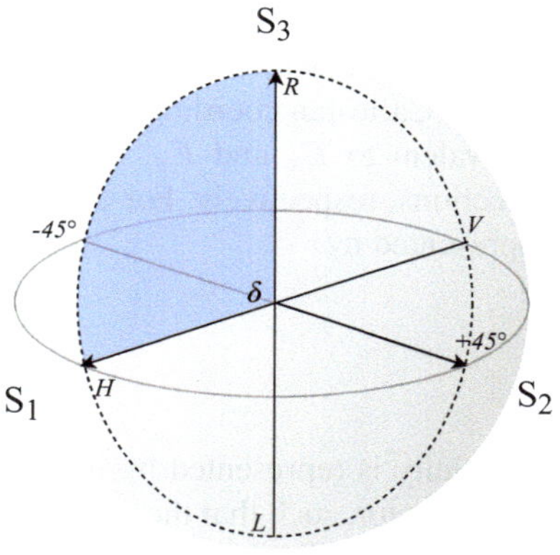

Fig. 2.2 A rotation around the sphere can represent the effect of a retardance. In this figure, δ represents a $\pi/2$ retardance that converts horizontally polarised light to right-circular

If we consider the same change in polarisation using the Poincaré sphere, the starting point of the rotation will be "H", representing linear horizontally polarised light. The effect of the retarder will be equivalent to a clockwise rotation of magnitude δ; the axis of rotation will be $\pm 45°$ axis of the sphere. The states that are reachable constitute the circle represented by the dotted line in Fig. 2.2. If the retarder is a quarter waveplate, δ is equivalent to $\pi/2$, and the final state of polarisation will be right-circular. This transformation is shown by the shaded region in Fig. 2.2.

This analysis shows that a single variable retardance at a fixed orientation can introduce a variable rotation on the Poincaré sphere around a fixed axis. The fixed axis of rotation will correspond to the orientation of the fast axis of the retarder. In order to reach any point on the sphere, effectively generating any polarisation state, a second variable retardance is needed. The axis of rotation of this retardance should be at 45° with respect to the first axis of rotation. We will discuss some experimental examples of the implementation of this concept in the generation of polarisation states in Sect. 2.2.2.

2.1.4 *Jones Calculus*

Another widely used mathematical formalism used to describe polarisation states and the polarisation effect of an optical element is Jones calculus. Description of this calculus can be found in texts such as Shurcliff and Ballard (1964) and Goldstein (2011). As is the case in Mueller calculus, optical elements are represented by matrices, and polarisation states are represented as vectors. However, Jones calculus uses 2×2 matrices and 2 element vectors. Contrary to Stokes parameters and Mueller matrices, Jones calculus considers the complex amplitude of the field of the electromagnetic wave, instead of the time-averaged intensity of the light source. Jones vectors and matrices can therefore include information on both the absolute phase, or wavefront, of a beam of radiation. It will also include information on the relative phase, which determines the ellipticity of the polarisation state.

In a Cartesian coordinate system, the two elements of a Jones vector are usually equivalent to E_x and E_y, the complex amplitudes of the field in the x- and y-directions, respectively. For example, radiation polarised in only the x-direction is represented by

$$\mathbf{j} = \begin{bmatrix} 1 \\ 0 \end{bmatrix}. \tag{2.19}$$

Ellipticity is represented by introducing complex terms to the individual elements of the vector, such that the appropriate phase difference exists between the two. For example, left-circular polarisation is represented by

$$\mathbf{j} = \frac{1}{\sqrt{2}} \begin{bmatrix} i \\ 1 \end{bmatrix}. \tag{2.20}$$

An advantage of the use of Jones calculus over Mueller is that Jones can include description of the absolute phase, while Mueller algebra cannot. Jones calculus, however, assumes that the light is fully polarised and coherent. It cannot be used to represent depolarising or scattering elements or partially polarised states (Shurcliff and Ballard 1964). Additionally, any orthogonal basis set can be chosen; in a polar coordinate system the field will be resolved into E_p and E_s, the radial and azimuthal components. In some instances it is advantageous to use the two circularly polarised states as the basis set, resolving the field into E_R and E_L.

These two formalisms are useful in differing situations due to the features mentioned above. For example, when measuring a polarisation state it is best to use Mueller matrices and Stokes parameters, as it is likely that a certain degree of unpolarised light will be present. However, when doing theoretical calculations, full polarisation is often assumed and it can be advantageous to consider the absolute phase of the light.

2.2 Polarimetry

A polarimeter is an instrument used to measure a polarisation state, or the effect of a sample on any state of polarisation. Stokes polarimeters are used to measure polarisation states and Mueller matrix polarimeters can use an appropriate set of Stokes measurements to construct the Mueller matrix of a component. Astronomical polarimetry involves the measurement of the polarisation state of a target, therefore in many instances a Stokes polarimeter will suffice as an instrument. Mueller matrix polarimeters can also be useful in astronomical applications, as they allow a more extensive and accurate calibration routine. While they are significantly more complex, Mueller matrix polarimeters are useful in many applications that require a high accuracy of measurement. In this section we describe some typical types of Stokes and Mueller polarimeters. We then discuss some of the available methods for optimisation and calibration of these systems.

2.2.1 Stokes Polarimeters

A Stokes polarimeter consists of an experimental system knows as a polarisation state analyser (PSA). In order to reconstruct the full Stokes parameters, a PSA should be able to measure intensities of at least 4 linearly independent polarisation states of the electromagnetic radiation incident on the system. These can correspond to four out of the six intensities referred to in the definition of the Stokes parameters (Eq. (2.1)), or any other linearly independent set of polarisation states. Stokes polarimeters achieve this by splitting the signal into a number of different components either in amplitude or time. If the full Stokes parameters are not required, fewer intensity measurements can be used.

Stokes polarimeters fall into two broad categories: division-of-time and division-of-amplitude. A division-of-time polarimeter measures the required intensities sequentially in time. One of the simplest setups for this type of Stokes polarimeter uses a single rotating polariser, with or without a waveplate, in front of a detector (or array of detectors). This kind of system is robust and compact, and has been commonly utilised (Collados 2003; Lites 1987; Elmore et al. 1992). Time-resolution, however, is sacrificed in these systems which is detrimental to measurements of unstable or rapidly changing targets.

Division-of-time polarimetry can also be carried out by using electro-optics such as liquid crystal optics (Alvarez et al. 2011; Pust and Shaw 2006; Tyo 2000), photo-elastic modulators (Diner et al. 2010; Kemp 1982; Wang et al. 2002), or Pockels' cells (Baur et al. 1980). These types of systems use a combination of variable retardances and polarisers to select a desired polarisation state for measurement. In contrast to rotating polariser polarimeters, electro-optical methods offer the flexibility of measuring the intensity of any polarised state without any moving parts. However, their behaviour may also change when the ambient temperature fluctuates, and these components generally work best with monochromatic or quasi-monochromatic light.

Division-of-amplitude polarimeters use beam-splitting components to direct separate portions of the beam onto spatially separated detectors. Splitting can be carried out using beamsplitters or custom prisms (Compain and Drévillon 1998; Collins et al. 2013). These components are most readily available in the UV, optical, and infrared regions of the electromagnetic spectrum. Advantages of this system are that they can image broadband sources, can operate at high-speed (Azzam 2016), and the full resolution of the image is maintained. The use of separate paths, however, means that these systems are more susceptible to stability issues. These systems are generally more complex than division-of-time polarimeters and careful pixel-to-pixel registration of the intensities should be carried out if imaging polarimetry is required (Mudge et al. 2009).

Two other subsets of division-of-amplitude polarimeters have also been developed extensively: division-of-aperture and division-of-focal plane. These offer improved stability and compactness compared to a conventional division-of-amplitude polarimeter. Division-of-focal plane polarimeters use micro-polariser or micro-retarder arrays placed in front of the detector array in the system (Andreou and Kalayjian 2002). These components consist of many μm-scale retarders or polarisers in a regular pattern such as 2×2, and are somewhat analogous to the Bayer filter used in colour imaging. Instead of individual pixels consisting of colour filters, each pixel will have a polariser or retarder corresponding to the intensity measurements that can be used to reconstruct the Stokes parameters. As these arrays must be match to the pixels on the used some loss of the resolution of the system will be required. They have the advantage of simultaneous measurements of all polarisation channels, which provides better stability than a conventional division-of-amplitude polarimeter. These types of PSAs have found many uses in wide-field imaging (Kudenov et al. 2011) and remote sensing (Tyo et al. 2006).

Division-of-aperture polarimetry uses a micro-lens array in the Fourier plane of the system to split the image into four images, which can be incident on the same detector, for example, Pezzaniti and Chenault (2005). Again there is a loss of resolution of at least a factor of 2 compared to a division-of-amplitude polarimeter, and registration of pixels on the four images is required. They do, however, have advantages in compactness and stability over a conventional division-of-amplitude polarimeter.

2.2.1.1 High-Energy Techniques

While components such as beamsplitters, waveplates, and polarisers are readily available for spectral ranges from the UV to the far infrared, special consideration is needed for polarimeters operating with high energy radiation, such as X-rays and γ-rays. The development of polarimeters in this spectral range is driven by the invention and improvement of detectors at these wavelengths (Bellazzini and Muleri 2010). In order for an X-ray photon to be detected, it must undergo an interaction with a material that results in some measurable quantity such as scattering or the photoelectric effect. The type of interaction that takes places will depend on the photon energy. Some of these interactions, such as the photoelectric effect, Compton scattering, and Bragg diffraction, have an intrinsic polarisation dependence.

For X-ray photons with energy on the order of keV, these detectors can be based on the photoelectric effect (Costa et al. 2001). A high energy photon incident on an atom will eject an electron from an inner orbital; the energy of the electron will be the difference between the photon energy and the initial binding energy. The scatter direction of the electron will have a dependence on the orientation of the electric field of the incident radiation, and this scatter can be used to estimate the polarisation state of the incident radiation. The average of many measurements of photons can be used to find I_H, I_V or other intensities that can be used to reconstruct the Stokes parameters. Detectors such as gas pixel detectors have been developed based on this effect (Bellazzini et al. 2007), and have been implemented in a number of polarimeters such as Tagliaferri et al. (2010), Costa et al. (2010), Soffitta et al. (2013).

For higher energy γ-rays of tens of keV, detectors based on Compton scattering has been used to estimate the polarisation state (Bloser et al. 2009; Kishimoto et al. 2007; Yonetoku et al. 2011). In these detectors the incident X-ray scatters off a solid target usually made of a low atomic number material. As is the case for the photoelectric effect, the direction of the Compton scatter will depend on the polarisation of the input radiation.

Bragg diffraction is another technique that can be useful for the measurement of the polarisation states of X-rays with energies of a few keV (Silver et al. 1989; Angel and Weisskopf 1970). This is because the Bragg reflectivity at incident angles of 45° is highly polarisation dependent. Rotation of the Bragg reflector may be carried out to obtain sufficient measurements to reconstruct the Stokes parameters (Marx et al. 2013). For further details on X-ray and γ-ray see Chaps. 13 and 14 of this publication.

2.2.2 *Mueller Polarimeters*

A polarisation state generator (PSG) is a system that can be used to change the polarisation state of a beam of radiation into a number of polarisation states. If a PSG is used in conjunction with a PSA, then the system is referred to as a Mueller matrix polarimeter and it can be used to measure the Mueller matrix of a specimen. A simple schematic of a Mueller matrix polarimeter is shown in Fig. 2.3. In this system a light source is first incident on the PSG, which can sequentially change the polarisation state of the source to some number of polarisation states. The beam is then incident on the sample, which will change the state of polarisation according to its Mueller matrix. The polarisation state of the light after the specimen is then measured using the PSA.

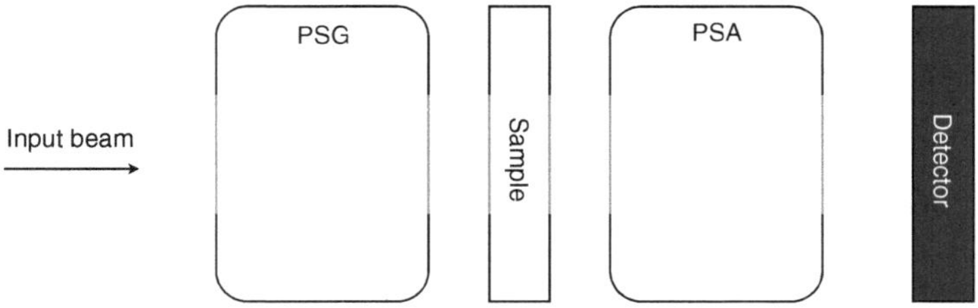

Fig. 2.3 A Mueller matrix polarimeter is used to measure the Mueller matrix of a sample. It includes a light source, polarisation state generator (PSG), polarisation state analyser, and detector

The PSG in a Mueller matrix polarimeter should be capable of generating at least four linearly independent polarisation states. In turn, each of these is incident on the sample, and the PSA records the Stokes parameters for every incident state. This set of Stokes parameter measurements can be used to reconstruct the Mueller matrix of the sample. PSGs have been constructed in a variety of ways such as rotating polarisers and waveplates (Azzam 1978; Hauge 1987), with Pockels cells (Rodríguez et al. 2010; Lara and Dainty 2006; Delplancke 1997), and liquid crystal optics (Kenny et al. 2012; Moreno et al. 2012; Delplancke 1997). The PSA can be the same as any of those used as Stokes polarimeters, previously described in Sect. 2.2.1.

2.2.3 Calibration Techniques

Calibration of a polarimeter should be undertaken if reliable quantitative measurements are desired. Small polarisation errors can arise in the instrument due to effects such as multiple reflections, imperfect components, or misalignments (Goldstein and Chipman 1990; Nee 2003), and these can add up to significant system errors in the instrument, which will affect measurements. There are a number of strategies for finding these errors, ranging from measurement of standard targets to a full system Mueller matrix, to a more involved calibration of the system using a polarisation state generator and considering the system as a full Mueller matrix polarimeter.

2.2.3.1 Polarisation Standards

As a baseline calibration tool, some standard targets can be measured when an instrument is on-sky. These are astronomical targets with accurate polarimetric measurements and have been shown to be relatively constant in the polarisation properties (Schmidt et al. 1992; Tinbergen 1979). They have, however, been shown to exhibit some variability (Hsu and Breger 1982; Whittet et al. 1992), and as such should not be the sole method used to calibrate an instrument. As they are astronomical sources they have the advantage of providing some information on the entire optical path from source to detector, and can therefore give some useful information on the polarisation effects of the entire optical path including atmosphere, telescope, and instrument.

2.2.3.2 Stokes Calibration Techniques

More sophisticated calibration and optimisation techniques for Stokes polarimeters have been developed, and these techniques are summarised in Boulbry et al. (2007). A Stokes polarimeter can extract the Stokes parameters from a set of intensity measurements using a matrix, $\mathbf{W}$, which is known as the data reduction matrix, in the following way:

$$\mathbf{s} = \mathbf{W}I, \tag{2.21}$$

where $\mathbf{s}$ are the Stokes parameters, and I is a vector comprised of the intensity measurements. If we measure six intensities corresponding to the 6 extrema of the Poincaré sphere, then $\mathbf{W}$ will be

$$\mathbf{W} = \begin{pmatrix} 1 & 1 & 0 & 0 & 0 & 0 \\ 1 & -1 & 0 & 0 & 0 & 0 \\ 0 & 0 & 1 & -1 & 0 & 0 \\ 0 & 0 & 0 & 0 & 1 & -1 \end{pmatrix} \tag{2.22}$$

$\mathbf{W}$, however, is not a unique matrix and other valid forms can be found that will also give the Stokes parameters. It has been shown that optimisation of the condition number of $\mathbf{W}$ will result in better performance due to the minimisation of the propagation of uncertainties and a reduction in sensitivity to noise. This optimal $\mathbf{W}$ has a condition number of $\sqrt{3}$, and was found by Tyo (2002):

$$\mathbf{W} = \begin{pmatrix} 1/3 & 1/3 & 1/3 & 1/3 & 1/3 & 1/3 \\ 1 & -1 & 0 & 0 & 0 & 0 \\ 0 & 0 & 1 & -1 & 0 & 0 \\ 0 & 0 & 0 & 0 & 1 & -1 \end{pmatrix}, \tag{2.23}$$

Even after this optimisation of the PSA, a measured set of Stokes parameters, $\mathbf{s}_i$, will still deviate from the true Stokes parameters, which we will term $\mathbf{s}$, by some amount due to the various polarisation aberrations that can occur in the system.

A second matrix, $\mathbf{A}$, is used to represent any errors in the PSA of a Stokes polarimeter. $\mathbf{A}$ is known as the sensitivity matrix, and each row of $\mathbf{A}$ represents a single analysis channel in the Stokes polarimeter. $\mathbf{A}$ relates the intensity measurements to the true Stokes parameters in the following way:

$$I = \mathbf{A}\mathbf{s} \tag{2.24}$$

According to Eq. (2.21), $\mathbf{A}$ should also be equivalent to the pseudoinverse of $\mathbf{W}$, the data reduction matrix. $\mathbf{A}$ has been conventionally estimated using the specifications of the components, however, this does not provide a true picture of the errors in the system and because of this, methods for finding $\mathbf{A}$ have been developed that do not make any assumptions about the components of the system (Boulbry et al. 2007).

2.2.3.3 Mueller Calibration Techniques

If the polarimeter is made more complex by the addition of a polarisation state generator, a more accurate calibration of the polarimeter can be carried out. After calibration, the PSG can be removed and the PSA can be used as a calibrated standalone Stokes polarimeter.

Mueller matrix polarimeters can be calibrated using a method known as the eigenvalue calibration method (ECM) (Compain et al. 1999). This centres on obtaining the two experimental matrices that represent the operation of the PSG and PSA arms of the setup. These are usually termed **W** and **A**, respectively. A comprehensive explanation of ECM was given by de Martino et al. (2004), and a summary of the method will be given in this section. ECM has been used in a wide variety of polarimeters, to measure both Stokes parameters (Collins et al. 2013; Mazumder et al. 2012) and Mueller matrices (Lara and Dainty 2006; Rodríguez et al. 2010; Kenny et al. 2012).

A number of calibration samples, such as polarisers and waveplates, are needed for this calibration procedure. One of these samples should simply be free space, and discussion of the optimum samples to use can be found in de Martino et al. (2003) and Hu et al. (2014). The calibration procedure involves using the polarimeter to measure the Mueller matrix of each sample. A minimisation procedure is then used to estimate **W**, **A**, along with the calibrated Mueller matrices of each calibration sample.

In the following analysis theoretical or ideal matrices will be referred to in upper case italics, while experimental Mueller matrices will use upper case bold type. The measurement of the Mueller matrix of the ith component using a polarimeter can be represented as follows:

$$\mathbf{B_i} = \mathbf{A} \cdot M_i \cdot \mathbf{W}, \tag{2.25}$$

where $\mathbf{B_i}$ is the measured Mueller matrix of the sample and M_i is the true Mueller matrix of the sample. **W** and **A** describe the effects of the PSG and PSA, respectively, and when **W** and **A** are known, the polarimeter is said to be calibrated. The true Mueller matrix of any element, M, can be extracted from a measurement using the inverses of the calibration matrices:

$$M_i = \mathbf{A}^{-1} \cdot \mathbf{B_i} \cdot \mathbf{W}^{-1}. \tag{2.26}$$

For $n = 0$, the experimental measurement, B_0, will be the polarimetric measurement of free space, the Mueller matrix of which is the identity matrix. We can therefore say that

$$\mathbf{B}_0 = \mathbf{A} \cdot \mathbf{W} \tag{2.27}$$

and therefore

$$\mathbf{W} = \mathbf{A}^{-1}\mathbf{B}_0. \tag{2.28}$$

The next step in the calibration is to calculate the set of products $\mathbf{C}_i$, for the remaining calibration samples ($i = 1, 2, 3$):

$$\mathbf{C}_i = (\mathbf{B}_0)^{-1} \cdot \mathbf{B}_i, \qquad \text{where } i > 0. \tag{2.29}$$

Substitution of $\mathbf{B}_0$ and $\mathbf{B}_i$ using Eqs. (2.27) and (2.26) yields

$$\mathbf{C}_i = \mathbf{W}^{-1} \cdot \mathbf{A}^{-1} \cdot \mathbf{A} \cdot M_i, \cdot\mathbf{W} \tag{2.30}$$

$$\mathbf{C}_i = \mathbf{W}^{-1} \cdot M_i . \cdot \mathbf{W} \tag{2.31}$$

The products $\mathbf{C}_i$ have the same eigenvalues as their corresponding "true" Mueller matrices, $\mathbf{M}_i$, since the eigenvalues of a product of matrices do not depend on the order of the product. This means that properties of $\mathbf{M}_i$ such as the retardance, angle of rotation, and diattenuation can be extracted from the calculation of $\mathbf{C}_i$.

Looking at Eq. (2.28), the problem of calibrating the system is reduced to finding either $\mathbf{W}$ or $\mathbf{A}$, since $\mathbf{B}_0$ is measured directly by the system. $\mathbf{W}$ is the unique solution of the set of simultaneous equations (de Martino et al. 2003):

$$M_i \cdot \mathbf{W} - \mathbf{W} \cdot \mathbf{C}_i = 0. \tag{2.32}$$

After $\mathbf{W}$ has been determined, $\mathbf{A}$ can be calculated by rearranging Eq. (2.28). ECM has the advantage of calculating the errors in the PSA and PSG in two distinct matrices, $\mathbf{W}$ and $\mathbf{A}$. If one wishes to use the instrument as a Stoke polarimeter, the PSG can be removed from the calibrated Mueller matrix polarimeter, and the PSA can be used alone to measure the calibrated Stokes parameters of any target.

2.3 Conclusion

In this chapter we have discussed the two most common mathematical formalisms used to describe polarised light and polarimeters. These are Mueller and Jones algebra. These both use vectors to represent polarisation states, and matrices to represent the polarisation effect of system components. The main difference between the two is that Mueller algebra can be used to describe partially or totally unpolarised light while Jones algebra cannot. Jones algebra, however, contains information about the complex amplitude of the electromagnetic field.

Polarimetric instruments can be divided into a number of subtypes. A Stokes polarimeter is used to measure the Stokes parameters of a source or target, while a Mueller matrix polarimeter measures the polarisation effects of a specimen. Polarimeters can further be categorised depending on how they obtain polarimetric information. A division-of-amplitude polarimeter will measure a number of intensities simultaneously, while a division-of-time polarimeter measures intensities sequentially in time.

Calibration and optimisation of a polarimeter should be carried out extensively so that accurate results can be obtained. Methods have been developed that do not require any assumptions about the system or its components. Finally, we described how a more extensive calibration of a Stokes polarimeter can be carried out if a polarisation state generator is introduced to the system and it is calibrated in the same way as a Mueller matrix polarimeter.

Appendix 1: Table of Stokes Parameters and Jones Vectors

Polarisation state	Stokes parameters	Jones vector
Linear horizontal	$\begin{pmatrix}1\\1\\0\\0\end{pmatrix}$	$\begin{pmatrix}1\\0\end{pmatrix}$
Linear vertical	$\begin{pmatrix}1\\-1\\0\\0\end{pmatrix}$	$\begin{pmatrix}0\\1\end{pmatrix}$
Linear +45°	$\begin{pmatrix}1\\0\\1\\0\end{pmatrix}$	$\frac{1}{\sqrt{2}}\begin{pmatrix}1\\1\end{pmatrix}$
Linear −45°	$\begin{pmatrix}1\\0\\-1\\0\end{pmatrix}$	$\frac{1}{\sqrt{2}}\begin{pmatrix}1\\-1\end{pmatrix}$
Right-circular	$\begin{pmatrix}1\\0\\0\\1\end{pmatrix}$	$\frac{1}{\sqrt{2}}\begin{pmatrix}1\\-i\end{pmatrix}$
Left-circular	$\begin{pmatrix}1\\0\\0\\-1\end{pmatrix}$	$\frac{1}{\sqrt{2}}\begin{pmatrix}1\\i\end{pmatrix}$
Unpolarised	$\begin{pmatrix}1\\0\\0\\0\end{pmatrix}$	N/A

Appendix 2: Commonly Used Mueller and Jones Matrices

Component	Mueller matrix	Jones matrix
Free space	$\begin{pmatrix} 1 & 0 & 0 & 0 \\ 0 & 1 & 0 & 0 \\ 0 & 0 & 1 & 0 \\ 0 & 0 & 0 & 1 \end{pmatrix}$	$\begin{pmatrix} 1 & 0 \\ 0 & 1 \end{pmatrix}$
Isotropic absorbing material	$\begin{pmatrix} k & 0 & 0 & 0 \\ 0 & k & 0 & 0 \\ 0 & 0 & k & 0 \\ 0 & 0 & 0 & k \end{pmatrix}$	$\begin{pmatrix} k & 0 \\ 0 & k \end{pmatrix}$
Horizontal polariser	$\frac{1}{2}\begin{pmatrix} 1 & 1 & 0 & 0 \\ 1 & 1 & 0 & 0 \\ 0 & 0 & 0 & 0 \\ 0 & 0 & 0 & 0 \end{pmatrix}$	$\begin{pmatrix} 1 & 0 \\ 0 & 0 \end{pmatrix}$
Vertical polariser	$\frac{1}{2}\begin{pmatrix} 1 & -1 & 0 & 0 \\ -1 & 1 & 0 & 0 \\ 0 & 0 & 0 & 0 \\ 0 & 0 & 0 & 0 \end{pmatrix}$	$\begin{pmatrix} 0 & 0 \\ 0 & 1 \end{pmatrix}$
+45° polariser	$\frac{1}{2}\begin{pmatrix} 1 & 0 & 1 & 0 \\ 0 & 0 & 0 & 0 \\ 1 & 0 & 1 & 0 \\ 0 & 0 & 0 & 0 \end{pmatrix}$	$\frac{1}{2}\begin{pmatrix} 1 & 1 \\ 1 & 1 \end{pmatrix}$
−45° polariser	$\frac{1}{2}\begin{pmatrix} 1 & 0 & -1 & 0 \\ 0 & 0 & 0 & 0 \\ -1 & 0 & 1 & 0 \\ 0 & 0 & 0 & 0 \end{pmatrix}$	$\frac{1}{2}\begin{pmatrix} 1 & -1 \\ -1 & 1 \end{pmatrix}$
Right-circular polariser	$\frac{1}{2}\begin{pmatrix} 1 & 0 & 0 & 1 \\ 0 & 0 & 0 & 0 \\ 0 & 0 & 0 & 0 \\ 1 & 0 & 0 & 1 \end{pmatrix}$	$\frac{1}{2}\begin{pmatrix} 1 & -i \\ i & 1 \end{pmatrix}$
Left-circular polariser	$\frac{1}{2}\begin{pmatrix} 1 & 0 & 0 & -1 \\ 0 & 0 & 0 & 0 \\ 0 & 0 & 0 & 0 \\ -1 & 0 & 0 & 1 \end{pmatrix}$	$\frac{1}{2}\begin{pmatrix} 1 & i \\ -i & 1 \end{pmatrix}$

(continued)

Component	Mueller matrix	Jones matrix
Ideal depolariser	$\begin{pmatrix} 1 & 0 & 0 & 0 \\ 0 & 0 & 0 & 0 \\ 0 & 0 & 0 & 0 \\ 0 & 0 & 0 & 0 \end{pmatrix}$	N/A
Quarter waveplate with fast axis at 0°	$\begin{pmatrix} 1 & 0 & 0 & 0 \\ 0 & 1 & 0 & 0 \\ 0 & 0 & 0 & 1 \\ 0 & 0 & -1 & 0 \end{pmatrix}$	$\begin{pmatrix} e^{-i\pi/4} & 0 \\ 0 & e^{i\pi/4} \end{pmatrix}$
Quarter waveplate with fast axis at 90°	$\begin{pmatrix} 1 & 0 & 0 & 0 \\ 0 & 1 & 0 & 0 \\ 0 & 0 & 0 & -1 \\ 0 & 0 & 1 & 0 \end{pmatrix}$	$\begin{pmatrix} e^{i\pi/4} & 0 \\ 0 & e^{-i\pi/4} \end{pmatrix}$
Quarter waveplate with fast axis at 45°	$\begin{pmatrix} 1 & 0 & 0 & 0 \\ 0 & 0 & 0 & -1 \\ 0 & 0 & 1 & 0 \\ 0 & 1 & 0 & 0 \end{pmatrix}$	$\frac{1}{\sqrt{2}} \begin{pmatrix} 1 & i \\ i & 1 \end{pmatrix}$
Quarter waveplate with fast axis at 45°	$\begin{pmatrix} 1 & 0 & 0 & 0 \\ 0 & 0 & 0 & 1 \\ 0 & 0 & 1 & 0 \\ 0 & -1 & 0 & 0 \end{pmatrix}$	$\frac{1}{\sqrt{2}} \begin{pmatrix} 1 & -i \\ -i & 1 \end{pmatrix}$
Half waveplate with fast axis at 0° or 90°	$\begin{pmatrix} 1 & 0 & 0 & 0 \\ 0 & 1 & 0 & 0 \\ 0 & 0 & -1 & 0 \\ 0 & 0 & 0 & -1 \end{pmatrix}$	$\begin{pmatrix} 1 & 0 \\ 0 & -1 \end{pmatrix}$
Half waveplate with fast axis at ±45°	$\begin{pmatrix} 1 & 0 & 0 & 0 \\ 0 & -1 & 0 & 0 \\ 0 & 0 & 1 & 0 \\ 0 & 0 & 0 & -1 \end{pmatrix}$	$\begin{pmatrix} 0 & 1 \\ 1 & 0 \end{pmatrix}$

References

Alvarez-Herrero A, Uribe-Patarroyo N, Parejo PG, Vargas J, Heredero RL, Restrepo R, Martínez-Pillet V, del Toro Iniesta JC, López A, Fineschi S, Capobianco G (2011) Proceedings of SPIE, vol 8160, 81600Y

Andreou AG, Kalayjian ZK (2002) IEEE Sensors A 2:566

Angel JRP, Weisskopf MC (1970) Astron J 75:231

Azzam RMA (1978) Opt Lett 2:148

Azzam RMA (2016) J Opt Soc Am A 33:1396

Azzam RMA, Bashara NM (2003) Ellipsometry and polarized light, 5th edn. Elsevier, Amsterdam, pp 45–56

Baur TG, House LL, Hull HK (1980) Sol Phys 65:111

Born M, Wolf E (1999) Principles of optics: electromagnetic theory of propagation, interference and diffraction of light, 7th edn. Pergamon Press, New York, pp 45–56

Bellazzini R, Muleri F (2010) Nucl Instrum Methods Phys Res A 623:766

Bellazzini R, Spandre G, Minuti M, Baldini L, Brez A, Latronico L, Omodei N, Razzano M, Massai MM, Pesce-Rollins M, Sgró C (2007) Nucl Instrum Methods Phys Res A 579:853

Bloser PF, Legere JS, McConnell ML, McConnell JR, Bancroft CM, Connor TP, Ryan JM (2009) Nucl Instrum Methods Phys Res A 600:424

Boulbry B, Ramella-Roman JC, Germer TA (2007) Appl Opt 46:8533

Collados MV (2003) Proceedings of SPIE 4843

Collett E (2005) Field guide to polarization. SPIE, Bellingham

Collins P, Kyne G, Lara D, Redfern M, Shearer A, Sheehan B (2013) Exp Astron 36:479

Compain E, Drévillon B (1998) Appl Opt 37:5938

Compain E, Poirier S, Drévillon B (1999) Appl Opt 38:3490

Costa E, Soffitta P, Bellazzini R, Brez A, Lumb N, Spandre G (2001) Nature 411:662

Costa E, Bellazzini R, Tagliaferri G, Matt G, Argan A, Attinà P, Baldini L, Basso S, Brez A, Citterio O, Di Cosimo S (2010) Exp Astron 28:137

De Martino A, Kim YK, Garcia-Caurel E, Laude B, Drévillon B (2003) Opt Lett 28:616

De Martino A, Garcia-Caurel E, Laude B, Drévillon B (2004) Thin Solid Films 455:112

Delplancke F (1997) Appl Opt 36:5388

di Serego Alighieri S (2017) Exp Astron 43:19

Diner DJ, Davis A, Hancock B, Geier S, Rheingans B, Jovanovic V, Bull M, Rider DM, Chipman RA, Mahler AB, McClain SC (2010) Appl Opt 49:2929

Elmore DF, Lites BW, Tomczyk S, Skumanich A, Dunn RB, Schuenke JA, Streander KV, Leach TW, Chambellan CW, Hull HK, Lacey LB (1992) Proceedings of SPIE 1746

Goldstein DH (2011) Polarized light, 3rd edn. CRC Press, Boca Raton

Goldstein DH, Chipman RA (1990) J Opt Soc Am A 7:693

Hauge PS (1987) J Opt Soc Am 68:1519

Hsu JC, Breger M (1982) Astron J 262:732

Hu H, Garcia-Caurel E, Anna G, Goudail F (2014) Opt Lett 39:418

Kemp JC (1982) Proceedings of SPIE 0307

Kenny F, Lara D, Rodríguez-Herrera OG, Dainty C (2012) Opt Express 20:14015

Kishimoto Y, Gunji S, Ishigaki Y, Kanno M, Murayama H, Ito C, Tokanai F, Suzuki K, Sakurai H, Mihara T, Kohama M (2007) IEEE Trans Nucl Sci 54:561

Kudenov MW, Escuti MJ, Dereniak EL, Oka K (2011) Appl Opt 50:2283

Lara D, Dainty C (2006) Appl Opt 45:1917

Lites BW (1987) Appl Opt 26:3838

Marx B, Schulze KS, Uschmann I, Kämpfer T, Lötzsch R, Wehrhan O, Wagner W, Detlefs C, Roth T (2013) Phys Rev Lett 110:254801

Mazumder N, Qiu J, Foreman MR, Romero CM, Hu CW, Tsai HR, Török P, Kao FK (2012) Opt Express 20:14090

Moreno I, Davis JA, Hernandez TM, Cottrell DM, Sand D (2012) Opt Express 20:364

Mudge J, Virgen M, Dean P (2009) Proceedings of SPIE 7461
Nee SMF (2003) Opt Soc Am A 20:1651
Pezzaniti JL, Chenault DB (2005) Proceedings of SPIE 5888
Pust NJ, Shaw JA (2006) Appl Opt 45:5470
Rodríguez-Herrera OG, Lara D, Dainty C (2010) Opt Express 18:5609
Schmidt GD, Elston R, Lupie OL (1992) Astron J 104:1563
Shurcliff WA, Ballard SS (1964) Polarized light. D. Van Nostrand Company, Princeton, pp 45–56
Silver EH, Simonovich A, Labov SE, Novick R, Kaaret PE, Martin C, Hamilton T, Weisskopf MC, Elsner RF, Beeman JW, Chanan GA (1989) Proceedings of SPIE 1160
Soffitta P, Barcons X, Bellazzini R, Braga J, Costa E, Fraser GW, Gburek S, Huovelin J, Matt G, Pearce M, Poutanen J (2013) Exp Astron 36:523
Tagliaferri G, Argan A, Bellazzini R, Bookbinder J, Catalano O, Cavazzuti E, Costa E, Cusumano G, Fiore F, Fiorini C, Giommi P (2010) Proceedings of SPIE 7732
Tinbergen J (1979) Astron Astrophys Suppl Ser 35:325
Tyo JS (2000) Opt Lett 25:1198
Tyo JS (2002) Appl Opt 41:619
Tyo JS, Goldstein DL, Chenault DB, Shaw JA (2006) Appl Opt 45:5453
Wang BB, List J, Rockwell RR (2002) Proceedings of SPIE 4819
Whittet DCB, Martin PG, Hough JH, Rouse MF, Bailey JA, Axon DJ (1992) Astron J 386:562
Yonetoku D, Murakami T, Gunji S, Mihara T, Sakashita T, Morihara Y, Kikuchi Y, Takahashi T, Fujimoto H, Toukairin N, Kodama Y (2011) Publ Astron Soc Jpn 63:625

Chapter 3
Optical Polarimetry: Methods, Instruments and Calibration Techniques

Andrei Berdyugin, Vilppu Piirola, and Juri Poutanen

Abstract In this chapter we present a brief summary of methods, instruments and calibration techniques used in modern astronomical polarimetry in the optical wavelengths. We describe the properties of various polarisation devices and detectors used for optical broadband, imaging and spectropolarimetry, and discuss their advantages and disadvantages. The necessity of a proper calibration of the raw polarisation data is emphasised and methods of the determination and subtraction of instrumental polarisation are considered. We also present a few examples of high-precision measurements of optical polarisation of black hole X-ray binaries and massive binary stars made with our DiPol-2 polarimeter, which allowed us to constrain the sources of optical emission in black hole X-ray binaries and measure orbital parameters of massive stellar binaries.

3.1 Introduction

The first measurements of polarisation (in the optical wavelengths)[1] were performed by Dominique Francois Jean Arago and they date back to the first half of the nineteenth century. However, optical polarimetry became a mainstream method of

[1]The terms *optical light* or *optical wavelengths* are usually understood as the wavelength range from 300 to 1000 nm. This is the range covered with the widely used UBVRI Johnson/Cousins photometric systems.

A. Berdyugin (✉) · V. Piirola
Department of Physics and Astronomy, University of Turku, Turku, Finland
e-mail: andber@utu.fi; piirola@utu.fi

J. Poutanen
Department of Physics and Astronomy, University of Turku, Turku, Finland

Nordita, KTH Royal Institute of Technology and Stockholm University, Stockholm, Sweden

Space Research Institute of the Russian Academy of Sciences, Moscow, Russia
e-mail: juri.poutanen@utu.fi

R. Mignani et al. (eds.), *Astronomical Polarisation from the Infrared to Gamma Rays*, Astrophysics and Space Science Library 460,
https://doi.org/10.1007/978-3-030-19715-5_3

astrophysical observations only in the end of 1960s. This happened due to the grown use of sensitive photoelectric detectors, such as photomultiplier tubes (PMTs) and charge coupling devices (CCDs) in optical photometry and spectroscopy.

PMTs and CCDs are able to provide a sufficient S/N for measuring optical polarisation even for low fluxes received from astronomical objects. CCD cameras, which are *panoramic* or *multi-cell* detectors, are capable of measuring polarisation in many elements of image or spectrum simultaneously. With the introduction of the new detectors, powerful polarisation devices and new techniques of polarisation measurements have been developed. Currently, optical polarimetry is a well-established observational tool, widely used on small and large optical telescopes for studying a large variety of astrophysical objects, from planets to active galaxies. As also emphasised at the EWASS 2018 meeting, about 4000 refereed publications on optical polarimetry have been published since year 2000.

3.1.1 Description of Polarisation: Stokes Parameters, Linear and Circular Polarisation

Polarisation can be described by the *normalised Stokes parameters* q, u and v defined as:

$$q = Q/I, \quad u = U/I, \quad v = V/I. \tag{3.1}$$

Here Q, U and V are the *absolute Stokes parameters*:

$$Q = I_P \cos 2\psi \cos 2\chi, \quad U = I_P \sin 2\psi \cos 2\chi, \quad V = I_P \sin 2\chi, \tag{3.2}$$

I is the total intensity of the partially polarised light and I_P is the intensity of its *elliptically* polarised fraction. The angle ψ, which is also often denoted in the literature as θ (Theta) or PA (polarisation Angle), is the *azimuthal* angle of the *polarisation ellipse* (see Fig. 3.1). In astronomical polarimetry it is measured counter-clockwise from the direction to the celestial North in the equatorial coordinate system. The angle χ is a measure of the *eccentricity* of the polarisation ellipse, so that $\tan\chi$ is the ratio of its minor and major axes. From the normalised Stokes parameters the *degree* P_L (or PD_L) and *direction* θ (PA) of linear polarisation, and the degree of circular polarisation, P_C (or PD_C), can be found:

$$P_L = \sqrt{q^2 + u^2}, \quad \theta = \frac{1}{2}\arctan\frac{u}{q}, \quad P_C = v. \tag{3.3}$$

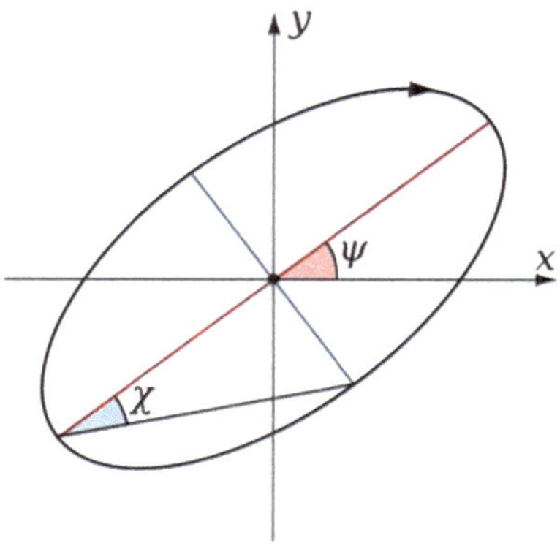

Fig. 3.1 Polarisation ellipse of fully elliptically polarised monochromatic electromagnetic wave. This ellipse is swept by the tip of the electric vector **E** on the plane which is orthogonal to the direction of propagation. Note: in radio astronomy the magnetic vector **B** is chosen for characterisation of polarisation

3.1.2 Mechanisms Producing Polarisation of Astrophysical Objects

There are several physical processes which may be responsible for producing substantial amount of linear and/or circular polarisation in the optical part of the spectrum. Those processes are:

1. Reflection of light from a solid surface (e.g., the surface of a planet or asteroid).
2. Scattering of light on free electrons, atoms, molecules or dust particles.
3. Propagation of light in the presence of the magnetic field (Zeeman and Paschen–Back effects).
4. Cyclotron and synchrotron radiation of accelerated free electrons moving in the magnetic field.
5. Interstellar polarisation of distant stars due to optical dichroism of non-spherical interstellar dust grains aligned by interstellar Galactic magnetic field.

Through the processes of light emission, reflection, scattering and absorption, optical polarimetry can provide us with important information on the properties of matter, such as spatial distribution, density, particle composition and size. It also helps us to derive the properties of stellar and interstellar magnetic fields, e.g., the field strength and geometry.

3.2 Polarisation Modulators, Analysers and Detectors for Optical Polarimetry

Most of optical polarimeters employ a polarisation modulator and polarisation analyser. Polarisation modulator is the optical device which *modulates* the state of polarisation of the incoming light beam with a given frequency. Polarisation analyser is the unit which separates the two orthogonally polarised components of light for measuring their intensities with the *detector*, synchronised with the modulation frequency.

3.2.1 *Modulators: Retarders with Constant Phase Shift (Wave Plates)*

A wave plate is a plane parallel plate made of a uniaxial crystal cut in the way that its optical axis (the direction in which refraction index n_e is minimum) is parallel to the plate surface. The light polarised *along* this axis moves *slower* than the light polarised in the orthogonal direction. Such plate introduces a phase shift, i.e., *retardation* τ between the orthogonally polarised components of the light beam entering the plate (Serkowski 1974):

$$\tau = 2\pi \Delta/\lambda, \quad \Delta = (n_e - n_o)s, \tag{3.4}$$

where s is the thickness of the retarder, λ is the wavelength, n_e and n_o are the refractive indices for the light components polarised parallel and perpendicular to the optical axis of the crystal, respectively, and Δ is the *path difference*.

There are two types of retarders widely used in optical polarimetry: a retarder with $\tau = \pi/2$ (or $\Delta = \lambda/4$) is called *quarter wave plate* (QWP), and with the $\tau = \pi$ (or $\Delta = \lambda/2$), called *half wave plate* (HWP). The QWP changes the circularly polarised light with $I = V$ into linearly polarised light with $I = (Q^2 + U^2)^{1/2}$. The HWP *rotates* the plane of polarisation of the linearly polarised light beam. Thus, if ϕ and θ are the directions of HWP axis and polarisation of the incoming light beam measured with respect to the northern celestial pole, the direction of polarisation plane of the beam emerging from the HWP becomes $2\phi - \theta$.

For a good optical retarder, the retardance should change as little as possible over the optical wavelengths. Currently available *superachromatic* optical wave plates are close to this requirement. For example, the multi-component superachromatic half and quarter wave plates offered by Bernhard Halle have the maximum deviation of retardation $\pm 1.3\%$ (HWP)—$\pm 4\%$ (QWP) and variation of the axis direction $\pm 1.5°$ (QWP)—$\pm 2.0°$ (HWP) over the spectral range 310–1100 nm. They are made of quartz and magnesium fluoride and available in various apertures, up to 50 × 50 mm. Polymer-made superachromatic wave plates offered by Astropribor closely match the optical properties of the best crystal-made retarders, although over the slightly narrower wavelength range. They are more durable and available in a larger set of apertures, up to 60 mm.

A QWP rotated in steps of 90° and placed in front of a polarisation analyser is frequently used to measure circular polarisation v. A HWP rotated by steps of 22°.5 is used for linear polarisation (Stokes q and u) measurements (see Sect. 3.3.1.1). Moreover, a QWP rotated with the steps of 22°.5 can be used to measure all Stokes parameters q, u and v. However, in the latter case the intensity modulation of the light beam due to linear polarisation has twice smaller amplitude, which reduces the efficiency of linear polarimetry by 50% (see e.g., Serkowski 1974).

3.2.2 Modulators: Retarders with Variable Phase Shift (PEMs and FLCs)

Photoelastic or piezoelectric modulator (PEM) is made of non-birefringent slab of material which is stressed using piezoelectric transducer at the natural slab frequency f_0. This allows to greatly reduce the power needed to sustain the standing wave in the slab. PEM operated at the base frequency f_0 acts as a *variable* QWP introducing the periodic phase shift $\lambda/4$ between orthogonally polarised components of the incoming light beam. Placed in front of the analyser, a PEM can be used for direct measurement of circular polarisation. If the PEM is driven at the frequency $2f_0$, the phase shift $\lambda/2$ is introduced and linear polarisation can be directly measured. However, in the latter case, either the PEM itself or the lower section of polarimeter (below the PEM) must be rotated to two orientations separated by 45° to allow measurement of the complete set of Stokes q and u (see Kemp 1981; Hough et al. 2006).

Nominal modulation frequencies of PEM are tens of kHz, well above fluctuations produced in the Earth's atmosphere which makes it well suitable for high-precision polarimetry. Among other advantages are the large useful aperture and possibility to use it in the fast optical beams (up to cone angles of $\leq$50°). The PEMs for optical polarimetry are usually made of fused silica with the anti-reflecting coating optimised for the desired photometric passband.

Ferro-electric liquid crystal (FLC) modulators are made of a thin layer of liquid crystal material, sandwiched between two glass plates. FLCs used in optical polarimetry have a fixed $\lambda/2$ retardation with switchable orientation of the optical axis controlled by applying drive voltage. Thus, they can be used for measuring linear polarisation in a way similar to that with the PEMs. The main advantage of the FLC modulator over the PEM is the nearly square-wave modulation, which makes it more efficient.

FLCs are operated at modulation frequencies in the range from hundreds of Hz to a few kHz which is fast enough to get rid of effects due to atmospheric intensity fluctuations. The FLCs are temperature sensitive devices and must be used in a temperature controlled environment to provide internal stability of the switching rate and the switching angle. FLCs are also known for producing a large instrumental polarisation from multiple internal reflections in the birefringent material between the plates. The magnitude of this polarisation can be in the range of 0.1–0.3% (Bailey et al. 2015) and is highly variable with the wavelength. Special measures must be taken to properly calibrate and remove it.

Similarly to PEMs, FLCs currently available on the market are optimised for different spectral regions. However, unlike the constant phase-shift wave plate which can be made superachromatic, the change of passband from blue to red in the polarimeter equipped with a variable phase-shift retarder requires changing the modulator unit.

3.2.3 Analysers for Optical Polarimetry: Single and Double-Beam Units

A single *polaroid* is probably the very first analyser ever used in optical polarimetry. A simple polaroid is made of hematite crystals, embedded in a polymeric transparent thin film, or layered on a glass surface. Crystals are aligned in one direction. Polaroid possesses a property which is called *optical dichroism*: a strong absorption of light linearly polarised in the direction parallel to the orientation of crystal axis. Thus a rotated polaroid can be used as the simplest *single-beam* analyser for linearly polarised light.

Obvious disadvantage in using polaroid is that half of the intensity of the incoming light beam is lost. Because in the most astronomical polarimetry applications the accuracy is determined by the number of available photons, it is important to retain as much light as possible. For this reason, polaroids and other single-beam analysers (i.e., Nicol and Glan–Thompson prisms) are rarely employed for optical polarimetry presently. However, in certain cases, the usage of polaroids can be justified, e.g., for the imaging polarimetry of extended objects (reflecting nebulae) as, for example, is done in the case of HST polarimeter.

A *double-beam* analyser, or *polarisation beam-splitter*, has the advantage of splitting the incoming light beam into two orthogonally polarised beams in such a way that both of them can be measured simultaneously. Such analyser utilises the property found in some uniaxial crystals, which is called a *birefringence*. In these crystals the orthogonally polarised components of the electromagnetic wave travel with different speed, i.e., have different refractive indices. The birefringence degree Δn is defined as a difference between the refraction indices of the *extraordinary* and the *ordinary* rays $n_e - n_o$. The extraordinary (e-ray) is polarised in the direction parallel to the crystal *principal axis*, and the ordinary (o-ray) is polarised in the perpendicular direction. There are crystals with negative Δn (beryl, calcite) and positive Δn (magnesium fluoride, quartz).

3.2.3.1 Double-Beam Polarisation Analysers for Optical Polarimetry: Plane-Parallel Calcite Plate, Savart Plate and Wollaston Prism

Because a double-beam polarisation analyser is the key element in many existing optical polarimeters, it is worth to mention the most important requirements which a good optical polarisation beam-splitter should meet:

1. High transparency in the optical spectral range.
2. Constant birefringence Δn over the optical wavelengths.
3. No aberrations are introduced to the telescope exit pupil.
4. The same optical path length for the e and o-rays.
5. High polarisation efficiency, i.e., the completely non-polarised light beam is split in two 100% orthogonally polarised beams with intensities I_e and I_o so that $I_e = I_o$.

There are three types of double-beam polarisation analysers most frequently used in optical broadband and spectropolarimetry. The most simple double-beam polarisation analyser is a *plane-parallel calcite plate*. It is made of calcite crystal cut parallel to the plate of cleavage (Fig. 3.2). Such plate splits stellar image in two by introducing a shift of the e-ray. The image separation (for $\lambda = 550\,\text{nm}$) is $d = 0.109 \times h$, where h is the calcite thickness. Being simple, this polarisation device can be manufactured up to the required thickness and aperture at a reasonable cost. Because it is a single-piece unit, it is not so fragile in comparison with other devices and can be handled with less caution. However, the optical path for the e and o-rays is not the same, and both images cannot be focused precisely on the same plane. Moreover, the light losses by reflection on calcite surface for the e and o-rays are different. Single calcite plate introduces a noticeable *dispersion* into the e-stellar image in the blue—near UV wavelengths. Despite of these shortcomings, the plane-parallel calcite plate is often used in applications where the image quality is not critical.

Most of the inconvenient features of a single calcite plate are avoided in *Savart plate* which is made of two plane-parallel calcite plates cemented with their principal axes at 45° to the surface normal and rotated through 90° with respect

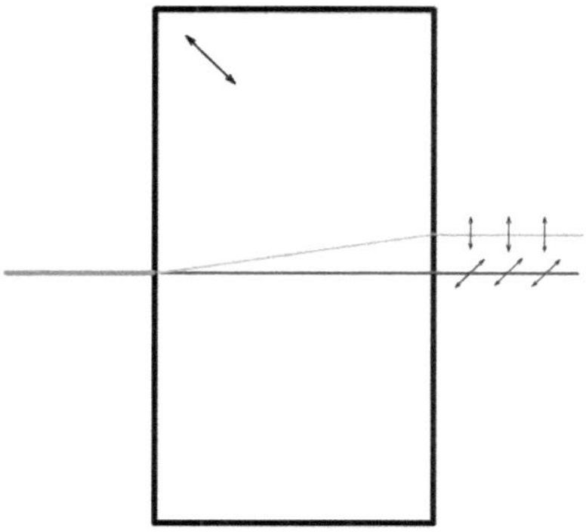

Fig. 3.2 Plane-parallel calcite plate. The principal axis has the angle of 45° to the plate normal. The direction of travel of the o-ray (blue) is not changed, but the e-ray (green) is shifted in the parallel direction

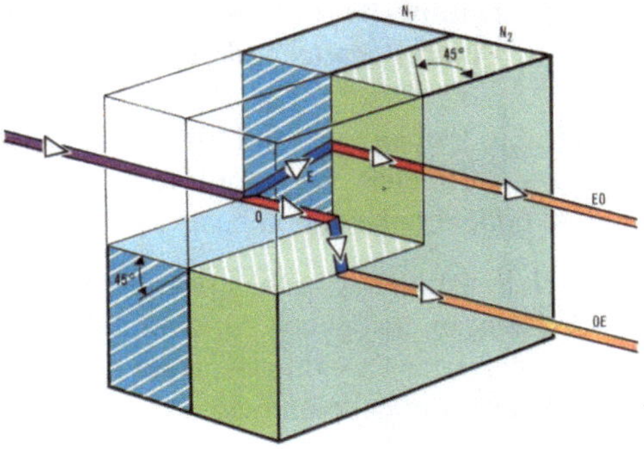

Fig. 3.3 Savart plate. When entering into the second crystal, the o-ray produced by first one becomes the e-ray and is displaced in the direction perpendicular to the first beam displacement. The result is two rays displaced along a diagonal. The optical path difference for the o- and e-rays is zero for normal incidence

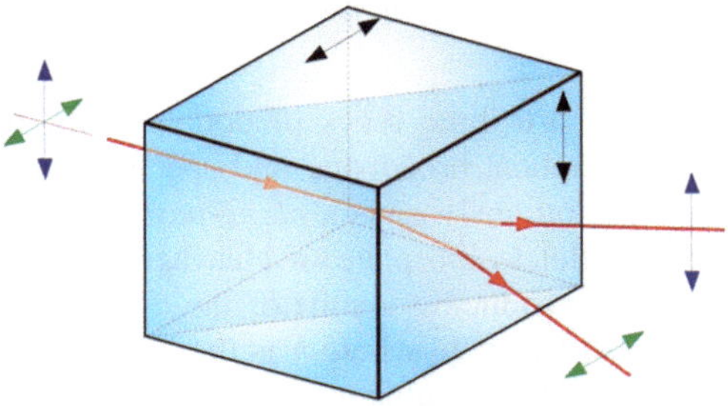

Fig. 3.4 Wollaston prism made of two cemented calcite wedges with orthogonally directed optical axes. The angle of separation of the o- and e-rays can be within 5°–20°

to each other (Fig. 3.3). Savart plate seems to be the best analyser for the stellar polarimetry with panoramic (i.e., CCD camera) detectors. Nevertheless, accurate measurements of polarisation might not be possible in crowded regions of the sky due to overlapping of the e- and o-images of the target and the nearby field stars. This problem can be solved by mounting the polarimeter on a rotator which can be turned at certain angle to change the direction of the stellar images split on the sky in a way to avoid the overlapping. Another solution is to put a stripe mask in the focal plane of the telescope, but this requires using re-imaging optics (collimator and camera lenses).

If polarimetry is done with photomultipliers, a large separation between the beams is convenient. The best double-beam analyser which provides a large image separation and high efficiency is the *Wollaston prism*. The *two-wedged* Wollaston prism (Fig. 3.4) is made of two cemented (calcite) or optically contacted (quartz, magnesium fluoride) wedges with optical axes orthogonal to each other and perpendicular to the direction of propagation of the incoming light beam. The angular separation between the outgoing e- and o-rays in the Wollaston prism depends on the number of wedges N, construction angle of the outer wedges A and birefringence of material as:

$$\alpha \simeq 2 \arctan[(N-1)\ |n_o - n_e|\ \tan A]\ . \tag{3.5}$$

In the optical wavelengths, calcite is the preferred material, because it is transparent from 350 to 2200 nm and has a high birefringence. Thus, two-wedged calcite Wollaston prisms offered by Bernhard Halle can provide the beam separation up to 20°. If a higher beam separation is required, the *three-wedged* calcite Wollaston prism can be used. It provides image separation up to 28° and gives the best image quality, approaching closely the ideal analyser for the optical polarimetry (Serkowski 1974). If a wider spectral range is needed, i.e., with the near UV and near-IR coverage, quartz or magnesium fluoride must be used. However, due to the much lower birefringence of these crystals, the beam separation is by an order of magnitude smaller. In comparison with the single calcite plate, the Wollaston prism is a more complex and expensive optical device and must be handled with care.

3.2.4 Detectors for Optical Polarimetry: CCDs, PMTs and APDs

There are three types of detectors used in optical polarimetry: CCD cameras, photomultiplier tubes (PMTs) and avalanche photodiodes (APDs). Each type of detector has its own advantages, disadvantages and areas of applications. There are also emerging detectors for the optical and NIR spectral ranges, namely MKIDs and Saphira APD arrays. These are likely to play an important role in future polarimeters, particularly Saphira based systems for NIR (Finger et al. 2014).

3.2.4.1 Detectors for Optical Polarimetry: CCD Cameras

CCD cameras are currently the most widely used detectors for photometry and spectroscopy at the optical wavelengths. Among their main advantages are the high quantum efficiency (QE), great versatility and convenience of data acquisition. Moreover, being panoramic (multi-cell) registration devices, they are very well suited for spectroscopic and imaging applications. This feature is also very useful for *double-beam polarimetry*, because it allows us to record two orthogonally polarised stellar images or spectra simultaneously on the same detector, so that variations in atmospheric seeing and transparency will have exactly the same effect on both. Moreover, the intensity of background sky is registered automatically on the pixels surrounding stellar images and this eliminates a necessity to measure intensity and polarisation of the sky separately.

The CCD cameras employed in optical polarimetry can be roughly divided into two classes: smaller units with the thermoelectric cooling and pixel area up to about 2048 × 2048 (used mostly for broadband polarimetry) and large pixel area (or mosaic) units with cryogenic cooling (used for spectropolarimetry, imaging polarimetry and in multi-functional instruments with optional *"polarimetry mode"*). The cameras of the first class can be connected to the computer via standard USB

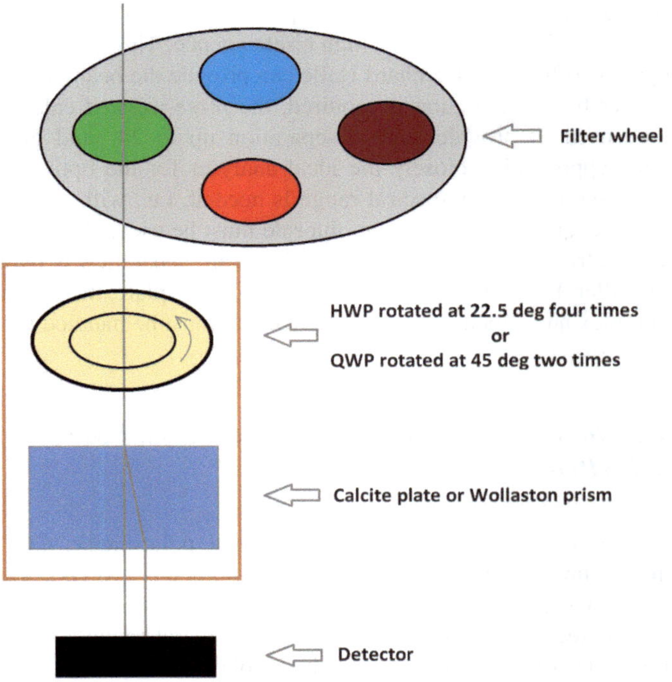

Fig. 3.5 The layout of the CCD optical polarimeter. Polarisation components can be installed between the CCD detector and filter wheel. After passing the beam-splitting analyser, the incoming light ray is split parallel in two orthogonally polarised rays. Instead of plane-parallel calcite shown on the figure, the other analyser, e.g., Wollaston prism can be used

port used for the camera control and data acquisition. They are often equipped with the high-grade, low noise CCD detector with sensitivity optimised either for the blue-visual or visual-red spectral ranges with the maximum QE $\geq 90\%$. Many such cameras are available currently on the market at a reasonably low cost. Thus, they are well suitable for building a low-budget instrument for stellar polarimetry at the optical wavelengths. The principal design of such CCD polarimeter is shown in Fig. 3.5.

In broadband CCD polarimetry the data reductions can be done with *aperture photometry* method which normally includes calibration (subtraction of dark, bias and flat-fielding), sky subtraction, determination of intensities of e- and o-stellar images and converting them to the *intensity ratios* $Q_i = I_e/I_o$. The Q_i must be determined for the each orientation of the wave plate $i = 0°, 22\overset{\circ}{.}5, 45°, \ldots$ The data reduction in CCD spectropolarimetry is done in the similar way: the e- and o-spectra must be extracted and, after wavelength calibration, the intensity ratios $Q_i(\lambda)$ are computed. The Stokes parameters of linear or circular polarisation are obtained from the intensity ratios Q_i with the formulae given in Sect. 3.3.1.1.

The CCD detectors, while being convenient for optical polarimetry, have also some shortcomings, e.g., the relatively small full-well capacity ($\leq$200000e^-/pixel) which results in fast pixel saturation (overexposure) of the images of the bright stars. The image download time for the small-sized CCD camera is typically around of $\simeq$1–2 s, while for the large, mosaic-type CCD detector it can be in range from few tens of seconds up to $\simeq$1 min. Thus, CCD polarimetry of bright targets, necessarily done with short exposures, may become very inefficient, because most of the time is spent on image downloads.

The download time problem is removed in CCD cameras with *fast frame transfer*. The CCD chip in such a camera has two areas, one is for the image exposure and another for the temporal image storage. As soon as the exposure is taken, the image is swiftly shifted vertically from the exposure area into the masked storage area from where it is transferred to the camera readout register. Thus, for example, ANDOR iXon Ultra 897 EMCCD camera can provide an impressive frame rate up to 56 full-frame images/sec. Even higher frame rate is achieved in the sub-frame readout mode. The frame-transfer EMCCDs are rather expensive and available only with a relatively small pixel area, up to 1024 $\times$ 1024. Nevertheless, such camera can be considered as a very good option for building an efficient and versatile CCD polarimeter operating in the optical wavelengths.

3.2.4.2 Detectors for Optical Polarimetry: Photomultiplier Tubes

Although most of the existing astronomical polarimeters employ CCD cameras, *photomultiplier tubes (PMTs)* are still used in optical polarimetry. The main advantages of PMTs, making them a good choice as detectors for certain applications, are the instant readout, wide dynamical range and ability to faithfully register high fluxes. Thus, a typical PMT retains linear response up to the level of 10^6e^-/s in the pulse counting mode and up to $\leq 10^8e^-$/s in a direct photo-current amplification mode.

Polarimetry is a *photon hungry* observational technique. The accuracy is critically dependent on the amount of registered photons or ADUs as:

$$\sigma_{q,u,v} = kN^{-1/2}, \tag{3.6}$$

where k is the analyser efficiency and $N = N_e + N_o$ is the total number of registered ADUs or counts. Thus, in order to achieve the precision at the level of 10^{-6} (part per million or ppm), 10^{12} ADUs must be collected. The early pixel saturation, a bottleneck for the CCD polarimeter when it is used for observation of the bright targets, normally is not an issue for the polarimeters equipped with a PMT detector. Moreover, the "instant" readout allows one to use PMT with the fast polarisation modulator such as PEM or FLC. Not surprisingly, the PMT detectors have been used in the most precise polarimeters such as HIPPI (Bailey et al. 2015) and POLISH2 (Wiktorowicz and Nofi 2015).

The major disadvantage of the PMT detectors is their relatively low quantum efficiency which is typically in the range 10–30%. The best in their class, super bialkali (SBA) and ultra bialkali (UBA) PMTs have QE $\simeq$ 40% with the peak near 400 nm. Thus, the polarimeters with PMT detectors, while being able to achieve an accuracy level of few ppm for bright stars, still cannot compete with the CCD polarimeters in the low flux domain. Because the PMT is a single-cell detector, it is suitable only for single target polarimetry of point-like objects. All existing PMT polarimeters utilise a diaphragm in the focal plane of the telescope, where the measured star is placed. The intensity and polarisation of sky background are measured separately by either pointing the telescope away from the star to the clear area of the sky or with a separate sky registration channel.

3.2.4.3 Detectors for Optical Polarimetry: Avalanche Photodiodes

The *avalanche photodiode* or APD is the semiconductor analog to the PMT. In an APD, as with any other photodiode, incoming photons produce electron-hole pairs; however, the APD is operated with a large reverse *bias voltage* (up to 2 kV), which accelerates photoelectrons. Those electrons collide with atomic lattice, releasing additional electrons via secondary ionisation. The secondary electrons are also accelerated, which results in an *avalanche* of charge carriers, hence the name.

Like the PMTs, APDs can work with high modulation frequencies (up to a few MHz) and can tolerate even higher fluxes, up to $\leq 10^9 e^-/\mathrm{s}$. They have a higher QE (up to 80% in the near IR) in comparison with the PMTs, but smaller or comparable QE in the blue wavelengths. The APDs have been used as detectors in PlanetPol (Hough et al. 2006), POLISH (Wiktorowicz and Matthews 2008) and OPTIMA (Kanbach et al. 2008) polarimeters. Although the APD detectors may be better suitable for high-precision polarimetry of the bright targets in the near IR wavelengths, they do not offer significant advantage over the PMTs in the blue and visual spectral ranges. In comparison with the PMT (and CCD) detectors, the APDs have a higher dark current and are more noisy. Another disadvantage is the small effective photosensitive area, typically a few mm^2, which requires a very precise positioning of the telescope exit pupil on the detector surface.

It seems that APDs are becoming less popular detectors for high-precision optical polarimeters. The HIPPI (a successor to PlanetPol) and POLISH2 (a successor to POLISH) are both employing PMTs.

3.3 Broadband, Imaging and Spectropolarimetry in the Optical Wavelengths

The various methods of measuring polarisation in optical wavelengths can be divided into three distinct groups: *broadband polarimetry*, i.e., polarimetry of stellar-like objects with the use of broadband optical filters; *imaging polarimetry* of extended objects, like nebulae, usually made also with filters; and *spectropolarimetry*, i.e., measuring polarisation over the regions of the optical spectrum with certain *spectral resolution*. Imaging polarimetry can be also done for the stellar-like targets. This method of observations has the advantage of simultaneous registration of sky background and comparison object which effectively shortens any exposure time.

3.3.1 Instruments for Broadband Optical Polarimetry

There are many instruments in this group utilising different approaches and focused on achieving different goals. We give here a brief description of the most commonly used techniques and designs currently employed in optical broadband polarimetry.

3.3.1.1 Double-Image CCD Polarimeters

The layout of the most commonly used broadband optical CCD polarimeter is shown in Fig. 3.5. It consists of four parts: (1) detector (CCD camera); (2) filter wheel; (3) rotatable half wave (HWP) or quarter wave plate (QWP); and (4) double-beam polarisation analyser such as plane-parallel calcite (or Savart) plate or Wollaston prism. The rotation of the HWP (QWP) is normally done with the use of a high-precision *stepper motor*.

In order to measure *linear polarisation*, a series of CCD images must be taken at the four positions of the HWP: 0°, 22°.5, 45° and 67°.5. Next step is to measure the intensities of the two orthogonally polarised stellar images I_e and I_o and compute their ratio $Q = I_e/I_o$. Then the normalised Stokes parameters q and u can be computed as:

$$Q_m = Q_0 + Q_{22.5} + Q_{45} + Q_{67.5}, \tag{3.7}$$

$$q = (Q_0 - Q_{45})/Q_m, \tag{3.8}$$

$$u = (Q_{22.5} - Q_{67.5})/Q_m. \tag{3.9}$$

Circular polarisation can be measured with the QWP in the following way: two CCD images are taken at the positions of 45° and 135° (or −45° and 45°) and the degree of circular polarisation $P_C = v$ is computed as:

$$Q_m = Q_{45} + Q_{135}, \tag{3.10}$$

$$v = 0.5\,(Q_{45} - Q_{135})/Q_m. \tag{3.11}$$

The setup shown in Fig. 3.5 can be used as a polarimetric *add-on* to a CCD photometer, a low resolution spectrograph or a multi-functional instrument. The

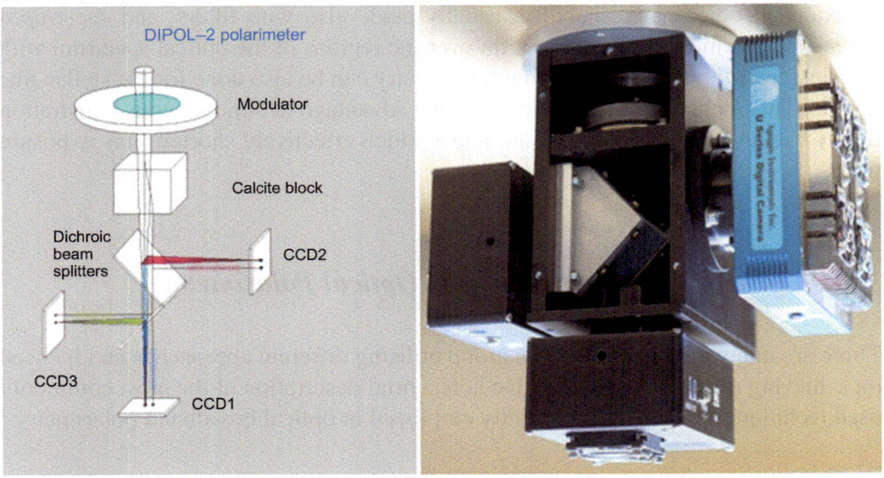

Fig. 3.6 Left: Schematic layout of the main components of the DiPol-2 polarimeter. Right: photo of DiPol-2 taken with the front cover removed. Polarisation unit consisting of calcite, exchangeable retarder and stepper motor is installed in the upper section. Dichroic mirrors are installed in the lower section. Apogee Alta U47 CCD (B-band) is on the right side, two ST-402ME CCDs are on the left (R-band) and on the bottom (V-band)

rotated wave plate and two-beam polarisation analyser can be moved into the optical path and removed when they are not necessary. This method of measurements, being simple, is probably the most efficient for the broadband and low-resolution spectropolarimetry of faint point-like sources. Simultaneous registration of two polarised images (or spectra) helps to maximise efficiency and to avoid most of systematic errors. Currently it is used in several existing instruments, e.g., ALFOSC at the NOT, ORM (La Palma) and EFOSC and FORS2 at the ESO (Chile).

The layout shown above can be easily expanded and enhanced. For example, the DiPol-2 polarimeter shown in Fig. 3.6 does not have a filter wheel, but employs two dichroic beam-splitters installed after the polarisation unit. The light beam is split into three passbands (B, V and R) registered with three CCD cameras simultaneously. Because DiPol-2 uses a plane-parallel calcite plate as a polarisation analyser, without having a focal plane mask, the o- and e-components of the sky overlap everywhere on the CCD image. Thus, the sky is not split into polarised components and sky polarisation automatically cancels out when the total sky intensity $I_e + I_o$ is subtracted from the stellar images in data reductions (Piirola et al. 2014). DiPol-2 has been built in three copies and used for high-precision polarimetry of the low-mass X-rays binaries (Kosenkov et al. 2017; Veledina et al. 2019) (see Sect. 3.5.1), early type binaries (Berdyugin et al. 2016, 2018) (see Sect. 3.5.2), novae (Harvey et al. 2018) and study of the interstellar polarisation in the vicinity of the Sun (Frisch et al. 2015). DiPol-UF, a successor to DiPol-2, currently being built at the University of Turku, employs three ANDOR iXon Ultra 897 EMCCD cameras as detectors and has a retractable polarisation unit which allows operation of the instrument also as a fast three-band photometer (simultaneous in BVR).

3.3.1.2 High-Precision Double-Image Polarimeters with High-Frequency Modulators

There are several examples of currently existing instruments employing the techniques of fast modulation. One example is the UBVRI polarimeter *TurPol* (Piirola 1973, 1988), the principal layout is shown in Fig. 3.7. In TurPol, the modulation of the incoming light, split by the calcite plate into two parallel beams, is done with a rotating mechanical chopper. The frequency of modulation is 25 Hz, which is enough to eliminate most of atmospheric effects. TurPol is equipped with a discretely rotated interchangeable $\lambda/2$ and $\lambda/4$ wave plates and is able to measure linear or circular polarisation of the stellar-like objects with an accuracy up to a

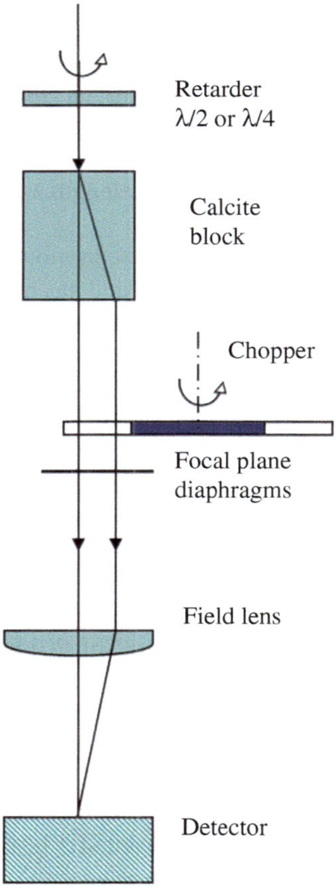

Fig. 3.7 The schematic layout of the TurPol polarimeter. The rotating chopper is installed after calcite block and in front of one of the selectable double diaphragms placed in the diaphragm wheel. After the double diaphragm, the field lens forms the exit pupil on the PMT detector. The registration of the intensities of e- and o-images of the star by the PMT is synchronised with the chopper rotation. In the real instrument, the light is split in five (U, B, V, R and I) beams by the four dichroic mirrors installed after the diaphragm section and registered simultaneously with the five PMTs

few times per 10^{-5} in the five UBVRI passbands simultaneously. The effect of sky polarisation is cancelled out due to the overlapping of the orthogonally polarised sky images in each of the two diaphragms. The total sky intensity $I_e + I_o$ is measured separately by pointing the telescope at the nearby clear area of the sky. TurPol has been made in several copies and still used for observations of polarisation of various types of variable stars, studies of interstellar polarisation and even for observations of exoplanets (Berdyugina et al. 2011). One of the main applications has been circular and linear polarimetry of magnetic cataclysmic variables (e.g., Piirola et al. 1987, 1990, 1993, 2008), with orbital periods of a few hours and rapid variability on time scales from a fraction of a second to minutes. Simultaneous measurements in the different passbands are therefore essential.

Another example of a high-precision broadband instrument which was able to reach an accuracy in stellar optical polarimetry of the order of few ppm is the PlanetPol (Hough et al. 2006). This instrument uses PEM as a polarisation modulator, three-wedged Wollaston prism as polarisation analyser and APD as detector. The working frequency of the PEM is $2f_0 = 40\,\mathrm{kHz}$ and it operates as the $\lambda/2$ retarder. To measure both Stokes parameters of linear polarisation, the whole instrument is rotated by 45°. Moreover, to minimise systematic errors, the lower section of the instrument with analyser and detectors is rotated from $-45°$ to $+45°$. The PlanetPol has two independent registration channels, one for the star (on axis) and one for the sky (off axis). The instrument has been designed to measure polarisation in a wide wavelength range 470–910 nm, centred at the $\lambda = 690$ nm. The effective wavelength λ_{eff} is dependent on the spectral type and lies in the range of 735–805 nm.

PlanetPol has been mounted in the Cassegrain focus of the 4.2 m WHT at the ORM (La Palma) and has measured polarisation of a number of bright stars ($m_V \leq 5\overset{\mathrm{m}}{.}0$) with very impressive accuracy of few ppm (Bailey et al. 2010).

HIPPI, a successor to PlanetPol (Bailey et al. 2015), employs the FLC modulator, operated at the frequency of 500 Hz, and PMT detectors. Unlike PlanetPol, it is optimised for high-precision polarimetry in the blue wavelengths, in the spectral region $\sim$ 400–700 nm. Mounted on the 3.9 m Anglo-Australian Telescope, it was used for high-precision optical polarimetry of the nearby FGK dwarfs in the brightness range $m_V \simeq 0\overset{\mathrm{m}}{.}4$–$6\overset{\mathrm{m}}{.}0$. The achieved accuracy of polarisation measurements (depending on brightness) is in the range of 1.5–10 ppm (Cotton et al. 2017).

POLISH2 polarimeter uses two PEM modulators operated at different frequencies to measure three Stokes parameters q, u and v in the UBV passbands (Wiktorowicz and Nofi 2015). Reported accuracy of the instrument for observations of the bright "naked-eye" stars with the 3 m Shane telescope of the Lick Observatory is at the level of few ppm.

Such an impressive precision of polarimeters with fast modulators can be only achieved with careful calibration procedure which is necessary to take into account and minimise all sources of systematic errors. The accuracy at the level of 10^{-5}–10^{-6} is only possible for bright stars and on telescopes with the aperture size $\geq$3 m. Thus, the primary goal of these instruments is limited to high-precision optical polarimetry of bright targets. An example of such application is a study of the interstellar polarisation and direction of the local galactic magnetic field in the vicinity of the Sun (Frisch et al. 2015; Cotton et al. 2019). The intrinsically non-polarised stars of A, F and G spectral types located at the distance <40 pc are bright, but show very low ($P \leq 0.01\%$ or $\leq 10^{-4}$) degree of interstellar polarisation which can only be measured with the instruments providing precision at the level of 10^{-5} or better.

3.3.1.3 High-Speed Broadband Polarimeters for the Optical Wavelengths

Measuring polarisation in the optical wavelengths with time resolution $\leq$1 s is a challenging task. An instrument capable to do this must be able to modulate the polarisation of incoming light beam so as to have all Stokes parameters recorded in a short exposure time, and also to maintain a high light throughput. This is rather difficult to achieve in practice. The general approach for fast optical polarimetry is a continuous rotation of the wave plate or polaroid with the frequency of few to 10 Hz.

HIPPO polarimeter, designed in the South African Astronomical Observatory (SAAO) is an example of an instrument built for high-speed photo-polarimetry of rapidly varying polarised astronomical sources, with particular interest in polarimetry of magnetic cataclysmic variables (mCVs). The layout of the instrument is shown in Fig. 3.8. HIPPO makes a single polarisation measurement every 0.1 s (Potter et al. 2010). The acquired raw data, if necessary, can be binned to any integer multiple of 0.1 s.

Among the results obtained with HIPPO are the discovery of polarised QPOs in polar IGR J14536-5522 (Potter et al. 2010) and white dwarf pulsar AR Sco (Potter and Buckley 2018). In both cases, the observations have been done with the 1.9 m telescope of the SAAO. The pulse periods, seen also in brightness variations, are 5.2 and ~2 min, respectively.

RINGO3[2] mounted on the robotic 2.0 m Liverpool Telescope at the ORM (La Palma) employs a rotated polaroid in front of the collimator, after which the light is split by two dichroic mirrors onto three Andor iXon 3 EMCCDs. After the year 2013, a Lyot depolariser has been installed after the polaroid to get rid of the large instrumental polarisation arising due to the dichroic mirrors. Initially the frequency of polaroid rotation was set to 1 Hz, which has allowed to measure the linear polarisation two times per second. Currently the rotation frequency is set at ~0.4 Hz.

The main shortcomings of the RINGO3, apart from rather large instrumental polarisation, are significant overheads for initialising the instrument for the exposure and the loss of 50% of light intensity in the polaroid. Thus, the eight images acquired during one rotation cycle must be stacked over many cycles in order to achieve a

[2]http://www.telescope.livjm.ac.uk/TelInst/RINGO3/.

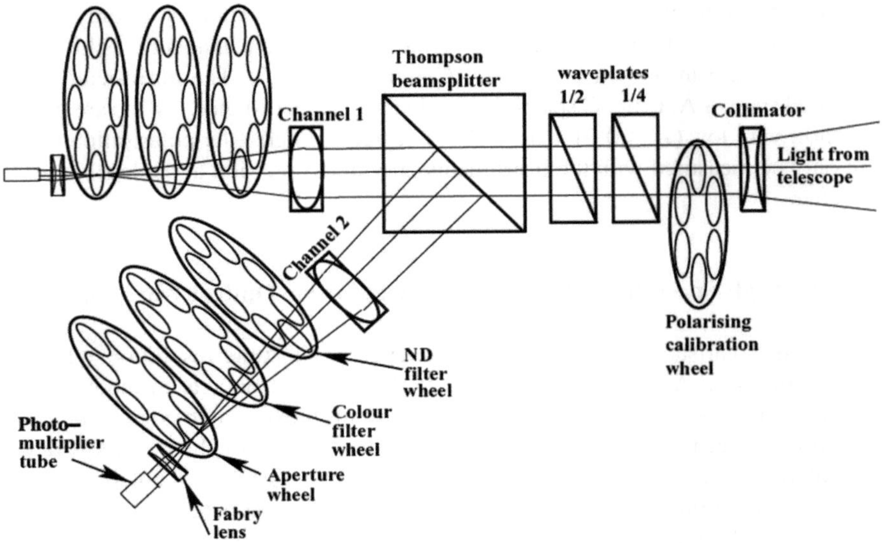

Fig. 3.8 The schematic layout of the HIPPO polarimeter. The HWP and QWP, placed in the collimated beam are contra-rotated at frequency of 10 Hz modulating the polarisation of the incoming light beam split by Thomson prism in e- and o-rays and registered separately by the two PMTs placed in Channels 1 and 2. Linear and circular only modes of polarimetry are possible and each channel can be used for independent measurements of three Stokes parameters with different filters. From Potter et al. (2010)

reasonable S/N. The recommended minimum integration time is 20 s (Słowikowska et al. 2016).

The GASP polarimeter (Collins et al. 2013; Kyne et al. 2016) is a DOAP (division of amplitude polarimeter) which uses two EMCCDs with a timing resolution down to 500 μs. It measures both linear and circular polarisation simultaneously with no moving parts.

We note that although the best high-speed polarimeters like HIPPO and GASP can sample the polarisation with time resolution $\leq$0.1 s, the acquired Stokes parameters are usually binned over time intervals $\geq$10 s to achieve sufficient S/N. Thus, in the case of AR Sco, binning over 10 s has been made (Potter and Buckley 2018). The binning interval for the polarisation data collected for IGR J14536-5522 is not mentioned in the original paper, but it can be roughly estimated from the plots shown in Figs. 6 and 7 of the paper by Potter et al. (2010) as $\sim$30–60 s. Both stars show rather strong (few percent of polarisation) signal which is strictly periodic and this helps to reveal it from the raw data accumulated over the sufficiently long time span. Because the accuracy of polarimetry is critically dependent on the count rate, the small number of ADUs registered over short time interval is generally the problem in detecting weak short-term polarisation signals.

It is worth to mention that for time resolution of ≥10 s the "traditional" CCD polarimeter with discretely rotated wave plate and fast readout CCD camera offers comparable, if not better performance. The single turn of the wave plate is done by stepper motor in 0.2–0.3 s. Thus, the time lost in wave plate rotation can be ≤0.5 s and ≤1.0 s for one cycle of circular and linear polarisation measurements, respectively. Stepper motors are compact, precise and provide reliable long-term operation. Moreover, modern EMCCD cameras offer frame rates of few tens per second and are able to "stack" images, thus allowing to shorten the sampling time interval considerably.

3.3.2 Instruments for Optical Imaging Polarimetry

3.3.2.1 Optical Imaging Polarimetry with Polaroids

The simplest way to measure polarisation of extended objects is to take a series of CCD images with a polaroid filter. Normally, four images are taken at 0°, 45°, 90° and 145° orientations of the polaroid. Because the images for the polarimetry cycle are not taken simultaneously, any variations in the atmospheric transparency will give spurious polarisation signal. Hence, this method of measurement can be applied only in strictly photometric nights. The calibration flat field must be obviously taken separately for each orientation of the polaroid. This method of imaging polarimetry has been used at the NOT (La Palma) with the ALFOSC (see, for example, Piirola et al. 1992; Harjunpää et al. 1999).

3.3.2.2 Optical Imaging Polarimetry with a Double-Beam analyser and Stripe Mask

The problems with atmospheric transparency variations and light losses in the polaroid can be solved with the use of a double-beam analyser and a *stripe mask*. The FOcal Reducer and low dispersion Spectrograph (FORS2) mounted on the Cassegrain focus of the 8.2 m UT1 telescope at the ESO (Chile) is equipped with a mask that can be inserted in the focal plane before the collimator, to avoid overlapping of the two orthogonally polarised beams produced by the Wollaston prism on the CCD. With this mask, a full scan of the imaging field is achieved by taking two series of frames displaced by $22''$.[3] This allows one to do imaging polarimetry of extended objects and crowded stellar fields.

Unfortunately, FORS has a strong field dependent off-axis instrumental linear polarisation which seriously compromises the performance of the instrument for

[3]http://www.eso.org/sci/facilities/paranal/instruments/fors/doc/VLT-MAN-ESO-13100-1543_v83.pdf.

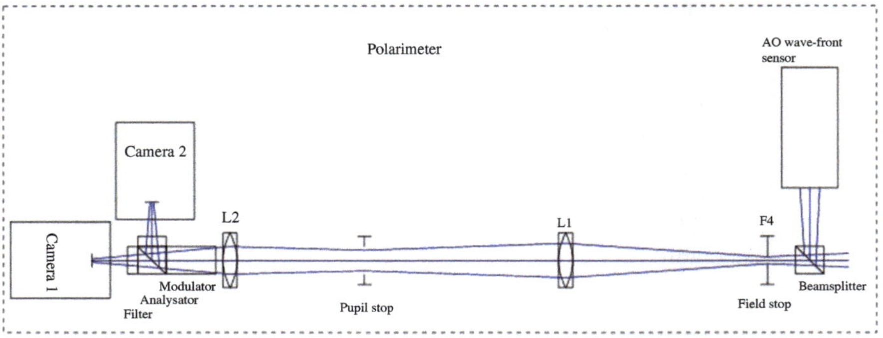

Fig. 3.9 The schematic layout of the GPP imaging polarimeter. The field stop (iris diaphragm) and pupil stop are used to block the unwanted part of the light. L1 and L2 are collimating and imaging lenses. After L2 the modulator is installed, followed by PBC unit. The filters can be inserted either in front of modulator or after analyser. From Gisler et al. (2016)

multi-target or extended object polarimetry.[4] Hence, this instrument is mostly used for broadband and spectropolarimetry of single targets in the centre of the field of view.

3.3.2.3 Optical Imaging Polarimetry with Fast-Frequency Modulators and Fast-Readout CCDs

The idea of using fast-readout EMCCD cameras in combination with the FLC modulators for optical imaging polarimetry is attractive. An example of this approach is the GREGOR Planet Polarimeter (GPP, Gisler et al. 2016). This design eliminates the necessity to use the stripe mask and allows recording of the two orthogonally polarised images simultaneously and synchronously with the fast modulation and two EMCCD cameras.

The layout of GPP is shown in Fig. 3.9. GREGOR is a 1.5 m solar telescope located at the Observatorio del Teide, Tenerife. This telescope, which is a double Gregory-type design, can be also employed for night-time observations. The polarimeter is located on the stationary optical bench after the 50/50 cube beam-splitter which reflects part of the light to the wave front sensor. The transmitted light is used by the polarimeter. The camera lens (see Fig. 3.10) images the focal plane to the CCD cameras with a resolution of $0.''086$/pixel. The polarisation unit consists of modulator and analyser. The modulator unit has two FLCs and two wave plates which allows to measure all Stokes parameters. A polarising beam-splitter cube (PBC) is used as the analyser. Two interchangeable modulators optimised for the blue-visible and visible—read spectral regions can be used. The modulation

[4]http://www.eso/org/sci/facilities/paranal/instruments/fors/inst/pola.html.

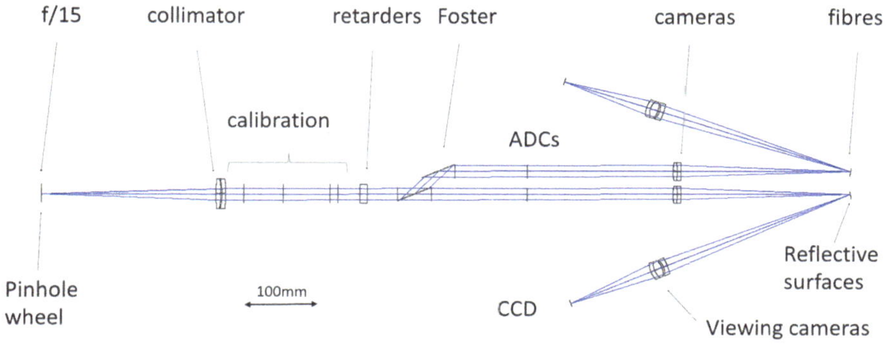

Fig. 3.10 Optical design of the PEPSI spectropolarimeter. The light enters from the left and exit into the fibres to the spectrograph section on right. The fibres are positioned on the central pinholes in the reflective surfaces. The individual optical components are marked. ADCs are the atmospheric dispersion correctors. The calibration components and QWP unit are retractable. From Strassmeier et al. (2015)

frequency is determined by the ANDOR iXon Ultra 897 camera frame rates. For the full frame images it is 14 Hz. For the sub-frame imaging mode, e.g., used for polarimetry of Uranus and Neptune, with the frame size of 128×128 pixels, the modulation frequency is 150 Hz (see Gisler et al. 2016).

The polarimeter can measure all Stokes parameters either with a single or two CCD cameras at the same time. Due to multiple reflections in telescope mirrors and alt-azimuth mount, the telescope instrumental polarisation is high and variable. To account for it, GREGOR has a special polarimetric calibrating unit installed in the telescope before the first diagonal mirror. The calibration procedure consists of taking calibration measurements at the same position as the object observed at the sky. Currently only relative calibration of polarisation within the image is possible, but the artificial unpolarised flat field light source will be implemented in the future. This should allow an absolute calibration of polarisation image.

3.3.3 Instruments for Optical Spectropolarimetry

In many cases, spectropolarimetry in the optical wavelengths can be done by adding the polarisation modulator and analyser to the optical path of the spectrograph. This method is well suitable for low and intermediate resolution spectropolarimetry and has been implemented in many existing instruments (see Sect. 3.3.1.1). The processes of data acquisition and data reduction do not differ much from those applied for traditional spectroscopy where only $I(\lambda)$ is registered (Sect. 3.2.4.1.)

3.3.3.1 Instruments for Low and Mid-Resolution Optical Spectropolarimetry

One example of a multi-functional instrument with spectropolarimetry observation mode implemented is FORS2. The polarisation unit consists of a discretely rotated wave plate (HWP or QWP), placed after the collimator into the parallel beam with the dedicated swing arm, and Wollaston prism, mounted in the uppermost filter wheel. With the inserted polarisation optics, spectropolarimetry can be done with most of the available grisms, with the spectral resolution $\lambda/\Delta\lambda$ from 450 to 2500. FORS2 employs superachromatic wave plate mosaics (3×3) and can be used for single and multi-object spectropolarimetry in the spectral range 330–1100 nm.

Spectropolarimetry can be done with the two CCD mosaics, either with the two $2k \times 4k$ MIT CCDs optimised for the red part of spectrum or with two $2k \times 4k$ E2V CCDs particularly sensitive in the blue range. Because FORS2 is mounted in the Cassegrain focus, it benefits from the very low telescope polarisation (in the centre of the field of view) and thus is well suitable for the absolute linear and circular spectropolarimetry of faint objects, being one of the best instruments in this class. FORS1 (a retired twin of FORS2) has been extensively used for the study of stellar magnetic fields with spectropolarimetry (see, for example, Bagnulo et al. 2002, 2006; Vornanen et al. 2010).

Spectropolarimetry with ALFOSC on the 2.52 m Nordic Optical Telescope (NOT) at the ORM (La Palma) is done with the HWP or QWP plates mounted in the converging beam (before the focal plane) and calcite beam-splitter mounted in the aperture wheel. The spectral resolution in the wavelength range 320–1100 nm can be selected from 210 (with the low-res, wide spectral range grism) up to 10,000 (high-res grism, centred at the Hα).

3.3.3.2 Instruments for High-Resolution Optical Spectropolarimetry

The most important application of high-resolution spectropolarimetry is for study of magnetic fields in solar-type stars. Spectral resolution higher than ~50,000 normally requires an échelle spectrograph.

Good example of a polarimeter implemented to a high-resolution échelle spectrograph is PEPSI (Potsdam Echelle Polarimetric and Spectroscopic Instrument) built for the 8.2 m large binocular telescope (LBT). The design allows to measure all four Stokes parameters (I, Q, U and V) with two identical independent polarimeters at the spectral resolution of 130,000 (Strassmeier et al. 2015). These polarimeters are installed at the straight-through f/15 Gregorian foci of the LBT. The entire PEPSI light feed is split into the *integral light* part with the permanent focus and *polarimetric-light* part which is available after dismounting other instruments (see Strassmeier et al. 2015, for further details).

The polarimeter units are of double-beam type with a Foster prism beam-splitter used as the polarisation analyser. For measuring linear polarisation the analyser unit is rotated. For measuring circular polarisation, a superachromatic QWP is inserted into the light beam. The principal layout of the instrument is shown in Fig. 3.10. Detailed description of the instrument can be also found in Ilyin et al. (2011). Spectropolarimetry with high-resolution spectrographs which always employ a significant number of reflecting optical elements, échelle grating, multiple mirrors, fibres etc., requires careful calibration of the instrumental polarisation (e.g., Ilyin 2012). The calibration process involves modelling the transformation of the Stokes parameters in the telescope and polarimeter optics with the Mueller matrices. The precise calibration is normally done with the use of polarised calibration light sources inserted into the optical path.

3.4 Calibration Techniques for Optical Polarimetry

The advantage of polarimetry over photometry is that the measurement is based on the *relative* intensity of two perpendicularly polarised components of light, rather than the total intensity measured in photometry. That helps in eliminating effects from variable atmospheric transmission and the throughput of the instrument. However, in contrast is the very minute amount of polarisation seen in astronomical objects, rarely exceeding a few percent and in the most demanding applications detection sensitivity at the ppm level (10^{-6}) is required.

3.4.1 Calibration of Linear Polarisation

In case of linear polarisation measurements, the calibration has to be carried out for:

1. Polarisation scale,
2. Zero-point of polarisation angle,
3. Instrumental polarisation produced by the polarimeter,
4. Polarisation produced by the telescope optics.

A straightforward way of polarisation scale calibration (1) is to insert a high quality polariser, e.g., a Glan prism with a nearly 100% efficiency, in front of the polarimeter. The scale calibration is particularly important for the high frequency modulators, where the efficiency can be as low as ~70% and strongly dependent on the wavelength, as the modulators are not achromatic. Double image polarisers, like calcite plate and Wollaston prism, have practically 100% efficiency. If the whole polarimeter is rotated at 45° steps, virtually no polarisation scale calibration is then needed. Superachromatic retarder plates with excellent performance in the 310–1100 nm range are available, but scale correction factors up to 1.02–1.03 may be encountered also with these devices. In practice, if no scale calibration optical component is available, one has to trust on published high polarisation standards to correct for the polarimeter efficiency factor.

For the angle calibration (2) one may use a high quality polariser oriented accurately in a known, e.g., equatorial, frame of references. The double image polarisers, calcite and Wollaston prism, are particularly advantageous in this regard, because they provide images with known polarisation orientation, readily fixed in the celestial coordinate system by the line connecting the two images. For that reason a rotatable (45° steps) double image polarimeter is an ideal instrument for establishing high polarisation standards, both for the scale and the angle calibration. Superachromatic retarder plates are less favourable for the highest precision angle calibration, because their optical axis typically varies in a complex way by several degrees over the wavelength regions applied. Therefore, it is also crucial not to use too broad passbands (<100 nm) if the angle calibration is critical. Differences in spectral energy distribution could easily introduce noticeable shifts in the effective wavelength and thereby errors in the corrections applied.

The best way of eliminating the instrumental polarisation produced by the polarimeter ($q_{\rm in}$, $u_{\rm in}$) is to rotate the whole instrument through 360° by 45° steps for one complete measurement. Typically, rotation effects (optics misalignments etc.) create a sinusoidal spurious modulation, but since the stellar polarisation gives a *double* sinusoid, the instrumental effects are largely cancelled in the reductions. Rotating the whole polarimeter, however, is often undesirable and a rotated retarder is used instead. Similarly, the half-wave plate must be rotated full 360° cycles ($22\overset{\circ}{.}5$ steps) to get the best accuracy and avoid effects from, e.g., dust particles on the retarder, non-parallelism of rotating components, etc.

Eliminating the spurious polarisation produced by the telescope optics (4) is perhaps the most difficult part to establish without leaving residual systematic errors, if the highest precision (a few ppm level) is required. Such a source like an "unpolarised star" probably does not exist. Stellar chromospheric activity may lead to detectable, and variable, intrinsic polarisation even for "normal" A–G main sequence stars. For a reliable determination it is necessary to observe a sample of (5–20) nearby stars (d < 25 pc, if possible) and obtain the normalised Stokes parameters of the telescope polarisation, ($q_{\rm tel}$, $u_{\rm tel}$) as the average of the (q, u) of the observed sample of stars. In this way also small effects of interstellar polarisation present in each of the observed stars will tend to cancel out.

Alt-az telescopes provide an additional way of determining ($q_{\rm tel}$, $u_{\rm tel}$). Because the field angle of the telescope optics continuously changes on the sky, this gives polarisation modulation: a double cosine curve over one full rotation. Fitting curve to this modulation yields the amplitude and angle of the telescope polarisation in the telescope optics coordinate system. These can then be rotated to the equatorial frame of references for the moment of the observation and subtracted from the stellar observation.

There are numerous ways, in addition to curve fitting, to determine the ($q_{\rm tel}$, $u_{\rm tel}$) for an alt-az telescope. During observations the polarimeter is rotated with respect to the telescope to keep it in the equatorial system. The observed (q, u) of zero-polarisation standard stars can be rotationally transformed to the telescope optics system. After the transformation, the ($q_{\rm tel}$, $u_{\rm tel}$) values are obtained by averaging the (q, u) values of the standard stars in the telescope frame of references. Also iterative methods can be applied to get both $(q_{\rm in}, u_{\rm in})$ and $(q_{\rm tel}, u_{\rm tel})$ from the same sample of calibration measurements.

Equatorial mounted telescope is convenient in the sense that the telescope polarisation does not rotate on the sky. It only gives a constant shift in (q, u). This is particularly important for searching small short-term periodic variations in binary and multiple stars, exoplanet systems, etc. Any residual telescope polarisation does not change the shape of the observed phase-locked patterns of the normalised Stokes parameters (q, u). The curves are only shifted by the amount of uncorrected $(q_{\rm tel}, u_{\rm tel})$, similarly to the effect of interstellar polarisation.

3.4.2 Calibration of Circular Polarisation

The telescope *circular* polarisation ($v_{\rm tel}$) is of an order of magnitude smaller than linear and, therefore, can be ignored in many applications. However, the effect of *cross-talk*, i.e., the transformation of high linear polarisation into circular which may occur in the telescope or polarimeter optics, must be carefully investigated and taken into account. The sign of the circular polarisation can be calibrated with the observations of the magnetic white dwarfs with the strong and non-variable circular polarisation, e.g., Grw+70 8247 which has $P_C \simeq -4.0\%$ in the B-band.

3.4.3 Calibration of the CCD Polarimetry: Flatfielding

Standard CCD reductions, bias, dark and flat field calibrations are obviously applied in the first part of the reduction pipeline scripts. Our experience with double image polarimeters having a rotatable superachromatic retarder (DiPol-1 and DiPol-2), and spectropolarimeters (ALFOSC at the NOT, EFOSC and FORS at the ESO), is that it is best not to apply different flat fields for each of the different position angle of the retarder, but use the same flat field for all (16) retarder positions. Individual flats fail to improve the precision, on the contrary. Common flat field is enough to suppress pixel-to-pixel variations in the CCD sensitivity. Measuring sequences consisting of cycles made over full rotation of the retarder make the polarimeter rather insensitive to flat field imperfections.

ESO staff recommends to take sky flats for calibration of polarimetry images acquired with the FORS with the *polarisation optics removed*. The reason for this is that the sky flat itself can be polarised. To avoid possible errors, it is advisable always to take sky flats in the direction of sky opposite to the rising or setting Sun.

For the highest S/N measurements it is advantageous to strongly defocus images to spread light over a very large number of pixels (Piirola et al. 2014). In this way we can expose up to 10^8 electrons in one stellar image without saturating the CCD pixels. Even if there are minor shifts in the position of the two perpendicularly polarised images of the target, vast majority of the pixels remain the same over the full measurement cycle (16 exposures). Minor imperfections in the flat field are eliminated in the reductions, since the ratio of the o- and e-beam transmission and efficiency, if constant, is automatically cancelled in the reduction algorithm. This provides inherently very stable instrument and detection sensitivity better than 10^{-5} ($<$10 ppm) in ~1 h for sufficiently bright stars. In fact, e.g., DiPol-2 polarimeter is photon-noise limited down to these very low polarisation signal levels.

3.5 High-Precision CCD Polarimetry of Binary Stellar Systems

One important application of optical polarimetry is to study interacting binary stars. The interaction, which involves transfer of matter from one component to another, very often gives rise to variable linear polarisation due to the light scattering on the circumstellar material such as gaseous streams, discs, jets and non-spherical envelopes. In binary systems with accreting black hole component, linear polarisation may also arise due to synchrotron radiation emitted in the jet. Thus, polarimetry of these objects can provide useful information on their physical properties. Below we provide examples of high-precision measurements of polarisation in black hole X-ray binaries and in early type binaries, made with the DiPol-2 instrument.

3.5.1 Polarimetry of Black Hole X-Ray Binaries

The low-mass black hole X-ray binary V404 Cyg went into outburst in 2015 after being 26 years in quiescence. In the peak of the outburst, the object showed very erratic behaviour with flares reaching a few tens of Crab in the hard X-ray domain, which for the known distance of 2.4 kpc corresponds to the Eddington luminosity for a 10 $M_{\odot}$ black hole (Rodriguez et al. 2015; Motta et al. 2017). The object also showed strong variability in the optical (see Kimura et al. 2016, and also lower left panel of Fig. 3.11). In order to understand the nature of the optical emission we performed simultaneous three-colour (BVR) polarimetric observations during and after the outburst (see Kosenkov et al. 2017, for details). We detected small but statistically significant change of the linear PD by ~1% between the outburst and the quiescence (see left panels of Fig. 3.11). We also found that the polarisation of V404 Cyg in the quiescent state agrees well with that of the visually close ($1\farcs4$) companion, as well as those of the surrounding field stars, indicating that it is predominantly of interstellar origin. From the observed variable polarisation during the outburst we showed that the intrinsic polarisation component peaks in the V band with $\mathrm{PD_V} = 1.1 \pm 0.1\%$, and in the R band it is factor of two smaller $\mathrm{PD_R} = 0.46 \pm 0.04\%$, while the polarisation PA $(\theta) = -7^{\circ} \pm 2^{\circ}$ is

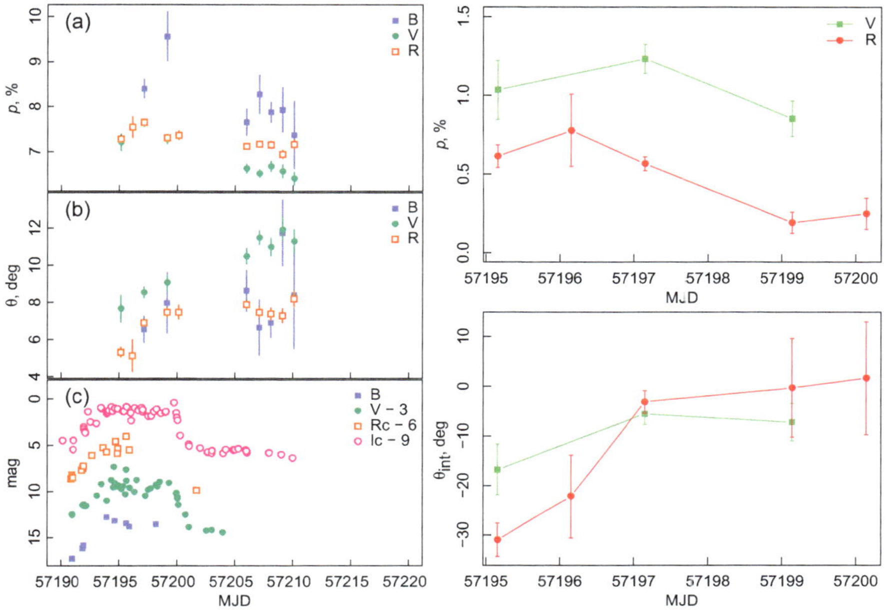

Fig. 3.11 The 2015 June–July outburst of the black hole X-ray binary V404 Cyg. *Left*: panels (**a**) and (**b**) show the observed linear PD (p) and PA (θ) in three bands; and panel (**c**) gives the optical light curve. *Right*: variations of the intrinsic linear PD (p) and PA (θ) in the V and R bands over the outburst. From Kosenkov et al. (2017)

similar in all three passbands. The observed wavelength dependence of the intrinsic polarisation does not support non-thermal synchrotron emission from a jet as a plausible mechanism, but is in better agreement with the combined effect of electron (Thomson) scattering and absorption in a flattened plasma envelope or outflow surrounding the illuminating source. Interestingly, we also found that the PA of the intrinsic polarisation is nearly parallel to the jet direction (i.e., perpendicular to the accretion disc plane) as determined by the VLBI observations.

Another black hole X-ray transient MAXI J1820+070 went into a spectacular outburst in March 2018. In the X-rays the source flux exceeded 3 Crabs and in the optical reached magnitude m_V = 12–13. This unusually bright event allowed detailed investigation of multi-wavelength spectral and timing properties and the source was monitored from radio to the γ-rays. Fast variability and powerful flares in the optical and infrared as well as optical and X-ray quasi-periodic oscillations were detected in the source. To disentangle sources of optical emission we performed BVR polarisation measurements during March–April 2018 (see Veledina et al. 2019, for details). We detected small, ∼0.7–0.9%, but statistically significant linear polarisation at PA of about 50° in all filters. We also detected a significant change in the PD after 2018 April 14 (see right panels in Fig. 3.12). The change is of the order of 0.1% and is most pronounced in the R band. By

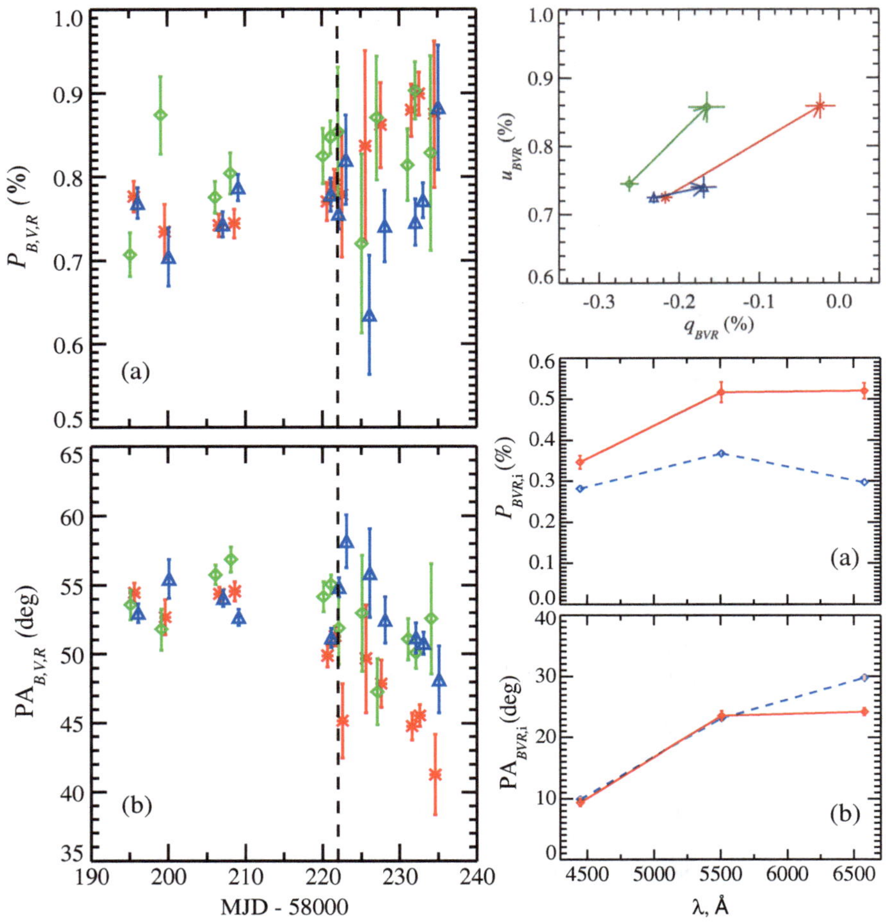

Fig. 3.12 Polarisation properties of black hole X-ray binary MAXI J1820+070. Left panels: Evolution of (**a**) PD (P) and (**b**) PA in different filters: B (blue triangles), V (green diamonds) and R (red crosses). The vertical dashed line marks MJD 58222 (2018 April 14), after which we detected a significant increase of polarisation. Right upper panel: Observed polarisation in three filters (B—blue triangles, V—green diamonds and R—red crosses) in the Stokes (u, q)-parameter plane before (lower left symbols) and after MJD 58222 (upper right symbols). Arrows show the direction of polarisation change. Right two lower panels: Wavelength dependence of the *intrinsic* (**a**) PD (P) and (**b**) PA. Weighted averages are shown with the blue dashed and red solid lines for observations before and after MJD 58222, respectively. Adapted from Veledina et al. (2019)

analysing a set of nearby field stars, we were able to determine the contribution of the interstellar polarisation. This allowed us to obtain the intrinsic polarisation of MAXI J1820+070 at the level of 0.3–0.5%, depending on the filter (see middle right panel in Fig. 3.12). The change of the source Stokes vector occurred simultaneously with the drop of the observed V magnitude and a slow softening of the X-ray spectrum, which we interpreted as a signature of an evolution towards the soft state. The increased polarisation may be a result of the decreasing contribution of the non-polarised component which we can associate with the hot flow or a jet. Its low polarisation likely results from the tangled geometry of the magnetic field or from the Faraday rotation in the dense, ionised and magnetised medium close to the black hole. We argued that the polarised optical emission is produced by the irradiated disc or scattering of its radiation in the optically thin outflow.

3.5.2 Polarimetry of Early Type Binaries

In massive early type binaries and various interacting systems (e.g., Algol-type), orbitally phase-locked linear polarisation may arise due to electron (Thomson) scattering. A detailed analysis of the polarisation variability yields independent estimates of the orbital orientation Ω and inclination i (e.g., Brown et al. 1978). The amplitude of the variability depends on the amount of the light-scattering material and usually is $\leq 10^{-3}$. In order to detect this variability, a high precision, at the level of 10^{-4} or better, is necessary. Examples of applications of CCD polarimetry on such binary systems are given in Figs. 3.13 and 3.14.

HD48099 (Fig. 3.13) is a massive, detached early type binary consisting of O5.5 V and O9 V stars. The orbital period is ~3.08 days. There is evidence of colliding stellar winds the system (e.g., Mahy et al. 2010). This binary is non-

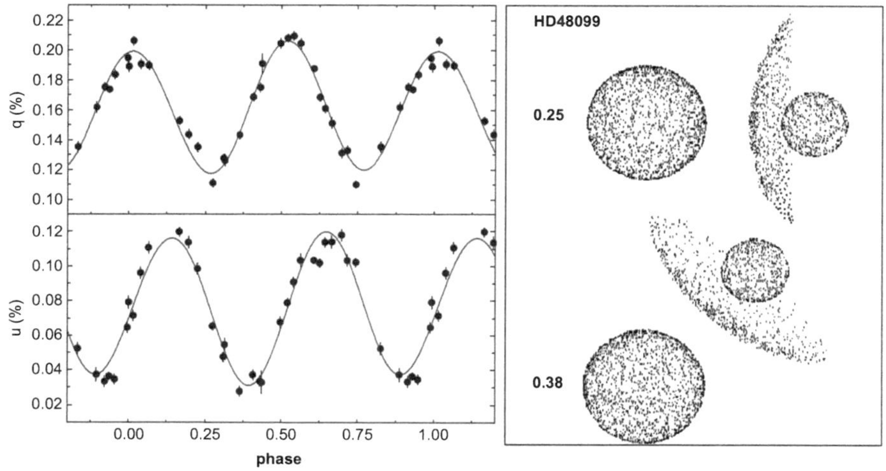

Fig. 3.13 Polarisation variability in HD48099. *Left*: Observed Stokes parameters q and u in the B-band, plotted over the phase of the orbital period. The vertical bars show 2σ errors. *Right*: The scattering model showing O-type components and the light scattering cloud produced by colliding stellar winds at the different phases of orbital period. Adapted from Berdyugin et al. (2016)

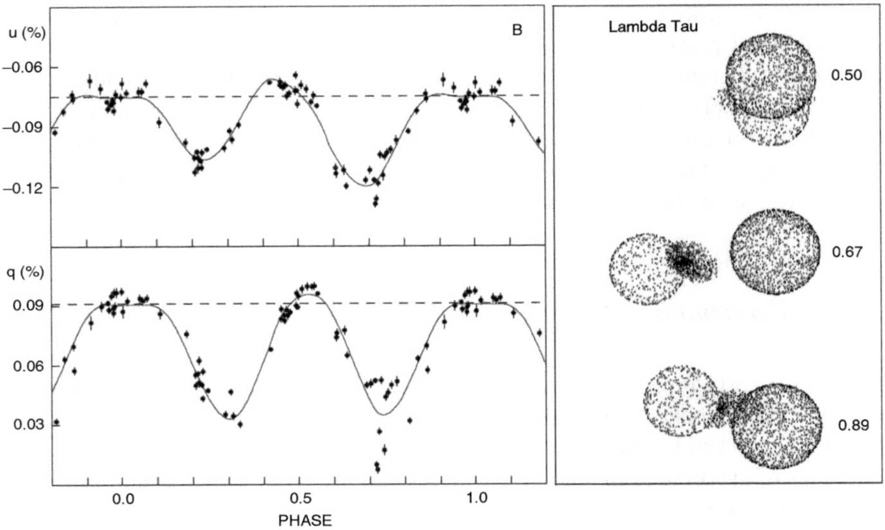

Fig. 3.14 Polarisation variability in λ Tauri. *Left*: Observed Stokes parameters q and u in the B-band plotted over the phase of the orbital period. The vertical bars show 2σ errors. *Right*: The scattering model showing primary and secondary components with the light scattering cloud between them at the different phases of orbital period. Adapted from Berdyugin et al. (2018)

eclipsing, and an independent estimate of the orbital inclination, provided by polarimetry, is important for constraining the component masses.

In Fig. 3.14 we show our results for λ Tau, a "classic" Algol-type triple system with the inner binary consisting of a more evolved and less massive mid-A type secondary filling its Roche lobe and a more massive B3 V primary star. The orbital period of the inner pair is ∼3.95 days. The unseen tertiary star (most probably a K dwarf) orbits around the inner binary with the period of ∼33 days (e.g., Fekel and Tomkin 1982).

In both cases, small-amplitude variations of polarisation (∼0.1% for HD48099 and ∼0.05% for λ Tau) have been revealed with DiPol-2 polarimetry. Modelling the polarisation data yielded the determination of the orbital parameters and the location of light scattering material (Berdyugin et al. 2016, 2018).

3.6 Summary

In this chapter we reviewed the methods, instruments and calibration techniques used in modern astronomical optical polarimetry. We described the properties of various polarisation devices and detectors used for optical broadband, imaging and spectropolarimetry, and discussed their advantages and disadvantages. We stressed the necessity to properly calibrate the raw polarisation data, and discussed the methods of the determination of instrumental polarisation and its subtraction. Finally, we presented a few examples of high-precision measurements of optical polarimetry, Fig. 3.14, X-ray binaries and on the orbital parameters of massive stellar binaries.

Acknowledgements We acknowledge support from the ERC Advanced Grant HotMol ERC-2011-AdG-291659 (AVB). The DiPol-2 was built in cooperation between the University of Turku, Finland, and the Kiepenheuer Institut für Sonnenphysik, Germany, with support from the Leibniz Association grant SAW-2011-KIS-7.

References

Bagnulo S, Szeifert T, Wade GA, Landstreet JD, Mathys G (2002) Measuring magnetic fields of early-type stars with FORS1 at the VLT. Astron Astrophys 389:191–201

Bagnulo S, Landstreet JD, Mason E, Andretta V, Silaj J, Wade GA (2006) Searching for links between magnetic fields and stellar evolution. I. A survey of magnetic fields in open cluster A- and B-type stars with FORS1. Astron Astrophys 450:777–791

Bailey J, Lucas PW, Hough JH (2010) The linear polarization of nearby bright stars measured at the parts per million level. Mon Not R Astron Soc 405:2570–2578

Bailey J, Kedziora-Chudczer L, Cotton DV, Bott K, Hough JH, Lucas PW (2015) A high-sensitivity polarimeter using a ferro-electric liquid crystal modulator. Mon Not R Astron Soc 449:3064–3073

Berdyugin A, Piirola V, Sadegi S, Tsygankov S, Sakanoi T, Kagitani M, Yoneda M, Okano S, Poutanen J (2016) High-precision broad-band linear polarimetry of early-type binaries. I. Discovery of variable, phase-locked polarization in HD 48099. Astron Astrophys 591:A92

Berdyugin A, Piirola V, Sakanoi T, Kagitani M, Yoneda M (2018) High-precision broad-band linear polarimetry of early-type binaries. II. Variable, phase-locked polarization in triple Algol-type system λ Tauri. Astron Astrophys 611:A69

Berdyugina SV, Berdyugin AV, Fluri DM, Piirola V (2011) Polarized reflected light from the exoplanet HD189733b: first multicolor observations and confirmation of detection. Astrophys J 728:L6

Brown JC, McLean IS, Emslie AG (1978) Polarisation by Thomson scattering in optically thin stellar envelopes. II. Binary and multiple star envelopes and the determination of binary inclinations. Astron Astrophys 68:415–427

Collins P, Kyne G, Lara D, Redfern M, Shearer A, Sheehan B (2013) The Galway astronomical Stokes polarimeter: an all-Stokes optical polarimeter with ultra-high time resolution. Exp Astron 36:479–503

Cotton DV, Marshall JP, Bailey J, Kedziora-Chudczer L, Bott K, Marsden SC, Carter BD (2017) The intrinsic and interstellar broad-band linear polarization of nearby FGK dwarfs. Mon Not R Astron Soc 467:873–897

Cotton DV, Marshall JP, Frisch PC, Kedziora-Chudzer L, Bailey J, Bott K, Wright DJ, Wyatt MC, Kennedy GM (2019) The wavelength dependence of interstellar polarization in the Local Hot Bubble. Mon Not R Astron Soc 483:3636–3646

Fekel Jr, FC, Tomkin J (1982) Secondaries of eclipsing binaries. IV. The triple system Lambda Tauri. Astrophys J 263:289–301

Finger G, Baker I, Alvarez D, Ives D, Mehrgan L, Meyer M, Stegmeier J, Weller HJ (2014) SAPHIRA detector for infrared wavefront sensing. In: Proceedings of SPIE (Adaptive Optics Systems IV), vol 9148, p 914817

Frisch PC, Berdyugin A, Piirola V, Magalhaes AM, Seriacopi DB, Wiktorowicz SJ, Andersson BG, Funsten HO, McComas DJ, Schwadron NA, Slavin JD (2015) Charting the interstellar magnetic field causing the interstellar boundary explorer (IBEX) ribbon of energetic neutral atoms. Astrophys J 814:112

Gisler, D., Berkefeld, T., and Berdyugina, S. (2016) Planet imaging polarimetry with the solar telescope GREGOR. In: Proceedings of SPIE (Ground-based and Airborne Telescopes VI), vol. 9906, p 99065E

Harjunpää P, Kaas AA, Carlqvist P, Gahm GF (1999) Linear polarization and molecular filamentary clouds. Astron Astrophys 349:912–926

Harvey EJ, Redman MP, Darnley MJ, Williams SC, Berdyugin A, Piirola VE, Fitzgerald KP, O'Connor EG (2018) Polarimetry and spectroscopy of the "oxygen flaring" DQ Herculis-like nova: V5668 Sagittarii (2015). Astron Astrophys 611:A3

Hough JH, Lucas PW, Bailey JA, Tamura M, Hirst E, Harrison D, Bartholomew-Biggs M (2006) PlanetPol: a very high sensitivity polarimeter. Publ Astron Soc Pac 118:1302–1318

Ilyin I (2012) Second-order error propagation in the Mueller matrix of a spectropolarimeter. Astron Nachr 333:213

Ilyin I, Strassmeier KG, Woche M, Dionies F, Di Varano I (2011) On the design of the PEPSI spectropolarimeter for the LBT. Astron Nachr 332:753

Kanbach G, Stefanescu A, Duscha S, Mühlegger M, Schrey F, Steinle H, Slowikowska A, Spruit H (2008) OPTIMA: a high time resolution optical photo-polarimeter. In: Phelan D, Ryan O, Shearer A (eds) High time resolution astrophysics (Astrophysics and Space Science Library), vol 351. Springer, Berlin, p 153

Kemp JC (1981) Photoelastic-modulator polarimeters in astronomy. In: Trapani G (ed) Proceedings of SPIE (Polarizers and applications), vol 307), p 83–88

Kimura M, Isogai K, Kato T, Ueda Y, Nakahira S, Shidatsu M, Enoto T, Hori T, Nogami D, Littlefield C, Ishioka R, Chen Y-T, King S-K, Wen C-Y, Wang S-Y, Lehner MJ, Schwamb ME, Wang J-H, Zhang Z-W, Alcock C, Axelrod T, Bianco FB, Byun Y-I, Chen W-P, Cook KH, Kim D-W, Lee T, Marshall SL, Pavlenko EP, Antonyuk OI, Antonyuk KA, Pit NV, Sosnovskij AA, Babina JV, Baklanov AV, Pozanenko AS, Mazaeva ED, Schmalz SE, Reva IV, Belan SP, Inasaridze RY, Tungalag N, Volnova AA, Molotov IE, Miguel ED, Kasai K, Stein WL, Dubovsky PA, Kiyota S, Miller I, Richmond M, Goff W, Andreev MV, Takahashi H, Kojiguchi N, Sugiura Y, Takeda N, Yamada E, Matsumoto K, James N, Pickard RD, Tordai T, Maeda Y, Ruiz J, Miyashita A, Cook L. M, Imada A., Uemura M (2016) Repetitive patterns in rapid optical variations in the nearby black-hole binary V404 Cygni. Nature 529:54–58

Kosenkov IA, Berdyugin AV, Piirola V, Tsygankov SS, Pallé E, Miles-Páez PA, Poutanen J (2017) High-precision optical polarimetry of the accreting black hole V404 Cyg during the 2015 June outburst. Mon Not R Astron Soc 468:4362–4373

Kyne G, Lara D, Hallinan G, Redfern M, Shearer A (2016) An investigation of the eigenvalue calibration method (ECM) using GASP for non-imaging and imaging detectors. Exp Astron 41:43–66

Mahy L, Rauw G, Martins F, Nazé Y, Gosset E, De Becker M, Sana H, Eenens P (2010) A new investigation of the binary HD 48099. Astrophys J 708:1537–1544

Motta SE, Kajava JJ, Sánchez-Fernández C, Beardmore AP, Sanna A, Page KL, Fender R, Altamirano D, Charles P, Giustini M, Knigge C, Kuulkers E, Oates S, Osborne JP (2017) Swift observations of V404 Cyg during the 2015 outburst: X-ray outflows from super-Eddington accretion. MNRAS 471:1797–1818

Piirola V (1973) A double image chopping polarimeter. Astron Astrophys 27:383–388

Piirola V (1988) Simultaneous five-colour (UBVRI) photopolarimeter. In: Coyne GV, Magalhaes AM, Moffat AF, Schulte-Ladbeck RE, Tapia S (eds) Polarized radiation of circumstellar origin, Vatican City State/Tucson, AZ. Vatican Observatory/University of Arizona Press, pp 735–746

Piirola V, Reiz A, Coyne GV (1987) Simultaneous five-colour (UBVRI) polarimetry of EF ERI. Astron Astrophys 186:120–128

Piirola V, Coyne GV, Reiz A (1990) Simultaneous UBVRI polarimetry of VV Puppis during an active phase. Astron Astrophys 235:245–254

Piirola V, Scaltriti F, Coyne GV (1992) Circumstellar disks deduced from sub-arcsecond polarization observations of two young stars. Nature 359:399–401
Piirola V, Hakala P, Coyne GV (1993) The discovery of variable polarization over the 13.9 minute spin period of the intermediate polar RE 0751 + 14. Astrophys J 410:L107–L110
Piirola V, Vornanen T, Berdyugin A, Coyne GV (2008) V405 aurigae: a high magnetic field intermediate polar. Astrophys J 684:558–568
Piirola V, Berdyugin A, Berdyugina S (2014) DIPOL-2: a double image high precision polarimeter. In: Proceedings of SPIE (Ground-based and Airborne Instrumentation for Astronomy V), vol. 9147, p 91478I
Potter SB, Buckley DAH (2018) Time series photopolarimetry and modelling of the white dwarf pulsar in AR Scorpii. Mon Not R Astron Soc 481:2384–2392
Potter SB, Buckley DAH, O'Donoghue D, Romero-Colmenero E, O'Connor J, Fourie P, Evans G, Sass C, Crause L, Still M, Butters OW, Norton AJ, Mukai K (2010) Polarized QPOs from the INTEGRAL polar IGRJ14536-5522 (=Swift J1453.4-5524). Mon Not R Astron Soc 402:1161–1170
Rodriguez J, Cadolle Bel M, Alfonso-Garzón J, Siegert T, Zhang X-L, Grinberg V, Savchenko V, Tomsick JA, Chenevez J, Clavel M, Corbel S, Diehl R, Domingo A, Gouiffès C, Greiner J, Krause MGH, Laurent P, Loh A, Markoff S, Mas-Hesse JM, Miller-Jones JCA, Russell DM, Wilms J (2015) Correlated optical, X-ray, and γ-ray flaring activity seen with INTEGRAL during the 2015 outburst of V404 Cygni. Astron Astrophys 581:L9
Serkowski K (1974) Polarization techniques. In: Carleton NP (ed) Astrophysics. Part A: optical and infrared. Academic Press, New York, pp 361–414
Słowikowska A, Krzeszowski K, Zejmo M, Reig P, Steele I (2016) Calibration of the Liverpool Telescope RINGO3 polarimeter. Mon Not R Astron Soc 458:759–771
Strassmeier KG, Ilyin I, Järvinen A, Weber M, Woche M, Barnes SI, Bauer S-M, Beckert E, Bittner W, Bredthauer R, Carroll TA, Denker C, Dionies F, DiVarano I, Döscher D, Fechner T, Feuerstein D, Granzer T, Hahn T, Harnisch G, Hofmann A, Lesser M, Paschke J, Pankratow S, Plank V, Plüschke D, Popow E, Sablowski D (2015) PEPSI: the high-resolution échelle spectrograph and polarimeter for the Large Binocular Telescope. Astron Nachr 336:324
Veledina A, Berdyugin AV, Kosenkov IA, Kajava JJE, Tsygankov SS, Piirola V, Berdyugina SV, Sakanoi T, Kagitani M, Kravtsov V, Poutanen J (2019) Evolving optical polarization of the black hole X-ray binary MAXI J1820+070. Astron Astrophys 623:A75
Vornanen T, Berdyugina SV, Berdyugin AV, Piirola V (2010) GJ 841B—the second DQ white dwarf with polarized CH molecular bands. Astrophys J 720:L52–L55
Wiktorowicz SJ, Matthews K (2008) A high-precision optical polarimeter to measure inclinations of high-mass X-ray binaries. Publ Astron Soc Pac 120:1282
Wiktorowicz SJ, Nofi LA (2015) Simultaneous linear and circular optical polarimetry of asteroid (4) vesta. Astrophys J 800:L1

Chapter 4
X-Ray Polarimetry (Instrument/Techniques/Calibration)

Paolo Soffitta and Enrico Costa

Abstract X-ray polarimetry is an "old" field in Astronomy which is rapidly evolving, thanks to the development of sensitive instrumentation and to mission proposals arrived at a various stages of implementation. It was recognised, since the dawn of X-ray Astronomy, as a formidable tool for the comprehension of the emission mechanisms and of the geometry at work in extreme gravitational and/or magnetic fields. We will review the subject of X-ray polarimetry spanning from 0.1 to 100 keV describing the different major principles of detection. We will show their practical implementation following the historical evolution of this subject from the past experiment to future (at the time of writing) expectations.

4.1 Statistics of Polarimetry in X-Rays

No realistic technique for measuring circular polarisation in a space experiment has been so far proposed, therefore we only discuss technique to measure linear polarisation. As we will see in the next chapters this measurement is based on the distribution of the carrier of the polarisation information in a realistic set-up. Such carriers in X-rays can be the diffracted photons (Sect. 4.2.1), the scattered photons (Sect. 4.2.2) or the photoelectrons and their emission directions (Sect. 4.5.1).

A conventional polarimeter is based on the probability to "pass" a characteristic axis. From basic principle this probability is proportional to $\cos^2(\phi)$ where ϕ (the azimuthal angle) is the angle between the direction of the electric field and its axis. As we will see, the integrated reflectivity of a crystal, the cross section for Compton effect and photoelectric effect are, indeed, dependent on $\cos^2(\phi)$. In the practical implementation such function need to be modified including the response of the instrument. The usual modification is the addition of a constant term that describes the not perfect capability of the instrument to respond to a polarised

---

P. Soffitta (✉) · E. Costa
IAPS/INAF, Roma, Italy
e-mail: paolo.soffitta@inaf.it; enrico.costa@inaf.it

R. Mignani et al. (eds.), *Astronomical Polarisation from the Infrared to Gamma Rays*, Astrophysics and Space Science Library 460,
https://doi.org/10.1007/978-3-030-19715-5_4

beam. Other possible additional (Fourier) components can be present. The most critical ones are those with a (2π) period because they are related to the presence of systematics. A "good" polarimeter will preserve, as much as possible, the above $\cos^2(\phi)$ dependence.

Historically the modulation amplitude for X-ray polarimetry derives from the modulation curve (the angular distribution of the carriers of the polarisation information with respect to the azimuthal angles) as:

$$A = \frac{N_{\max} - N_{\min}}{N_{\max} + N_{\min}} \tag{4.1}$$

where N is the number of counts in each bin of the modulation curve. The fitting function is

$$C(\phi) = a + b\cos^2(\phi - \phi_0) \tag{4.2}$$

Where ϕ_0 is the phase of the polarisation beam. Applying the fitting function to the modulation curve we derive that the Eq. (4.1) can be written as:

$$A = \frac{b}{2a + b} \tag{4.3}$$

and the polarisation of a source is derived as:

$$P = \frac{A}{\mu} \tag{4.4}$$

where μ is the modulation factor which is the amplitude A for a 100% polarised beam. Another fitting function is the following:

$$C(\phi) = A_0(1 + A(\cos(2(\phi - \phi_0)) \tag{4.5}$$

This fitting function naturally provides the modulation amplitude A (and therefore also μ) regardless the possible covariance between a and b present in the formula (4.3). We note that in X-rays, as in other wavelength, we measure a polarisation from a distribution of angles. We cannot measure the polarisation of a single photon.

The above formalism, actually, is perfectly equivalent to that of the Stokes parameter used in radio, infrared and optical polarimetry. Indeed the Eq. (4.2) can be expressed as:

$$C(\phi) = I + U\cos(2\phi) + Q\sin(2\phi) \tag{4.6}$$

where: $Q = \frac{b}{2}\sin(2\phi_0)$; $U = \frac{b}{2}\cos(2\phi_0)$ $I = a + \frac{b}{2}$ and $\phi_0 = \frac{1}{2}\arctan\left(\frac{Q}{U}\right)$. To cope with the fact that in X-rays we deal with individual photons, Kislat et al. (2015)

defined, for each carrier of the X-ray polarisation information (electron or photon) its Stokes parameter (u_i and v_i). Exploiting their additivity:

$$U = \Sigma u_i;\ V = \Sigma v_i \tag{4.7}$$

See Kislat et al. (2015) for a complete description of the tools for using the Stokes parameter in X-ray astronomy. Other authors explore different properties of the modulation curve comparing them with Stokes analysis (Massaro et al. 2018).

A further important parameter is the sensitivity of an X-ray polarimeter called minimum detectable polarisation (MDP). The MDP is the measured polarisation level that excludes the hypothesis of null polarisation in the data given a confidence level. Indeed as shown, for example, in Vaillancourt (2006); Elsner et al. (2012); Kislat et al. (2015) it is possible to derive, from the statistical properties of the Stokes parameters, the joint probability of the polarisation degree and of the polarisation angle given their expected values. From this equation, integrating over the polarisation angle, the Rice distribution is derived providing the probability for a measure of P given its expected value. Using this equation we find the MDP assuming an expected zero polarisation and a given confidence level (usually assumed to be 99%). The general expression for the MDP given a certain confidence level CL is

$$\mathrm{MDP_{CL}} = \frac{2\sqrt{-\ln(1-\mathrm{CL})}}{\mu S}\sqrt{\frac{S+B}{T}} \tag{4.8}$$

where S is the source counting rate, B is the background counting rate, T is the integration time μ is the modulation factor and CL is the confidence level accepted. In case of a CL = 99%, the MDP is

$$\mathrm{MDP_{99}} = \frac{4.29}{\mu S}\sqrt{\frac{S+B}{T}} \tag{4.9}$$

The question now is the relation between $\mathrm{MDP_{99}}$ and the significance of a measure of a linear polarisation degree ($\mathrm{PD_L}$) and of a polarisation angle (PA) that is explored in Weisskopf et al. (2010); Elsner et al. (2012); Strohmayer et al. (2013). Such relation, indeed, depends on whether the polarisation angle is a free fitting parameter (2D fitting) or it is known and fixed (1D fitting). Table 4.1 summarises this relation in terms of needed increase of the number of counts (and therefore of the observing time) to reach a given significance in the measurement $\beta = \frac{\sigma(P)}{P} = \frac{\sigma(A)}{A}$ with respect to the counts needed to get $P = \mathrm{MDP_{99}}$. As can be seen from Table 4.1, in the 1D case the number of counts providing $P = \mathrm{MDP_{99}}$ (the same counts are necessary to get $A = \mathrm{MDA_{99}}$) corresponds to a significance on P of $3\sigma_{1D}$. We need, instead, more counts to get the same significance for a joint fit. For example, for low modulation amplitude, $P = \mathrm{MPD_{99}}$ corresponds to a $2\sigma_{2D}$ measurement. This is not unexpected because, for a 2D fit, σ_{2D} is larger. Lastly we mention that in Strohmayer et al. (2013) the significance of a measurement on the angle as a

Table 4.1 Increase of the number of counts with respect to that necessary to get $P = MDP_{99}$ (from Strohmayer et al. (2013))

No. free parameter	A	$\frac{N}{N_{MDP}}$
1D (A)	All	$\frac{\beta_{1D}^2}{9.2}$
2D (A,ϕ)	$<\frac{1}{3}$	$\frac{\beta_{2D}^2}{4.1}$
	$\frac{1}{3}$	$\frac{\beta_{2D}^2}{4.4}$
	$\frac{1}{2}$	$\frac{\beta_{2D}^2}{5.1}$
	$\frac{2}{3}$	$\frac{\beta_{2D}^2}{6.1}$

function of β_{2D} is derived as:

$$\sigma(\phi) = \frac{28.5^\circ}{\beta_{2D}} \tag{4.10}$$

Therefore for a $3\sigma_{2D}$ measurement on P the error on the polarisation angle is slightly below 10°.

4.2 Classical Techniques to Measure Polarisation in X-Rays

4.2.1 Bragg Diffraction

Bragg diffraction in a crystal or in a multi-layer structure can be exploited for polarimetry of a photon beam. The general condition for the reflection is derived from a constructive interference of waves generated in the crystal lattice and is

$$\sin(\theta_B) = \frac{n\lambda}{2d} \tag{4.11}$$

where d is the lattice plane distance, λ is the wavelength of the radiation and n is the order of the refraction. In X-ray polarimetry crystals are oriented at 45° and the polarisation vector of the reflected crystal is normal to the incidence plane that is the plane containing both the incoming direction and the normal to the crystal plane. The radiation with opposite polarisation is, instead, absorbed. Naively, if we consider the lattice points as dipoles oscillating parallel to the electric vector of the incoming beam, we obtain the maximum of the emission perpendicularly to the vibration direction. If the vibration is parallel to the incidence plane the radiation is absorbed. More quantitatively we define a quantity called integrated reflectivity as:

$$R_C = \int_0^{\frac{\pi}{2}} R(\lambda;\theta)d\theta \tag{4.12}$$

where $R(\lambda;\theta)$ is the coefficient of reflection of the X-ray radiation at the wavelength λ and angle θ. It represents the reflected power $J(\theta)$(in energy unit time^{-1} unit area^{-1}unit) for an incident spectrum I(λ) (in energy unit time^{-1} unit wavelength^{-1}unit area^{-1}unit). R_c is a characteristics of the material. If we consider the component σ of the radiation as the radiation with polarisation perpendicular to the incidence plane and π the radiation parallel to the incidence plane we can define the quantity:

$$k = \frac{R_{c\pi}}{R_{c\sigma}} \tag{4.13}$$

where $R_{c\pi}$ and $R_{c\sigma}$ are the component π and the component σ of the integrated reflectivity. The reflected power (per unit area and time is) is

$$J(\theta) = I(\lambda)\lambda\cot(\lambda)R_c \tag{4.14}$$

where R$_c$ is the average integrated reflectivity for the two polarisation state σ and π.

We can rotate the crystal maintaining constant the incidence angle around the direction of a fixed source placed at infinite distance. The polarised component of a partially polarised radiation will be therefore alternatively σ and π and the power $J(\theta)$ will vary cyclically with a characteristic frequency of twice the rotation frequency of the crystal. The variation of the measured power is defined as:

$$V = \frac{J_{\max} - J_{\min}}{J_{\max} + J_{\min}} \tag{4.15}$$

It is possible to demonstrate (Evans et al. 1977) that the degree of polarisation of the incoming radiation is

$$P = \frac{V}{\frac{1-k}{1+k}} \tag{4.16}$$

where the denominator of Eq. (4.16) is the modulation factor of the polarimeter.

In the ideal condition $k = 0$ and in the case of a perfect crystal it is shown (Evans et al. (1977)) that $k =\mid \cos(2\theta) \mid$ so the maximum modulation requires $\theta = \frac{\pi}{4}$). In the real case the incidence angle is not known with infinite accuracy, also it is very difficult to make a perfectly parallel beam and the lattice cannot be perfectly regular so the modulation factor can be smaller than 100%. In case of a perfectly parallel beam the Bragg condition provides the wavelength of the diffraction at 45 deg:

$$\lambda = \frac{2d}{n}\sin\left(\frac{\pi}{4}\right) \tag{4.17}$$

or in the usual physical unities in X-ray astronomy:

$$\text{Energy(keV)} = \frac{n \times 12.3984}{2d(\text{Å}) \sin(\frac{\pi}{4})} \tag{4.18}$$

The energy band for the validity of Eq. (4.18) is much smaller than 1 eV and therefore perfect crystals cannot be used in X-ray astronomy characterised by continuum spectra and paucity of photons. In order to enlarge the energy band, even if at the expense of the modulation factor, one can use a mosaic crystal, instead. A mosaic crystal is composed of minuscule crystal tiles with thickness smaller than the absorption length of the photon and slightly misaligned one respect to the other. The advantage is that the band of reflection of a continuum X-ray source is larger than that of a perfect crystal. Indeed the radiation that does not satisfy the Bragg condition for a particular tile will encounter another tile oriented in such a way that the Bragg diffraction is satisfied. The energy band can so be increased up to 100 eV for the graphite crystal. The two kinds of set-up for an experiment based on Bragg diffraction are shown in Fig. 4.1.

The integrated reflectivity of an ideal mosaic crystal depends on the polarisation angle as Silver et al. (1990):

$$\Delta\Theta(E_B) = \frac{N^2 F^2 r_0^2}{2\mu} \left[\frac{hc}{E_B}\right]^3 \left[\frac{1}{\sin(2\theta_B)} - \frac{\sin(2\theta_B)}{2}(1 + P\cos(2\phi)\right] \tag{4.19}$$

Here N is the number of scattering cell per unit volume, F is the crystal structure factor (that is the electrons for unit cell), r_0 is the classical radius of an electron, θ_B is the Bragg angle (45° in our case); ϕ is the azimuthal angle which is the angle between the incidence plane and the plane formed by the incoming direction of the photon beam and its polarisation vector. The rate ($\frac{\text{counts}}{s}$) of X-ray from a continuous source reflected by a mosaic crystal onto a detector derives from Eq. (4.14). This expression is

$$S(E_B) = I(E_B)\epsilon(E_B)A(E_B)\cot\theta_B\Delta(E_B) \tag{4.20}$$

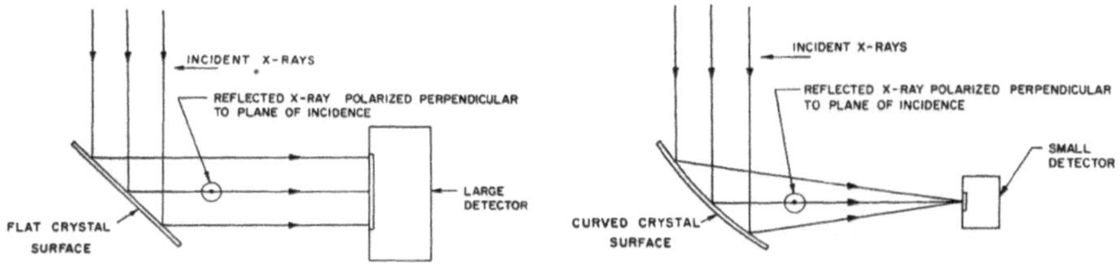

Fig. 4.1 (**Left**) The Bragg diffraction from a flat crystal. This configuration maximises the spectroscopical capabilities. (**Right**) The Bragg diffraction from a parabolic surface. This configuration maximises the polarimetric capabilities. Mosaic crystal in addition allows for a larger bandwidth and better sensitivity (figures from Silver et al. (1990))

Table 4.2 Maxima integrated reflectivity (from Silver et al. (1990))

Crystal	E_{45° (keV)	$\Delta\theta \times 10^{-4}$
Mica	0.880	0.2
WS_2	1.40	15.0
MoS_2	1.43	8.3
PET	2.01	4.3
Graphite	2.61	15.9
Calcite	2.89	3.5
LiH	4.30	35.0
LiF	4.35	5.2

where I is the intensity ($\frac{ph}{\mathrm{s\ cm^2\ keV}}$), ϵ is the detector efficiency (including the possible effect of the window transparency), A is the geometrical area of the crystal, Δ is the integrated reflectivity, E_B is the Bragg energy (which is the energy at 45° reflection in our case).

We show below a table with the theoretical integrated reflectivity of some useful crystals.

During 1970s scientists exploited Bragg diffraction from graphite crystals (Schnopper and Kalata 1969) for X-ray polarimetry on-board of rockets and satellite like *OSO-8* and *Ariel 5* described below. Graphite crystals were preferred to other kind of crystals because the experimental reflectivity matched its high theoretical value. This was not the case for LiH that, also, reacts very heavily with H_2O and must be protected from moisture. With graphite the energy for making polarimetry is a narrow band around 2.6 keV (I refraction order) and 5.2 keV (II refraction order). Other orders were precluded by the paucity of celestial photons (Table 4.2).

4.2.2 Compton Scattering

Compton effect cross section depends on polarisation as described by the Klein–Nishina formula:

$$\frac{d\sigma}{d\Omega} = \frac{r_0^2}{2}\frac{E'^2}{E^2}\left[\frac{E}{E'} + \frac{E'}{E} - 2\sin^2(\theta)\cos^2(\phi)\right] \tag{4.21}$$

where θ is the polar scattering angle and ϕ is the azimuthal angle or the angle with respect to the polarisation vector of the incoming photon and:

$$\frac{E'}{E} = \frac{1}{1 + \frac{E}{m_e c^2}(1 - \cos(\theta))} \tag{4.22}$$

At low energies (compared to mc^2) photons 100% polarised scattered at polar angles around 90° from the original direction are preferentially scattered at azimuth

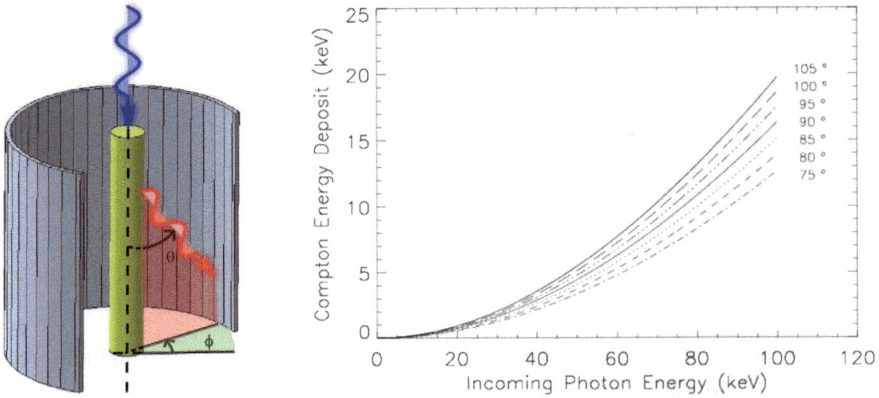

Fig. 4.2 (**Left**) A configuration for an X-ray scattering polarimeter (possibly at a focus of an X-ray optics). (**Right**) The energy deposited in the scattering element as function of the incoming photon energy for different scattering angle around 90° (figures from Fabiani et al. (2013))

angles perpendicular to electric vector. On other extreme of the distribution the photons scattered backward and forward are isotropic in azimuth.

A polarimeter based on the Compton effect near the Thomson limit is usually designed so that most of photons scattered perpendicular to the beam direction are collected with detectors providing an adequate coverage of the solid angle to achieve a good efficiency and avoiding to collect photons scattered forward and backward. It is evident that one of the major design trade-off is the accepted range of polar scattering angles, namely a trade-off of efficiency and modulation.

Every Compton polarimeter (see Fig. 4.2, left) is based on two interactions. A first interaction is the Compton scattering with a release of energy to the electron (see Fig. 4.2, right). The residual energy is carried by the new photon and in a Compton telescope is subject to a new interaction. If this is a photoelectric absorption (possibly followed by the emission and a near local re-absorption of a fluorescence photon) all the energy of the primary photon is lost in the instrument. If the second interaction is another Compton scattering, not followed by a local further interaction, a fraction of the energy escapes the system and is not measured by the instrument. In any case the azimuth angular distribution of the photons scattered at polar angles near to 90° is the basis of the measurement of polarisation. At low energies the distribution is near to $\cos^2$ At higher energies, far from the Thomson limit, the modulation is reduced and somehow bent forward.

How to convert these distributions into an actual instrument is a matter of heavy trade-off between mass, encumbrance, band, statistical sensitivity, control of systematics. We sketch here some of the major alternatives.

4.2.2.1 High Systematics

An important peculiarity of scattering polarimetry is that the control of systematics is very challenging. An efficient detection of photons requires low atomic number materials of a thickness of the order of $10\,\mathrm{g/cm^2}$. For typical low Z materials this means also thicknesses of 10 cm to achieve a high efficiency of detection. The photons that carry the highest information on polarisation are those scattered around 90°. But in the path from the scattering to the absorption point they can be scattered or absorbed in the scatterer itself or in the passive mounting structures and, since this cannot be totally independent from the azimuth angle, systematics are unavoidably there. Moreover the response to polarisation depends on the polar scattering angle. To derive the polarisation of the beam from the azimuth angular distribution a good knowledge of the polar angular distribution is needed. Given the thickness of the scatterer this is not easy.

A further source of systematics is the geometry. The probability that a pair of events at a certain angle is detected does not depend only on the cross section but also on the angular distribution of pair of detector pixels available. For instance, on a detector of square geometry the angles parallel to diagonals will be better covered than those parallel to sides.

Another source of systematics is the difference in thresholds of the different pixels. These differences are non-necessarily constant with time. The gain of inorganic scintillators depends on temperature and thence the threshold expressed in keV can vary during the flight and also within the same measurement. A good energy calibration possibly during flight affording for a good trimming of thresholds is very important for the robustness of results.

4.2.2.2 One Material vs Two Materials

For a certain material the cross section for photo-absorption decreases according to a very steep power law, while the cross section for Compton scattering is almost constant. Thence if the scatterer and the absorber are of the same material, a reasonable probability of a first Compton interaction combined with a reasonable probability of a photo-absorption of the scattered photon can be only achieved on a narrow range, around the energy where the two absorption coefficients cross. Given that the Compton scattering is a broad band process, a scattering polarimeter based on a scatterer and a detector of the same material will make, by first principles, a poor exploitation of the mass employed. On the contrary if a material of low atomic number is used as scatterer and a material of high atomic number as absorber, the instrument can operate on a much broader range. For instance, in the case of an array of long sticks of parallel scintillators, it has been demonstrated that the sensitivity is much better with a combination of plastic scintillators and inorganic scintillators (Costa et al. 1995).

4.2.2.3 Passive Scatterer vs Active Scatterer

As stated above the material selected for the scatterer should be of low atomic number. Hydrogen and Helium are very light gasses and in practice could be used as scatterers only at very high pressure or in the condensed phase. But the structure to keep the pressure or the low temperature would be very massive and irregular itself so that nothing like this is in practice viable. The first useful element is Lithium, which has a density of $0.53\,\mathrm{g/cm^3}$ and the scattering prevails on photo-absorption at 9 keV. LiH is slightly better mainly because of the greater density ($0.82\,\mathrm{g/cm^3}$). Unfortunately Lithium is mechanically soft and both materials are highly hygroscopic and subject to the attack of Oxygen. Therefore they can only be used if contained in a sealed can of a harder material such as beryllium and the effective range will be somehow higher. Solid beryllium is viable at higher energies since the scattering prevails around 12 keV. Given these numbers scattering polarimetry is definitely confined to hard X-rays. Any application at lower energies, although sometimes declared, will be marginal.

By further increasing the atomic number a new possibility is viable. The scatterer instead of a passive material can be a detector itself. Plastic scintillators are basically made of carbonium with some hydrogen. In the first case the fact that the radiation is following the sequence scattering—absorption must be guaranteed by a focussing optics and/or by shielding or collimation system. If the scatterer is a detector the coincidence of the signals allows to associate the two events with an improvement of the kinematic reconstruction and a dramatic improvement in the background rejection.

Unfortunately the energy transferred to the Compton electron at low energies is low (see again Fig. 4.2, right). For a scattering at polar angle of 90° the energy transferred is 420 eV at 15 keV, 750 eV at 20 keV and 1.16 keV at 25 keV, very low for a detection in a plastic scintillator. Moreover the scatterer to have a reasonable efficiency must be long a few centimetres (5 to 10). Light photons following the electron detection will be significantly self-absorbed by the scintillator itself in the transfer to the optical device (Vacuum or Silicon photomultiplier). Therefore the detection of these events is the most challenging task and determines the band pass of an active a scattering polarimeter. Given that spectra of sources and band pass of optics are a fast decreasing function of the energy, in many cases the lower threshold has a great impact on the sensitivity. Some studies have been performed to assess this threshold (Fabiani et al. 2013; Chattopadhyay et al. 2014; Kislat et al. 2018).

We could conclude stating that scattering polarimetry is viable with passive scatterers above 5 keV and with active scatterers above 20 keV.

4.2.2.4 Focal Plane vs Collimation vs Wide Field

Non-solar X-Ray Astronomy was born on 1962 with a rocket-borne experiment based on proportional counters with collimation scanning the sky with rocket spinning. This scheme was followed, with various improvements until 1969 when the *Einstein* satellite was launched, using a grazing incidence optics and various focal plane detector. The optics provides imaging and a dramatic improvement in sensitivity because of the effective background reduction. In fact photons from a weak source are focussed in a small spot of the detector and are compared only to the background events in the same region.

A scattering focal plane polarimeter is basically a stick of some low atomic number material surrounded with a well of detectors to detect the photons scattered all around. The difference between a focal plane polarimeter based on photoelectric effect and another one based on scattering is relevant.

If we put a photoelectric polarimeter in the focal plane, the imaging property holds and the increase of sensitivity will also apply to the polarimetry. For a scattering device the improvement is much smaller. In fact after collecting the photons from the source towards a focal point we scatter them by means of a scatterer of a certain length and we collect them with a set of detectors surrounding the scatterer. The instrument is not imaging and the reduction in the background is still there but it is not that dramatic. The background will be roughly proportional to the area of the whole detectors acting as absorbers.

These limits are evident if we consider the Stellar X-Ray Polarimeter (SXRP, see Fig. 4.3) (Kaaret et al. 1989, 1994). This was built to be mounted in one of the focal plane slits of the *Soviet Danish X-Ray Telescope* (*SODART*) aboard the *Spectrum X-Ray Gamma* (*SRG*) mission (Schnopper 1990). SXRP was built, integrated, calibrated and accepted. It was stored for years while the *SRG* mission was slowly delayed and, eventually dropped, in the general sinking of the Soviet system. Each of the *SODART* telescopes had a consistent collecting surface of

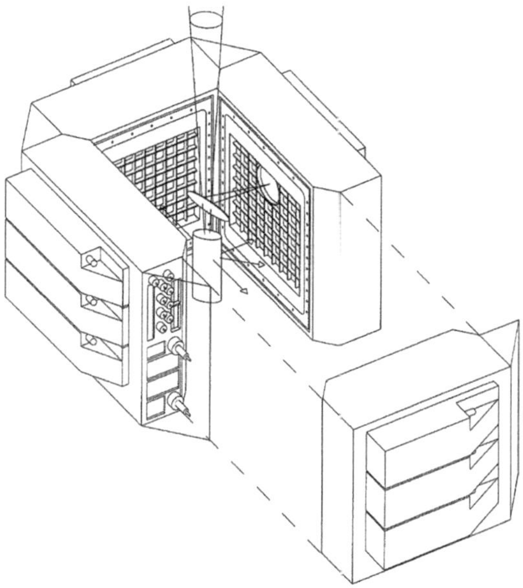

Fig. 4.3 SXRP experiment. It was supposed to fly on-board Spectrum X-Gamma. It consisted of a graphite mosaic crystal stacked with a lithium rod encapsulated in beryllium and encircled by four imaging proportional counters (figure from Soffitta et al. (1998))

the order of 1600 cm^2 and an angular resolution of 2 arcmin. Above the focal plane a scatterer of mosaic graphite 50 μm thick diffracted the 2.6 and 5.2 keV photons on a secondary focal plane inside an imaging proportional counter. The crystal was transparent to higher energy photons that impinged on a lithium scatterer encapsulated in beryllium positioned on the primary focus. The scattered photons were detected from a well of four proportional counters filled with a mixture of Xe, Ar and CO_2. The polarimeter was the most sensitive with current technologies. Yet due to the alignment tolerances, the pointing error and Point Spread Function of the optics, the diameter of the scatterer was of 30 mm and the systematics due to the self- absorption with an off-axis beam were not negligible. Moreover in order to minimise the difference in acceptance of solid angle at different azimuth angles, the detectors are mounted at a distance of 90 mm from the scatterer. Due to this large surface the expected background was of the order of 300 mCrab. Therefore the instrument was a real breakthrough only for bright sources while for weak sources the Bragg polarimeter was much more competitive.

In the design of a focal plane scattering polarimeter the possibility to position exactly on the axis the scattering element must be a driver. Actually in the *X-Calibur* Experiment (described below, see Sect. 4.4.3), although the focal length of the telescope is very similar to that of *SODART*, the diameter of the scatterer is the minimum compatible with the divergence of the beam and the well detectors are positioned very close to the scatterer, which also is compatible with an anti-coincidence.

But if the scatterer is a detector, the coincidence will allow to reject a large fraction of these background counts. A focal plane scattering polarimeter is suitable only for a telescope with a significant band-pass above 20 keV, typically a multi-layer optics with focal lengths of the order of 8–10 m. Given that the band pass is in any case determined by the scattering and that these multi-layer optics are very ambitious and demanding in terms of satellite mission, scattering polarimeters of large area with a collimator in the front can be competitive and feasible aboard smaller size satellites. An example of such a mission is in an advanced state of development in India. *POLIX* (see Fig. 4.4) is based on a lithium/beryllium scatterer in a well of four proportional counters. A collimator of honeycomb shape allows to pass only the photons inside a field of view of $3° \times 3°$ with a flat top of 2°. The shape of the scatterer is still to be defined, but it will be conical and axis symmetric. The expected modulation factor is of around 40%. The band pass is 5 to 50 keV. The compensation of systematics will be performed through the rotation of the whole satellite around the axis. The expected sensitivity is 2–3% MDP for a 50 mCrab sources with 1 Ms exposure. The extragalactic sources are out of reach (except exceptional outbursts) but a good sample of binaries should be studied with an instrument of the total mass of 125 kg. Further information are available in Rishin et al. (2010), Maitra et al. (2011).

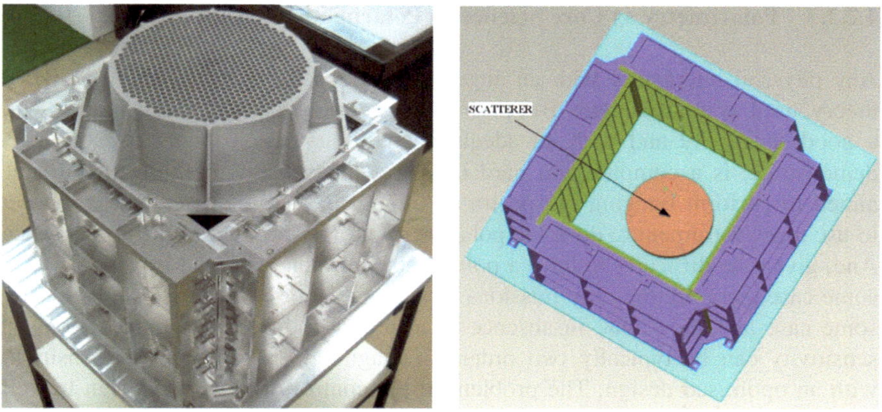

Fig. 4.4 (**Left**) The *POLIX* experiment with the collimator to minimise the impact of the diffuse emission. (**Right**) The layout of *POLIX* with the scatterer surrounded by four imaging proportional counters. Figures from: http://www.rri.res.in/~bpaul/polix.html

The need for a large area and the need to control systematics following the self-absorption of scattered photons results in the need of a fine subdivision. A collimated scattering polarimeter is a bunch of long thin detectors, possibly distributed on a honeycomb pattern.

If the fine subdivision is adequate, systematics can be kept under control. The residual effects can be compensated by rotating the whole assembly around the observing axis. A special case is that of polarimeters devoted to measure polarisation of sources from a wide field of view. In general the sensitivity of Wide Field Instruments is much lower than that of Narrow Field Instruments, basically because a large fraction of the background is proportional to the field of view. Yet in some application and, in particular, to perform polarimetry of sources of previously unknown direction, only a Wide Field Instrument can work. Beside the high background these instruments must face the problem that the systematics cannot be averaged by rotating the instrument around the view axis. In fact there is no view axis previously known. Therefore the control of systematics is all based on a good knowledge of the response of the instrument at different directions on the sky. And this relies on detailed simulations and on accurate calibrations. Furthermore the latter cannot be used in a straightforward manner because in the Hard X-ray, soft gamma-ray band the effects of backscattering are relevant and on ground and in space the conditions may be significantly different. A ground calibration must be corrected for the effects of the laboratory environment (pavement, walls) while in space the effects of the satellite and of earth atmosphere (included delays) must be accounted.

4.2.2.5 Polarimetry as Core Science vs Polarimetry as a By-product

Any physical interaction with an angular dependence on polarisation can be, in theory, used as a polarimeter. But performing polarimetry of celestial sources with a good sensitivity and with an adequate credibility, as 50 years of experience demonstrate, is a complex and hard task. Yet many teams building instruments aimed to perform imaging or spectra or timing of X-ray sources have proposed to use their instrument to do some polarimetry as a by-product of the core science. And, given the difficulty to have a polarimetry mission approved by an agency, in some case these are the only missions producing data on polarisation. We discuss some case and show that, in absence of a design optimised for polarisation, the sensitivity can be typically two orders of magnitude worse than the sensitivity with an optimised design. The problem of the control of systematics can be very serious as well. In general the discrepancy is reduced at increasing energies. A good example is that of pixellated detectors for focal plane of optics. We discuss the idea to use contiguous pixels fired on focal plane CCDs (see Sect. 4.5.1.2) to search for polarisation of the beam due to the photoelectron created within one pixel and propagated to the pixel nearby. This idea never arrived to practical implementation because of the poor sensitivity. A similar concept, based on the scattering, was developed. Multi-layer optics focus on a broad band. To make an efficient use of this band in many proposed instruments a stack of detectors was assumed. A Si detector in front to detect photons up to 10 keV and a CdTe (or CZT) detector in the back to detect photons of higher energies passing through the Si wafer. For some of these missions it was foreseen to perform some polarimetry by detecting a scattering event in the Si followed by an absorption in CdTe. It has been demonstrated (Muleri et al. 2012) that this would be at least a factor 30 less sensitive than a dedicated polarimeter

4.3 Past Experiments of X-Ray Polarimetry with Classical Techniques

4.3.1 Rockets

The first evidence of polarisation from a celestial source came from a rocket experiment on-board of a Aerobee 350 (Novick et al. 1972; Weisskopf et al. 1972) launched on 22nd February 1971 from the base of Wallops Island in USA. The payload consisted of two kinds of polarimeters, one based on Bragg diffraction and the other based on Thomson scattering. The polarisation is derived from the modulation of the counting rate at 2 times the frequency ω, of the rotation of the rocket. Four panels opened in-flight hosting mosaic graphite crystal onto a parabolic surface to focus the radiation on a small detector reducing the contribution of the background with the addition of hexagon honeycomb collimators. The modulation factor for incidence angles between 40° and 50° was 97%.

Additionally Thomson scattering section of the polarimeter consisted of 28 metallic lithium and 52 beryllium-window proportional counters placed at the side of the blocks. Non- diffused X-ray background contribution was reduced using the technique of pulse shape discrimination, anti-coincidence and pulse amplitude. With an observing time of 330 s the minimum detectable polarisation was 10%.

4.3.1.1 Systematics Effects

Some systematics were expected during flight:

- Misalignment of the rotation axis with the direction of the X-ray source. This misalignment is responsible for the variation of the exposed area of the panels and variation of the counting rate due to uniformities of the detector
- Background induced polarisation. This is mainly due to east-west effects and the derived anisotropy of the particle background,

It turned out that the main systematics found in the data was a 1-ω effect due to the variation of the exposed area (3° of misalignment corresponded to 3% of spurious modulation at 1ω and 1% at 2ω). Other systematics were kept below 1% (Novick et al. 1972).

4.3.1.2 Results

Using both the Thomson scattering and Bragg diffraction portion of the instrument (unfortunately half of the data were lost in transmission for each section) the polarisation from the Crab Nebula was measured as (18.2 ± 6.1)% which is a 3-$sigma$ result (Novick et al. 1972) later confirmed with higher precision with a measurement on-board of a satellite (see Sect. 4.3.2.1). Such measurement allowed to definitely confirm that the emission of the Nebula is synchrotron up to X-rays requiring a way to continuously accelerate electrons up to 10–100 TeV.

4.3.2 *Satellites*

Flown polarimeter was based on Bragg diffraction either as secondary mode of a spectroscopic measurement or as dedicated experiment. The spectrometer on-board of *Ariel 5* (Griffiths et al. 1976; Gowen et al. 1977) (see Fig. 4.5) was capable to measure only a coarse 3-σ upper limit on Sco X-1 of 7.7% in 10 days of source observation (and 3 days of background acquisition). Such spectrometer was designed as a pair of perfect flat crystals of 234 cm^2 of LiF and Graphite

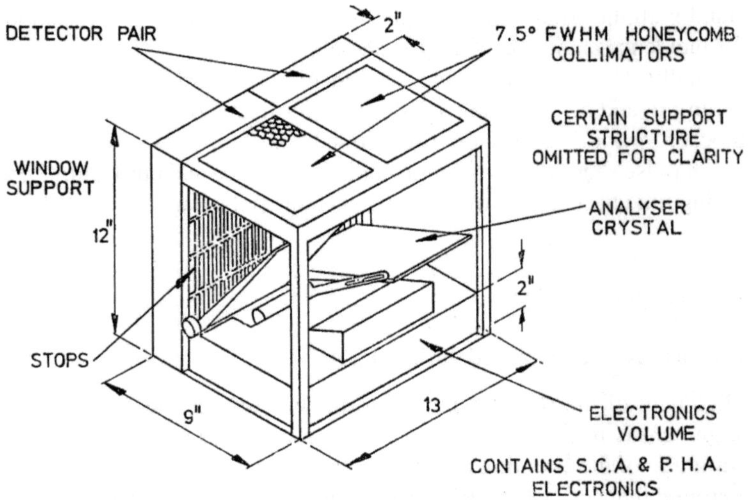

Fig. 4.5 *Ariel 5* was primarily a mission of spectroscopy with two flat crystals: one with LiF and another one with graphite. When mounted at 45° the graphite crystal was in polarimetry mode. However only coarse upper limit on Sco X-1 was measured (figure from Griffiths et al. (1976))

adjacent crystals. It needed an adequately large proportional counter with resulting poor performances in terms of source-to-background ratio notwithstanding the high modulation factor of 97.3.

Instead the Bragg diffraction experiment on-board of *OSO-8* satellite was designed to be a polarimeter (see Fig. 4.6). The 2 × 468 mosaic crystals were placed on two parabolic surface of 140 cm^2 each one with Bragg angle spanning between 40° and 50°. The modulation factor obtained was 94%. However the proportional counter was sensitively smaller than the one in *Ariel 5* and so the background level. Both *Ariel 5 OSO-8* were spinning satellites so polarisation was searched in the data modulated at twice the satellite spin frequency. The better sensitivity of *OSO-8* was indeed mainly due to its superior source-to-background performances (a factor of 10 for the first diffraction order). The comparison of *Ariel 5* with *OSO-8* is a clear picture of the different performances of a dedicated polarimeter and a by-product polarimeter.

4.3.2.1 Results from *OSO-8* Polarimeter

The only positive measurement to-date was indeed obtained by the *OSO-8* polarimeter in 71.2 h of net observation out a total of 12 days selecting portion of the orbit with background that was just 10% of the signal. Exploiting the lunar occultation technique, precision polarimetry (19.22 ± 0.92)%, combined at 2.6 keV and 5.2 keV was obtained (Weisskopf et al. 1978) (see Fig. 4.6, right). Indeed it was excluded

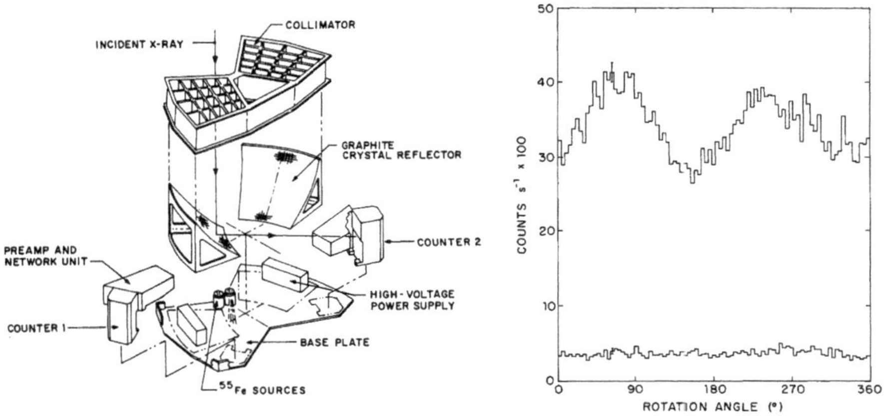

Fig. 4.6 (**Left**) The Bragg diffraction polarimeter on board of *OSO-8*. Graphite mosaic crystals are placed on two parabolic surfaces to increase the source-to-background ratio (figure from Weisskopf (1976)). (**Right**) The modulation curve obtained observing the Crab Nebula showing, also, the contribution of the background (negligible). The asymmetry is mainly due to the misalignment between the direction of the Crab Nebula and the spin axis of the satellite (figure from Weisskopf et al. (1978))

Table 4.3 Upper limits on Sco X-1 and Cyg X-1 with *OSO-8* (Long et al. 1979, 1980)

Sco X-1	$PD_L = (0.39 \pm 0.20)\%$	$PA = 29° \pm 10°$	2.6 keV
	$PD_L = (1.31 \pm 0.40)\%$	$PA = 57° \pm 6°$	5.2 keV
Cyg X-1	$PD_L = (2.44 \pm 1.07)\%$	$PA = 162° \pm 13°$	2.6 keV
	$PD_L = (5.3 \pm 2.5)\%$	$PA = 155° \pm 14°$	5.2 keV

in the analysis the data corresponding to the portion of the lightcurve with high contribution of the pulsar. However only coarse upper limits were found for the pulsar (Silver et al. 1978).

More stringent upper limits were found in particular for Sco X-1 and Cyg X-1 with their spectral states unknown at the time of the measurement. The results are reported in Table 4.3.

Coarse upper limits on other celestial sources were found (Hughes et al. 1984) providing evidence that, besides the Crab Nebula, they are polarised at level not compatible with this experiment. However if we compare the 3-σ upper limit obtained with *OSO-8* at 2.6 keV Cygnus X-1 results are not compatible with the results of the SPI measurement (Jourdain et al. 2012) at higher energies.

4.4 Flown Hard X-Ray Polarimeters

4.4.1 *IKAROS/GAP*

The Gamma-Ray Burst Polarimeter (GAP) (Yonetoku et al. 2011) is a small experiment, aboard the Small Solar Power Sail Demonstrator IKAROS, devoted to polarimetry of Gamma-Ray Bursts. The scatterer is a dodecagon of plastic scintillator, 6 cm thick, acting as scatterer. The diameter is 140 mm. In the bottom the scintillator block is tapered to a photomultiplier. In front of each of the twelve sides, a crystal of CsI(Tl), long as the scatterer and 5 mm thick has its own read out photomultiplier each pair of detectors shares a radioactive calibration source of Am^{241}. Another source is calibrating the scatterer. The sole photomultiplier cannot localise the site of scintillation. Therefore the angle is given only by the absorbing crystal. There is also a certain probability of a double interaction within the plastic block. The advantage is the high uniformity of light collection from the plastic, the good circular symmetry and the good calibration of thresholds. The modulation factor is almost constant around 30% from 50 to 300 keV and the efficiency has a peak of 17% at 130 keV. The detection of polarisation is very robust at small angles from the observation axis. At large incidence angles the spurious polarisation becomes relevant and the detection of polarisation becomes heavily correction dependent. GAP has detected high polarisation in three GRBs (Yonetoku et al. 2012).

4.4.2 *POLAR*

POLAR is a compact wide-field instrument to detect polarisation of Gamma Ray Bursts, or other transients, in the energy range from 50 keV to 500 keV. It consists of a 40 × 40 array of plastic scintillator bars used as a detection material. The bars dimensions are 4 × 4 × 4 × 200 mm^3. They are grouped in 25 detector modules, each read with a multi-anode photomultiplier. Each bar is optically separated from the others. The typical interactions are a Compton scattering in one bar and a second Compton scattering or a photoelectric absorption in another bar. A lower and an upper limit to the detected energy limit the energy range from 50 to 500 keV. As in the case of any polarimeter made with one material only the energy of each photon is poorly constrained and energy resolved polarimetry not easy. On the other side the selected design is feasible with the well-established technology of pixel PMTs and an area of almost 400 cm^2 was arrived. The response at large angles from the axis requires a huge simulation activity. POLAR was calibrated in great detail and the simulation mass model was verified. Also an on-board calibration with weak Na22 sources was performed. Nevertheless the not uniformity of the threshold of each bar in terms of energy required a severe activity of calibration and equalisation on ground.

POLAR was launched aboard the Chinese space-lab TG-2 for a 1 year observing time (Xiao et al. 2018).

4.4.3 *X-Calibur*

X-Calibur is the only actual implementation of the concept of focal plane scattering polarimetry (see Fig. 4.7). It is based on the InFOCμS telescope of 8 m focal length with a Wolter I design and a multi-layer coating (Pt/C). *X-Calibur* is definitely aimed to a satellite mission, but two flights aboard balloons have been already performed. The focal plane instrument has been adjusted for these flights. In a first test-flight of 1 day from Fort Summer in 2014, the basic functionality of the system was verified and the background was measured. The scatterer was a cylinder of plastic scintillator. The absorber was a well of 32 CZT detectors surrounding the scatterer on four sides. The whole had around a passive shield. The coincidence between the scatterer and the absorber detectors helped to reject 90% of the background.

Following this first test the experiment was significantly re-designed. The efficiency of the active scattering system was relatively low. In the new design the scatterer is made of solid beryllium, 16 cm long, with a diameter of 12 mm. The top surface of the beryllium rod is positioned on the focal plane of the telescope. The swap from plastic to beryllium increased the efficiency of 45%. Also to compensate the increase in background due to the drop of active scatterer in coincidence, the shield was replaced with an additional CsI anti-coincidence plate (Kislat et al. 2018).

X-Calibur with the improved design has been flown again aboard a long duration balloon flight from Antarctica. Its main target is Vela X-1 and GX301-2.

For the satellite version a first scatterer of LiH followed by a second stage of beryllium is foreseen.

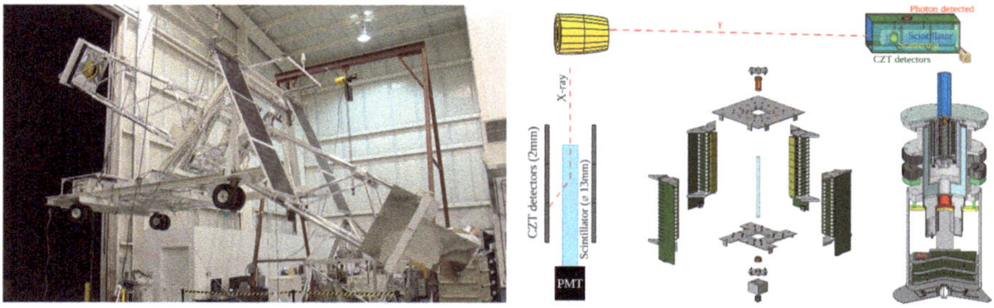

Fig. 4.7 (**Left**) The *X-Calibur* experiment at the launch base. (**Right**) The detection principle based on Compton scattering (figure from Kislat et al. (2017))

4.5 Modern Techniques

4.5.1 Photoelectric Effect and Its Exploitation

An ideal polarimeter is based on a physical interaction whose angular distribution is of the form:

$$P(\phi, \theta) \propto \cos^2(\phi)\sin^2(\theta) \tag{4.23}$$

where θ is the polar angle, namely the angle with respect to the photon direction and ϕ is the azimuth angle with respect to the electric field.

This is the case of the angle of ejection of s photoelectrons extracted from the photo-absorption (see Fig. 4.8). Above the K edge this is the dominant interaction. Anyway also the angular distribution of L photoelectrons is modulated to a reasonable level. Beside the two s electrons also the p electrons are ejected with a good modulation.

The basic physics of the interaction of polarised X-rays with matter was first assessed by Auger, Kirkpatrick, Kirchner and Richardson, in the years spanning from 1925 to 1930 (Compton and Ellison 1935). The angular distribution of *s* photoelectrons is of the type:

$$P(\phi, \theta) \propto \frac{\cos^2(\phi)\sin^2(\theta)}{(1 - \beta \times \cos(\theta))^4} \tag{4.24}$$

where β is the velocity of the electron divided by that of light.

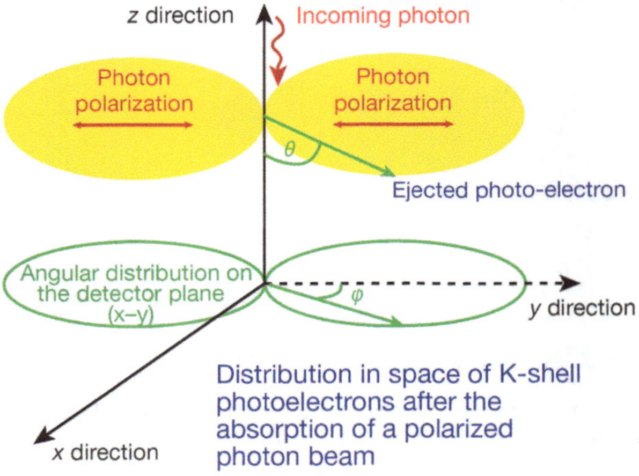

Fig. 4.8 The photoemission properties of s-photoelectrons. Their emission direction is maximally modulated by the polarisation of the incoming beam (figure from Costa et al. (2001)). Indeed at 90° from the electric field of the incoming beam, no photoelectrons are ejected. A smaller modulation is foreseen for p-photoelectrons (Afanas'ev et al. 1988)

As to the polar angle the maximum ejection is in the plane perpendicular to the beam. The term at denominator accounts for the forward bending that increases with energy, to preserve the momentum, but, at the energies of interest this is not relevant and in practice the Eq. (4.24) applies. The angular distribution of photoelectrons around the direction of the incoming X-ray beam can, therefore, be used to perform polarimetry. The information on the linear polarisation is encoded in the direction of the photoelectrons ejected in the interaction. But after the first ejection the electron is scattered, slowed by ionising and non-ionising collisions, and eventually stopped. Ionisation events from the first creation of the electron to the end is named Track. At these energies, in the low energy side of the Bethe-Bloch curve, the energy loss $\frac{dE}{dx}$ is inversely proportional to β^2 and, in practice, to the kinetic energy, and the scattering probability is inversely proportional to the square of the kinetic energy. The track is straight and *pale* at the beginning, then it is continuously increasing the linear energy loss and thence the ionisation density. The beginning of the track is straight then nuclear scattering changes suddenly the direction here and there. At the end the direction is totally randomised and most of the energy is deposited. The information on polarisation is all contained in the other side of the track near the absorption point, which sometimes is strengthened by an Auger electron. If we want to exploit this information, we must visualise the ionisation track, distinguish between the start and the end of the track and derive the direction of the very beginning before the first scattering changes it. These studies were performed in the 1920s by means of cloud chambers. The most extensive studies were performed by Pierre Auger who also used stereoscopic chambers to visualise the tracks in 3-dimensions (Auger 1920). Incidentally, while studying the photoelectrons from X-rays and the dependence on the track shape on the gas atomic number, Auger discovered the existence of the, since then, named Auger electrons.

4.5.1.1 Anti-Coincidence and Rise-Time in Proportional Counters

With the rise of space age, after the World War Two, the focus of technology moved to low mass, small encumber devices, compatible with probe rockets. After the discovery of X-rays of cosmic origin on 1962, and after the detection of X-ray emission from the Crab Nebula, that in the radio showed evidence of synchrotron radiation, theoreticians wondered whether this emission could extend up the X-ray range. In theory X-ray polarimetry was conceived soon after the start of X-ray Astronomy. In practice the task came out to be more difficult than hoped. For sure nobody was able to launch a cloud chamber that is massive, cannot be triggered and uses a photographic readout. Since the beginning of X-ray astronomy Thomson Scattering and Bragg diffraction were the techniques candidate to perform this measurement. But all the early X-ray astronomy was performed with detectors based on the photoelectric effect and a few attempts to perform photoelectric polarimetry were also performed.

The baseline detector for spectra and timing was the multi-wire proportional counter. The sensitive volume was divided into cells centred on an anode wire operated at high voltage and contoured by grounded cathode wires. Therefore nearby cells are simply delimited by the geometrical surface defined by the locations where the electric field points to one anode or to another one. In these conditions a photoelectron absorbed near the border between two cells could generate a track with a part in one cell and the remnant on the other. In this case one should have two signals from the two anodes in temporal coincidence. If a flux of polarised X-rays impinges on a detector in a direction parallel to the wires we will have a certain number of coincident signals that will be larger for pairs of anodes parallel to the electric field. The rate of these coincidences would be modulated with the phase of the spinning of the satellite. No conclusive results were achieved or, at least, no results encouraging to prosecute documented by publications. The only paper of this kind in the literature is Riegler et al. (1975) of is dated 1975.

Another detector for X-ray Astronomy aimed to perform polarimetry exploiting the Photoelectric effect is described in a paper of 1970 by Sanford et al. (1970). This team of researchers from Mullard Space Science Laboratory (MSSL) tried a method based on pulse shape discrimination in proportional counters used for background rejection. It was known that the rise-time of a proportional counter is shorter if the track is parallel to the anode and slower when the track is perpendicular. They found a difference in the rise-time for polarisations parallel and perpendicular to the anode, but the effect was little (see Fig. 4.10) and the research was stopped there.

4.5.1.2 X-Ray CCDs

More recently, charge-coupled devices (CCDs) (Tsunemi et al. 1992; Holland et al. 1995; Kotthaus et al. 1998) have been tested with polarised X-rays and simulated. At energy below 10 keV optics for X-ray astronomy are very effective. In this energy range the actual pixel size of a CCD is still larger than the photoelectron track length and single pixel events outnumbers two pixels events, called pairs, and moreover multi-pixels events for a true electron tracking. In this energy range therefore this approach is still not very effective, as only the few photons absorbed close to the pixel edge produce pairs whenever an electron leaks from one pixel to the adjacent. The orientation of these pairs is modulated with the polarisation angle. However these events are extremely sensitive to systematic effects, when the pixels are not uniformly illuminated. This is because the probability of a photoelectron switching on two contiguous pixels could depend more on the probability of being absorbed in the frontier zone at a particular phase than on the polarisation itself. These effects can be controlled and calibrated reliably in laboratory (Buschhorn et al. 1994). The control and calibration of these effects are much more difficult in an astronomical observation because of the satellite pointing instability and because the point spread function of the optics can be of the same size as the CCD pixels. Moreover for this device, track imaging is possible only at high energy (>20–30 keV) where the efficiency becomes extremely small.

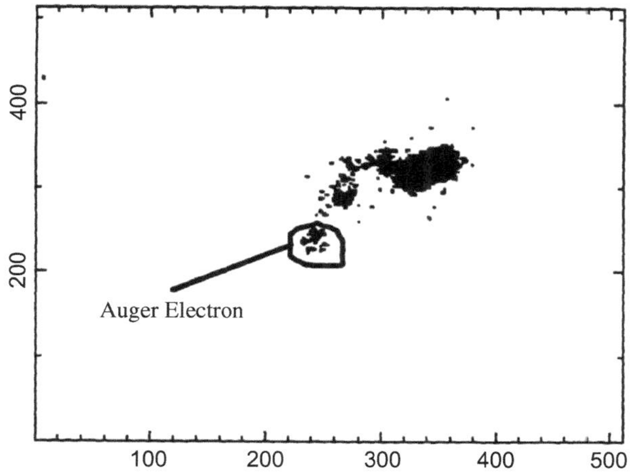

Fig. 4.9 The image of a photoelectron from Austin et al. (1994). The photoimager was filled with 2 Atm of Ar-TMA mixture with a 50 cm drift length. The energy threshold for polarimetry was set 30 keV due to the drift length and mixture used

4.5.1.3 Gas Imagers

Another approach was followed by Austin (Austin et al. 1994). He built a device that imaged the light from a drifting track (see Fig. 4.9) produced by a hard X-ray photon by using a Quartz lens and a CCD. He reported a modulation measurement of 30% for 60 keV photons by using an X-ray imager filled with Ar-TMA. The choice of such a gas establishes a threshold in energy estimated by the authors to be around 30 keV above which it is possible to optically image the track. This is mainly due to the reduced range of the photoelectrons with respect to low-Z gas mixture (Iskef et al. 1983) and to the long drift region (50 cm). A different kind of photoelectron imager, more oriented to lower energy photons, was proposed as an X-ray polarimeter (LaMonaca et al. 1998; Sakurai et al. 2004). The aim here was to get a modulation at energies below 10 keV, where the grazing incidence optics can focus X-rays.

4.5.2 Time Projection Chambers

Inspired by the discovery that it was possible to image the track by the Gas Pixel Detector (GPD, see Sect. 4.5.3), and to measure polarisation with good modulation factor and quantum efficiency, it was devised a polarimeter using the technique of Time Projection Chamber (Black et al. 2007). Such device forms 2-dimensional images of the photoelectron track in gas using both the 1-D spatial information from micropattern gas proportional counter and the time history of the signal of each strip.

The latter timing information is converted in space coordinate by knowing precisely the drift velocity of the charge cloud in the gas. Photons enters in the gas cell along the direction of the strips, they are converted in a photoelectron track and they drift perpendicularly to the incoming direction towards the strips of the micropattern. A Gas Electron Multiplier is placed in front of it at a short distance. Each strip of the micropattern is coupled to an ASIC equipped with a charge sensitive preamplifier which, in turn, is connected to a continuously sampling analog-to-digital converter (ADC). Due to the orthogonality of the drift with respect to the absorption direction, such design allows in principle for decoupling the modulation with respect to the efficiency at the expenses of imaging. Indeed for one coordinate it is not known the conversion time in the gas, for the other coordinate the practical implementation requires inclined beam from an X-ray optics in a long device which impacts on the 1-d reconstruction capability. A major weakness of this technique is that by drifting the electrons of the track in a direction perpendicular to the optical axis the symmetry around this axis is lost. Azimuth-dependent systematics are introduced and can only be compensated through the rotation of the instrument (or of the satellite). Indeed such detectors were proposed and developed for *GEMS* (*Gravity and Extreme Magnetism SMEX*) NASA mission (Swank et al. 2010; Jahoda et al. 2014) and for *PRAXyS* (*Polarimeter for Relativistic Astrophysical X-ray Sources*) (Jahoda et al. 2016). *GEMS* was discontinued for programmatic reasons in 2012 while *PRAXyS* performed phase A as a SMEX mission which study ended in 2017 with the selection of *IXPE* the *Imaging X-ray Polarimetry Explorer* based on Gas Pixel Detector technology. Due to the particular geometry of this device, the TPC polarimeter can be coupled in a stack together with a hard X-ray polarimeter based on Compton scattering. Such detector configuration requires two beryllium windows at the X-ray entrance and exit. Low energy X-rays are absorbed in the TPC polarimeter while high energy X-rays are absorbed by the rear Compton polarimeter. Such disposition is at the base of a design of a wide energy polarimeter for the proposed, at the time of writing, X-ray Polarimetry Probe (XPP) (Krawczynski et al. 2019).

4.5.3 Gas Pixel Detectors

With the aim to arrive at a device capable to measure polarisation in the *classic* energy band, 2 to 10 keV, by exploiting the photoelectric effect, we detected, for the first time, sensitivity to polarisation below 10 keV with a 1-dimensional detector, of a high space resolution, and (Soffitta et al. 1995, 2001) using a low atomic number gas mixture (see Fig. 4.10). The choice of Neon was intended to operate well above the K edge and, thence to have the maximum energy transfer to the an s-photoelectron and to minimise, compared to Ar and Xe, the elastic scattering that unavoidably blurs the information on the original direction. The results are shown in Fig. 4.10 (right).

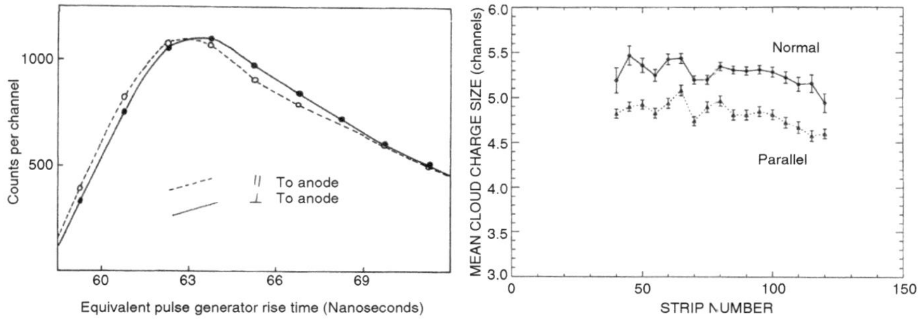

Fig. 4.10 (**Left**) The difference in rise rime for two different orientation of the X-ray polarisation obtained by a single wire proportional counter filled with Xenon (Sanford et al. 1970) at 16.5 keV. (**Right**) The difference in the mean number of strips (mean cloud charge size) in a 1-d Microgap detector filled with Neon-DME mixture at 8.04 keV which demonstrated the feasibility of photoelectric polarimetry below 10 keV by 1-D imaging (Soffitta et al. 1995, 2001)

The publishing process in refereed journal was slow and the paper was printed only a few months before the publication of the paper on the Gas Pixel Detector. However, the paper includes all the basic concepts and analysis of physics, statistics, simulations, the astrophysics rationale for the future development of the GPD are already there.

In theory such detector, combined with rotation, would allow to perform polarimetry, but the target was more ambitious. In X-ray astronomy the passage from collimated spinning or rocking instruments to focal plane imaging detectors had the consequence that the need for a spinning satellite or subsystem was dropped. Most of bi-dimensional gas detectors at the epoch were based on cross-wire or cross-strip technique. This technique is not viable for polarimetry. Hosting a polarimeter in the focus of a telescope, possibly to be alternated with an imaging or a spectrometer on a sliding device, would be much easier with a real bi-dimensional detector, namely with a full pixel geometry. A rotation would be not needed or kept as a back-up to compensate possible residual systematics.

This concept drove the second dramatic passage with a major step forward in technology that was performed by the Pisa/INFN group, lead by Ronaldo Bellazzini, that in 2001 arrived to the first prototype, which started a new era for astronomical X-ray Polarimetry (Costa et al. 2001; Bellazzini et al. 2006a, 2007). The development of GPD is well documented in the literature. Here we give an overview outlining the three major steps. All three versions have some relevant commonalities:

- A gas cell with a ceramic body and a Be frontal window. Contrary to the TPC the gas in contact with the window is fully sensitive
- An electric field parallel to the optical axis to drift the electrons of the track to a multiplication stage.
- A Gas Electron Multiplier (GEM) to perform a proportional multiplication of the electrons to amplify the track while preserving its shape.

- An electric field, also directed as the optical axis, to drift the multiplied electrons from the bottom of the GEM to a sense plane.
- A plane of metal pads distributed on a honeycomb pattern to act as anodes to collect the charge multiplied by the GEM.
- An electronic chain independent for the charge collected by each pad.

4.5.3.1 From Printed Circuit Board to a Self-Triggered Sealed Detector

One of the major difficulties was to route the charge collected from the pad to the electronic chain, also given that the pads form a compact plane with only a thin separation. The first prototype was based on trigger by GEM and collection plane built with multi-layer printed circuit technology. The pad pitch was 200 μm. The GEM pitch was 90 μm (Costa et al. 2001). Signals were passed to lower layers of the PCB by vias and, for each plain, routed horizontally to a set of application specific integrated circuit (ASIC) chips, relatively far from the gas cell. However, it should be clear that the technology applied had serious limitations that would make a flight device still cumbersome and with severe limitations.

This technique of multi-layer PCB would never allow for a number of pixels larger than $\simeq$1000 and anode pads below 100 μm. Given the length of the tracks, this would limit the field of view of any application of the GPD and, in practice, to point-like sources and, from the physics point of view, a finer readout would give much better performances.

Also, while the small dimension of the pad would provide a very small capacity and a very low noise, the long tracks would add a large capacity and an increase of the noise.

But data from this prototype showed that imaging photoelectric polarimetry with a full pixel detector was, at last, feasible. The modulations found were adequate to provide a dramatic step forward in sensitivity and the control of systematics (Costa et al. 2001). Such experimental results determined a new bunch of theoretical papers on X-ray Polarimetry.

The next step performed by the Pisa team was the development of a very large-scale integration (VLSI) ASIC chip that solved together the problem of minimising the routing noise, to reduce the pixel size and to increase the total effective area. The first layer of the ASIC hosted the metal collecting pads. In the inferior layers, below the projection of each pad an analog and digital front end electronic chain handles the signal from each pad. Under trigger from the GEM, the preamplified and shaped analog signal is held. The content of all pixels is Analog to Digital (A/D) converted. The digitised image is analysed to detect the cluster of pixels of the track with a charge content, identify the starting point and determine the emission direction of the photoelectron. The revolution of this device is that everything collapses inside the same cell containing the gas. The dimensions become incredibly small. The electronics contiguous to the pad has a very low noise. The whole system benefits from the stability and the radiation hardness of the high integration VLSI. The chips of this stage are two, one with 2000 pixels (Bellazzini et al. 2004) and the other with 22,000 pixels (Bellazzini et al. 2006b). The pitch of the pad was reduced to 80 μm.

These two devices are still based on a trigger from the Gas Electron Multiplier starting the download and the conversion of the content of all the pixels of the chip and an analysis of the image of the whole detector to identify the track. The maximum frequency of the conversion cannot exceed 10 MHz. This resulted into an unavoidable dead time. The double target of reducing the pixel size and increasing the total area of the detector would result in a significant increase of the number of pixels and in an unacceptable dead time.

The last passage was a further increase of the area with a chip of 105,600 pixels, with a pitch of 50 μm, for a total of 2.25 cm^2 (Bellazzini et al. 2006a) and this requested to introduce a system of self-triggering, leaving to the GEM the only function of multiplier without making any use of the signal from it. A Region of Interest is defined around the pixels which have triggered. Only the content of the pixels inside the region of Interest are downloaded and converted. Therefore, for X-rays in the range, the number of pixels downloaded is typically less than 1% of the total. The three generations of ASIC are shown in Fig. 4.11.

This new chip was mounted inside a sealed detector with a thin (50 μm) beryllium window (Bellazzini et al. 2007) (see Fig. 4.11). The very new technology was, therefore, combined with a very traditional set-up of detector manufacture for long duration operations. These first sealed prototypes had an effective area of 2.25 cm^2 and a total mass of 50 g. The three generations of the chip are shown in figure.

In Rome our team improved the testing and measurement capability, by building a set of monochromatic X-ray sources polarised to near 100% by Bragg diffraction and very compact capillary plates as collimators (Muleri et al. 2007). The sources combined with collimators, diaphragms and micrometric computer-controlled stages allow for a detailed characterisation of the detectors. The absence of systematics was tested with a Fe^{55} source which is unpolarised by prime principles.

Fig. 4.11 (**Left**) The three different generation of the ASIC which collects the charge of the photoelectron track. The ASIC employed has self-trigger capability and features the Region of Interest determination and its download (Bellazzini et al. 2004, 2006a). (**Right**) The first sealed Gas Pixel Detector (Bellazzini et al. 2007)

Table 4.4 Main characteristics of the low energy polarimeter produced until December 2018

Size	4 cm × 4 cm
	9 cm × 9 cm
Active area	1.5 cm × 1.5 cm
Drift	1 cm
Spacer	Macor
Window	beryllium (50 μm)
Filling	Ne80-DME20 (1 bar)
	He20-DME80 (1 bar)
	DME100 (0.8 bar)
	DME100 (0.4 bar)
Dead Time	1 msec
Number of pixels	105,600
Pitch	50 μm hexagonal
GEM	Kapton
	Liquid crystal polymer
Thickness	100 μm or 50 μm
Hole pitch	80 μm or 50 μm hexagonal

This detector was eventually ready to be the long foreseen focal plane polarimeter and has been proposed to various agencies. Eventually IXPE the *Imaging X-ray Polarimetry Explorer* was approved on 2017 by NASA to be launched in the middle of 2021. It includes three telescopes each with a GPDs in the focus. It will be the first satellite mission totally dedicated to low energy X-ray polarimetry ever.

4.5.3.2 Low Energy Gas Device for X-Ray Astronomy

The Gas Pixel detector for low energy application was devised in order to be used at the focus of a conventional X-ray optics. We opted for a low-atomic number mixture and 1 cm drift depth after intensive Monte Carlo simulations. The characteristics of the GPD are shown in Table 4.4. A version of the GPD with a flight-like design is shown in Fig. 4.12.

4.5.3.3 Medium Energy Gas Device for Hard X-Ray Astronomy and Solar Flares

Simulations show that GPD can work effectively as polarimeters with Argon based mixtures, with a pressure up to 3 atm and an absorption gap of the order of 30 mm (Fabiani et al. 2012a). The team built a first prototype of such a medium energy polarimeter and had to face a problem of uniformity of electric field. To have more freedom to increase the thickness of the absorption gap they had to make wider electrodes. Thus, they changed the design. Nowadays, this larger body is also adopted for lower energies, included the detectors for *IXPE*. A second prototype was

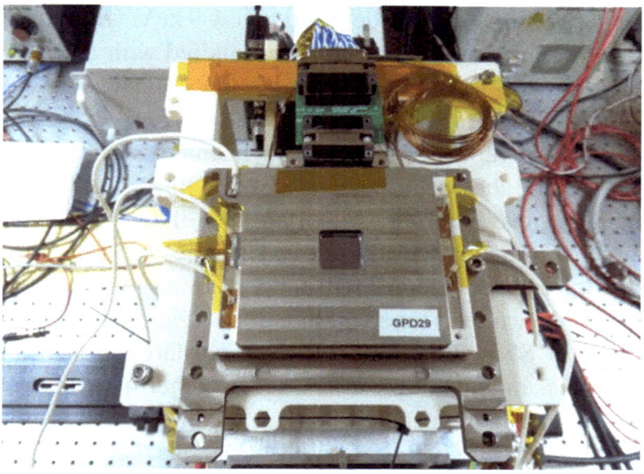

Fig. 4.12 The flight version of the Gas Pixel Detector. The dimensions are larger (9 cm × 9 cm) than the first design but the active area and the drift thickness is the same. This design reduces the impact of the background because the walls are more distant from the active reagin. Moreover it increases the uniformity of the electric field especially at the edge of the active area, thanks to a larger guard ring

built, based on this new geometry. First, testing at IAPS showed that it was working as predicted by simulation, but a leak problem turned the prototype into something not operative in a few weeks (Fabiani et al. 2012a). If we consider that:

- The credibility of simulations increases with energy.
- In gas detectors, there is much more experience in the Argon based mixtures than in pure Dimethylether (DME).
- Testing of the last prototype has only failed under the aspect of leak tightness, a problem not trivial but faced and solved many times in the past, also for detectors more sensitive to gas purity as Gas Scintillation Proportional Counters.

We can assume, with reasonable confidence, that this extension of the GPD performance to higher energies, should work, and arriving to the Readiness Level needed for a mission proposal should be feasible in a reasonable time. Now, a good question is whether it is worthwhile to conceive a mission hosting at least one of these telescopes that the team names Medium Energy Polarimeter (MEP). In fact, the change of the band is more a shift than an extension. The Auger electron in Argon, which has an isotropic distribution, has an energy of about 3 keV. The photoelectron must, at least, exceed this energy to preserve a minimum of the anisotropy that carries the polarimetric information. Let us say that 8–25 keV is a practical range of operation. There are a few, but not trivial at all, topics where the MEP would represent a great step forward:

- The Supernova Remnants (SNRs) at energies below 8 keV are dominated by a huge unpolarised thermal emission. For sure at higher energies, the study of the non-thermal emission and thence of the shock front will be much more effective.

- No cyclotron line has been so far detected below 9 keV. Only for a very fortunate combination of circumstances, one could be studied with *IXPE*. A good fraction of them (from one third to one half) falls below 30 keV. They are an excellent target for a MEP.
- A significant role of reflection is foreseen in many scenarios of High Energy Astrophysics. Beside the spectral signatures (Fe lines, Compton bump) and the timing signatures (reverberation), important and distinct polarisation signatures have been predicted in binaries and Active Galactic Nuclei. Some measurements for a few of the brightest sources will be performed by *IXPE*. However, this is the domain of an instrument tuned on higher energies.
- As in the case of SNR, the spectrum of solar flares is also dominated by a huge thermal emission but above 10 keV the non-thermal emission is prevailing. Fabiani et al. (2012b) demonstrated that even a single MEP, without any optics, would be more sensitive to polarisation of solar flares than any of the experiments flown so far.

4.5.4 Bragg Diffraction from Multi-Layer Mirrors

It is interesting and rewarding, for some kind of classes of sources (Blazars, Isolated Pulsars), to measure polarisation below 1 keV. In this energy band polarimetry can be accomplished either by using specific organic crystals or by means of multi-layer mirrors. By means of vacuum deposition technology, structures consisting of alternating layers of high- and low-Z materials, with individual layers having thicknesses of the order of nanometers, can be fabricated on suitable substrates. These structures act as multi-layer interference reflectors for X-rays, soft X-rays and extreme ultraviolet (EUV) light. By choosing suitable graded separations the bandwidth can be increased up to tens of eV. Such technique is being developed for hard X-ray mirrors and soft X-ray polarimeters. A design like *OSO-8* would require large pieces of such kind of mirrors and it would be therefore difficult to build. Disposing smaller mirrors with suitable graded spacing to reflect radiation dispersed by a grating system and focussed by an X-ray optics could be a more viable design.

4.5.5 PolarLight a GPD in Orbit On-Board of a Cubesat

On October 29, 2018 a GPD polarimeter, filled with a 0.8 atm of DME and equipped with a micro-channel type collimator to reduce diffuse background, sized as a 1U Cubesat embedded in a 6U Cubesat, was launched in a low-earth Sun-synchronous orbit. On December 18, 2018 the HVs were successfully turned on. The characteristics of the experiments are discussed in Feng et al. (2019). This is the first time for a GPD technology in orbit and builds confidence on its use in space.

PolarLight is expecting to gather data from Sco X-1, the Crab Nebula and GRS 1915 + 105 with very long exposure times. More Cubesat launches are expected during 2019.

4.6 Future X-Ray Missions

4.6.1 Rocket Experiment Demonstration of a Soft X-Ray Polarim REDSox

The *Rocket Experiment Demonstration of a Soft X-ray Polarimeter* (*REDSoX* polarimeter) (Marshall et al. 2018) is a polarimeter based on the technique of multi-layer mirrors for sensitive X-ray polarimetry in the soft X-ray energy band. This experiment is designed for a first flight on-board of a rocket. In this experiment 6 gratings diffract X-ray coming from a conventional optics onto three laterally graded mirrors (LGMLs) that in turn reflect at 45° the radiation onto a CCD detector. Such mirrors are clocked at 120° for a complete determination of the Stokes parameters. The focal length is 2.5 m. The geometry is chosen such that the modulation factor is about 90% and, thanks to the lateral graded coating, the energy band is 0.2–0.8 keV. The Critical Angle Transmission (CAT) gratings have an efficiency greater than 20% (Heilmann et al. 2016) while the LGMLs have a reflectivity better than 10% (Marshall et al. 2015). *REDSoX* can observe Mkn 421 in a 300 s flight with a MDP of 11% (Marshall et al. 2018). The design of *REDSoX* experiment and its layout are shown in Fig. 4.13.

Grating A Critical Angle Transmission (CAT) grating is fabricated at MIT. The dispersion of the grating at the first (+1) order matches the peak reflectivity of the LGML. The dispersion rule is given by:

$$m\lambda = P\sin(\phi) \tag{4.25}$$

where m is the grating order of interest, P is the period of the grating and ϕ is the dispersion angle. Matching the Eq. (4.25) with the diffraction condition for the multi-layer mirror is possible to derive, by knowing the lateral dispersion of the LGML which set the energy band, the distance of the CAT with respect to the respective LGML and therefore its geometrical design.

LGML LGML are fabricated by Reflective X-ray Optics (RXO) and the Centre for X-ray Optics (CXRO). Multi-layer is deposited on highly polished silicon wafers. The multi-layer has a linear spacing variation of 0.88 $\frac{\text{Å}}{\text{mm}}$ to reflect the radiation from 170 to 730 eV. The material pairs are Cr/SC, C/CrCo, LA/B_4C and their Reflectance has been measured at 45° of incidence.

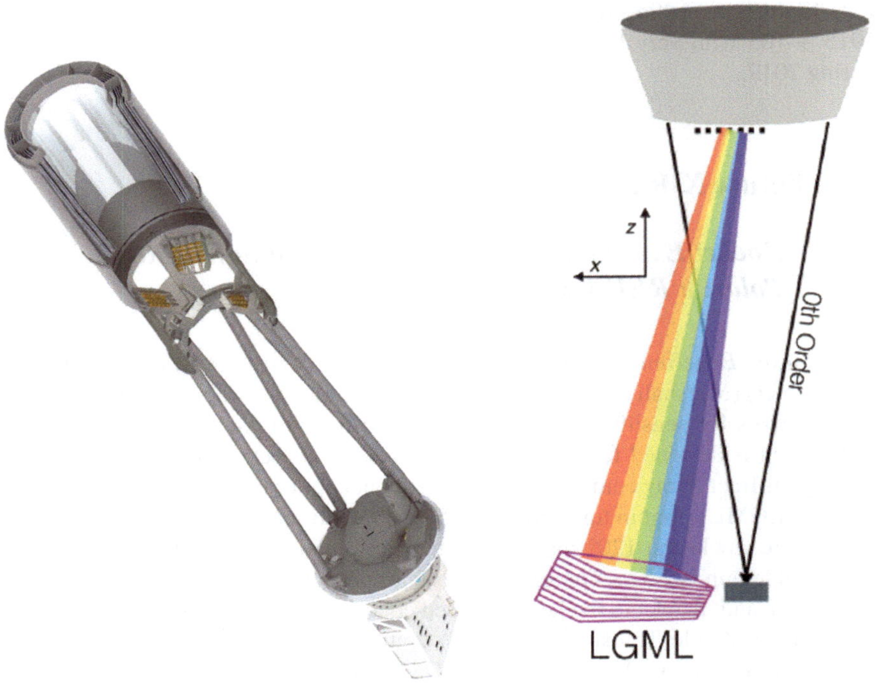

Fig. 4.13 (**Left**) Cutaway rendering of the engineering design of *REDSoX* Polarimeter (Marshall et al. 2018). Photons reflected by the mirror shells (on top figure) are diffracted by 6 gratings clocked at 60° (gold colour structures) and impinge onto three lateral graded multi-layer mirrors that are the polarisation analyser. In turn these reflected photons are detected by three X-ray CCD. (**Right**) Layout of the *REDSoX* polarimeter showing the Laterally Graded Multi-layer

Detector The detectors are commercially available CCD cameras back-illuminated with 80–90% quantum efficiency in the energy range 150–500 eV in a surface of about 25 mm^2 and 1500 × 1500 pixels.

4.6.2 Imaging X-Ray Polarimetry Explorer, IXPE

The *Imaging X-ray Polarimetry Explorer* (*IXPE*) (Weisskopf et al. 2016) has been selected in January 2017 as a Small Explorer NASA mission (SMEX) for a flight in April 2021.

Italian Space Agency (ASI) funds the Instrument composed of three Detector Units (DUs) plus one spare and the Detector Service Unit. ASI also provides the Malindi Ground Station and part of the scientific reduction pipeline through its Space Science Data Centre (SSDC). Marshall Space Flight Centre (MSFC) is the PI (Martin C. Weisskopf) institution. MSFC provides the optics, the calibration facility and the Science Operation Centre. IAPS/INAF manages the provision of the instrument, provides the filter and calibration set and the facility for the DU calibration and instrument integration. INFN provides the GPD, the data acquisition board and integrates the DU. The Mission Operation Centre is at LASP (Boulder, Co), OHB-Italia supplies the Filter and Calibration Wheel (FCW), the High Voltage Power Supply and the Detector Service Unit. Ball Aerospace provides the spacecraft and key payload elements. Orbital ATK supplies the extendible boom.

IXPE is the first mission devoted entirely to X-ray polarimetry (see Fig. 4.14) and it will provide the answer related to the NASA first question: "how the Universe works ?". The detection of polarisation is by means of Photoelectric effect with the GPD as enabling technology. Three low energy GPDs are placed at the focus of three Mirror Modules assembly with 4 m focal length. The GPD with the associated back-end electronics and the FCW are sub-units of the DU (see Fig. 4.14). The DUs are oriented at 120° to minimise possible residual systematics.

The key questions to which *IXPE* will answer are:

- What are the emission mechanisms at work in cosmic X-ray sources
- Are General Relativity and Quantum Electrodynamics still valid in extreme environments?
- What are the geometries of active galactic nuclei (AGN) and microquasars?
- What are the geometries and the magnetic field strength in magnetars?
- What are the geometries and the origin of X-ray radiation in radio pulsars?
- What are the geometries in neutron stars binary systems?
- What are the acceleration mechanisms in Pulsar Wind Nebulae and in (shell-type) Supernova Remnants (SNR)?

4.6.2.1 The Mission

The main characteristics of *IXPE* mission are described in Table 4.5.

Background is negligible for all the observation due to the brightness of the sources providing enough statistics for sensitive polarisation measurement. X-ray shields around the optics and detector baffles allow for source dominated measurement also when observing faint extended molecular clouds in the galactic centre regions. Data (both tracks and processed information on energy time, position and photoelectron emission direction) are in free distribution through NASA's High-Energy Astrophysics Science Archive Research Centre (HEASARC) at GSFC.

4.6.2.2 The On-Ground and In-Flight Calibration

IAPS built a set of monochromatic X-ray sources polarised to near 100% by Bragg diffraction (see Fig. 4.15). Lines and continuum produced by different X-ray tubes are reflected at 45° by suitable crystals by means of very compact capillary plates as collimators (Muleri et al. 2008). These sources, combined with diaphragms and

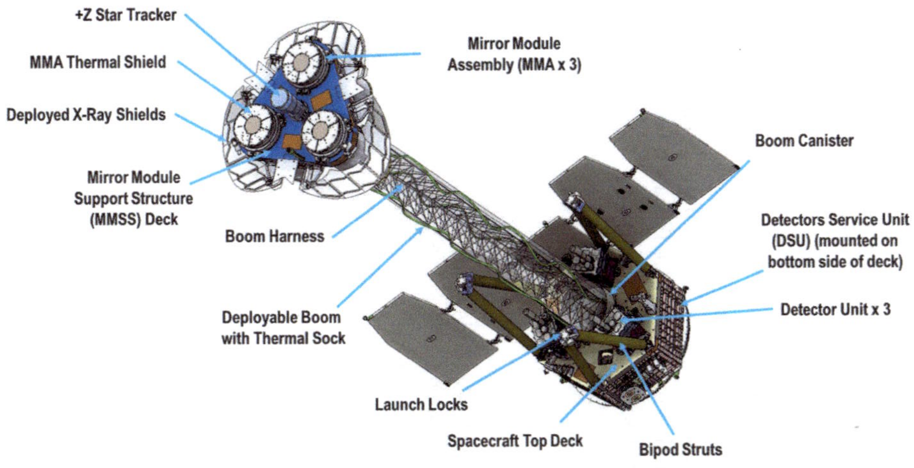

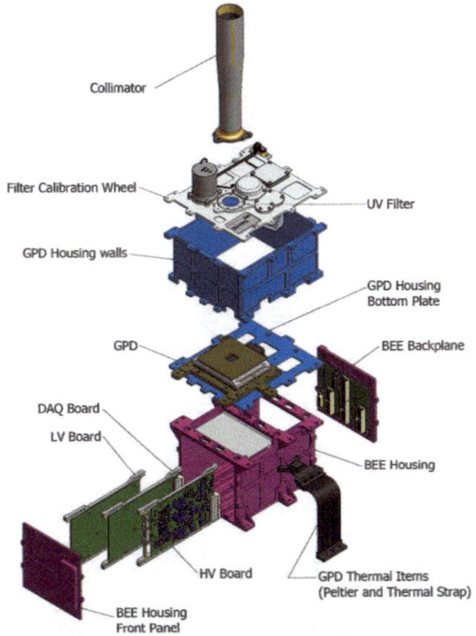

Fig. 4.14 (**Left**) The *IXPE* mission (figure from O'Dell et al. (2018)). The payload elements are labelled. (**Right**) The Detector Unit hosting the Gas Pixel Detector. Elements are labelled

Table 4.5 Characteristic of *IXPE* mission

Category	NASA astrophysics small explorer
Operation	2021 launch, 2 years of operation plus possible extension
Orbit	Circular at 540–620 km, equatorial, Malindi (ASI) primary ground station
Spacecraft	Three axis stabilised, with GPS and no controlled reentry
Payload	3 identical X-ray telescopes, 4.0 m focal length with a co-aligned star tracker
Optics	3 × 24 monolithic Wolter-1 nested design as electroformed NiCo shells
Optics area	700 cm^2 (3 keV, detector efficiency not included)
Detectors	3 × Gas pixel detectors (low energy polarimeter)
Polarisation sensitivity	$MDP_{99} < 5.5\%$ at 0.5 mCrab, 10 days of observation scalable as $\frac{1}{\sqrt{S \times T}}$
Spurious modulation	<0.3%
Angular resolution	<30-arcsec half power diameter (HPD)
Field of view	≃10 arcmin (overlapped region)
Spectroscopic capability	2–8 keV energy band, 20% energy resolution at 5.9 keV
Timing	20 μs accuracy, 1 ms dead time
X-ray calibration	At detector unit level, mirror unit level and at telescope level. In-orbit calibration system available

In the polarisation Sensitivity S is the source flux in mCrab (which corresponds to $2.010^{-11} ergs^{-1} cm^{-2}$ in the 2–8 keV energy band) and T is the observing time in days

micro-metric linear and rotary stages, allow for a detailed characterisation of the Detector Units. On-ground calibration at IAPS (see Fig. 4.15) measures:

- Detection efficiency and attachment coefficient
- Energy resolution and spatial resolution as a function of energy
- Modulation factor as a function of energy and position
- Spurious modulation as a function of energy and position
- Accuracy of the polarisation angle determination
- Gain as a function of position
- Dead time fraction as a function of rate

X-ray mirrors will be calibrated at MSFC at the 100 m X-ray stray-light test facility. In particular the Point Spread Function (PSF) and the effective area will be measured at different energies and orientation. The possible presence of ghost images will be, also, checked. Using the Electrical Ground Support Equipment at MSFC, the Telescope will be also calibrated to check the consistency between the results obtained at IAPS at DU level and those at mirror level at MSFC. A calibration system will be available during *IXPE* flight. Such system will be employed during source occultation to:

- Measure the modulation factor at 2.9 keV and 5.9 keV
- Measure the level of absence on spurious modulation at 5.9 keV

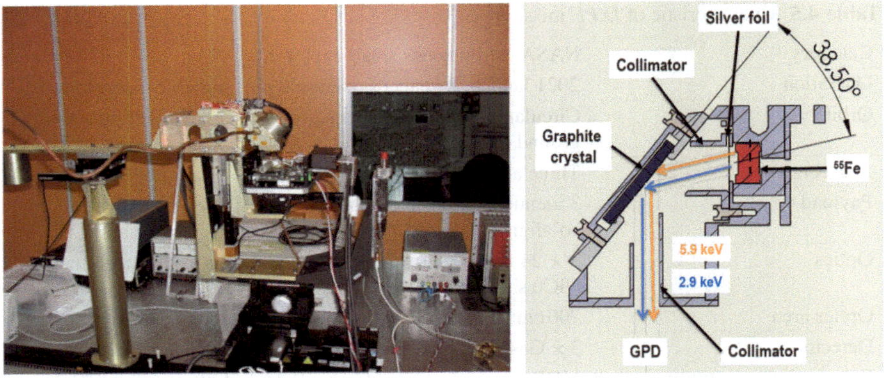

Fig. 4.15 (**Left**) An X-ray test facility has been developed at IAPS for the production of polarised and monochromatic radiation. Linear and rotary stages are completely computer controlled and equipped with collimator, diaphragms, X-ray tubes, suitable crystals, capillary plates, X-ray CCD, SDD and Si-PiN. (**Right**) The sketch of the polarised X-ray source to be used in flight as generator of two polarised X-ray lines

- Providing the gain map at 5.89 keV and 1.7 keV
- Providing the energy resolution at 1.7 keV, 2.98 keV and 5.89 keV

Four calibration sources, included one polarised (see Fig. 4.15), are mounted onto a filter and calibration wheel (FCW). The polarised X-ray source (Muleri et al. 2018), shown in Fig. 4.15) exploits the Bragg diffraction of photons produced by a commercial ^{55}Fe (5.89 keV). This source is partially absorbed by a silver thin sheet which, in turn, emits L-lines (2.98 keV). Photons of both energies are diffracted at 38.7° and 38.3°, respectively, from a single graphite mosaic crystal at the first and at the second order. Derived polarisation of the diffracted photons are 69% and 67%. Counts from these two energies will be discriminated on-ground by using pulse amplitude and exploiting the energy resolution of the detector. The other X-ray calibration sources are conventional ^{55}Fe radioactive capsules. A collimated high activity source will be used to check for the absence of spurious modulation. A large-aperture source will allow for measuring the gain. Lastly another source extracts K-lines from a Silicon target at 1.7 keV to measure the gain at low energy. Upon a time-tagged telecommand the FCW rotates and places the relevant source above the GPD.

4.6.3 Enhanced X-Ray Timing and Polarimetry Mission, eXTP

The *eXTP* mission (Zhang et al. 2019) is devoted to study the matter under extreme condition of density, gravity and magnetism. The *eXTP* international consortium is led by the Institute of High Energy Physics (IHEP) of the China Academy of

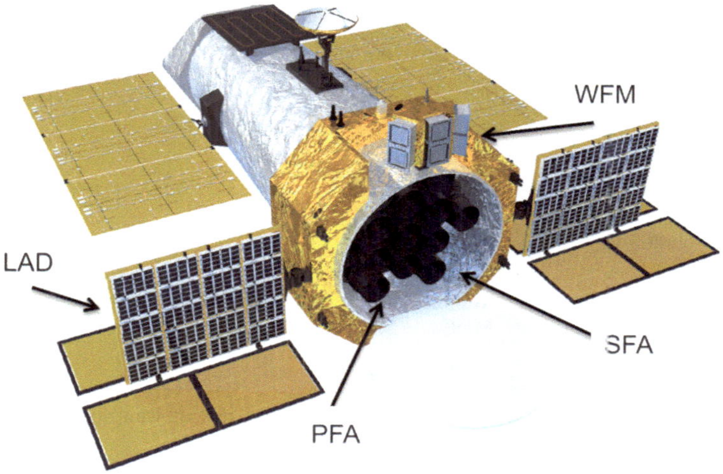

Fig. 4.16 Rendering of the *eXTP* mission (Zhang et al. 2019) in developing phase in China. *eXTP* is equipped with a set of 4 telescopes with polarimetric capability based on the GPD technology and pointing in the same direction of the timing and spectroscopy instruments

Science (CAS). It includes several institutions of CAS and Chinese Universities and institution from 10 European countries. The extended phase A study is completed at the end of 2018 for a launch around 2025. The scientific payload of *eXTP* consists (see Fig. 4.16) of four instrument array, the Spectroscopic Focusing Array (SFA), the Large Area Detector (LAD), the Polarimetric Focusing Array (PFA) that share the same observing direction and the Wide Field Monitor (WFM). The PFA consists of 4 identical telescopes for sensitive X-ray imaging polarimetry in the 2–8 keV energy band. These telescopes work in synergy with all the narrow field instruments with better angular resolution helping to discriminate sources in the field of view. The GPD is built and calibrated by the Tsinghua University group. The optics consist of X-ray grazing-incidence Wolter-I made by electro-forming Nickel shells with Gold and Carbon deposition. The characteristics of the PFA are shown in Table 4.6 The sensitivity of the PFA is about a factor of 2 better than *IXPE* because of the larger collecting area of the optics and the number of telescopes. The PFA takes advantage of the synergy with spectroscopy and timing instruments for detailed time-resolved polarimetry. For example, polarimetry of Quasi Periodic Oscillations in black hole candidates will be possible, thanks to the simultaneous observations with the LAD (Ingram et al. 2015; Ingram and Maccarone 2017). Also the PFA will measure the geometrical parameters of the accretion in pulsating neutron stars. Such information is crucial when fitting the LAD timing data of such pulsars (Salmi et al. 2018; Watts et al. 2019) allowing an accurate determination of the equation of state of the neutron

Table 4.6 Characteristic of the PFA on-board *eXTP* mission

Mixture	Pure DME 0.8 atm
Window	50 μm
Energy range	2–8 keV
Effective area (inc. det. eff.)	915 cm^2 at 2 keV
	495 cm^2 at 3 keV
	216 cm^2 at 6 keV
Energy resolution	<18% at 6 keV
Field of view	<8 arcmin
Polarisation sensitivity	Minimum detectable polarisation (99% confidence level) $MDP_{99} = 2.6\%$ at 0.5 mCrab, 10 days of observation

stars in such extreme density regime. We described in this chapter the hard X-ray polarimeters flown or to be flown.

For the soft gamma polarimeter we refer the reader to Chap. 5.

Acknowledgements The GPD project was funded by ASI, INAF and INFN during many years of development. Presently Paolo Soffitta and Enrico Costa acknowledge ASI for granting *IXPE* according to Accordo Attuativo n. 2017-12-H.0 with INAF.

References

Afanas'ev RM, Imanove RM, Pashaev EM, Peregudov VE, Abdullaev MI (1988) Sov Phys Solid State 30(6):1004

Auger MP (1920) Sur l' effect photoelectrique composé. J Phys Radium 6:205–208

Austin RA, Minamitani T, Ramsey BD (1994) Development of a hard x-ray imaging polarimeter. In: Fineschi S (ed) X-Ray and ultraviolet polarimetry, Proceedings of Society of Photo-Optical Instrumentation Engineers, vol 2010, pp 118–125

Bellazzini R, Angelini F, Baldini L, Bitti F, Brez A, Ceccanti M, Latronico L, Massai MM, Minuti M, Omodei N, Razzano M, Sgro C, Spandre G, Costa E, Soffitta P (2004) Reading a GEM with a VLSI pixel ASIC used as a direct charge collecting anode. Nucl Inst Methods Phys Res A 535:477–484

Bellazzini R, Spandre G, Minuti M, Baldini L, Brez A, Cavalca F, Latronico L, Omodei N, Massai MM, Sgro' C, Costa E, Soffitta P, Krummenacher F, de Oliveira R (2006) Direct reading of charge multipliers with a self-triggering CMOS analog chip with 105 k pixels at 50 μm pitch. Nucl Inst Methods Phys Res A 566:552

Bellazzini R, Angelini F, Baldini L, Bitti F, Brez A, Cavalca F, Del Prete M, Kuss M, Latronico L, Omodei N, Pinchera M, Massai MM, Minuti M, Razzano M, Sgro C, Spandre G, Tenze A, Costa E, Soffitta P (2006) Gas pixel detectors for X-ray polarimetry applications. Nucl Inst Methods Phys Res A 560:425–434

Bellazzini R, Spandre G, Minuti M, Baldini L, Brez A, Latronico L, Omodei N, Razzano M, Massai MM, Pesce-Rollins M, Sgró C, Costa E, Soffitta P, Sipila H, Lempinen E (2007) A sealed gas pixel detector for X-ray astronomy. Nucl Inst Methods Phys Res A 579:853

Black J, Baker RG, Deines-Jones P, Hill JE, Jahoda K (2007) X-ray polarimetry with a micropattern TPC. Nucl Inst Methods Phys Res A 551:755–760

Buschhorn G, Kotthaus R, Kufner W, Rössl W, Rzepka M, Schmidt KH, Genz H, Gräf HD, Hoffmann-Stascheck P, Nething U, Richter A, Dix W-R, Illing G, Lohmann M, Pflüger J, Reime B, Schildwächter L (1994) X-ray polarimetry using the photoeffect in a CCD detector. Nucl Inst Methods Phys Res A 346:578–588

Chattopadhyay T, Vadawale SV, Shanmugam M, Goyal SK (2014) Measurement of low energy detection efficiency of a plastic scintillator: implications on the lower energy limit and sensitivity of a hard X-Ray focal plane compton polarimeter. Astrophys J Suppl Ser 212:12

Compton A, Allison SK (1935) X-rays in theory and experiment. D. Van Nostrand, New York

Costa E, Cinti MN, Feroci M, Matt G, Rapisarda M (1995) Design of a scattering polarimeter for hard X-ray astronomy. Nucl Inst Methods Phys Res A 366:161–172

Costa E, Soffitta P, Bellazzini R, Brez A, Lumb N, Spandre G (2001) An efficient photoelectric X-ray polarimeter for the study of black holes and neutron stars. Nature 411:662

Elsner RF, O'Dell SL, Weisskopf MC (2012) Measuring x-ray polarization in the presence of systematic effects: known background. In: Society of Photo-Optical Instrumentation Engineers (SPIE) Conference Series, vol 8443

Evans KD, Hall R, Lewis M (1977) The calibration of Bragg X-ray analyser crystals for use as polarimeters in X-ray astronomy. Space Sci Instrum 3:163

Fabiani S, Bellazzini R, Berrilli F, Brez A, Costa E, Minuti M, Muleri F, Pinchera M, Rubini A, Soffitta P, Spandre G (2012a) Performance of an Ar-DME imaging photoelectric polarimeter. In: Space telescopes and instrumentation 2012: ultraviolet to gamma ray, Society of Photo-Optical Instrumentation Engineers (SPIE) Conference Series, vol 8443, p 84431C

Fabiani S, Costa E, Bellazzini R, Brez A, di Cosimo S, Lazzarotto F, Muleri F, Rubini A, Soffitta P, Spandre G (2012b) The gas pixel detector as a solar X-ray polarimeter and imager. Adv Space Res 49:143–149

Fabiani S, Campana R, Costa E, Del Monte E, Muleri F, Rubini A, Soffitta P (2013) Characterization of scatterers for an active focal plane Compton polarimeter. Astropart Phys 44:91–101

Feng H, Jiang W, Minuti M, Wu Q, Jung A, Yang D, Citraro S, Nasimi H, Yu J, Jin G, Huang J, Zeng M, An P, Baldini L, Bellazzini R, Brez A, Latronico L, Sgrò C, Spandre G, Pinchera M, Muleri F, Soffitta P, Costa E (2019) PolarLight: a CubeSat X-ray polarimeter based on the gas pixel detector. Exp Astron 47(1–2):225–243

Gowen RA, Cooke BA, Griffiths RE, Ricketts MJ (1977) An upper limit to the linear X-ray polarization of SCO X-1. Mon Not R Astron Soc 179:303–310

Griffiths RE, Ricketts MJ, Cooke BA (1976) Observations of the X-ray nova A0620-00 with the Ariel V crystal spectrometer/polarimeter. Mon Not R Astron Soc 177:429–440

Heilmann RK, Bruccoleri AR, Kolodziejczak J, Gaskin JA, O'Dell SL, Bhatia R, Schattenburg ML (2016) Critical-angle x-ray transmission grating spectrometer with extended band pass and resolving power >10,000. In: Space telescopes and instrumentation 2016: ultraviolet to gamma ray, Proceedings of Society of Photo-Optical Instrumentation Engineers, vol 9905, p 99051X

Holland AD, Short ADT, Fraser GW, Turner MJL (1995) The X-ray polarisation sensitivity of CCDs. Nucl Inst Methods Phys Res A 355:526–531

Hughes JP, Long KS, Novick R (1984) A search for X-ray polarization in cosmic X-ray sources. Astrophys J 280:255

Ingram AR, Maccarone TJ (2017) An observational method for fast stochastic X-ray polarimetry timing. Mon Not R Astron Soc 471:4206–4217

Ingram A, Maccarone TJ, Poutanen J, Krawczynski H (2015) Polarization modulation from lense-thirring precession in X-Ray binaries. Astrophys J 807:53

Iskef H, Cunningham JW, Watt DE (1983) Projected ranges and effective stopping powers of electrons with energy between 20 eV and 10 keV. Phys Med Biol 28:535–545

Jahoda KM, Black JK, Hill JE, Kallman TR, Kaaret PE, Markwardt CB, Okajima T, Petre R, Soong Y, Strohmayer TE, Tamagawa T, Tawara Y (2014) X-ray polarization capabilities of a small explorer mission. In: Space telescopes and instrumentation 2014: ultraviolet to gamma ray, Proceedings of Society of Photo-Optical Instrumentation Engineers, vol 9144, p 91440N

Jahoda K, Kallman TR, Kouveliotou C, Angelini L, Black JK, Hill JE, Jaeger T, Kaaret PE, Markwardt CB, Okajima T, Petre R, Schnittman J, Soong Y, Strohmayer TE, Tamagawa T,

Tawara Y (2016) The polarimeter for relativistic astrophysical X-ray sources. In: Space telescopes and instrumentation 2016: ultraviolet to gamma ray, Proceedings of Society of Photo-Optical Instrumentation Engineers, vol 9905, p 990516

Jourdain E, Roques JP, Chauvin M, Clark DJ (2012) Separation of two contributions to the high energy emission of cygnus X-1: polarization measurements with INTEGRAL SPI. Astrophys J 761:27

Kaaret P, Novick R, Martin C, Hamilton T, Sunyaev R, Lapshov I, Silver E, Weisskopf M, Elsner R, Chanan G, Manzo G, Costa E, Fraser G, Perola GC (1989) SXRP. A focal plane stellar X-ray polarimeter for the SPECTRUM-X-Gamma mission. In: Hoover RB (ed) Society of Photo-Optical Instrumentation Engineers (SPIE) Conference Series, vol 1160, pp 587–597

Kaaret PE, Schwartz J, Soffitta P, Dwyer J, Shaw PS, Hanany S, Novick R, Sunyaev R, Lapshov IY, Silver EH, Ziock KP, Weisskopf MC, Elsner RF, Ramsey BD, Costa E, Rubini A, Feroci M, Piro L, Manzo G, Giarrusso S, Santangelo AE, Scarsi L, Perola GC, Massaro E, Matt G (1994) Status of the stellar x-ray polarimeter for the spectrum-X-Gamma mission. In: Fineschi S (ed) Society of Photo-Optical Instrumentation Engineers (SPIE) Conference Series, vol 2010, pp 22–27

Kislat F, Clark B, Beilicke M, Krawczynski H (2015) Analyzing the data from X-ray polarimeters with Stokes parameters. Astropart Phys 68:45–51

Kislat F, Beheshtipour B, Dowkontt P, Guarino V, Lanzi RJ, Okajima T, Braun D, Cannon S, de Geronimo G, Heatwole S, Hoorman J, Li S, Mori H, Shreves CM, Stuchlik D, Krawczynski H (2017) Design of the telescope truss and gondola for the balloon-borne X-ray polarimeter X-Calibur. J. Astron. Instrum. 6:1740003

Kislat F, Abarr Q, Beheshtipour B, De Geronimo G, Dowkontt P, Tang J, Krawczynski H (2018) Optimization of the design of X-Calibur for a long-duration balloon flight and results from a one-day test flight. J Astron. Telescopes Instrum. Syst. 4(1):011004

Kotthaus R, Buschhorn G, Rzepka M, Schmidt KH, Weinmann PM (1998) Hard x-ray polarimetry exploiting directional information of the photoeffect in a charge coupled device. In: Fineschi S (ed) X-Ray and ultraviolet spectroscopy and polarimetry II, Proceedings of Society of Photo-Optical Instrumentation Engineers, vol 3443, pp 105–116

Krawczynski H, Matt G, Ingram AR et al (2019) Using X-ray polarimetry to probe the physics of black holes and neutron stars. arXiv e-prints, arXiv:1904.09313

La Monaca A, Costa E, Soffitta P, di Persio G, Manzan M, Martino B, Patria G, Cappuccio G, Zema N (1998) A new photoelectron imager for X-ray astronomical polarimetry. Nucl Inst Methods Phys Res A 416:267–277

Long KS, Chanan GA, Ku W, Novick R (1979) The linear X-ray polarization of Scorpius X-1. Astrophys J 232:L107–L110

Long KS, Chanan GA, Novick R (1980) The X-ray polarization of the Cygnus sources. Astrophys J 238:710–716

Maitra C, Paul B (2011) Prospect of polarization measurements from black hole binaries in their thermal state with a scattering polarimeter. Mon Not R Astron Soc 414:2618–2628

Marshall HL, Schulz NS, Windt DL, Gullikson EM, Craft M, Blake E, Ross C (2015) The use of laterally graded multilayer mirrors for soft x-ray polarimetry. In: Society of Photo-Optical Instrumentation Engineers (SPIE) Conference Series, Proceedings of Society of Photo-Optical Instrumentation Engineers, vol 9603, p 960319

Marshall HL, Günther HM, Heilmann RK, Schulz NS, Egan M, Hellickson T, Heine SNT, Windt DL, Gullikson EM, Ramsey B, Tagliaferri G, Pareschi G (2018) Design of a broadband soft x-ray polarimeter. J. Astron. Telescopes Instru. Syst. 4(1):011005

Massaro E, Fabiani S, Campana R, Costa E, Del Monte E, Muleri F, Soffitta P (2018) Correlation methods for the analysis of X-ray polarimetric signals. Nucl Inst Methods Phys Res A 885:7–14

Muleri F, Soffitta P, Bellazzini R, Brez A, Costa E, Fabiani S, Frutti M, Minuti M, Negri MB, Pascale P, Rubini A, Sindoni G, Spandre G (2007) A very compact polarizer for an X-ray polarimeter calibration. In: Proceedings of Society of Photo-Optical Instrumentation Engineers, vol 6686, p 668610

Muleri F, Soffitta P, Bellazzini R, Brez A, Costa E, Frutti M, Mastropietro M, Morelli E, Pinchera M, Rubini A, Spandre G (2008) A versatile facility for the calibration of x-ray polarimeters with polarized radiation. In: Proceedings of Society of Photo-Optical Instrumentation Engineers, vol 7011-84

Muleri F, Bellazzini R, Brez A, Costa E, Fabiani S, Minuti M, Pinchera M, Rubini A, Soffitta P, Spandre G (2012) A new design for the gas pixel detector. In: Society of Photo-Optical Instrumentation Engineers (SPIE) Conference Series, Society of Photo-Optical Instrumentation Engineers (SPIE) Conference Series, vol 8443

Muleri F, Lefevre C, Piazzolla R, Morbidini A, Amici F, Attina P, Centrone M, Del Monte E, Di Cosimo S, Di Persio G, Evangelista Y, Fabiani S, Ferrazzoli R, Loffredo P, Maiolo L, Maita F, Primicino L, Rankin J, Rubini A, Santoli F, Soffitta P, Tobia A, Tortosa A, Trois A (2018) Calibration of the IXPE instrument. In: Space telescopes and instrumentation 2018: ultraviolet to gamma ray, Society of Photo-Optical Instrumentation Engineers (SPIE) Conference Series, vol 10699, pp 106995C

Novick R, Weisskopf MC, Berthelsdorf R, Linke R, Wolff RS (1972) Detection of X-Ray polarization of the crab nebula. Astrophys J 174:L1

O'Dell SL et al. (2018) The imaging X-ray Polarimetry Explorer (IXPE): technical overview. In Society of Photo-Optical Instrumentation Engineers (SPIE) Conference Series, vol 10699, p 106991

Riegler GJ, Garmire GP, Moore WE, Stevens JC (1970) A low energy X-ray polarimeter. Bull Am Phys Soc 15:635

Rishin PV et al (2010) Development of a Thomson X-ray polarimeter, p 83

Sakurai H, Tokanai F, Gunji S, Motegi S, Toyokawa H, Suzuki M, Hirota K, Kishimoto S (2004) Measurement of X-ray polarization using optical imaging detector with capillary plate. Nucl Inst Methods Phys Res A 525:6–11

Salmi T, Nättilä J, Poutanen J (2018) Bayesian parameter constraints for neutron star masses and radii using X-ray timing observations of accretion-powered millisecond pulsars. Astron. Astrophys. 618:A161

Sanford PW, Cruise AM, Culhane JL (1970) Techniques for improving the sensitivity of proportional counters used in X-Ray astronomy. In: Gratton L (ed) Non-solar X- and gamma-ray astronomy, IAU Symposium, vol 37, p 35

Schnopper HW (1990) SODART telescope on spectrum-roentgen-gamma and its instrumentation. In: Kondo Y (ed) IAU Colloq. 123: observatories in earth orbit and beyond, Astrophysics and Space Science Library, vol 166, pp 119–128

Schnopper HW, Kalata K (1969) Polarimeter for celestial X rays. Astronom J 74:854

Silver EH, Kestenbaum HL, Long KS, Novick R, Wolff RS, Weisskopf MC (1978) Search for X-ray polarization in the Crab pulsar. Astrophys J 225:221

Silver E, Holley J, Ziock K, Novick R, Kaaret P, Weisskopf M, Elsner R (1990) Bragg crystal polarimeters. Opt Eng 29:759–766

Soffitta P, Costa E, Morelli E, Bellazzini R, Brez A, Raffo R (1995) Sensitivity to x-ray polarization of a microgap gas proportional counter. In: Fineschi S (ed) X-Ray and EUV/FUV spectroscopy and polarimetry, Proceedings of Society of Photo-Optical Instrumentation Engineers, vol 2517, pp 156–163

Soffitta P, Costa E, Kaaret P, Dwyer J, Ford E, Tomsick J, Novick R, Nenonen S (1998) Proportional counters for the stellar X-ray polarimeter with a wedge and strip cathode pattern readout system. Nucl Inst Methods Phys Res A 414:218–232

Soffitta P, Costa E, di Persio G, Morelli E, Rubini A, Bellazzini R, Brez A, Raffo R, Spandre G, Joy D (2001) Astronomical X-ray polarimetry based on photoelectric effect with microgap detectors. Nucl Inst Methods Phys Res A 469:164–184

Strohmayer TE, Kallman TR (2013) On the statistical analysis of X-ray polarization measurements. Astrophys J 773:103

Swank J et al (2010) Gravity and extreme magnetism SMEX (GEMS). In: Bellazzini R, Costa E, Matt G, Tagliaferri G (eds) X-ray polarimetry: a new window in astrophysics by Ronaldo

Bellazzini, Enrico Costa, Giorgio Matt and Gianpiero Tagliaferri. Cambridge University Press, Cambridge, p. 251, ISBN: 9780521191845
Tsunemi H, Hayashida K, Tamura K, Nomoto S, Wada M, Hirano A, Miyata E (1992) Detection of X-ray polarization with a charge coupled device. Nucl Inst Methods Phys Res A 321:629–631
Vaillancourt JE (2006) Placing confidence limits on polarization measurements. Publ Astron Soc Pac 118:1340–1343
Watts AL, Yu W, Poutanen J, Zhang S, Bhattacharyya S, Bogdanov S, Ji L, Patruno A, Riley TE, Bakala P, Baykal A, Bernardini F, Bombaci I, Brown E, Cavecchi Y, Chakrabarty D, Chenevez J, Degenaar N, Del Santo M, Di Salvo T, Doroshenko V, Falanga M, Ferdman RD, Feroci M, Gambino AF, Ge M, Greif SK, Guillot S, Gungor C, Hartmann DH, Hebeler K, Heger A, Homan J, Iaria R, Zand Ji, Kargaltsev O, Kurkela A, Lai X, Li A, Li X, Li Z, Linares M, Lu F, Mahmoodifar S, Méndez M, Coleman Miller M, Morsink S, Nättilä J, Possenti A, Prescod-Weinstein C, Qu J, Riggio A, Salmi T, Sanna A, Santangelo A, Schatz H, Schwenk A, Song L, Šrámková E, Stappers B, Stiele H, Strohmayer T, Tews I, Tolos L, Török G, Tsang D, Urbanec M, Vacchi A, Xu R, Xu Y, Zane S, Zhang G, Zhang S, Zhang W, Zheng S, Zhou X (2019) Dense matter with eXTP. Sci China Phys Mech Astron 62:29503
Weisskopf MC, Berthelsdorf R, Epstein G, Linke R, Mitchell D, Novick R, Wolff RS (1972) A graphite crystal polarimeter for stellar X-ray astronomy. Rev Sci Ins 43:967
Weisskopf MC, Cohen GG, Kestenbaum HL, Long KS, Novick R, Wolff RS (1976) Measurement of the X-ray polarization of the Crab Nebula. Astrophys J 208:L125
Weisskopf MC, Silver EH, Kestenbaum HL, Long KS, Novick R (1978) A precision measurement of the X-ray polarization of the Crab Nebula without pulsar contamination. Astrophys J 220:L117
Weisskopf MC, Elsner RF, O'Dell SL (2010) On understanding the figures of merit for detection and measurement of x-ray polarization. In: Society of Photo-Optical Instrumentation Engineers (SPIE) Conference Series, vol 7732
Weisskopf MC, Ramsey B, O'Dell SL, Tennant A, Elsner R, Soffita P, Bellazzini R, Costa E, Kolodziejczak J, Kaspi V, Mulieri F, Marshall H, Matt G, Romani R (2016) The imaging X-ray polarimetry explorer (IXPE). In: Space telescopes and instrumentation 2016: ultraviolet to gamma ray, Proceedings of Society of Photo-Optical Instrumentation Engineers, vol 9905, p 990517
Xiao H, Hajdas W, Wu B, Produit N, Bao T, Bernasconi T, Cadoux F, Dong Y, Egli K, Gauvin N, Kole M, Kramert R, Kong S, Li L, Li Z, Liu J, Liu X, Marcinkowski R, Rybka DK, Pohl M, Shi H, Song L, Sun J, Xiong S, Szabelski J, Socha P, Wang R, Wen X, Wu X, Zhang L, Zhang P, Zhang S, Zhang X, Zhang Y, Zwolinska A (2018) In-flight energy calibration of the space-borne Compton polarimeter POLAR. Astropart Phys 103:74–86
Yonetoku D, Murakami T, Gunji S, Mihara T, Sakashita T, Morihara Y, Kikuchi Y, Takahashi T, Fujimoto H, Toukairin N, Kodama Y, Kubo S, Ikaros Demonstration Team (2011) Gamma-ray burst polarimeter (GAP) aboard the small solar power sail demonstrator IKAROS. Publ Astron Soc Jpn 63:625–638
Yonetoku D, Murakami T, Gunji S, Mihara T, Toma K, Morihara Y, Takahashi T, Wakashima Y, Yonemochi H, Sakashita T, Toukairin N, Fujimoto H, Kodama Y (2012) Magnetic structures in gamma-ray burst jets probed by gamma-ray polarization. Astrophys J 758:L1
Zhang S et al (2019) The enhanced X-ray Timing and Polarimetry mission-eXTP. Sci China Phys Mech Atron 62:029502

Chapter 5
Gamma-Ray Polarimetry

Vincent Tatischeff, Mark L. McConnell, and Philippe Laurent

Abstract Gamma-ray polarimetry can provide a powerful diagnostic of the high-energy physics at work in the most extreme sources of the universe: active galactic nuclei, gamma-ray bursts, black hole binaries, pulsars, etc. The linear polarisation information of gamma-ray sources is measured using Compton scattering up to a few MeV and the conversion of high-energy photons to electron–positron pairs at higher energies. Thanks to the development of gamma-ray detectors with high position and energy resolution, several balloon-borne and satellite experiments are getting important results with this new astronomical tool.

5.1 Introduction

Polarimetry has not yet become a standard tool for γ-ray astronomy (which we take here to be the energy range from roughly 50 keV up to several hundred GeV). Low flux levels, coupled with relatively high background levels, often make this a challenging regime in which to do polarimetry, especially when the polarisation measurement itself is already quite challenging. High-energy polarimeters must be carefully designed to avoid systematic effects that either degrade an existing

V. Tatischeff (✉)
Centre de Sciences Nucléaires et de Sciences de la Matière (CSNSM), CNRS/IN2P3, Univ. Paris-Sud, Université Paris-Saclay, Orsay, France
e-mail: vincent.tatischeff@csnsm.in2p3.fr

M. L. McConnell
Space Science Center, University of New Hampshire, Durham, NH, USA

Department of Earth, Oceans, and Space, Southwest Research Institute, Durham, NH, USA
e-mail: mark.mcconnell@unh.edu

P. Laurent
CEA/DRF/IRFU/DAp, CEA Saclay, Gif sur Yvette, France

Laboratoire APC, Paris Cedex 13, France
e-mail: philippe.laurent@cea.fr

R. Mignani et al. (eds.), *Astronomical Polarisation from the Infrared to Gamma Rays*, Astrophysics and Space Science Library 460,
https://doi.org/10.1007/978-3-030-19715-5_5

polarisation signal or masquerade as a polarisation signal when none exists. Despite these difficulties, there is considerable interest in developing the capability of measuring γ-ray polarisation from astrophysical sources.

In this chapter, we briefly review the scientific motivation for high-energy polarisation measurements and then review the two basic principles (Compton scattering and pair production) on which γ-ray polarimeters are based. We then follow this with a review of polarimeters that have been demonstrated with ground-based measurements, those that have been flown on high-altitude balloons, and those that have been flown on orbit. Along the way, we summarise some of the observations that have been made already, observations that hint at the importance of measurements that have been made to date, and the future potential value of γ-ray polarimetry.

5.2 Science Motivation

Gamma-ray polarimetry can provide valuable information on the geometrical configuration, magnetisation, and high-energy emission mechanism of an astrophysical source, which may not be accessible via spectral, timing, or imaging observations.

Measurements of the γ-ray polarisation of **blazars** can provide definitive insight into the presence of accelerated hadrons in the jets of these objects, since hadronic models for their high-energy radiation predict a much higher level of polarisation than leptonic models (Zhang et al. 2016). Revealing the nature of the emitting particles in these active galactic nuclei is of utmost importance to understand the origin of ultra-high-energy cosmic rays (Böttcher 2019).

Gamma-ray polarimetry can also reveal the nature of the central engines producing ultra-relativistic jets in **gamma-ray bursts** (GRBs), as well as the physical properties of these jets, the energy dissipation sites, and the radiation mechanisms (see, e.g., Toma et al. (2009); Tatischeff et al. (2018)). Joint γ-ray polarisation and gravitational wave observations of a short GRB would be key to obtain a more complete knowledge of the source, as these two observables provide complementary information (de Angelis et al. 2018).

Polarimetric measurements of γ-ray **pulsars** can shed new light on the magnetic field structure around the compact object, the processes of particle acceleration and pair cascades in the pulsar magnetosphere, as well as on the γ-ray emission mechanisms (curvature radiation or synchrotron radiation) (Harding and Kalapotharakos 2017).

In **microquasars** (i.e., black hole binaries featuring jets), γ-ray polarimetry can disentangle the jet emission, which can be highly polarised, from the Compton scattering of disk photons on thermal electrons, which is expected to be only weakly polarised (Laurent et al. 2011b), thus probing the poorly known physics of the jets in the black hole vicinity.

In **solar flares**, beams of accelerated electrons impinging on the photosphere generate electron bremsstrahlung radiation. The polarisation of this radiation can provide important insights into the extent to which the accelerated electrons are beamed (i.e., the pitch-angle distribution). This, in turn, places constraints on acceleration models for solar flares (Kontar et al. 2011; McConnell et al. 2002a).

Measuring γ-ray polarisation from cosmological sources (GRBs, blazars) is also relevant for **fundamental physics**. Thus, such measurements (e.g., Laurent et al. (2011a)) currently provide the most stringent test of the violation of Lorentz invariance arising from vacuum birefringence, which is predicted from the fuzzy nature of space time in quantum gravity. The resulting helicity dependence of the photon propagation velocity can be best constrained by measuring how the polarisation direction of a γ-ray beam of cosmological origin changes as function of energy (Gambini and Pullin 1999). Gamma-ray polarisation of distant GRBs can also be used to probe the existence of axion-like particles, which are predicted by various extensions of the standard model of particle physics (Bassan et al. 2010). In addition, γ-ray polarimetric observations of the Galactic Center could perhaps enable the indirect detection of dark matter annihilation in this region (Huang and Ng 2018).

5.3 Measurement Principle

The linear polarisation information of incident γ-rays is partially preserved by both Compton scattering and electron–positron pair creation. The sensitivity of the photoelectric effect to polarisation is better suited to the X-ray domain (see Chap. 4 on X-ray polarimetry by Soffitta and Costa in this book).

5.3.1 Compton Scattering

The Compton effect is used for polarimetry in the hard X-ray range (see Soffitta and Costa in this book) and the soft γ-ray domain up to a few MeV. The Klein–Nishina differential cross section for Compton scattering of a linearly polarised photon by an electron at rest is given by Heitler (1954):

$$\left(\frac{d\sigma}{d\Omega}\right)_{\mathrm{KN}} = \frac{r_e^2 \epsilon^2}{2}(\epsilon + \epsilon^{-1} - 2\sin^2\theta\cos^2\varphi), \tag{5.1}$$

where r_e is the classical electron radius, θ the polar angle between the incident photon direction and the scattered photon direction, φ the azimuthal angle of the scattered photon with respect to the electric vector of the incident photon, and $\epsilon = E'_\gamma/E_\gamma$ the ratio of the scattered photon energy to the incident photon energy, which is related to the scatter angle by the Compton equation:

$$\epsilon = \left(1 + \frac{E_\gamma}{m_e c^2}(1 - \cos\theta)\right)^{-1}. \quad (5.2)$$

Here, $m_e c^2$ is the electron rest mass energy. Equation (5.1) shows that linearly polarised photons tend to scatter perpendicularly to the incident polarisation vector (minimising the term $2\sin^2\theta\cos^2\varphi$). The polarisation information of a source of polarised photons is reflected in the probability density function (PDF) of scattering with a given azimuthal angle φ:

$$P(\varphi; E_\gamma, \theta) = \frac{1}{2\pi}[1 + \mu_0(E_\gamma, \theta)\cos 2(\varphi - \varphi_0 + \pi/2)]\,, \quad (5.3)$$

which can be readily derived from Eq. (5.1). Here, φ_0 is the polarisation angle of the source, whose polarisation fraction is $\Pi = \mu_0/\mu_{100}$, where μ_{100} is the modulation expected to be measured for a 100% polarised source.

The expected modulation factor for an ideal polarimeter can be obtained from Eqs. (5.1)–(5.3):

$$\mu_{100}(E_\gamma, \theta) = \frac{P_{\max} - P_{\min}}{P_{\max} + P_{\min}} = \frac{\sin^2\theta}{\epsilon + \epsilon^{-1} - \sin^2\theta}, \quad (5.4)$$

where $P_{\max}$ and $P_{\min}$ are represented in Fig. 5.1a. The ideal modulation as a function of θ is shown in Fig. 5.1b for six values of E_γ in the range 0.1–10 MeV. We see that

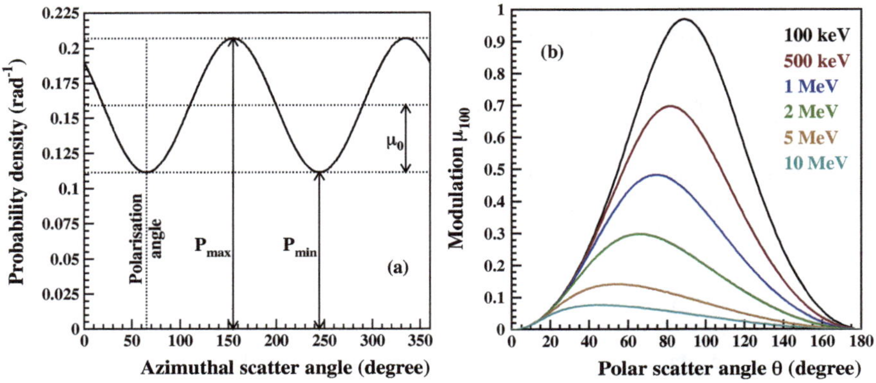

Fig. 5.1 (**a**) Representation of a polarisation signal (also called polarigram) from a Compton polarimeter; the source polarisation angle is measured from the position of the minimum of the distribution (here $\varphi_0 = 65°$), the source polarisation level being obtained from the amplitude $\mu_0 = (P_{\max} - P_{\min})/2$ compared to the ideal modulation factor for a 100% polarised γ-ray source, μ_{100}. (**b**) μ_{100} as a function of the Compton scatter angle, for six values of the incident photon energy E_γ

photons that are forward or backward scattered ($\theta \approx 0°$ or $\approx 180°$) carry almost no information about the polarisation of the incident photon beam. We also see that μ_{100} strongly decreases with energy, from a maximum value of $\approx$1 for $\theta = 90°$ at low energies, i.e., when Thomson scattering applies in good approximation, to a maximum value of 0.078 for $\theta = 46°$ at 10 MeV. This fundamental loss of sensitivity of Compton polarimeters at high energies (see also Bernard (2015)) explains why there is still no measurement of celestial source polarisation above 2 MeV (Sects. 5.6 and 5.7). In practice, the modulation response of any polarimeter is lower than the ideal modulation calculated above due to several factors, such as (i) the spectral and position resolution of the detectors, (ii) the presence of passive material, and (iii) the imperfect Compton event reconstruction (see, e.g., Lei et al. (1997)).

The first step in Compton polarisation data analysis is generally to select the detected events (in terms of incident photon energy, Compton scatter angle, distance between the interactions, etc.) that provide the lowest minimum detectable polarisation (MDP):

$$\mathrm{MDP} = \frac{n_\sigma}{\mu_{100} S}\sqrt{\frac{S+B}{T}}, \tag{5.5}$$

where n_σ is related to the desired confidence level of the detection (e.g., $n_\sigma = 4.29$ for a confidence level of 99%), S and B are the source and background count rates, and T is the observation time. The standard approach is then to extract the parameters μ_0 and φ_0 from a fit of a generalised form of Eq. (5.3) to a measured histogram of azimuthal scattering angles. However, combining the Compton scatter angle and photon energy measurements with the φ measurement in an unbinned, maximum likelihood analysis can improve the MDP (i.e., the polarisation sensitivity) by ~20% (Lowell et al. 2017a). A similar improvement can be obtained with a moments method aiming at computing a simple estimator of the polarisation fraction that makes optimal use of all the kinematic information (Bernard 2015).

5.3.2 Electron–Positron Pair Production

Above the threshold energy $2 \times m_e c^2 = 1.022$ MeV, the linear polarisation of an astrophysical source can also be measured by using the conversion of an incident γ-ray to an electron–positron pair in the electromagnetic field of target nucleus ($\gamma Z \rightarrow e^+e^- Z$). The polarisation information (i.e., the polarisation fraction Π and angle φ_0) can be extracted from a probability distribution of the pair azimuthal angle of the same form as Eq. (5.3):

$$P(\varphi; E_\gamma) = \frac{1}{2\pi}[1 + A_{\mathrm{eff}}(E_\gamma)\Pi \cos 2(\varphi - \varphi_0)], \tag{5.6}$$

where $A_{\text{eff}} = D \times A$ is the effective polarisation asymmetry, D being the (polarimeter-dependent) dilution factor, and A the ideal asymmetry, which is obtained by integration of the full differential cross section of pair production over all the variables that describe the final e^+–e^- state, except φ (see Gros and Bernard (2017) and the references therein). Noteworthy, in the case of nuclear conversion, the e^+–e^- pair is preferentially emitted in the plane of polarisation of the incident photon, such that the polarisation angle φ_0 corresponds to the position of the maximum of the azimuthal angle distribution.

However, as the final e^+–e^- state is determined by five variables, e.g., the polar and azimuthal angles of the electron and positron ($\theta_+, \varphi_+, \theta_-, \varphi_-$), plus the relative energy of one of the two leptons, the reference azimuthal angle can be defined in several ways, including (see Gros and Bernard (2017)):

- the azimuthal angle of the pair plane, ω (see Fig. 5.2a and, e.g., Wojtsekhowski et al. (2003)),
- the azimuthal angle φ_p of the pair momentum $\vec{p}_p = \vec{p}_+ + \vec{p}_-$, which is equivalent to the azimuthal angle of the recoiling nucleus φ_r, as $\varphi_p = \varphi_r \pm \pi$,
- the bisector angle of the pair $\phi = (\varphi_+ + \varphi_-)/2$ (Fig. 5.2a).

Using an exact, five-dimensional (5D) event generator for pair creation by linearly polarised photons (Bernard 2013; Gros and Bernard 2017) have recently shown that the bisector angle ϕ provides a significantly higher asymmetry near the reaction

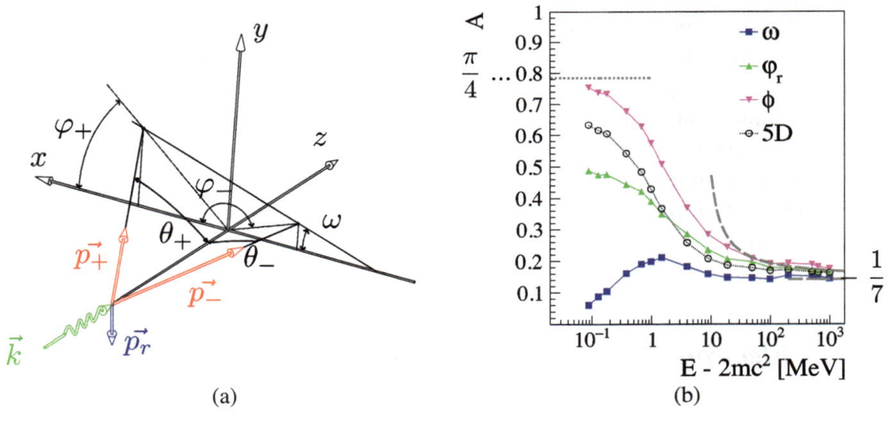

Fig. 5.2 (**a**) Representation of a γ-ray conversion. $\vec{k}$ (along the z axis), $\vec{p}_+$, $\vec{p}_-$, and $\vec{p}_r$ are the momenta of the incident photon, positron, electron, and recoil nucleus, respectively; the azimuthal angles φ_+ (positron) and φ_- (electron) are defined with respect to the x axis. (**b**) Amplitude of the modulation for the conversion of a 100% polarised γ-ray source as a function of the photon excess energy relative to 1.022 MeV (two times the electron rest mass), for various definitions of the azimuthal angle: blue squares—ω; green triangles—φ_r; magenta triangles—ϕ (see text). Also shown (black circles) is the performance of a *5D estimator* (Gros and Bernard 2017) built on the whole set of variables that describe the final state using the moments' method. Adapted from Gros and Bernard (2017)

threshold than, e.g., the commonly used pair plane azimuthal angle ω (see Fig. 5.2b). Thus, at $E_\gamma = 2$ MeV, the asymmetry computed with ϕ is about 2.9 times higher than that obtained with ω, and the same improvement factor applies to the polarisation sensitivity given by the MDP (replace μ_{100} by A in Eq. (5.5)). As shown in Fig. 5.2b, only the asymmetry computed with ϕ tends to the low-energy asymptote $A = \pi/4$ that can be derived from the Bethe–Heitler cross section (Gros and Bernard 2017).

The main limitation to polarimetry with pair creation comes from the effect of multiple Coulomb scattering (also called Molière scattering), which degrades the accuracy with which the electron and positron directions emanating from the pair vertex can be measured. The corresponding dilution factor can be written as Kotov (1988); Bernard (2013)

$$D = A_{\mathrm{eff}}/A = \exp(-2\sigma_\phi^2), \tag{5.7}$$

where σ_ϕ is the root-mean-square deviation of the azimuthal angle due to multiple scattering. For a pair production foil of thickness L made of a material of radiation length X_0, we have in first approximation $\sigma_\phi \approx 14\sqrt{L/X_0}$ rad (Mattox et al. 1990). Applying this formula to the silicon detectors of the *Fermi Large Area Telescope* (Atwood et al. 2009), for which $L = 400\,\mu$m and $X_0(\mathrm{Si}) = 9.37$ cm, we get $\sigma_\phi \approx 0.2$ rad and $D \approx 0.2$. Would such a dilution value be achievable, the polarisation of some of the brightest γ-ray sources (the Vela and Crab pulsars and the blazar 3C 454.3) might be detectable with *Fermi* LAT after 10 years of observation by selecting events of photon conversion in the Si detectors, if the polarisation level of these sources above 50 MeV is $\gtrsim$30–50% (Giomi et al. 2017). In the tungsten foil of the front converter of *Fermi* LAT ($L = 100\,\mu$m, $X_0(\mathrm{W}) = 0.35$ cm), the dilution is of the order of 10^{-5}, which is much too low to allow for any polarisation measurement.

Above the threshold energy $4 \times m_e c^2 = 2.044$ MeV, polarisation can also be measured by using triplet conversion, that is, the conversion of a photon to an e^+–e^- pair in the field of an electron ($\gamma e^- \rightarrow e^+ e^- e^-$). In this process, the target electron generally recoils with a large polar angle with respect to the incident photon direction, which in principle can make it easier to measure its azimuthal angle (Depaola and Iparraguirre 2009). However, the probability for pair production in the field of an electron is lower than that for pair production in the field of a nucleus of atomic number Z by a factor $f \approx (1\text{–}3)Z$, which depends on the incident photon energy (e.g., $f = 2.6Z$ for $E_\gamma = 6.5$ MeV and $1.2Z$ for $E_\gamma = 100$ MeV (Evans 1955)). In addition, the fraction of target electrons recoiling with a large enough energy to generate an additional track in the instrument is always very small (see Fig. 6 in Bernard (2013)). As a result, the sensitivity of any space polarimeter that would use triplet conversion should be very low.

The best way to overcome the effect of multiple scattering is to use a homogeneous active target as a photon converter that can record the three-dimensional trajectories of the e^+–e^- pairs. In the GRAINE project (Takahashi et al. 2016), the converter is made of a stack of nuclear emulsion films allowing the lepton trajectories to be tracked near the vertex with a spatial resolution of $\sim 1\,\mu$m. The polarisation sensitivity of this technology has been demonstrated using a polarised γ-ray beam in the range 0.8–2.4 GeV from the laser electron photon beamline at Spring-8 (LEPS) (Ozaki et al. 2016). The AdEPT (Hunter et al. 2014) and HARPO (Bernard et al. 2014) projects aim at measuring the polarisation of cosmic sources at lower energies, in the MeV range, where the polarisation asymmetry is higher (see Fig. 5.2b) and the γ-ray flux from cosmic sources is generally also higher (Bernard 2013).[1] Both AdEPT and HARPO use as their active target a time projection chamber (TPC) filled with a high-pressure gas (0.5–5 bar). Polarimetry in the MeV range using pair creation has recently been demonstrated for the first time with this technology (see Gros et al. (2018) and Sect. 6.4.2).

5.4 On-Ground Calibration

The modulation response of a space polarimeter (i.e., for a Compton polarimeter, the parameter $\mu_{100}(E_\gamma, \theta)$ in Eq. (5.3), or, for a pair-creation polarimeter, $A_{\text{eff}}(E_\gamma)$ in Eq. (5.6)) generally depends in a complicated way on the energy spectrum and position of the source with respect to the instrument line of sight. It must generally be obtained using Monte Carlo simulations, which need to be benchmarked beforehand with on-ground calibrations. An uncontrolled systematic error in the instrument response can result in a fake polarisation signal even if the incoming radiation is not polarised. On-ground calibration must therefore be performed with both polarised and non-polarised photon beams.

5.4.1 Compton Polarimeters

Compton polarimeters can be calibrated on ground with radioactive sources used in particular setups. A first calibration method is based on the partial linear polarisation acquired by an unpolarised γ-ray after Compton scattering. This method was used by, e.g., Kroeger et al. (1996) (see also Lei et al. (1997)) to measure the polarimetric performance of a germanium strip detector. The polarised photon beams were obtained by 90° scattering of γ-rays from a ^{57}Co source ($E_\gamma = 122$ keV) or from a ^{137}Cs source ($E_\gamma = 662$ keV) off a plastic scintillator, which was also used for tagging the scattered photons. With this setup, the authors (Kroeger et al. 1996) were able to calibrate their detector with a beam of $\sim$100 keV photons (see

[1] One can show from Eq. (5.6) that the uncertainty on the measurement of the polarisation level goes like $\sigma_\Pi \approx A_{\text{eff}}^{-1}\sqrt{2/N}$, where N is the number of events in the sample (Bernard 2013). So polarimetry always requires significant statistics, especially when the polarisation asymmetry is small.

Eq. (5.2)) polarised at ∼97% and another beam of ∼290 keV photons polarised at ∼ 60%.

Another method, which was used by the POLAR collaboration (Orsi et al. 2011), is based on the correlated polarisation of the two 511 keV photons emitted back-to-back in the annihilation of a positron and an electron. The measurement can be done with a ^{22}Na positron source placed at the centre of the setup. Then, one of the 511 keV photon from the source is detected in the polarimeter to calibrate, while the other is detected in coincidence by a relatively simple Compton polarimeter of known modulation response (see Fig. 3 in Orsi et al. (2011)).

A third calibration method given in the literature exploits the polarisation-direction correlation of the two γ-rays emitted in the decay of ^{60}Co. This radioactive nucleus mainly decays (by β^- emission) to the $J^\pi = 4^+$ excited state of ^{60}Ni at 2506 keV, which then decays to the ground state by successively emitting two γ-rays at 1173 and 1333 keV. By selecting the emission angle θ_γ between the two photons, the population of magnetic sub-states of the intermediate $J^\pi = 2^+$ state at 1333 keV is also selected, which leads to the linear polarisation of the decay γ-rays. Thus, for $\theta_\gamma = 90°$, the selected γ-rays are ∼17% polarised (see Lei et al. (1997) and the references therein).

If many Compton polarimeters (such as *GRAPE* and *COSI*, see Sect. 9.5) relied solely on radioactive calibration sources for ground-calibration, the use of tagged polarised photons produced at accelerator facilities can be essential to calibrate the instruments above 1.3 MeV. In the 1–10 MeV range, highly polarised γ-ray beam of low background can be best produced by Compton backscattering of laser optical photons against high-energy (∼1 GeV) electrons (Schaerf 2005), at facilities such as the NewSUBARU/Spring-8 electron storage ring[2] (Japan) or the High Intensity Gamma-Ray Source (HIGS) of the Duke Free Electron Laser Laboratory[3] (DFELL, NC, USA). In the sub-MeV γ-ray domain, fully polarised photon beams can also be produced as synchrotron radiation, e.g., at the European Synchrotron Radiation Facility[4] (ESRF, France). However, an issue with calibrations at accelerator facilities is that they usually deliver very fine pencil beams, whereas the full measurement of a space polarimeter response requires a broad beam source covering a large fraction of (if not all) the instrument. A way to mimic uniform irradiation is to place the polarimeter on a motorised table designed to provide a scan of the instrument during irradiation (Kole et al. 2017).

One of the first space-dedicated instruments to be thoroughly calibrated was the *Medium Energy Gamma-ray Astronomy* (*MEGA*) telescope, whose objective was to observe the sky in the energy band 0.4–50 MeV by the joint detection of photons in both the Compton (0.4–10 MeV) and pair-creation (>10 MeV) energy ranges (Kanbach et al. 2005). The instrument was calibrated with radioactive sources (^{22}Na, ^{137}Cs, ^{88}Y) and at the DFELL-HIGS facility with monoenergetic

[2]http://www.lasti.u-hyogo.ac.jp/NS-en/facility/bl01/.

[3]http://www.tunl.duke.edu/facilities/.

[4]http://www.esrf.eu/home/UsersAndScience/Experiments/StructMaterials/ID15A.html.

($\Delta E/E < 2\%$) and fully polarised pencil beams at different energies (0.7, 2, 5, 8, 10, 12, 17, 25, 37, and 49 MeV) and different angles of incidence (0°, 30°, 60°, 80°, 120°, and 180°). The measured instrument response, including its polarimetric capability, was then compared with detailed Monte Carlo simulations performed with the MEGAlib software (Zoglauer et al. 2006). This work showed that the *MEGA* instrument concept, which is at the origin of the current proposals for the next generation of γ-ray space observatories—*e-ASTROGAM* and *AMEGO*—can have excellent polarimetric capability in the Compton range.[5]

Figure 5.3 shows some results of an experiment conducted at the BL08W beamline of Spring-8 for the development of the soft gamma-ray detector (SGD) of the *Hitomi* (*ASTRO-H*) satellite (Tajima et al. 2018). In this setup, the primary beam consisted of highly polarised ($\Pi > 99.99\%$) photons of 250 keV delivered by a multipole wiggler. The primary photons were scattered at ~90° by an aluminium block placed in the beamline, in order to irradiate the Compton camera by a broad (diameter of 30 mm) secondary beam of reduced intensity (1.8×10^5 ph s^{-1} compared to 2×10^{10} ph s^{-1} for the primary beam) of 168.0 ± 1.4 keV photons (see Eq. (5.2)) polarised at $92.5 \pm 0.3\%$. The instrument was rotated around the beam axis several times during the experiment to measure its response for various polarisation angles. The top panel of Fig. 5.3b shows the measured distribution of reconstructed azimuth scatter angle for $\varphi_0 = 0°$. We see that this distribution is very different from a $\cos 2\varphi$ function (see Eq. (5.3)), which is due to the strong asymmetry of the detector azimuthal acceptance for unpolarised photons. The latter (middle panel of Fig. 5.3b) was estimated from a Monte Carlo simulation benchmarked with experimental data (see Takeda et al. (2010)). Once corrected for the detector response, the measured polarigram (bottom panel) can be well fitted by the theoretical function of Eq. (5.3). The fitted modulation factor, $\mu_0 = 0.82 \pm 0.01$, shows that the tested SGD prototype was an excellent Compton polarimeter (compare μ_0 with the ideal modulation μ_{100} shown in Fig. 5.1b).

[5]Both the *e-ASTROGAM* (Tatischeff et al. 2016; De Angelis et al. 2017) and *AMEGO* (Moiseev and Amego Team 2017) γ-ray telescopes are made up of three detection systems: a silicon tracker composed of a stack of double-sided strip detectors, in which the cosmic γ-rays undergo a Compton scattering or a pair conversion; a calorimeter to absorb and measure the energy of the secondary particles; and a coincidence system to veto the prompt-reaction background induced by charged particles. The *e-ASTROGAM* calorimeter is a pixelated detector made of a high-Z scintillation material, CsI(Tl), for efficient absorption of Compton scattered γ-rays and e^+–e^- pairs. The *AMEGO* calorimeter consists of two subsystems: a segmented CdZnTe (CZT) detector optimised for Compton events and a CsI(Tl) calorimeter similar to that of *Fermi* LAT. Thanks to the high position and energy resolution of these detectors, both *e-ASTROGAM* and *AMEGO* would be able to perform unprecedented polarimetry in the MeV range (see Tatischeff et al. (2018)).

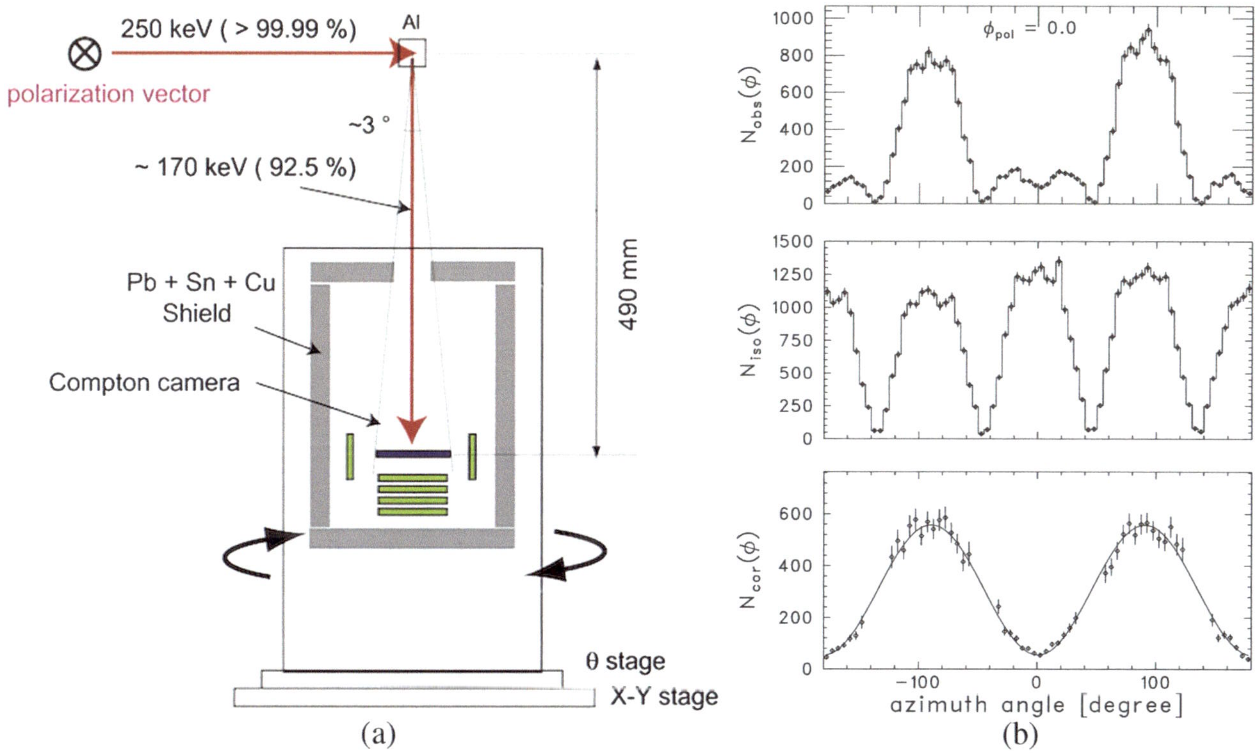

Fig. 5.3 (**a**) Schematic view of the experiment conducted at SPring-8/BL08W to study the polarimetric performance of the *Hitomi*-SGD prototype (adapted from Takeda et al. (2010)). (**b**) Polarisation signal obtained in this experiment for the polarisation angle $\varphi_0 = 0°$: *top panel*—measured azimuthal angle distribution $N(\varphi)$; *middle panel*—simulated $N(\varphi)$ for non-polarised γ-rays; *bottom panel*—measured $N(\varphi)$ corrected for the detector response, together with a fit to the data of a generalised form of Eq. (5.3), yielding $\mu_0 = 0.82 \pm 0.01$ and $\varphi_0 = 0.6° \pm 0.5°$. From Takeda et al. (2010)

5.4.2 Pair-Creation Polarimeters

Although no significant polarimetry of an astrophysical source has ever been achieved by a pair-creation polarimeter, recently two experiments dedicated to high-energy astronomy have demonstrated the detection of a linear polarisation signal in a γ-ray beam using pair conversion. The *GRAINE* (*Gamma-Ray Astro-Imager with Nuclear Emulsion*) project is a balloon-borne experiment dedicated to the observation of the γ-ray sky between 10 MeV and 100 GeV with an emulsion-based pair-creation telescope (Takahashi et al. 2016). The instrument was exposed for $\sim$1 s to a polarised ($\Pi = 66\%$) γ-ray beam of energy in the range 0.8–2.4 GeV from the Spring-8/LEPS facility (Ozaki et al. 2016). After manual checks of the emulsion films and track measurements using an optical microscope, 1372 events were selected to build a histogram of the pair-creation azimuthal distribution. The latter shows the detection at the 3σ significance level of a non-zero effective polarisation asymmetry: $A_{\text{eff}} = 0.21^{+0.11}_{-0.09}$.

The HARPO (Hermetic ARgon POlarimeter) detector is a demonstrator of the performance of a TPC for measuring the polarisation of astrophysical sources down to the threshold of pair creation (Bernard et al. 2014). Polarimetry in the 1–100 MeV range is important, because many high-energy cosmic sources have their peak luminosity in this energy domain. To overcome the effect of multiple scattering, the HARPO project uses as photon converter a TPC filled with a relatively low-density gas, an argon:isobutane gas mixture (95%:5%) at about 2 bar, which is immersed into an electric field. The electrons produced by the ionisation of the gas drift along the electric field toward the readout plane, which is segmented in X and Y so as to provide a 2D image of the fast particle tracks. The third coordinate of the tracks is provided by the measurement of the drift time (see Fig. 5.4a).

The HARPO detector was tested at the BL01 line of the NewSUBARU facility, where it was exposed to monochromatic γ-rays between 1.74 and 74 MeV (Gros et al. 2018). The polarimetric response of the instrument was measured with nearly fully polarised beams between 4 and 20 MeV. The detector azimuthal acceptance was obtained from data acquired with unpolarised γ-rays, as well as with results from a detailed Geant4 simulation of the experiment. Notably, this simulation work required the development of an exact 5D, polarised, Bethe–Heitler event generator for γ-ray conversion (Bernard 2018), which is now available to the community in the latest Geant4 release (Semeniouk and Bernard 2018). The polarigram finally measured at 11.8 MeV is shown in Fig. 5.4b. The fitted effective polarisation asymmetry, $A_{\text{eff}} = 10.5\% \pm 0.6\%$, corresponds to a dilution factor of about 0.4 (see the ideal asymmetry in Fig. 5.2). The HARPO in-beam calibration represents the first demonstration of polarimetry in the MeV range using photon conversions to e^+–e^- pairs.

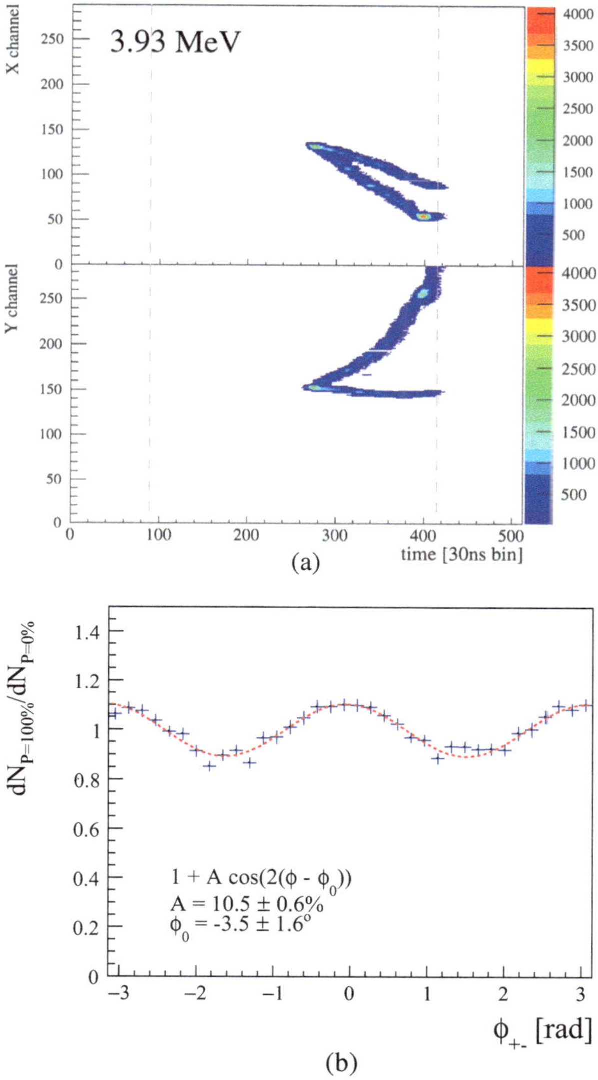

Fig. 5.4 (**a**) Pair conversion event recorded by the HARPO detector during the test experiment at NewSUBARU (Gros et al. 2018): transverse coordinates X and Y as a function of the drift time providing the longitudinal coordinate Z within the TPC. The energy of the incident polarised γ-ray is $E_\gamma = 3.93$ MeV. (**b**) Measured azimuthal angle distribution corrected for the detector response for $E_\gamma = 11.8$ MeV: ratio of fully polarised data to unpolarised data. The fit to the data yields $A = 10.5\% \pm 0.6\%$ and $\phi_0 = -3.5° \pm 1.6°$. From Gros et al. (2018)

5.5 Balloon-Borne Experiments

Balloon experiments have long provided an ideal platform for γ-ray instrument development, sometimes leading to subsequent space missions. In many cases, however, balloon experiments are themselves designed to provide significant scientific capability. Although there are many advantages for balloon payloads (most notably, larger mass and power budgets), there are also many challenges. The most significant of those challenges are the atmospheric absorption and scattering, and the limited exposure that is offered by current ballooning technology. In this section, we survey those balloon experiments that have most contributed (or are contributing) to the development of γ-ray polarimetry as a viable tool of the high-energy astronomer.

5.5.1 *NCT/COSI*

The *Compton Spectrometer and Imager* (*COSI*) (Yang et al. 2018; Sleator et al. 2017; Chiu et al. 2017; Kierans et al. 2014), previously known as the Nuclear Compton Telescope (NCT) (Boggs et al. 2005; Chang et al. 2007), is a balloon-borne, Ge-based Compton telescope that operates in the 200 keV–5 MeV energy band, with a field of view that covers about 25% of the sky. *COSI* uses an array of 12 cross-strip germanium detectors to track the path of incident photons (Fig. 5.5), where position and energy deposits from Compton interactions allow for a reconstruction of the source position in the sky, an inherent measure of the linear polarisation, and significant background reduction. The size of each of the 12 high-purity Ge crystals is $8 \times 8 \times 1.5$ cm. Orthogonal strip electrodes (with a pitch of 2 mm) are deposited on both the anode and cathode surface of each detector.

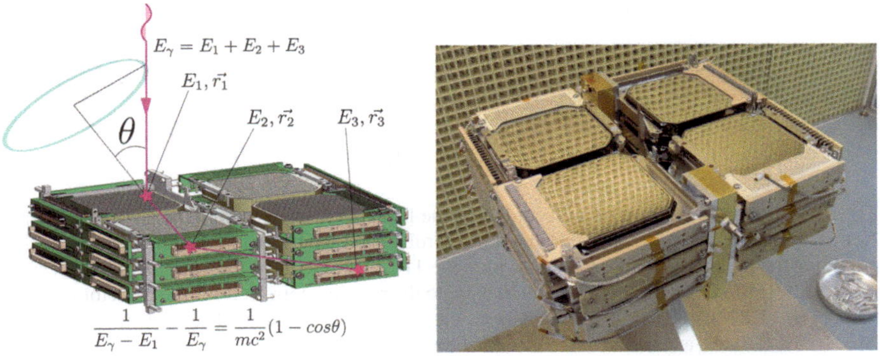

Fig. 5.5 A schematic view of the *COSI* detector array (left) shows the principle of Compton imaging, which is intrinsically sensitive to polarisation (Sect. 5.3.1). The photo (right) shows the Ge detectors in the lab

The detectors have very good position resolution (1.6 mm^3) and provide excellent spectral resolution (<1%). Although the effective area is rather small, ranging from $\sim$18 cm^2 at 300 keV to $\sim$6 cm^2 at 2 MeV, the imaging capability, coupled with the low background nature of the design, provides significant source sensitivity.

Polarimetry is one of the primary science goals of this experiment, which is designed to fly as an ultra-long duration balloon (ULDB) payload on NASA's Super-Pressure Balloon (SPB) system, with expected flight durations of $\sim$100 days. In its most recent (and most successful) balloon flight, *COSI* was launched from Wanaka, New Zealand, on May 16, 2016. The ensuing flight lasted 46 days, although problems with the stability of the balloon resulted in the payload spending much of its time at low altitudes (Kierans et al. 2016). During the flight it observed several steady sources (including the Crab, Cyg X-1, and Cen A), and one intense GRB (GRB 160530A) (Chiu et al. 2017). A polarisation analysis of the GRB data was conducted using two different analyses (one using binned data, one using unbinned data), resulting in an upper limit on the polarisation level of 46% (at the 90% confidence level) (Lowell et al. 2017a; Lowell et al. 2017b). Polarisation results from the other sources have not yet been reported. Future plans for the *COSI* payload include additional SPB balloon flights.

5.5.2 *PoGO*

The goal of the polarised *Gamma Ray Observer* (*PoGO*) programme (Blanford et al. 2003) is to study the polarisation of hard X-ray sources in the energy range of 25–200 keV. The PoGO design is based on an array of phoswich detector cells (PDCs), each of which is an independently collimated hexagonal unit (Fig. 5.6) that provides a FoV of $\sim 2.5° \times 2.5°$. Two types of slow scintillator are used for both the collimator and the rear anti-coincidence shield. The primary detector consists of a fast scintillator. The entire assembly is read out with a single 25 mm photomultiplier tube (PMT). A valid photon interaction is one that registers as a fast signal without any slow signal components. The full PDC can identify different types of events, as shown in (Fig. 5.7).

The initial flight version was referred to as the *PoGOLite Pathfinder* (Fig. 5.8) (Kiss and Pearce 2007; Kiss et al. 2008; Chauvin et al. 2015). It consisted of an array of 61 PDC elements surrounded by both a BGO anti-coincidence shield for rejecting photons and charged particles, and a polyethylene neutron shield to moderate atmospheric neutrons, which might otherwise mimic valid photon events in the PDC array. Heat generated within the PMT array was radiated away using a fluid-based cooling system. The Pathfinder instrument was flown on a circumpolar balloon flight in 2013, launched from Kiruna, Sweden, and landing 13 days later near the Russian town of Norilsk. The polarimeter control system failed after the first 2 days of flight (two observations totalling $\sim$10 h), after which observations were no longer possible. Although a Crab signal was detected using polarised events

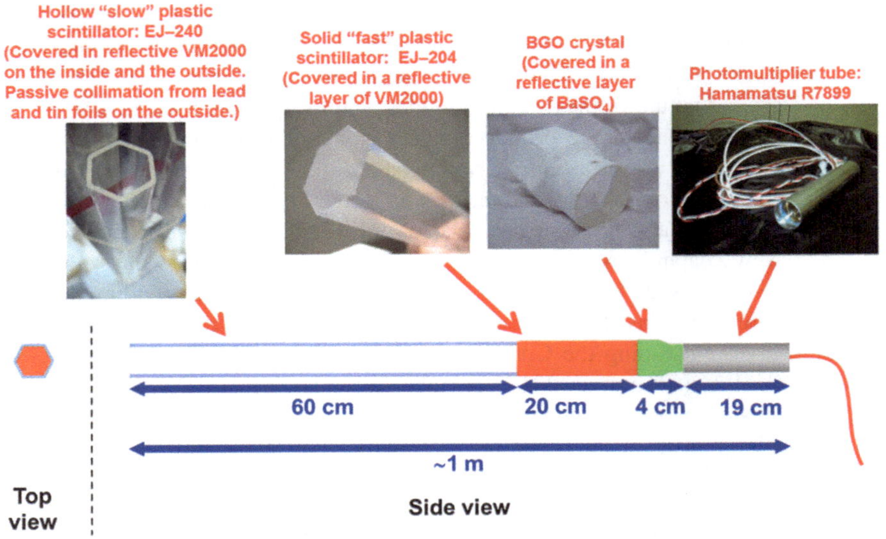

Fig. 5.6 A schematic view of single phoswich detector cell (PDC) as flown on the *PoGOLite* Pathfinder in 2013. For the *PoGO+* balloon flight in 2016, the active 60 cm collimator (the slow scintillator on the left) was replaced with a passive collimator and the length of the fast scintillator was reduced to 12 cm

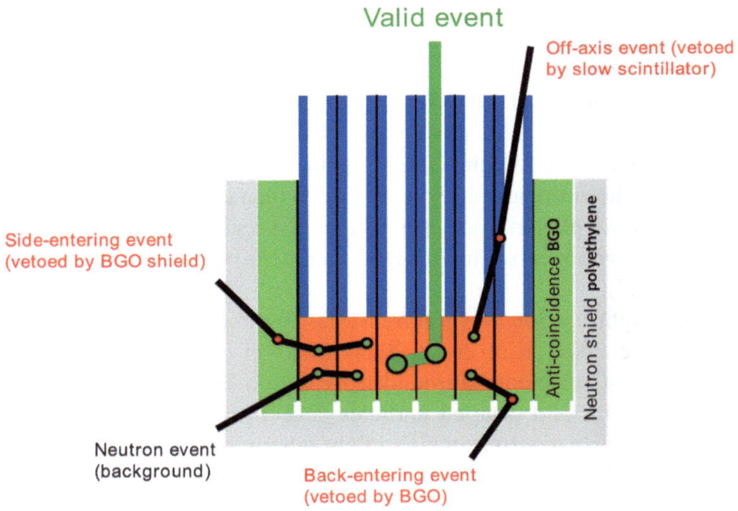

Fig. 5.7 The different types of events that can be registered in the *PoGOLite* Pathfinder

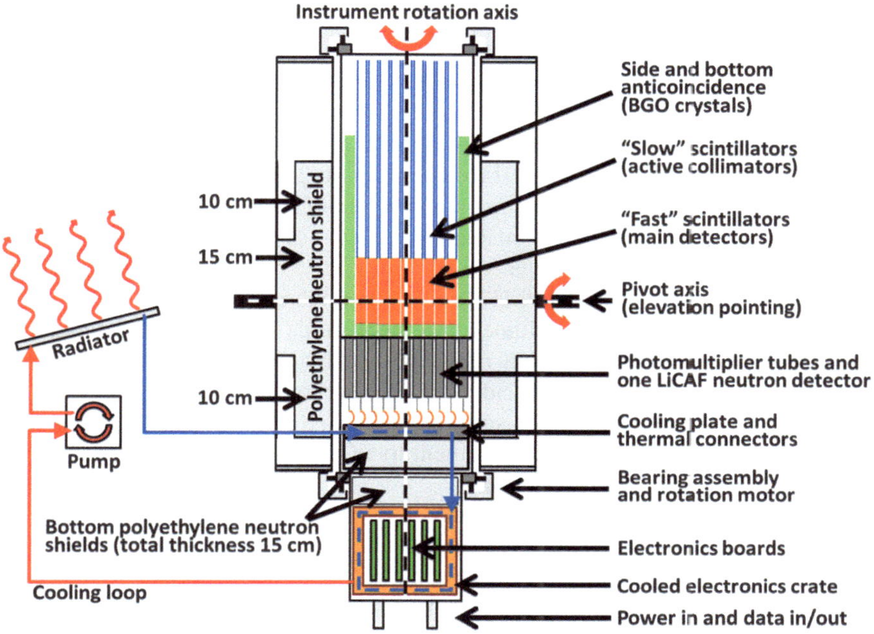

Fig. 5.8 The payload configuration of the *PoGOLite* Pathfinder

in the 25–110 keV energy range, the limited observation time resulted in a lower than expected MDP. An upper limit to the polarisation level of the integrated Crab signal was reported to be 42.4% at the 99% confidence level (Chauvin et al. 2016).

An improved version of the PoGOLite instrument was developed for a second balloon flight in 2016. Referred as *PoGO+*, this instrument incorporated several improvements with respect to the Pathfinder design. In particular, the active PDC collimators used in the Pathfinder (whose signal had been hard to discern from low-energy events in the central scintillators) were replaced with passive collimator material. In addition, the primary scintillator elements were reduced in length from 20 to 12 cm; this had little effect on the total effective area, but served to increase the light collection efficiency and reduce the (volume-dependent) background. Finally, improved reflective coatings on all scintillators were employed to increase the clarity of the scintillation signals. The second balloon flight lasted 7 days, as the balloon travelled from its launch site in Kiruna to Victoria Island, Canada. Polarisation results from both the Crab pulsar and nebula (Chauvin et al. 2017) and the Cygnus X-1 microquasar (Chauvin et al. 2018) have been reported. The polarisation level of the phase-integrated Crab emission (20–160 keV) was reported to be $(20.9 \pm 5.0)\%$, whereas the off-pulse polarisation fraction was measured at $(17.4 \pm 9.0)\%$. An upper limit of 8.6% (at the 90% confidence level) was reported for Cyg X-1 in the 19–181 keV energy range (Chauvin et al. 2018).

5.5.3 GRAPE

The *Gamma RAy Polarimeter Experiment* (*GRAPE*) is a balloon programme, whose goal is to conduct GRB observations on long duration balloon flights. *GRAPE* is a large field-of-view Compton polarimeter operating in the 50–500 keV energy range (McConnell et al. 2009b, 2013, 2014b; Bloser et al. 2009; McConnell et al. 1999, 2004, 2005). This energy range complements that of Compton telescope designs, which typically have thresholds at higher energies (around 200 keV in the case of COSI). The instrument design is a modular one, based on the use of a 64-channel multi-anode PMT (MAPMT) to read out a hybrid array of small scintillator elements, each of which has a volume of $5 \times 5 \times 50$ mm. *GRAPE* uses two scintillator types (plastic and CsI(Tl)), which provides moderate energy resolution and low susceptibility to neutron background. The scintillators are arranged with a central array of 6×6 plastic scintillators surrounded by CsI(Tl) scintillators. Ideal events are those that scatter from a plastic element to a CsI(Tl) element. Each module has an effective area of $\sim 2\,\mathrm{cm}^2$ at 150 keV.

To date, the *GRAPE* programme has conducted two balloon flights intended to validate the instrument design by performing a measurement of the Crab polarisation. In 2011, the payload was flown with an array of 16 collimated polarimeter modules. The detector assembly was placed within a pressure vessel and mounted on a rotating platform that was intended to correct for systematic effects within the polarimeter. Higher-than-expected background levels prevented a successful Crab measurement (McConnell et al. 2014b). A subsequent flight in 2014 was made with improved shielding and a larger array of 24 modules, but the flight did not last long enough for a full observation of the Crab.

The programme is currently working on a revised design in preparation for a series of long duration balloon flights (McConnell et al. 2018, 2016). In order to avoid optical crosstalk issues, the more recent design has moved away from the use of MAPMTs to the use of small individual scintillator elements, each with its own Si photomultiplier (SiPM) readout. Arranged in a cubic array (Fig. 5.9), this next-generation design will also provide imaging capability. Although the limited spatial and spectral resolutions provide only moderate angular resolution, it provides a significant increase in sensitivity by reducing the instrumental background for any given GRB measurement while maintaining the low-energy response provided by the low-Z scattering elements.

The *GRAPE* project has provided the basis for several space-borne GRB polarimeter proposals. *POET* (*POlarimeters for Energetic Transients*) was proposed in both 2008 (Hill et al. 2007; McConnell et al. 2009a) and 2014 (Bloser et al. 2014; McConnell et al. 2014a) as a NASA Small Explorer (SMEX) mission. In both cases, the payload consisted of a high-energy (50–500 keV) Compton scatter polarimeter and a low-energy (2–15 keV) photoelectric polarimeter. *PETS* (*Polarimetry of Energetic Transients in Space*) was proposed in 2012, to fly a high-energy polarimeter as an external payload on the International Space Station (ISS).

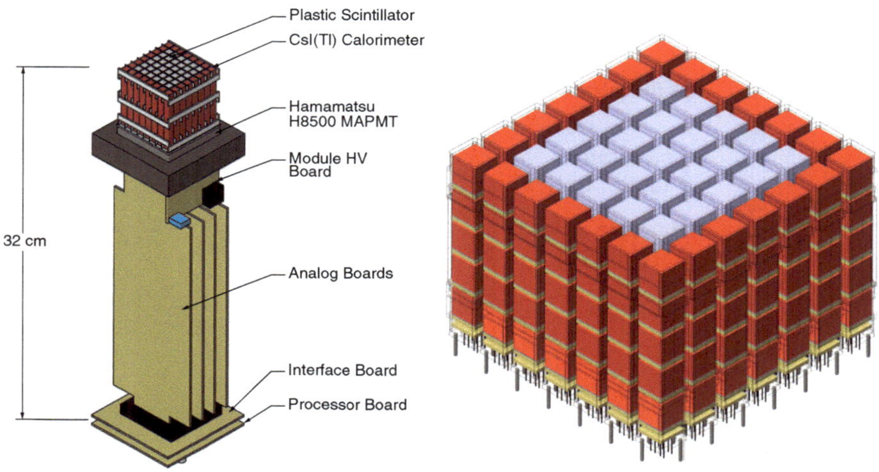

Fig. 5.9 A schematic view of single *GRAPE* module (left) showing the layout of the scintillator elements (plastic elements surround by CsI(Tl) elements) and the discrete electronics boards. An improved design for *GRAPE* (right) consists of a three-dimensional array of small scintillator elements (each a cube of 12.5 × 12.5 × 12.5 mm^3 read out by a SiPM) that acts as a Compton imager, but retains the advantages of low-Z scattering elements

A similar proposal concept, known as *LEAP* (*LargE Area burst Polarimeter*), was also submitted in 2016.

5.5.4 *PHENEX*

The development of the *PHENEX* (*Polarimetry for High ENErgy X-rays*) polarimeter was developed specifically for 40–200 keV polarisation studies of the Crab (Kishimoto et al. 2006). Although developed independently, the design of a single *PHENEX* "unit counter" is remarkably similar to that of the GRAPE polarimeter, consisting of an array of scintillator elements (both CsI and plastic) mounted on the front end of a 64-channel MAPMT. Each unit counter (Fig. 5.10) incorporates its own passive collimator assembly, which constrains the FoV to 4.8°. Although each plastic scintillator element is mounted directly to the MAPMT, the CsI elements are mounted to the MAPMT via the use of a fibre optic plate (FOP) to reduce optical crosstalk from the CsI to plastic elements.

The *PHENEX* instrument was flown on a 6 h balloon flight in 2006 (Kishimoto et al. 2009; Gunji et al. 2007, 2008; Kishimoto et al. 2007; Tanaka et al. 2007). The payload consisted of an array of four unit counters plus one "monitor counter" consisting entirely of CsI elements. As with *GRAPE*, the detector assembly was placed within a pressure vessel and mounted on a rotating platform to correct for systematic effects within the polarimeter. During the flight, problems with the

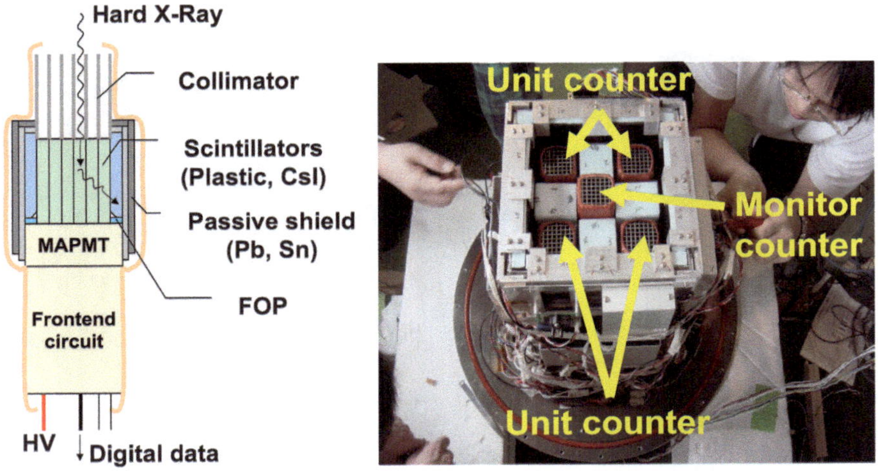

Fig. 5.10 A schematic view (left) of a single *PHENEX* unit counter (left) showing the layout of the scintillator array, the collimator, the passive shielding, and the front-end electronics. The photo (right) shows the balloon payload configuration being readied for its 2006 balloon flight

attitude control system (ACS) resulted in a wandering of the pointing direction about the Crab location. Only about 1 h of useful Crab data was obtained. Although the Crab was detected at a significance level of 8σ, no significant polarisation signal was measured due to the limited exposure and the limited size of the detector array. A subsequent balloon flight (with 8 unit counters, added passive shielding, improved low-energy performance, and improved ACS) was targeted for 2009 (Kishimoto et al. 2008, 2011; Gunji et al. 2008), although no results seem to have been reported in the literature.

5.5.5 *ASCOT*

The *Advanced Scintillator COmpton Telescope* (*ASCOT*) represents a Compton telescope design based on the use of modern scintillator and readout technologies (Bloser et al. 2016a,b, 2018). Similar to the concept behind the COMPTEL instrument of the *Compton Gamma-Ray Observatory*, it employs two detector layers—a low-Z scattering layer and a high-Z absorbing layer. The low-Z material is P-Terphenyl and the high-Z material is $CeBr_3$, both fast scintillator materials. Scintillator readout is provided by SiPMs that allow for a more closely packed detector arrangement. One of the goals of this design is to employ pulse-shape discrimination (PSD) and time-of-flight (ToF) measurements to reduce instrumental background, despite a more compact design. As with any Compton telescope, *ASCOT* is intrinsically capable of making polarisation measurements. A prototype instrument was flown out of Palestine, TX in 2018, with 5 h at float. The principle goal of this flight was to evaluate the in-flight background and to validate the instrument design by generating an image of the Crab. The small prototype did not have sufficient sensitivity for polarisation studies, but future (larger) versions of *ASCOT* may be more capable.

5.5.6 GRIPS

One balloon instrument has been developed specifically for fine imaging polarimetry of solar flares. The *Gamma-Ray Imager/Polarimeter for Solar flares* (*GRIPS*) is based on three-dimensional Ge technology, similar to that pioneered for *COSI* (Duncan et al. 2013; Shih et al. 2012). Operating in the 20 keV to 10 MeV energy range, it employs a combination of a Compton scatter detector with an indirect imaging technique to provide sub-arcminute angular resolution. The fine imaging is achieved using a coded-aperture style imaging system, a single-grid modulation collimator, the multi-pitch rotating modulator (MPRM). *GRIPS* is capable of angular resolution in the range of 12.5–162 arc s. The GRIPS payload was successfully flown on an 11-day long duration balloon flight from McMurdo in 2016 (Duncan et al. 2016). Solar activity was relatively low during the flight, although the Sun did produce 21 C-class flares during the flight. No useful results were obtained from these events (Duncan 2017).

5.5.7 GRAINE

The *Gamma-Ray Astro-Imager with Nuclear Emulsion* (*GRAINE*) experiment is a balloon-borne instrument designed to study the polarisation of gamma-ray sources in the pair-production regime (10 MeV–100 GeV) (Rokujo et al. 2018; Takahashi et al. 2015, 2018; Ozaki et al. 2016). *GRAINE* is based on the use of emulsion films for tracking the electron–positron pair. Since emulsion films must be optically scanned to identify and track particle interactions, their use would not be practical for an orbiting instrument. For a balloon payload, where the instrument is recovered post-flight, emulsions offer several unique advantages. Traditional approaches to pair production imaging (such as that employed in *Fermi* LAT) utilise high-Z converter materials interspersed with low-Z detector materials that track the electron–positron pairs. This type of design results in multiple scattering of the electron–positron pairs that degrades knowledge of the initial track direction. The multiple scattering effects limit not only the angular resolution, but also the ability to make polarisation measurements (Sect. 5.3.2). Emulsions, on the other hand, serve both as the conversion material (due to the high-Z composition), but also the tracking medium. With sub-mm spatial resolution, the emulsions are significantly more effective at discerning the initial track direction. Angular resolutions of

<0.1° can be achieved and polarimetry can be far more effective. GRAINE has demonstrated its polarimetric capability on ground with a 0.8–2.4 GeV γ-ray beam from Spring-8 (Sect. 4.2). Balloon flights of an emulsion telescope have been conducted in 2011 and 2015. During the 14 h balloon flight in 2015, calibration gamma-rays were generated from cosmic ray interactions with an aluminium plate that was located in front of the emulsion stack. Analysis of the flight data resulted in an image of the radiating plate. Future plans call for a series of longer duration (week-long) balloon flights using an emulsion array with an area of 10^2 m.

5.6 Satellite Experiments

5.6.1 The RHESSI Mission

The *RHESSI* mission was an observatory dedicated to the study of the Sun. It hosted on-board a spectrometer which was composed of nine cryogenically cooled germanium detectors, which produced Sun images, thanks to modulated fine grids. It could also measure γ-ray polarisation by studying Compton scattering between two of the Ge detectors in the 20–2000 keV energy range (McConnell et al. 2002a).

In 2002, *RHESSI* observed a very bright gamma-ray burst, GRB021206, and detected a clear polarisation signal (Fig. 5.11). The data analysis made by Coburn and Boggs (2003) was however rediscussed 2 years later by Rutledge and Fox (2004), who demonstrated that most of the "Compton events" used by Coburn and

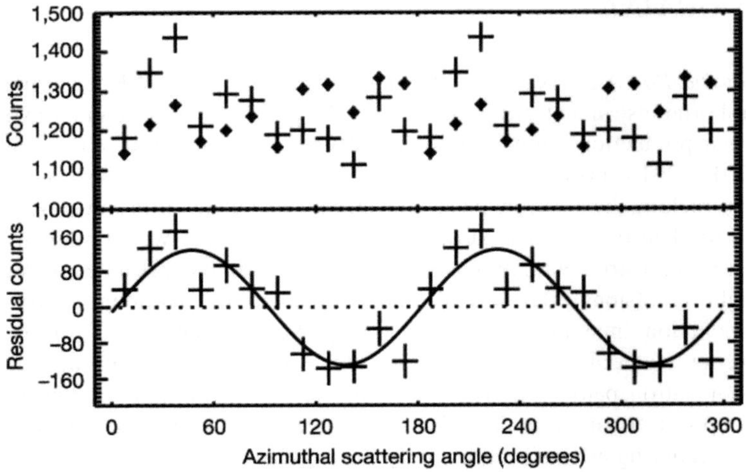

Fig. 5.11 Polarisation signal observed with *RHESSI* from GRB021206 in the 150 keV–2 MeV energy range. *Top panel*: measured data (crosses) with simulated data for an unpolarised source (points). *Bottom panel*: data obtained after subtraction of the simulated distribution. From Coburn and Boggs (2003)

Boggs were not real double events, but single events repeated by the *RHESSI* on-board electronics. They showed that the number of real double events was in fact a factor of ten below the one estimated by Coburn and Boggs, which implied a lower signal-to-noise ratio for the GRB. Considering this, they show finally that the published polarisation signal was in fact dominated by systematics (see also Wigger et al. (2004)). This example has demonstrated how the Compton polarisation data analysis is complex, and that strong efforts should be devolved to the study of systematics, which could mask or sometimes mimic a polarisation signal.

The ability of *RHESSI* to measure polarisation has also been used for measuring solar flare polarisation. Measurements reported to date for the 100–350 keV energy range (Boggs et al. 2006; Suarez-Garcia et al. 2006) have provided only marginal evidence for solar flare polarisation. None of the measurements have a statistical significance above 3σ.

5.6.2 *INTEGRAL*

5.6.2.1 The *INTEGRAL* Polarimetric Capabilities

INTEGRAL is the second medium-size mission (M2) of the Horizon 2000 ESA programme. Its payload comprises two main instruments, an imager, IBIS, and a spectrometer, SPI, dedicated to the observation of the γ-ray sky. They are both spectro-imagers working in the same range of energy. IBIS has been designed for the detection and accurate localisation (1 arcmin) of γ-ray point sources in the 15 keV–10 MeV energy range, measuring their spectrum, variability, and polarisation. These sources are mainly compact accreting stars, active galactic nuclei, pulsars, supernovae, supernova remnants (SNRs), and GRBs. The spectrometer SPI is dedicated to measuring very accurately the spectrum of γ-ray sources ($E/\Delta E \approx 500$), thanks to its nineteen cryogenically cooled Ge detectors. The characteristics of SPI make it particularly suitable for the measurement of galactic emission at 511 keV and in the ^{26}Al line (1.8 MeV), which require a large field of view and very long observation times. SPI is also optimised for measuring radioactive decay lines (^{56}Co, ^{60}Fe, ^{44}Ti, etc.). These measurements allow decisive progress in the field of nucleosynthesis and galactic evolution.

The SPI telescope can be used as a Compton polarimeter by selecting Compton scattering between different Ge detectors. The derivation of the polarimetric properties of γ-ray sources with SPI required very intensive Monte Carlo simulations and complex data analysis, as described in Dean et al. (2008).

Thanks to its two position-sensitive detectors, ISGRI (made of CdTe crystals and sensitive in the 15–1000 keV energy band) and PICsIT (made of CsI bars and sensitive in the 200 keV–10 MeV energy band), the IBIS telescope has been also used as a Compton polarimeter to study many compact objects. The procedure to measure the polarisation is described in Forot et al. (2008). It allowed to control

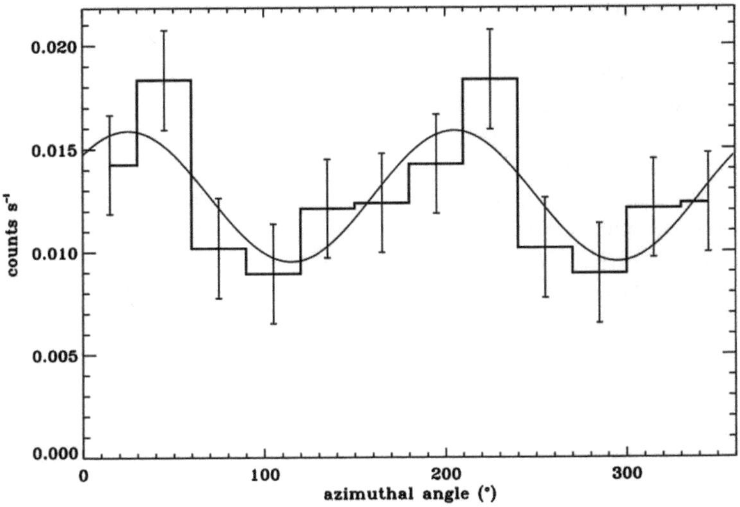

Fig. 5.12 Phase-averaged polarigram of the Crab pulsar and nebula obtained by *INTEGRAL*/IBIS in fall 2015 in the 300–450 keV energy band. From Moran et al. (2016)

systematic effects and to successfully detect a polarised signal from the Crab nebula. In this work, the source flux was derived as a function of the azimuthal scattering angle φ by dividing the Compton photon distribution into 6 bins of $\Delta\varphi = 30°$. The chance coincidence (i.e., photons interacting in both detectors, but not related to a Compton event) has been subtracted from each detector image following also the procedure described in Forot et al. (2008). The derived detector images were then deconvolved to obtain sky images, where the flux of the source in each azimuthal bin was measured by fitting the instrumental PSF to the source peak. The azimuthal profiles (i.e., the polarigrams, see Fig. 5.12) were fitted using a least squares technique to derive the polarisation fraction Π and angle φ_0.

5.6.2.2 *INTEGRAL* Observation of the Crab Pulsar and Nebula Polarisation

The Crab nebula is regularly observed since the *INTEGRAL* launch for calibration purposes. Polarisation results obtained with IBIS until 2014 were published in Forot et al. (2008), Moran et al. (2016). In Fig. 5.12, we show the polarigram obtained with IBIS using data obtained between 2015, August 18th and October 17th for a total exposure time of 500 ks. Optical polarimetric observations of the Crab were also performed at the same epoch at the William Herschel Telescope (WHT), using the GASP polarimeter (Gouiffes et al. 2016). Several observations have demonstrated

Table 5.1 *INTEGRAL*/IBIS polarisation results in the 300–450 keV band compared to optical measurements

Observation date	Position angle (°)	Polarisation fraction (%)	Optical position angle (°)	Reference
2003–2007	115 ± 11	96 ± 34	109.5 ± 0.7	Forot et al. (2008)
2012–2014	80 ± 12	98 ± 37	85.3 ± 1.4	Moran et al. (2016)
2015	125 ± 15	89 ± 28	130 ± 1.0	Gouiffes et al. (2016)

that the Crab polarisation angle has shifted during the 2003–2015 period, jointly in the optical and γ-rays (see Table 5.1).

5.6.2.3 Cygnus X-1 Polarisation

The X-ray binary Cygnus X-1 is the first Galactic source known to host a black hole with a mass of 14.8 ± 1.0 $M_{\odot}$ (Orosz et al. 2011). The donor star is the high mass O9.7 Iab star HDE 226868. The system is located at a distance of $d = 1.86 \pm 0.12$ kpc from Earth with an orbital period of 5.6 days.

Cygnus X-1 is observed in two main spectral states, the low hard state (LHS), characterised by a Comptonised thermal emission with a temperature around 50 keV and the high soft state (HSS), dominated by a bright X-ray thermal emission coming most probably from the accretion disk. Up to 2010, Cygnus X-1 was predominantly in the LHS. Then, its behaviour has changed and it spends most of its time in the HSS. It sometimes undergoes partial (failed) transitions from the LHS to the HSS, and can be found in a transitional or intermediate state. The detection of compact relativistic jets in the LHS (Stirling et al. 2001) places Cyg X-1 in the family of microquasars.

The presence of a MeV tail in the low hard state, first detected by COMPTEL (McConnell et al. 2002b), has recently been confirmed by the two main instruments on-board *INTEGRAL* (Jourdain et al. 2012; Rodriguez et al. 2015). Analysing SPI and IBIS data, Jourdain et al. (2012), Laurent et al. (2011b), respectively, have also shown that the Cygnus X-1 emission above 400 keV is polarised at a level of about 70%. They obtained a 20% upper limit on the degree of polarisation at lower energies. Both teams suggested that the polarised emission is due to synchrotron emission coming from a compact jet. The detection of polarised emission and measurement of its energy dependency give strong constraints on our understanding of accretion and ejection processes in microquasars. It helps to distinguish the different proposed emitting media (Comptonisation corona versus synchrotron jets) in these objects and provides important clues to the composition, energetics, and magnetic field of the jets.

5.6.2.4 GRB Observations and Constraints on Lorentz Invariance Violation

Up to now, three GRBs were sufficiently bright in the *INTEGRAL* field of view to enable a polarisation study (Kalemci et al. 2007; McGlynn et al. 2007, 2009; Götz et al. 2009, 2013, 2014): GRB140206A, GRB061122, and GRB041219A. The polarised γ-ray light of GRB041219A was measured with both SPI (McGlynn et al. 2007) and IBIS (Götz et al. 2009). In addition, thanks to deep infra-red multi-band ($YJHK_s$) imaging using the WIRCam instrument at the 3.6 m Canadian French Hawaiian Telescope (CFHT), Götz et al. (2011) were able to identify the host galaxy of GRB041219A, and constrain its photometric redshift to $z = 0.31^{+0.54}_{-0.26}$ (the errors are at the 1σ confidence level). This implies a luminosity distance interval of [0.222–5.406] Gpc, assuming standard cosmological parameters ($\Omega_m = 0.3$, $\Omega_\lambda = 0.7$, and $H_0 = 70$ km/s/Mpc).

The polarigram deduced from IBIS data during the observation of GRB041219A is shown in Fig. 5.13. It was computed by Laurent et al. (2011a) into two energy bands in order to detect a possible polarimetric angle shift with energy, reminiscent of Lorentz invariant violation ("LIV") effects (Sect. 5.2). The two bands were chosen to be [200–250 keV] and [250–325 keV], where the source has merely the same signal-to-noise ratio. The measured evolution of the polarimetric angle shift between these two energy bands, along the burst duration, is consistent with zero,

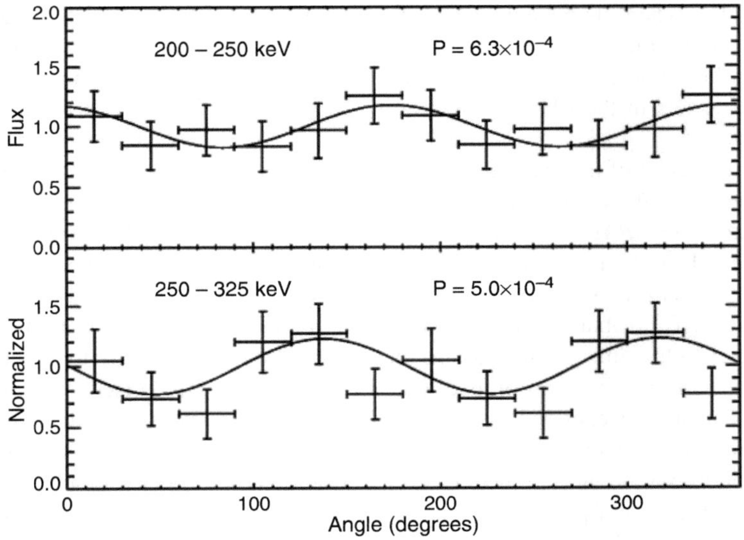

Fig. 5.13 Polarigram of GRB041219A in the 200–250 keV and in the 250–325 keV energy band. These distributions are consistent with a highly polarised signal. The chance probability of a non-polarised signal is reported in each panel. No statistically significant shift was observed during the whole duration of the gamma-ray burst. From Laurent et al. (2011a)

with a mean value of 21° ± 47°. This has put one of the strongest constraint ever derived on the existence of LIV effects.

5.6.3 *IKAROS/GAP*

The GAP experiment is a Japanese Compton polarimeter working in the hard X-ray/soft γ-ray domain (see also the chapter on X-ray polarimetry by Soffitta and Costa in this book). It was launched in May 21st 2010, on-board the *IKAROS* solar sail (Yonetoku et al. 2011). It is composed of an assembly of two coaxial detectors, made, respectively, of a plastic scintillator and CsI crystals. Its polarimetric capabilities have been tested on ground in polarised beams to calibrate systematic effects.

In 2010, GRB100826A was observed by GAP in the 70–300 keV energy band. The modulation curve was fitted by a model from a Monte Carlo simulation of the experiment (Yonetoku et al. 2011), and a polarised signal was marginally observed with a significance of 2.9σ (see Fig. 5.14). The polarisation angle was observed to vary by ~90° during the prompt emission, which can be interpreted as evidence for patchy emission, patchy magnetic field, or as photospheric emission from a variable jet. In 2011, the GAP instrument measured the polarisation of two more GRBs, GRB110301A and GRB110721A.

According to the Japan Aerospace Exploration Agency (JAXA), *IKAROS* finished all planned experiments in December 2010, but the instrument continued to work beyond that date. The mission entered in hibernation in 2012, and the GAP observations stopped, due to lack of power from the solar panels.

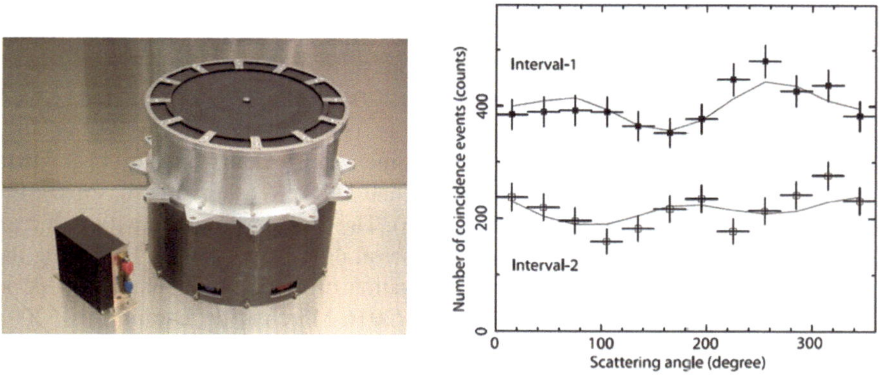

(a) (b)

Fig. 5.14 (**a**) View of the Ikaros/GAP polarimeter. (**b**) Polarised signal observed with GAP from GRB100826A in the 70–300 keV energy range (Yonetoku et al. 2011)

5.6.4 POLAR

POLAR is a Swiss/Chinese mission dedicated to GRB polarisation measurements. It has been launched on September 15th, 2016 and placed on the Chinese space station Tiangong 2 (Produit et al. 2005). This Compton telescope is made of 1600 bars of plastic scintillators read out in groups of 64 using an 8×8 MAPMT. Its design ensures a very wide field of view, one-third of the sky, and very good polarimetric capabilities ($\mu_{100} \approx 35\%$). Due to a failure of its power distribution unit, the telescope was switched off in 2017.

The *POLAR* telescope measured the light curves of 55 GRBs. Among these, five have been bright enough in the 10–1000 keV energy range to enable a polarisation analysis. This work has revealed that, on average over the GRB duration, the polarisation fraction is rather low, around 10% for the five bursts. However, when considering time resolved analysis for the brighter burst, GRB 170114A, this fraction is measured to be as high as 28%. This may indicate that the γ-ray light is mildly polarised, but the time-integrated polarisation degree is measured to be low because the polarisation angle changes during the burst (Zhang et al. 2019).

5.6.5 Hitomi

The JAXA project, *ASTRO-H*, renamed *Hitomi*, was the sixth Japanese satellite to observe the sky in the field of X-rays and low-energy γ-rays. It carried four types of telescopes operating in a band of energy ranging from 0.3 keV to 600 keV. Its main scientific objectives were to reveal the large-scale structures and the evolution of the Universe, as well as to understand the various high-energy sources.

The satellite was placed in a low Earth orbit (LEO) at 550 km altitude and 31° inclination on February 17th, 2016 with an H-IIA rocket from the Tanegashima Space Center, Kagoshima, Japan. Unfortunately, after an error of attitude, *Hitomi* switched to survival mode on March 26th, 2016. Then, on April 28th, JAXA declared the mission lost.

Hitomi was composed of four instruments including a Compton telescope, which acted as a Compton polarimeter between 5 and 600 keV (Tajima et al. 2010): the soft gamma-ray detector (SGD, see Fig. 5.15). The SGD has been heavily tested on ground, and shows very good polarimetric possibilities (Sect. 5.4.1). Following the start-up operations for the SGD, the observation of the Crab nebula with *Hitomi* was performed from 12:35 to 18:01 UT on 2016 March 25th. During this 5200 s observation, before the satellite failure, the SGD instrument has clearly detected the Crab polarisation between 60 and 160 keV.

The analysis has been done through deep analysis of Crab and background data, and comparison with Monte Carlo simulations. The obtained polarisation fraction of the phase-integrated Crab emission (sum of pulsar and nebula emissions) is $22.1\% \pm 10.6\%$, and the polarisation angle is $110.7° \pm 13°$. The polarisation angle

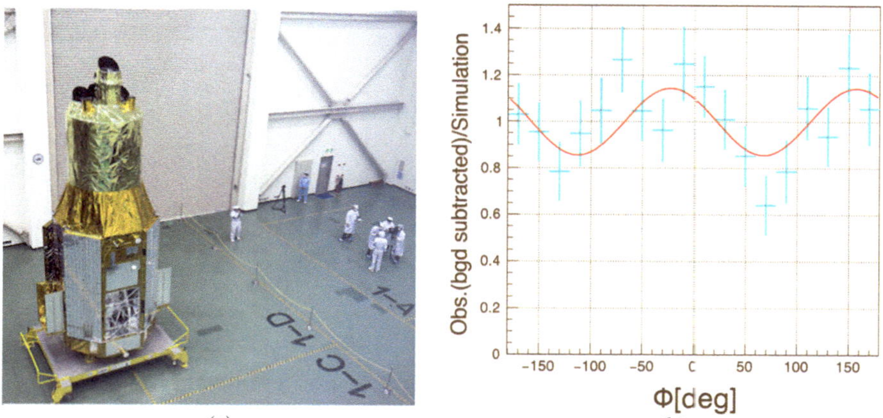

(a) (b)

Fig. 5.15 (**a**) View of the *Hitomi* satellite at JAXA before flight. SGD is the black box aside the satellite. (**b**) Polarigram obtained with SGD during the Crab observation. The data points show the ratio of the background-subtracted observation data to the unpolarised simulation data (Hitomi Collaboration et al. 2018)

measured by SGD is consistent with the projected spin axis of the pulsar (124°) (Hitomi Collaboration et al. 2018).

5.6.6 *ASTROSAT*

ASTROSAT is the first Indian satellite dedicated to astronomical observations. It has been launched on 2015 September 28 on a LEO by a PSLV Indian rocket. All instruments are operating successfully since then. One of its instruments, the Cadmium Zinc Telluride Imager (CZTI) is a coded mask telescope composed with a 1000 cm^2 array of CdZnTe detectors, working in the 20 to 150 keV energy range. This detector can be also used as a Compton polarimeter in the 100–300 keV energy range, taking into account photons which make nearly 90° scatterings into the detector plane. This capability has been heavily calibrated before flight (Vadawale et al. 2015).

In the first 1 year of operation (2015 October 6 to 2016 October 5), CZTI has detected a total of 47 GRBs, among which eleven were bright enough to compute their polarisation signature (Chattopadhyay et al. 2017). One example, GRB160821A, where a strong polarisation fraction has been measured, is shown in Fig. 5.16. In this polarigram, the data were corrected from the high modulation due to the CZTI non-uniform response, which was determined by Monte Carlo simulations. Most of the studied GRBs were highly polarised, implying most probably synchrotron radiation in a uniform magnetic field.

In 2016–2017, the CZTI team also measured the polarisation signature of the Crab pulsar and nebula. It has allowed to obtain the most precise determination

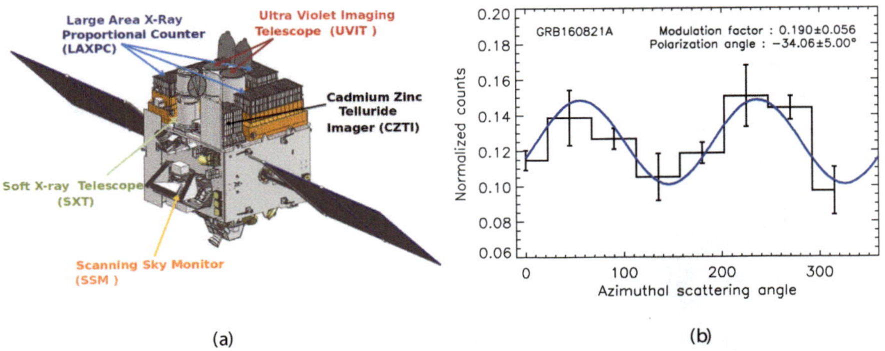

Fig. 5.16 (**a**) Sketch of the *ASTROSAT* satellite. (**b**) Polarigram for GRB 160821A. The data are corrected from CZTI non-uniformity effects. The blue solid line is the sinusoidal fit to the data. From Chattopadhyay et al. (2017)

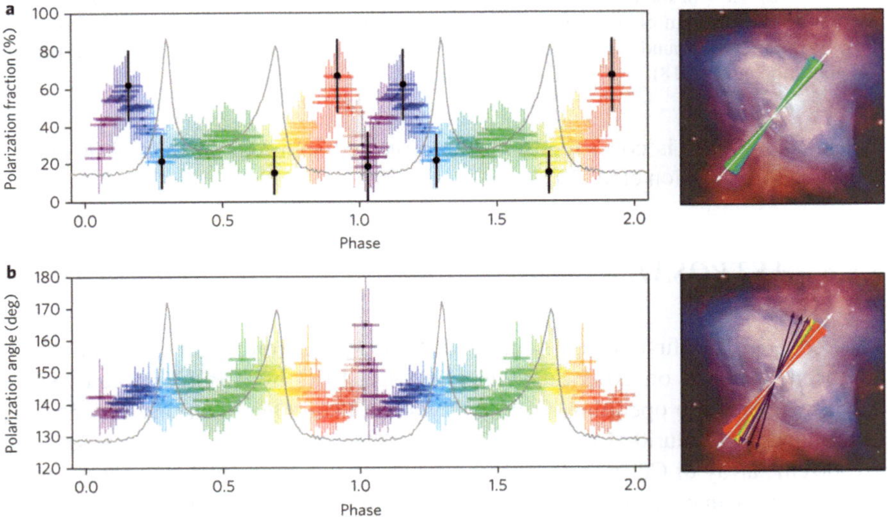

Fig. 5.17 Phase resolved polarimetry of the Crab pulsar and nebula. We see the evolution over the pulsar's phase of (**a**) the polarisation fraction and (**b**) the polarisation angle (from Vadawale et al. (2018))

of the pulsar and nebula polarisation properties over the phase range. What is noticeable is the detection of a strong polarisation fraction in the off-pulse region, associated with a shift of the polarisation angle. Also, the polarisation angle seems to shift along the rise and the decay of the pulsar's pulses (Vadawale et al. (2018); see Fig. 5.17), which cannot be fully explained by the current theoretical models of high-energy emission from pulsars.

5.7 Summary

Gamma-ray polarimetry is still in its infancy, but some polarisation results already obtained by balloon-borne and satellite experiments in the soft γ-ray domain show the importance that this new astronomical tool should take in the coming years. Thus, measurements of GRB polarisation have already placed interesting constraints on the nature of the ultra-relativistic outflow, its geometry and magnetisation, as well as on the high-energy radiation mechanism (e.g., Götz et al. (2009); Yonetoku et al. (2011); Lowell et al. (2017b); Chattopadhyay et al. (2017)). The GRB polarisation information has also been used to provide a strong constraint on the possible violation of Lorentz invariance arising from vacuum birefringence (Laurent et al. 2011a). Repeated polarisation observations of the Crab γ-ray emission have revealed a significant variation of the polarisation characteristics with the pulsar phase (Chauvin et al. 2017; Forot et al. 2008; Moran et al. 2016; Hitomi Collaboration et al. 2018; Vadawale et al. 2015), which is not well explained by the current pulsar models. The detection of highly polarised emission from Cygnus X-1 above 400 keV, whereas the lower-energy radiation is not significantly polarised, has helped to distinguish the emitting regions (Comptonisation corona vs. synchrotron jets) in this microquasar (Chauvin et al. 2018; Laurent et al. 2011b; Jourdain et al. 2012).

These topics will greatly benefit in the coming years from the development of several balloon-borne Compton telescopes with excellent polarimetric capability (e.g., COSI-X, PoGO+, ASCOT). Moreover, in the current proposals for the next generation of γ-ray space observatories, the γ-ray instruments are expressly designed for polarisation measurements, using detectors with excellent spectral and spatial resolution, and dedicated on-ground calibrations for the polarisation response of the space telescopes are foreseen (see, e.g., de Angelis et al. (2018)). This was not the case for previous γ-ray missions. Thus, both the *AMEGO* (Moiseev and Amego Team 2017) and *e-ASTROGAM* (Tatischeff et al. 2016; De Angelis et al. 2017) telescopes would be able to perform unprecedented polarisation measurements in the MeV range using Compton interactions (see, e.g., Tatischeff et al. (2018)).

The *AMEGO* and *e-ASTROGAM* instruments should also have some sensitivity to polarisation above 2 MeV, using γ-ray conversion in the Si detectors of the tracker. However, multiple scattering of the e^+–e^- pairs within these solid-state detectors should induce a significant dilution of the polarisation asymmetry. Space polarimetry with pair creation might be better achieved using as γ-ray converter a homogeneous active target capable of recording the three-dimensional trajectories of the e^+–e^- pairs, such as a gaseous time projection chamber (Hunter et al. 2014; Bernard et al. 2014). A demonstration of this concept in a balloon-borne experiment would be desirable.

Acknowledgements The authors would like to warmly thank Lorraine Hanlon and Denis Bernard for fruitful discussions and their constructive comments on the manuscript.

References

Atwood WB, Abdo AA, Ackermann M, Althouse W, Anderson B, Axelsson M, Baldini L, Ballet J, Band DL, Barbiellini G et al (2009) The large area telescope on the Fermi Gamma-ray Space Telescope mission. Astrophys J 697:1071–1102

Bassan N, Mirizzi A, Roncadelli M (2010) Axion-like particle effects on the polarization of cosmic high-energy gamma sources. J Cosmol Astropart Phys 5:010

Bernard D (2013) Polarimetry of cosmic gamma-ray sources above e^+e^- pair creation threshold. Nucl Inst Methods Phys Res A 729:765–780

Bernard D (2015) Compton polarimetry revisited. Nucl Inst Methods Phys Res A 799:155–158

Bernard D (2018) A 5D, polarised, Bethe-Heitler event generator for $\gamma \rightarrow e^+e^-$ conversion. Nucl Inst Methods Phys Res A 899:85–93

Bernard D, Bruel P, Frotin M, Geerebaert Y, Giebels B, Gros P, Horan D, Louzir M, Poilleux P, Semeniouk I, Wang S, Anvar S, Attié D, Colas P, Delbart A, Sizun P, Götz D (2014) HARPO: a TPC as a gamma-ray telescope and polarimeter. In: Space telescopes and instrumentation 2014: ultraviolet to gamma ray, Proceedings of Society of Photo-Optical Instrumentation Engineers, vol 9144, p 91441M

Bianford R, Chen P, Kamae T, Madejski GM, Ng J, Mizuno T, Tajima H, Thurston T, Barbier L, Bloser PF et al (2003) Large-area balloon-borne polarized gamma ray observer (PoGO). In: Nuclear Science Symposium Conference Record, 2003. IEEE, Piscataway, pp 1708–1713

Bloser PF, Legere JS, Bancroft CM, Ryan JM, McConnell ML (2016a) Balloon flight test of a Compton telescope based on scintillators with silicon photomultiplier readouts. Nucl Inst Methods Phys Res A 812(C):92–103.

Bloser PF, Legere JS, McConnell ML, Macri JR, Bancroft CM, Connor TP, Ryan JM (2009) Calibration of the gamma-ray polarimeter experiment (GRAPE) at a polarized hard X-ray beam. Nucl Inst Methods Phys Res A 600(2):424–433

Bloser PF, McConnell ML, Legere JS, Ertley CD, Hill JE, Kippen RM, Ryan JM (2014) A high-energy Compton polarimeter for the POET SMEX mission. In: Takahashi T, den Herder J-WA, Bautz M (eds) Proceedings of the Society of Photo-Optical Instrumentation Engineers, SPIE, Bellingham, pp 91441I–91441I–6

Bloser PF, Sharma T, Legere JS, Bancroft CM, McConnell ML, Ryan JM, Wright AM (2016b) The advanced scintillator Compton telescope (ASCOT) balloon project. Proc Soc Photo Opt Instrum Eng 9905:99056K–99056K–7

Bloser PF, Sharma T, Legere JS, Bancroft CM, McConnell ML, Ryan JM, Wright AM (2018) The advanced scintillator Compton telescope (ASCOT) balloon payload. In: den Herder J-WA, Nakazawa K, Nikzad S (eds) Proceedings of the Society of Photo-Optical Instrumentation Engineers. The University of New Hampshire, Durham, p 106995X

Boggs SE, Bandstra M, Bowen J, Coburn W, Lin RP, Wunderer C, Zoglauer A, Amman M, Luke P, Jean P, von Ballmoos P (2005) Performance of the nuclear Compton telescope. Exp Astron 20(1):387–394

Boggs SE, Coburn W, Kalemci E (2006) Gamma-ray polarimetry of two X-class solar flares. Astrophys J 638(2):1129–1139

Böttcher M (2019) Progress in multi-wavelength and multi-messenger observations of blazars and theoretical challenges. Galaxies 7:20

Chang H-K, Boggs SE, Chang Y-H (2007) The nuclear Compton telescope (NCT): scientific goals and expected sensitivity. Adv Space Res 40(8):1281–1287

Chauvin M, Florén H-G, Jackson M, Kamae T, Kawano T, Kiss M, Kole M, Mikhalev V, Moretti E, Olofsson G, et al (2015) The design and flight performance of the PoGOLite Pathfinder balloon-borne hard X-ray polarimeter. *Exp Astron* 55–25

Chauvin M, Florén H-G, Jackson M, Kamae T, Kawano T, Kiss M, Kole M, Mikhalev V, Moretti E, Olofsson G, et al (2016) Observation of polarized hard X-ray emission from the Crab by the PoGOLite Pathfinder. Mon Not R Astron Soc Lett 456(1):L84–L88

Chattopadhyay T, Vadawale SV, Aarthy E, Mithun NPS, Chand V, Basak R, Rao AR, Mate S, Sharma V, Bhalerao V, Bhattacharya D (2017) Prompt emission polarimetry of gamma ray bursts with ASTROSAT CZT-Imager. ArXiv e-prints. arXiv:1707.06595

Chauvin M, Florén H-G, Friis M, Jackson M, Kamae T, Kataoka J, Kawano T, Kiss M, Mikhalev V, Mizuno T, Ohashi N, Stana T, Tajima H, Takahashi H, Uchida N, Pearce M (2017) Shedding new light on the Crab with polarized X-rays. Sci Rep 7(1):7816

Chauvin M, Florén H-G, Friis M, Jackson M, Kamae T, Kataoka J, Kawano T, Kiss M, Mikhalev V, Mizuno T, Ohashi, N, Stana T, Tajima H, Takahashi H, Uchida N, Pearce M (2018) Accretion geometry of the black-hole binary Cygnus X-1 from X-ray polarimetry. Nat Astron 44:49–27

Chiu JL, Boggs SE, Kierans CA, Lowell A, Sleator C, Tomsick JA, Zoglauer A, Amman M, Chang H-K, Chu CY, Tseng CH, Yang CY, Lin CH, Jean P, von Ballmoos P (2017) The Compton Spectrometer and Imager (COSI). In: 35th International Cosmic Ray Conference, 10–20 July, vol 35, p 796

Coburn W, Boggs SE (2003) Polarization of the prompt γ-ray emission from the γ-ray burst of 6 December 2002. Nature 423:415–417

Dean AJ, Clark DJ, Stephen JB, McBride VA, Bassani L, Bazzano A, Bird AJ, Hill AB, Shaw SE, Ubertini P (2008) Polarized gamma-ray emission from the crab. Science 321:1183

de Angelis A, Tatischeff V, Tavani M, Oberlack U, Grenier I, Hanlon L, Walter R, Argan A, von Ballmoos P, Bulgarelli A, et al (2017) The e-ASTROGAM mission. Exploring the extreme Universe with gamma rays in the MeV - GeV range. Exp Astron 44:25–82

de Angelis A, Tatischeff V, Grenier IA, McEnery J, Mallamaci M, Tavani M, Oberlack U, Hanlon L, Walter R, Argan A et al (2018) Science with e-ASTROGAM. A space mission for MeV-GeV gamma-ray astrophysics. J High Energy Astrop 19:1–106

Depaola GO, Iparraguirre ML (2009) Angular distribution for the electron recoil in pair production by linearly polarized γ- rays on electrons. Nucl Inst Methods Phys Res A 611:84–92

Duncan N (2017) Gamma-ray observations of solar flares with RHESSI imaging spectroscopy and the GRIPS instrument. In: ProQuest Dissertations and Theses; Thesis (Ph.D.)–University of California

Duncan N, Shih A, Hurford GJ, Saint-Hilaire P, Zoglauer A, Bain H, Amman M, Boggs SE, Lin RP (2013) Detector and imaging systems for the gamma-ray imager/polarimeter for solar flares (GRIPS) instrument. In: Fineschi S, Fennelly J (eds) Proceedings of Society of Photo-Optical Instrumentation Engineers. Space Sciences Lab., Univ. of California, Berkeley, p 88620W

Duncan N, Saint-Hilaire P, Shih AY, Hurford GJ, Bain HM, Amman M, Mochizuki BA, Hoberman J, Olson J, Maruca BA, et al (2016) First flight of the Gamma-Ray Imager/Polarimeter for Solar flares (GRIPS) instrument. In: den Herder J-WA, Takahashi T, Bautz M (eds) Proceedings of Society of Photo-Optical Instrumentation Engineers. SPIE, Bellingham, pp 99052Q–18

Evans R (1955) The atomic nucleus. McGraw-Hill, New York

Forot M, Laurent P, Grenier IA, Gouiffès C, Lebrun F (2008) Polarization of the Crab Pulsar and Nebula as Observed by the INTEGRAL/IBIS Telescope. Astrophys J 688:L29

Gambini R, Pullin J (1999) Nonstandard optics from quantum space-time. Phys Rev D 59(12):124021

Giomi M, Bühler R, Sgrò C, Longo F, Atwood WB (2017) Estimate of the Fermi large area telescope sensitivity to gamma-ray polarization. In: 6th International Symposium on High Energy Gamma-Ray Astronomy, American Institute of Physics Conference Series, vol 1792, p 070022

Götz D, Laurent P, Lebrun F, Daigne F, Bošnjak Ž (2009) Variable polarization measured in the prompt emission of GRB 041219A using IBIS on board INTEGRAL. Astrophys J 695:L208–L212

Götz D, Covino S, Hascoët R, Fernandez-Soto A, Daigne F, Mochkovitch R, Esposito P (2011) A detailed spectral study of GRB 041219A and its host galaxy. Mon Not R Astron Soc 413:2173–2183

Götz D, Covino S, Fernández-Soto A, Laurent P, Bošnjak Ž (2013) The polarized gamma-ray burst GRB 061122. Mon Not R Astron Soc 431:3550–3556

Götz D, Laurent P, Antier S, Covino S, D'Avanzo P, D'Elia V, Melandri A (2014) GRB 140206A: the most distant polarized gamma-ray burst. Mon Not R Astron Soc 444:2776–2782

Gouiffes C, Laurent P, Shearer A, O'Connor E, Moran P (2016) New hard X-rays and optical polarimetric observations of the Crab nebula and pulsar. In: Proceedings of the 11th INTEGRAL Conference Gamma-Ray Astrophysics in Multi-Wavelength Perspective. 10–14 October 2016 Amsterdam, The Netherlands (INTEGRAL2016), p 38

Gros P, Bernard D (2017) γ-Ray polarimetry with conversions to e^+e^- pairs: polarization asymmetry and the way to measure it. Astropart Phys 88:30–37

Gros P, Amano S, Attié D, Baron P, Baudin D, Bernard D, Bruel P, Calvet D, Colas P, Daté S et al (2018) Performance measurement of HARPO: a time projection chamber as a gamma-ray telescope and polarimeter. Astropart Phys 97:10–18

Gunji S, Sakurai H, Tokanai F, Kishimoto Y, Kanno M, Ishikawa Y, Hayashida K, Anabuki N, Tsunemi H, Mihara T, et al (2007) Observation of Crab Nebula with hard x-ray polarimeter: PHENEX. In: Proceedings of Society of Photo-Optical Instrumentation Engineers, vol 6686, p 37

Gunji S, Kishimoto Y, Sakurai H, Tokanai F, Kanno M, Ishikawa Y, Hayashida K, Anabuke N, Tsunemi H, Mihara T, et al (2008) The PHENEX experiment result. In: Proceedings of "Polarimetry days in Rome: Crab status", p 5

Harding AK, Kalapotharakos C (2017) Multiwavelength polarization of rotation-powered pulsars. Astrophys J 840:73

Heitler W (1954) Quantum theory of radiation. Oxford University Press, Oxford

Hill JE, Barthelmy SD, Black JK, Deines-Jones P, Jahoda K, Sakamoto T, Kaaret PE, McConnell ML, Bloser PF, Macri JR, Legere JS, Ryan JM, Smith BRJ, Zhang B (2007) A burst chasing x-ray polarimeter. In: Proceedings of Society of Photo-Optical Instrumentation Engineers, vol 6686, p 29

Hitomi Collaboration, Aharonian F, Akamatsu H, Akimoto F, Allen SW, Angelini L, Audard M, Awaki H, Axelsson M, Bamba A, Bautz MW, et al (2018) Detection of polarized gamma-ray emission from the Crab nebula with the Hitomi Soft Gamma-ray Detector. Publ Astron Soc Jpn 70(6): 113

Huang W-C, Ng K-W (2018) Polarized gamma rays from dark matter annihilations. Phys Lett B 783:29–35

Hunter SD, Bloser PF, Depaola GO, Dion MP, DeNolfo GA, Hanu A, Iparraguirre M, Legere J, Longo F, McConnell ML, Nowicki SF, Ryan JM, Son S, Stecker FW (2014) A pair production telescope for medium-energy gamma-ray polarimetry. Astropart Phys 59:18–28

Jourdain E, Roques JP, Chauvin M, Clark DJ (2012) Separation of two contributions to the high energy emission of Cygnus X-1: polarization measurements with INTEGRAL SPI. Astrophys J 761:27

Kalemci E, Boggs SE, Kouveliotou C, Finger M, Baring MG (2007) Search for polarization from the prompt Gamma-Ray Emission of GRB 041219a with SPI on INTEGRAL. Astrophys J 169:75–82

Kanbach G, Andritschke R, Zoglauer A, Ajello M, McConnell ML, Macri JR, Ryan JM, Bloser P, Hunter S, Di Cocco G, Kurfess J, Reglero V (2005) Development and calibration of the tracking Compton/Pair telescope MEGA. Nucl Inst Methods Phys Res A 541:310–322

Kierans CA, Boggs SE, Lowell A, Tomsick J, Zoglauer A, Amman M, Chiu JL, Chang HK, Lin CH, Jean P et al (2014) Calibration of the Compton Spectrometer and Imager in preparation for the 2014 balloon campaign. Proc Soc Photo Opt Instrum Eng 9144:91443M–91443M–16

Kierans C, Boggs SE, Chiu JL, Lowell A, Sleator C, Tomsick JA, Zoglauer A, Amman M, Chang H-K, Tseng CH, Yang CY, Lin, CH, Jean P, von Ballmoos P (2016) The 2016 super pressure balloon flight of the Compton spectrometer and imager. In: Proceedings of the 11th INTEGRAL Conference Gamma-Ray Astrophysics in Multi-Wavelength Perspective. 10–14 October 2016 Amsterdam, p 75

Kishimoto Y, Gunji S, Ishigaki Y, Kanno M, Murayama H, Ito C, Tokanai F, Suzuki K, Sakurai H, Mihara T, et al (2006) Basic performance of PHENEX: a polarimeter for high ENErgy X rays. In Nuclear Science Symposium Conference Record, 2006. IEEE, Piscataway, pp 421–425

Kishimoto Y, Gunji S, Ishigaki Y, Kanno M, Murayama H, Ito C, Tokanai F, Suzuki K, Sakurai H, Mihara T, et al (2007) Basic performance of PHENEX: a polarimeter for high ENErgy X rays. IEEE Trans Nucl Sci 54:561–566

Kishimoto Y, Gunji S, Ishikawa Y, Takada M, Toukairin N, Tanaka Y, Tokanai F, Sakurai H, Mihara T, Sato T, et al (2008) Expected performance of the PHENEX polarimeter by enlargement of detection area. In: Nuclear Science Symposium Conference Record, 2008. NSS '08. IEEE, Piscataway, pp 722–726

Kishimoto Y, Gunji S, Ishikawa Y, Takada M, Toukairin N, Tanaka Y, Tokanai F, Sakurai H, Mihara T, Sato T, et al (2009) Observation of polarization in hard X-ray region with PHENEX polarimeter. In Astrophysics with All-Sky X-Ray Observations, p 380

Kishimoto Y, Gunji S, Ishikawa Y, Takada M, Mihara T, Hayashida K, Anabuki N, Saito Y, Kohama M, Suzuki M (2011) Observation of polarization for the Crab Nebula with PHENEX polarimeter. Trans Jpn Soc Artif Intell 8

Kiss M, Pearce M (2007) PoGOLite: opening a new window on the universe with polarized gamma-rays. Nucl Inst Methods Phys Res A 580:876–879

Kiss M, Larsson S, Arimoto M, Axelsson M, Bettolo CM, Bogaert G, Florén H-G, Fukazawa Y, Gunji S, Hjalmarsdotter L, et al (2008) The PoGOLite balloon-borne soft gamma-ray polarimeter. Cool Discs 1054:225–232

Kole M, Li ZH, Produit N, Tymieniecka T, Zhang J, Zwolinska A, Bao TW, Bernasconi T, Cadoux F, Feng MZ, et al (2017) Instrument performance and simulation verification of the POLAR detector. Nucl Inst Methods Phys Res A 872:28–40

Kontar EP, Brown JC, Emslie AG, Hajdas W, Holman GD, Hurford GJ, Kašparová J, Mallik PCV, Massone AM, McConnell ML, Piana M, Prato M, Schmahl EJ, Suarez-Garcia E (2011) Deducing electron properties from hard X-ray observations. Space Sci Rev 159(1):301–355

Kotov ID (1988) Methods of measurement of gamma-ray polarization. Space Sci Rev 49:185–195

Kroeger RA, Johnson WN, Kinzer RL, Kurfess JD, Inderhees SE, Phlips BF, Graham B (1996) Gamma-ray instrument for polarimetry, spectroscopy, and imaging (GIPSI). In: Ramsey BD, Parnell TA, (eds) Gamma-Ray and Cosmic-Ray Detectors, Techniques, and Missions, Proceedings of Society of Photo-Optical Instrumentation Engineers, vol 2806, pp 52–63

Laurent P, Götz D, Binétruy P, Covino S, Fernandez-Soto A (2011a) Constraints on Lorentz invariance violation using integral/IBIS observations of GRB041219A. Phys Rev D 83(12):121301

Laurent P, Rodriguez J, Wilms J, Cadolle Bel M, Pottschmidt K, Grinberg V (2011b) Polarized gamma-ray emission from the galactic black hole Cygnus X-1. Science 332:438

Lei F, Dean AJ, Hills GL (1997) Compton polarimetry in gamma-ray astronomy. Space Sci Rev 82:309–388

Lowell AW, Boggs SE, Chiu CL, Kierans CA, Sleator C, Tomsick JA, Zoglauer AC, Chang H-K, Tseng C-H, Yang C-Y, Jean P, von Ballmoos P, Lin C-H, Amman M (2017) Maximum likelihood Compton polarimetry with the Compton spectrometer and imager. Astrophys J 848:120

Lowell AW, Boggs SE, Chiu CL, Kierans CA, Sleator C, Tomsick JA, Zoglauer AC, Chang H-K, Tseng CH, Yang CY, Jean P, von Ballmoos P, Lin CH, Amman M (2017) Polarimetric analysis of the long duration gamma-ray burst GRB 160530A with the balloon borne Compton spectrometer and imager. Astrophys J 848(2):119

Mattox JR, Mayer-Hasselwander HA, Strong AW (1990) Analysis of the COS B data for evidence of linear polarization of VELA pulsar gamma rays. Astrophys J 363:270–273

McConnell ML, Macri JR, McClish MA, Ryan JM, Forrest DJ, Vestrand WT (1999) Development of a hard X-ray polarimeter for astrophysics. IEEE Trans Nucl Sci 46(4):890–896

McConnell ML, Ryan JM, Smith DM, Lin RP, Emslie AG (2002a) RHESSI as a hard X-ray polarimeter. Sol Phys 210:125–142

McConnell ML, Zdziarski AA, Bennett K, Bloemen H, Collmar W, Hermsen W, Kuiper L, Paciesas W, Phlips BF, Poutanen J, Ryan JM, Schönfelder V, Steinle H, Strong AW (2002b) The soft gamma-ray spectral variability of Cygnus X-1. Astrophys J 572:984–995

McConnell ML, Ledoux JR, Macri JR, Ryan JM (2004) Dedicated polarimeter design for hard x-ray and soft gamma-ray astronomy. Proc Soc Photo Opt Instrum Eng 5165:334–345

McConnell ML, Bloser PF, Legere JS, Macri JR, Narita T, Ryan JM (2005) A hard X-ray polarimeter designed for transient astrophysical sources. Nucl Sci Symp Conf Rec 1:462–466
McConnell ML, Angelini L, Baring MG, Barthelmy SD, Black JK, Bloser PF, Dennis BR, Emslie AG, Greiner J, Hajdas W, et al (2009a) GRB polarimetry with POET. AIP Conf Proc 1133(1):64–66
McConnell ML, Bancroft CM, Bloser PF, Connor TP, Legere JS, Ryan JM (2009b) GRAPE: a balloon-borne gamma-ray polarimeter. Proc Soc Photo Opt Instrum Eng 7435:13
McConnell ML, Bloser PF, Connor TP, Ertley C, Legere JS, Ryan JM, Wasti SK (2013) Plans for the next GRAPE balloon flight. In: Siegmund OH (ed) Proceedings of the Society of Photo-Optical Instrumentation Engineers. SPIE, Bellingham, pp 885909–885909–10
McConnell ML, Baring MG, Bloser PF, Dwyer JF, Emslie AG, Ertley CD, Greiner J, Harding AK, Hartmann DH, Hill JE, Kaaret PE, Kippen RM, Mattingly D, McBreen S, Pearce M, Produit N, Ryan JM, Ryde F, Sakamoto T, Toma K, Vestrand WT, Zhang B (2014a) POET: a SMEX mission for gamma ray burst polarimetry. Proc Soc Photo Opt Instrum Eng 9144:91440O–91440O–8
McConnell ML, Bloser PF, Ertley C, Legere JS, Ryan JM, Wasti SK (2014b) Current status of the GRAPE balloon program. Proc Soc Photo Opt Instrum Eng 9144:91443P–91443P–10
McConnell ML, Bloser PF, Legere JS, Ryan JM (2016) The development of a low energy Compton imager for GRB polarization studies. Proc Soc Photo Opt Instrum Eng 9905:99052O–99052O–9
McConnell ML, Bloser PF, Legere JS, Ryan JM, Hanlon LO, McBreen S, Uliyanov A (2018) The continued development of a low energy Compton imager for GRB polarization studies. In: den Herder J-WA, Nakazawa K, Nikzad S, (eds) Proceedings of the Society of Photo-Optical Instrumentation Engineers. International Society for Optics and Photonics, Bellingham, p 106992O
McGlynn S, Clark DJ, Dean AJ, Hanlon L, McBreen S, Willis DR, McBreen B, Bird AJ, Foley S (2007) Polarisation studies of the prompt gamma-ray emission from GRB 041219a using the spectrometer aboard INTEGRAL. Astron Astrophys 466:895–904
McGlynn S, Foley S, McBreen B, Hanlon L, McBreen S, Clark DJ, Dean AJ, Martin-Carrillo A, O'Connor R (2009) High energy emission and polarisation limits for the INTEGRAL burst GRB 061122. Astron Astrophys 499:465–472
Moiseev A, Amego Team (2017) All-sky medium energy gamma-ray observatory (AMEGO). Int Cosmic Ray Conf 35:798
Moran P, Kyne G, Gouiffès C, Laurent P, Hallinan G, Redfern RM, Shearer A (2016) A recent change in the optical and γ-ray polarization of the Crab nebula and pulsar. Mon Not R Astron Soc 456:2974–2981
Orosz JA, McClintock JE, Aufdenberg JP, Remillard RA, Reid MJ, Narayan R, Gou L (2011) The mass of the black hole in Cygnus X-1. Astrophys J 742:84
Orsi S, Haas D, Hajdas W, Honkimäki V, Lamanna G, Lechanoine-Leluc C, Marcinkowski R, Pohl M, Produit N, Rapin D, Suarez-Garcia E, Rybka D, Vialle J-P (2011) Response of the Compton polarimeter POLAR to polarized hard X-rays. Nucl Instrum Methods Phys Res Sect A 648:139–154
Ozaki K, Takahashi S, Aoki S, Kamada K, Kaneyama T, Nakagawa R, Rokujo H (2016) Demonstration of polarization sensitivity of emulsion-based pair conversion telescope for cosmic gamma-ray polarimetry. Nucl Instrum Methods Phys Res Sect A 833:165–168
Produit N, Barao F, Deluit S, Hajdas W, Leluc C, Pohl M, Rapin D, Vialle J-P, Walter R, Wigger C (2005) POLAR, a compact detector for gamma-ray bursts photon polarization measurements. Nucl Instrum Methods Phys Res Sect A 550:616–625
Rodriguez J, Grinberg V, Laurent P, Cadolle Bel M, Pottschmidt K, Pooley G, Bodaghee A, Wilms J, Gouiffès C (2015) Spectral state dependence of the 0.4-2 MeV polarized emission in Cygnus X-1 seen with INTEGRAL/IBIS, and links with the AMI radio data. Astrophys J 807:17
Rokujo H, Aoki S, Hamada K, Hara T, Inoue T, Ishiguro K, Iyono A, Kawahara H, Kodama K, Komatani R, et al (2018) First demonstration of gamma-ray imaging using a balloon-borne emulsion telescope. Prog Theor Exp Phys 2018(6):063H01

Rutledge RE, Fox DB (2004) Re-analysis of polarization in the γ-ray flux of GRB 021206. Mon Not R Astron Soc 350:1288–1300

Schaerf C (2005) Polarized gamma-ray beams. Phys Today 58(8):44–52

Semeniouk I, Bernard D (2018) C++ implementation of Bethe–Heitler, 5D, polarized, $\gamma \rightarrow e^+e^-$ pair conversion event generator. Nucl Instrum Methods Phys Res A. https://doi.org/10.1016/j.nima.2018.09.154

Shih AY, Lin RP, Hurford GJ, Duncan NA, Saint-Hilaire P, Bain HM, Boggs SE, Zoglauer AC, Smith DM, Tajima H, Amman MS, Takahashi T (2012) The gamma-ray imager/polarimeter for solar flares (GRIPS). In: Takahashi T, Murray SS, den Herder J-WA (eds) Proceedings of Society of Photo-Optical Instrumentation Engineers, SPIE, Bellingham, p 84434H–16

Sleator C, Boggs SE, Chiu J-L, Kierans C, Lowell A, Tomsick JA, Zoglauer A, Amman M, Chang H-K, Tseng C-H, Yang C-Y, Lin C-H, Jean P, vonBallmoos P (2017) Benchmarking COSI's detector effects engine. In: Proceedings of 11th INTEGRAL Conference Gamma-Ray Astrophysics in Multi-Wavelength Perspective — PoS(INTEGRAL2016). SISSA Medialab, Trieste, p 087

Stirling AM, Spencer RE, de la Force CJ, Garrett MA, Fender RP, Ogley RN (2001) A relativistic jet from Cygnus X-1 in the low/hard X-ray state. Mon Not R Astron Soc 327:1273–1278

Suarez-Garcia E, Hajdas W, Wigger C, Arzner K, Güdel M, Zehnder A, Grigis P (2006) X-Ray polarization of solar flares measured with RHESSI. Sol Phys 239:149–172

Tajima H, Blandford R, Enoto T, Fukazawa Y, Gilmore K, Kamae T, Kataoka J, Kawaharada M, Kokubun M, Laurent P, et al (2010) Soft gamma-ray detector for the ASTRO-H Mission. In: Space telescopes and instrumentation 2010: ultraviolet to gamma ray, Proceedings of Society of Photo-Optical Instrumentation Engineers, vol 7732, p 773216

Tajima H, Watanabe S, Fukazawa Y, Blandford R, Enoto T, Goldwurm A, Hagino K, Hayashi K, Ichinohe Y, Kataoka J, et al (2018) Design and performance of Soft Gamma-ray Detector onboard the Hitomi (ASTRO-H) satellite. J Astron Telesc Instrum Syst 4(2):021411

Takahashi S, Aoki S, Kamada K, Mizutani S, Nakagawa R, Ozaki K, Rokujo H (2015) GRAINE project: the first balloon-borne, emulsion gamma-ray telescope experiment. Prog Theor Exp Phys 2015(4):043H01

Takahashi S, Aoki S, Hamada K, Hara T, Inoue T, Ishiguro K, Iyono A, Kawahara H, Kodama K, Komatani R, et al (2016) GRAINE 2015, a balloon-borne emulsion γ-ray telescope experiment in Australia. Prog Theor Exp Phys 2016(7):073F01

Takahashi S, Aoki S, Collaboration G (2018) GRAINE project, prospects for scientific balloon-borne experiments. Adv Space Res 62(1):2945–2953

Takeda S, Odaka H, Katsuta J, Ishikawa S-n, Sugimoto S-i, Koseki Y, Watanabe S, Sato G, Kokubun M, Takahashi T, Nakazawa K, Fukazawa Y, Tajima H, Toyokawa H (2010) Polarimetric performance of Si/CdTe semiconductor Compton camera. Nucl Instrum Methods Phys Res Sect A 622:619–627

Tanaka T, Arimoto M, Axelsson M, Bjornsson C-I, Bogaert G, Carlson P, Cooney M, Craig W, Engdegård O, Fukazawa Y, et al (2007) Data acquisition system for the PoGOLite astronomical hard X-ray polarimeter. In: Nuclear Science Symposium Conference Record, 2007. NSS '07. IEEE, Piscataway, pp 445–449

Tatischeff V, Tavani M, von Ballmoos P, Hanlon L, Oberlack U, Aboudan A, Argan A, Bernard D, Brogna A, Bulgarelli A, et al (2016) The e-ASTROGAM gamma-ray space mission. In: Space telescopes and instrumentation 2016: ultraviolet to gamma ray, Proceedings of Society of Photo-Optical Instrumentation Engineers, vol 9905, p 99052N

Tatischeff V, De Angelis A, Gouiffès C, Hanlon L, Laurent P, Madejski GM, Tavani M, Ulyanov A (2018) e-ASTROGAM mission: a major step forward for gamma-ray polarimetry. J Astron Telesc Instrum Syst 4(1):011003

Toma K, Sakamoto T, Zhang B, Hill JE, McConnell ML, Bloser PF, Yamazaki R, Ioka K, Nakamura T (2009) Statistical properties of gamma-ray burst polarization. Astrophys J 698:1042–1053

Vadawale SV, Chattopadhyay T, Rao AR, Bhattacharya D, Bhalerao VB, Vagshette N, Pawar P, Sreekumar S (2015) Hard X-ray polarimetry with Astrosat-CZTI. Astron Astrophys 578:A73

Vadawale SV, Chattopadhyay T, Mithun NPS, Rao AR, Bhattacharya D, Vibhute A, Bhalerao VB, Dewangan GC, Misra R, Paul B, Basu A, Joshi BC, Sreekumar S, Samuel E, Priya P, Vinod P, Seetha S (2018) Phase-resolved X-ray polarimetry of the Crab pulsar with the AstroSat CZT Imager. Nat Astron 2:50–55

Wigger C, Hajdas W, Arzner K, Güdel M, Zehnder A (2004) Gamma-ray burst polarization: limits from RHESSI measurements. Astrophys J 613:1088–1100

Wojtsekhowski B, Tedeschi D, Vlahovic B (2003) A pair polarimeter for linearly polarized high-energy photons. Nucl Instrum Methods Phys Res Sect A 515:605–613

Yang CY, Lowell A, Zoglauer A, Tomsick J, Chiu JL, Kierans C, Sleator C, Boggs SE, Chang H-K, Jean P, et al (2018) The polarimetric performance of the Compton spectrometer and imager (COSI). In: den Herder J-WA, Nakazawa K, Nikzad S (eds) Proceedings of the Society of Photo-Optical Instrumentation Engineers, National Tsing Hua Univ. (Taiwan), SPIE, p 106992K

Yonetoku D, Murakami T, Gunji S, Mihara T, Toma K, Sakashita T, Morihara Y, Takahashi T, Toukairin N, Fujimoto H, Kodama Y, Kubo S, IKAROS Demonstration Team (2011) Detection of gamma-ray polarization in prompt emission of GRB 100826A. Astrophys J 743:L30

Zhang H, Diltz C, Böttcher M (2016) Radiation and polarization signatures of the 3D multizone time-dependent hadronic blazar model. Astrophys J 829:69

Zhang S-N, Kole M, Bao T-W, Batsch T, Bernasconi T, Cadoux F, Chai J-Y, Dai Z-G, Dong Y-W, Gauvin N, Hajdas W, Lan M-X, Li H-C, Li L, Li Z-H, Liu J-T, Liu X, Marcinkowski R, Produit N, Orsi S, Pohl M, Rybka D, Shi H-L, Song L-M, Sun J-C, Szabelski J, Tymieniecka T, Wang R-J, Wang Y-H, Wen X, Wu B-B, Wu X, Wu X-F, Xiao H-L, Xiong S-L, Zhang L-Y, Zhang L, Zhang X-F, Zhang Y-J, Zwolinska A (2019) Detailed polarization measurements of the prompt emission of five gamma-ray bursts. Nat Astron 3:258–264

Zoglauer A, Andritschke R, Schopper F (2006) MEGAlib the medium energy gamma-ray astronomy library. New Astron Rev 50:629–632

Chapter 6
Polarimetric Observations of the Sun

David Orozco Suárez

Abstract Probing the magnetism of the solar atmosphere requires measuring the polarisation state of the light in different wavelength bands. Since the physical conditions of the solar plasma change dramatically from the optically thick photosphere which is the main contributor to the emitted radiation in visible wavelengths to the optically thin outer solar atmosphere, several polarisation mechanisms contribute to the continuum and spectral line polarisation throughout the solar atmosphere. This chapter presents an overview of the main sources of light polarisation in the Sun along with some recent investigations of the polarisation produced in the solar atmosphere.

6.1 Polarimetry in Solar Physics

The study of magnetic fields is a topic of fundamental importance in many research areas, especially in the field of astrophysics. These magnetic fields are typically measured with polarimetry. In this context, and due to its accessibility, the Sun is the best laboratory for plasma physics studies—many solar phenomena having connections to plasma processes can be captured with solar telescopes and examined in detail. Early observations of magnetic fields in the Sun played a pioneering role in astrophysical polarimetry. Since the discovery of strong magnetic fields in sunspots by Hale et al. (1908)[1] the solar community has contributed to the development of sophisticated instrumentation for obtaining information about the polarisation state of the light. It also mastered calibration techniques needed to achieve the

[1]He employed a rudimentary but efficient polarisation analyser to separate the different polarisation states, pretty much the same strategy we follow nowadays.

D. O. Suárez (✉)
Instituto de Astrofísica de Andalucía, Glorieta de la Astronomía s/n, Granada, Spain
e-mail: orozco@iaa.es

R. Mignani et al. (eds.), *Astronomical Polarisation from the Infrared to Gamma Rays*, Astrophysics and Space Science Library 460,
https://doi.org/10.1007/978-3-030-19715-5_6

required polarisation sensitivities. Although the history of solar polarimetry is brief compared to that of spectroscopy[2] (Stenflo 2017; del Toro Iniesta 1996), these two observational techniques combined (i.e. spectropolarimetry) open the possibility of inferring the vector magnetic fields in astrophysical plasmas. Currently, magnetic fields are being routinely measured in the solar atmosphere using ground-based and space-borne telescopes, through the interpretation of polarised light. Most of the observations are taken in visible and near-infrared (below 1.5 μm) spectral lines.

The Sun activity and dynamics is driven by magnetic fields. Its atmosphere has been traditionally divided into four layers: the photosphere (at the bottom, where the photon mean free path is small), the chromosphere, the transition region, and the corona (the outermost layer). In standard uni-dimensional solar models, the gas density drops exponentially with height while the temperature drops in the photosphere until the temperature minimum region. Beyond that point it rises again, slowly through the chromosphere and much faster in the transition region until the low corona, where it increases more gently. Various physical phenomena related to magnetic fields take place in the photosphere. To name a few, we can mention the formation of tiny kilo-Gauss magnetic flux tubes, the emergence of magnetic loops through the solar surface, and the formation of active regions. The kilo-Gauss flux tubes are considered to be the building blocks of the solar surface magnetism. The ascent of loops of various sizes is a clear manifestation of interaction between the magnetic features with the non-magnetic plasma. They are seen to evolve by advection of the external convective motions. Active regions are responsible for the sudden and violent release of energy in flares and coronal mass ejections from the atmosphere to the heliosphere.

The surface of the Sun harbour ubiquitous magnetic fields: there is no region without magnetic fields covering all spatial and temporal scales. For that reason, solar polarisation signals have a wide range of variation. Photospheric lines formed in active regions display very strong polarisation signals. In turn, scattering processes in quieter areas produce much weaker signals, of the order of ten thousandths of the intensity of the continuum. Most of these signals are produced at very small spatial scales and usually are short-lived. This sets conditions on the required spectral, spatial, and temporal resolutions for spectropolarimetric observations. In spite of being counterintuitive to most of the astrophysical community, this fact leads to the so-called *photon starvation* problem in solar physics (Stenflo 2003). Current spectropolarimetric observations cannot comply with the spectral, spatial, and temporal resolution and the polarimetric sensitivity requirements due to the limited photon budget at the focal plane of the instruments. Hence, trade-offs between these observing parameters must be assumed. A further difficulty is that polarimetry is a differential technique. It requires a very good knowledge of systematic errors and precise instrumental calibrations to reach the required polarisation accuracy. This might be the reason why polarimetry has not been yet fully exploited in other fields of astrophysics.

[2]Already in 1802, William Wollaston detected solar Fraunhofer lines.

This chapter provides an overview about the generation of polarised radiation in the Sun and gives illustrative examples of the polarisation signals solar physicists are obtaining and interpreting nowadays.

6.2 Generation of Polarised Radiation in the Sun

There are several physical processes that can generate polarised radiation in the solar atmosphere. The two most relevant ones are the *Zeeman effect* (the energy splitting of atomic sub-levels in the presence of magnetic fields) and *scattering processes* (due to anisotropic illumination). The latter produce an excess or defect of population in certain atomic sub-levels, which leads to polarised light in spectral lines. The scattering polarisation is modified by magnetic fields through the *Hanle effect*. These mechanisms are responsible for most of the polarisation signals that have been measured in the solar spectrum, from the ultraviolet (Kano et al. 2017) to the visible (Stenflo and Keller 1997) and infrared (Trujillo Bueno et al. 2002b). A formal theoretical description of the above processes can be found in Landi Degl'Innocenti and Landolfi (2004) and del Toro Iniesta (2003). There are other sources of polarisation in the solar atmosphere, particularly in the corona:

- Thermal bremsstrahlung emission, responsible of the polarisation signals that can be measured in the millimetre and sub-millimetre spectral ranges.
- Thermal gyroresonance emission (cyclotron radiation) produces polarisation in the radio and microwave spectral ranges.
- Compton scattering and bremsstrahlung which are responsible for the production of linearly polarised X-rays during energy releases in solar flares.
- Thomson scattering, occurring in the K-corona.

The most convenient description of the radiation field is through the Stokes parameters. The Stokes vector $S^T = (I, Q, U, V)$, where ^{T} stands for matrix transposition, fully accounts for all polarisation states of light. Stokes I represents the intensity of the radiation field. Stokes Q is the difference between two orthogonal linearly polarised states, namely vertical and horizontal in the plane perpendicular to the direction of propagation. Stokes U is the intensity difference between linear polarisation at $+45°$ and $-45°$ in the same perpendicular plane. Finally, Stokes V is the intensity difference between two orthogonal (right-handed and left-handed) circularly polarised states. In solar polarimetry, it is customary to normalise the Stokes parameters by the value of the local continuum intensity or by $I(\lambda)$ to obtain the fractional polarisation at each wavelength. For a formal introduction to the Stokes formalism see, e.g., Landi Degl'Innocenti (1992) and del Toro Iniesta (2003).

6.2.1 Zeeman Effect

It is the most familiar polarisation mechanism in the solar atmosphere, due to the perturbation of atomic energy levels by an external magnetic field. Under the presence of a magnetic field, each atomic level of total angular momentum J splits into $2J + 1$ sub-levels characterised by their magnetic quantum numbers M. The amount of splitting is proportional to the corresponding Landé factors. For a specific line transition between angular momenta J and J', there will be as many spectral lines as transitions allowed by the appropriate selection rule. For the case of dipole transitions, those with $\Delta M_J = M - M' = 0$ correspond to the so-called π components and those with $\Delta M_J = \pm 1$ are the σ_b and σ_r components, respectively. The spectral line polarisation is then produced by the wavelength shifts between the σ and π components:

$$\Delta\lambda_B = 4.67 \times 10^{-13} g_{\text{eff}} \lambda^2 B\ , \tag{6.1}$$

where λ represents the wavelength in mÅ of the spectral line under consideration, B is the magnetic field strength in Gauss, and g_{eff} is the effective Landé factor. This equation shows that the Zeeman splitting of a given spectral line is proportional to $g_{\text{eff}}\lambda^2 B$. Therefore, the Zeeman effect in spectral lines can be detected more easily when the magnetic field is strong (see Fig. 6.4), for large values of g_{eff}, or at longer wavelengths. In solar plasmas, however, the sensitivity of a spectral line to the Zeeman effect depends also on its *Doppler width*, which depends on the plasma kinetic temperature, the microturbulent velocity,[3] and linearly on the wavelength. The sensitivity of a spectral line to the Zeeman effect is usually given by the ratio between the Zeeman splitting and the Doppler width. Typical spectral lines are the Fe I 630 nm pair of lines with $g_{\text{eff}} = 1.67$ and 2.5 currently used by the spectropolarimeter onboard the *Hinode* space-borne solar observatory (Kosugi et al. 2007; Tsuneta et al. 2008) or the Fe I 617.3 nm line with $g_{\text{eff}} = 2.5$ observed by the Helioseismic and Magnetic Imager (HMI) instrument onboard the *Solar Dynamic Observatory* (*SDO*) (Schou et al. 2012; Pesnell et al. 2012) and the Polarimetric and Helioseismic Imager of the *Solar Orbiter* mission (Solanki et al. 2015; Müller et al. 2013). Spectral lines in the infrared are a good choice because the splitting increases quadratically with wavelength although at the price of losing spatial resolution (e.g. the Fe I pair of lines at 1.565 μm with $g_{\text{eff}} = 2.5$ and 3). There are many other candidates. These particular ones have a clean local continuum which allows a better modelling.

When the magnetic field is aligned with the observer's line of sight (LoS), the only non-zero Stokes parameters are I and V. Likewise, when the magnetic field is perpendicular to the LoS, we only detect Stokes I, Q, and U signals. In the absence of velocity and magnetic field gradients along the line forming region the shape of

[3] Microturbulence accounts for plasma motions taking place at spatial scales smaller than the photon mean free path.

Stokes V is antisymmetric, with a well-defined zero-crossing point and two lobes of opposite sign, while the Stokes Q and U profiles are symmetric. The shape of the profiles can be easily guessed by looking at the expression of the emissivity for Stokes V and Q:

$$\epsilon_V \approx (\phi_r - \phi_b) \cos\theta, \tag{6.2}$$

$$\epsilon_Q \approx \left(\phi_\pi - \frac{\phi_r + \phi_b}{2}\right) \sin\theta^2 \cos 2\chi, \tag{6.3}$$

where ϕ_r, ϕ_b, and ϕ_π are the emissivity of the red and blue σ components, and the π components, respectively. The angle θ represents the inclination of the field with respect to the LoS and χ represents the angle of the transverse component of the magnetic field vector (contained in the plane perpendicular to the LoS) to the chosen Stokes $Q > 0$ direction. The $\phi_{r,b,\pi}$ profiles are given by a Voigt function. An important property of the Zeeman effect to bear in mind is that the Stokes V signal has a linear dependence with the magnetic field strength while this dependence is non-linear and of second order for the Stokes Q and U signals. As a consequence, in solar observations, Stokes V signals are systematically much more prominent over the noise than linear signals. This has a particular impact when trying to infer the magnetic field vector in many solar physical scenarios, as we will see later in this chapter.

6.2.2 *Scattering Polarisation and the Hanle Effect*

The mere detection of Zeeman signals implies the existence of magnetic fields. However, in the solar atmosphere, it is not always possible to record polarisation signals. The polarisation signals generated by fields with strengths below, e.g., 100 G often show very small amplitudes, on the order of 10^{-3} and smaller. Modern solar polarimeters are photon noise limited and hardly reach 10^{-4} in polarimetric sensitivity such that the weakest polarisation signals are buried in the noise. This situation worsens when individual magnetic fields are not resolved by the instrument, that is, when the structure that generates polarisation signals occupies only a fraction of the resolution element while the rest samples a nonmagnetic plasma. In this situation, photon noise becomes more important because the observed polarisation signals are proportional to their filling factor, hence smaller. The fact that in the solar atmosphere magnetic structures with different strength and orientations can coexist within the same resolution element worsen the situation: Zeeman signals with opposite orientations may cancel out. Researchers refer to those undetected signals as the *hidden* quiet Sun magnetic fields (Bellot Rubio and Orozco Suárez 2019).

Fortunately, other mechanisms for generating polarisation signals exist: *scattering polarisation* (Stenflo 1982; Trujillo Bueno 2001). Over the years, there has been a strong theoretical effort to understand the physics behind scattering polarisation and the Hanle effect so observers have now meaningful diagnostics for interpreting their measurements (see, e.g., the review by Trujillo Bueno 2010). To understand how these two mechanisms work, one should consider first the concept of *atomic level polarisation*. It occurs when certain atomic sub-levels "absorb" light more strongly than others. In this situation, it is easy to see that polarisation can be generated even if there is no wavelength shifts between the σ and π components of the spectral line transition—hence without the need of a magnetic field. Assume, for instance, a specific line transition between angular momenta $J_{\text{low}} = 1$ and $J_{\text{up}} = 0$. If we illuminate the atom with unidirectional and unpolarised light (parallel to the quantisation axis of total angular momentum), the atom must absorb ± 1 units of angular momentum from the light and hence no transitions can occur from the magnetic $m = 0$ sub-level of the ground state. The excited atom now has a probability of 1/3 of falling back to any of the ground-state sub-levels (neglecting other relaxation mechanisms) and therefore at some point, all atoms will be pumped into the $m = 0$ sub-level of the lower level. This mechanism is known as *depopulation pumping* (Happer 1972; Trujillo Bueno et al. 1997): a redistribution of atomic sub-levels populations through the absorption of light plus spontaneous emission. Depopulation pumping produces lower level polarisation and the emergent spectral line radiation is linearly polarised because of the ensuing selective absorption processes (Trujillo Bueno et al. 1997, 2002b; Trujillo Bueno 2001, 2014). Repopulation is the spontaneous decay of polarised excited levels. In multilevel atoms atomic level polarisation occurs as a combination of such two mechanisms.

The generation of linear polarisation due to lower-level polarisation can be clearly understood by looking at the general expressions of the absorption coefficient for Stokes Q for a magnetic field pointing to the observer:

$$\eta_Q \approx \left(\rho_{(0,0)} \phi_\pi - \frac{\rho_{(-1,-1)} \phi_r + \rho_{(1,1)} \phi_b}{2} \right) \sin^2 \theta, \tag{6.4}$$

where θ is the angle between the quantisation axis and the observer and $\rho_{(M,M)}$ represents the population of the lower-level sub-level with magnetic quantum number M. One can see that the absorption is non-zero even if the σ and π components are not split because of differences in the $\rho_{(M,M)}$ populations. They would cancel out in the absence of an external magnetic field if all of the magnetic substates are equally populated. It should be noted that the Stokes Q profile resembles the Stokes I shape because the $\phi_{r,b,\pi}$ are not split in wavelength but weighted differently.

In the Sun, the main driver for atomic level polarisation is the *anisotropic illumination* of the atoms. As discussed before, when atoms are illuminated anisotropically, atomic level polarisation can be generated whenever the depolarisation rates from elastic collisions are much lower.

Scattering processes due to anisotropic unpolarised illumination[4] produce a rich linear polarisation spectrum at the solar limb, which was not accessible until recently due to the very low signals. The "second solar spectrum", as it was called by Stenflo et al. (1983a,b, 2000), Stenflo and Keller (1997) and Gandorfer (2000, 2002) shows an overwhelming number of features in linear polarisation. It opened new research windows in solar physics and motivated the development of rigorous theories for the generation and transfer of polarisation in spectral lines. An example is, for instance, the enigmatic Na I D1 589.6 nm line whose linear polarisation signals have recently been understood (Belluzzi and Trujillo Bueno 2013; Belluzzi et al. 2015) after many previous theoretical attempts (Landi Degl'Innocenti 1998; Trujillo Bueno et al. 2002a). Note that, for symmetry reasons (Trujillo Bueno 2001), linear scattering polarisation signals are zero at disc centre. In 1-dimensional models of the solar atmosphere the scattering polarisation signals increase monotonically towards the limb. This is why researchers usually record their data at the limb, where the scattering signals are expected to be maximum.

Scattering polarisation combined with the Hanle effect represents a useful diagnostic in solar physics (Trujillo Bueno 2001). Discovered by Hanle in 1924 (Hanle 1924), it is the modification of the atomic level polarisation due to the presence of a magnetic field inclined with respect to the symmetry axis of the pumping radiation field. In particular, the presence of a magnetic field induces a depolarisation and a rotation of the direction of the linear polarisation, i.e., it reduces the Stokes Q amplitude signal and generates Stokes U. The theory of scattering polarisation and the Hanle effect can be found in Landi Degl'Innocenti and Landolfi (2004). The Hanle effect operates below the critical strengths $B_{\rm H}$ typically between mili-Gauss to tens of Gauss, depending on the Landé factor g and the natural width $t_{\rm life}$ of the particular line transition (e.g. Trujillo Bueno 2014):

$$B_{\rm H} = \frac{1.137 \times 10^{-17}}{t_{\rm life}\, g}. \tag{6.5}$$

The joint action of atom level polarisation and the Hanle effect is useful for determining the vector magnetic field in weakly magnetised solar (spicules, prominences, the quiet Sun, and the solar corona) and stellar plasmas. It is also of special interest for determining the field strength of unresolved magnetic fields, i.e., when the induced Zeeman signals are too small to be detected because of small filling factors or when they are partially cancelled out (Stenflo 1982; Trujillo Bueno et al. 2004). Scattering polarisation signals and the Hanle effect do not suffer from such cancellations and are, therefore, an important tool for inferring weak and/or tangled magnetic fields that are not accessible through the Zeeman effect.

An example of Stokes profiles showing the joint action of the Zeeman and atomic level polarisation and the Hanle effect is shown in Fig. 6.1. The signals correspond to the fine structure of the He I infrared triplet at 1083.0 nm taken in a

[4]The anisotropy is caused by the solar limb-darkening and local inhomogeneities.

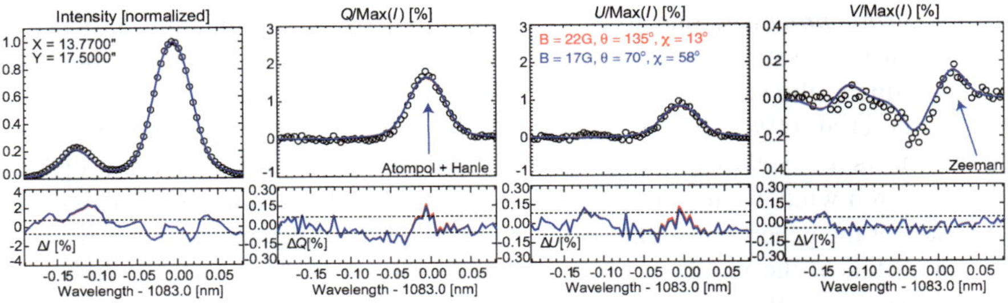

Fig. 6.1 Example of He I 1083.0 nm Stokes I, Q, U, and V profiles observed with the Tenerife Infrared Polarimeter instrument (Collados et al. 2007) installed at the 70 cm Vacuum Tower Solar Telescope (Izaña Observatory, Tenerife, Spain). The data were recorded in 90° scattering geometry with the spectrograph crossing a solar prominence located at the limb. The circular and linear signals are normalised to Stokes I maximum peak value. Open dots represent the observations while solid lines show the fitting result using a theoretical model. Differences between the observation and the model are shown in the bottom panels. Blue and red colour codes represent two different Van Vleck solutions. Image reproduced with permission from Orozco Suárez et al. (2014c), copyright by AAS

solar prominence located at the limb. There is a weak blue component at 1082.9 nm ($^3S_1 - ^3P_0$) separated about 0.12 nm from the two stronger blended components at about 1083.03 nm ($^3S_1 - ^3P_1$ and $^3S_1 - ^3P_2$). The Stokes Q signal shows the typical signature of scattering polarisation, i.e., a Stokes profile with the same shape as Stokes I. In this case, at 90° scattering, the signals are due to *selective emission* processes resulting from the atomic polarisation of the $^3P_1$ and $^3P_2$ upper levels. It should be noted that there is no Stokes Q signal around the blue component. Since the upper level $^3P_0$ has $J = 0$ it cannot be polarised, the only way to see polarisation in the blue component is through the selective absorption that results from the lower level $^3S_1$ polarisation (Trujillo Bueno et al. 2002b). This happens as a result of anisotropic illumination. The presence of Stokes U signals indicates the action of the Hanle effect caused by the presence of a magnetic field inclined with respect to the symmetry axis of the pumping radiation and having a well-defined orientation. Finally, Stokes V shows the typical shape of the Zeeman effect.[5] Hanle effect-based diagnostics in a spectral line allow to infer the strength and orientation of the field vector although with ambiguities and always assuming that the magnetic field vector has a fix orientation within the resolution element and there is no mixed polarities. The non-linear relationship between the magnetic field and the generated linear polarisation signals gives rise to the so-called 90° ambiguity of the Hanle effect. It is associated with the Van Vleck angles (54.74° and 125.26°) and the fact that Stokes Q and U can take the same values for different field orientations (Asensio Ramos et al. 2008; Casini et al. 2005, 2009; Merenda et al. 2006). The

[5]Notice that for strong fields the He I infrared triplet in the incomplete Paschen-Back regime (Socas-Navarro et al. 2004).

Hanle effect is also affected by the 180° azimuth ambiguity, which is similar to that of the Zeeman effect. In this case, the linear polarisation signals do not change when the magnetic field vector is rotated by 180° around the LoS. Since the Hanle effect shows up prominently in Stokes Q and U but not in Stokes V, it is important to detect Stokes V signals originated from the Zeeman effect as well to constrain the longitudinal component of the magnetic field vector.

6.2.3 Thermal Bremsstrahlung

Thermal bremsstrahlung consists in the emission of a photon by an electron due to its interaction with charged particles in the solar plasma. Radiation emitted at these frequencies, (sub-)millimetre wavebands, comes from plasmas in local thermodynamic conditions (Wedemeyer et al. 2016). The degree of circular polarisation in the millimetre spectral range depends directly on the free–free[6] emission brightness (thermal bremsstrahlung), on the observing frequency, and on the field strength and its orientation with respect to the LoS. Hence, bremsstrahlung radiation provides direct information on the magnetic field strength because the free–free opacity depends on the local magnetic field strength. This is remarkable because it is then relatively easy to obtain information about the plasma thermal and magnetic conditions from the observations. In the solar case, most radiation observed in the (sub-)millimetre band comes from the solar chromosphere, which makes observations in this spectral range very appealing for the scientific community. Thermal conditions and field strengths at different height can be inferred by combining different spectral band-passes (Bogod and Gelfreikh 1980). Moderate resolution images can be taken at, e.g., the RATAN-600 (Bogod 2011) or the Nobeyama Radio Heliograph (Nakajima et al. 1994) at short cm wavelengths. The most promising facility is the Atacama Large Millimetre/sub-millimetre Array (ALMA) Observatory, in operation since 2011 which can perform solar observations at different frequencies in this band. Moreover, it is able to measure the Stokes profiles (Loukitcheva et al. 2017). In principle, most of the solar structures and physical phenomena can be observed in polarised light by ALMA. It is not the first observatory able to observe polarisation in the millimetre range (Grebinskij et al. 2000) but promises to provide polarimetric observations at an unprecedented high spatial resolution.

6.2.4 Gyroresonance Emission in Radio Observations

Gyroresonance emission or cyclotron radiation originates from the acceleration of electrons as they spiral around magnetic field lines (White 2001, 2005; Brosius

[6]Mainly due to electron–ion and H^- free–free absorption.

et al. 2002; Brosius and White 2006). The circular polarisation signal associated with gyroresonance is linearly proportional to the magnetic field strength itself and shows up at radio frequencies. Hence, radio observations provide a direct way to diagnose the magnetic field strength in the solar corona. It can be possible to infer field orientations as well. Gyroresonance emission can also be detected in the microwave spectral range (White and Kundu 1997; Grebinskij et al. 2000). Gyroresonance emission in radio is usually employed to infer fields in coronal areas above active regions (Gary and Hurford 1994). In other areas Stokes V signals are typically around 0.1% and 10%. Bremsstrahlung emission also contributes to the microwave polarisation.

6.2.5 Compton Scattering and Bremsstrahlung in the X-ray Range

The generation and interpretation of X-ray linear polarisation signals due to Compton scattering and bremsstrahlung radiation in the solar atmosphere is rather more complex than those produced in the millimetre spectral range. They are mostly generated when the velocity of the electrons changes continuously in the solar plasma as a result of their spiral trajectory along the magnetic field lines (Emslie and Brown 1980). The amount of polarisation depends on the anisotropy of the radiation field, i.e., the electron beam characteristics, the observer viewing angle, and the magnetic field. Hence, these measurements are usually prone to large errors in the determination of the magnetic field strength and direction. Indeed, there are only few instruments capable of measuring polarisation in the range from 20 keV to 1 MeV, despite X-rays dominate the spectra of solar flares. Flares have been investigated through polarisation measurements in the X-ray spectral range using the *Reuven Ramaty High Energy Spectroscopic Imager* (*RHESSI*) satellite (e.g. Suarez-Garcia et al. 2006). X-ray polarisation can also be used to observe plasma heating processes in the keV spectral range (soft X-rays).

6.2.6 Thomson Scattering: Polarisation in the K-corona

The theory of Thomson scattering applied to solar physics is described in Quémerais and Lamy (2002). It basically consists in the scattering of radiation by electrons in the solar corona. It takes place at relatively long distances from the Sun, below 5 solar radii, in the so-called K-corona. The observations of while light polarisation in the corona can be interpreted by means of inversion methods (van de Hulst 1950). They are still applied to the determination of electron densities in coronal streamers (e.g. Thernisien et al. 2006).

6.3 Polarimetric Observations in the Photosphere

In the solar photosphere, the ratio of the gas pressure to the magnetic pressure β is larger than unity. Hence, gas flows govern the structuring and dynamics of magnetic fields there. The strongest fields can be clearly identified on the solar surface as dark structures surrounded by the brighter solar granulation. These structures, namely sunspots and pores forming active regions, follow the 11-year solar cycle. The rest of the surface is apparently devoid of magnetic fields when seen in imaging observations. Polarimetry has shown us that this is not the case. The so-called quiet Sun is permeated by magnetic fields forming two well-differentiated components: the network fields, a pattern of fields located preferentially in the borders of supergranular cells ($\sim$10,000 km) and covering the whole solar surface and the solar internetwork (IN) (Martin 1984; Lites et al. 1993). Network fields are not larger than a 1000 km in size and can be detected relatively easily in polarised light with current low-resolution instrumentation. These structures have field strengths of more than 1 kG. In contrast, IN fields are very weak (below 1 kG) and exhibit tiny polarisation signals.

This section reviews most important polarimetric observations carried out in the solar photosphere, showing several examples of the type of signals that can be obtained with current instrumentation.

6.3.1 Quiet Sun Magnetic Fields

For many years the kilo-Gauss slender magnetic flux tubes conforming the solar network were considered fundamental for understanding the solar surface magnetism. Of spatial sizes well below 100 km, they connect the different solar layers and are certainly essential ingredients of the quiet Sun. However, thanks to investigations based on the Hanle effect in the atomic Sr I line at 460.7 nm and the C_2 molecular lines, it is known that the quieter areas of the solar surface, the IN, contain significant amounts of energy in the form of weak magnetic fields (Trujillo Bueno et al. 2004). Since most Hanle measurements integrate the signal from long periods of time and have low resolution, these measurements cannot be used to describe how the fields are organised at the different spatial scales along the quiet Sun surface (Bellot Rubio and Orozco Suárez 2019). Many remarkable physical phenomena take place in the quiet Sun, such as magnetic cancellation, the emergence of magnetic flux from deeper layers, or magnetic field reconnection. Hence, a proper understanding the quiet solar magnetism needs an in-deep analysis of all these physical phenomena. To measure the weak polarisation signals generated by magnetic fields organised on very small spatial scales very high spatial and temporal resolution and high polarimetric sensitivity observations are required.[7]

[7]A full review of the quiet Sun magnetic fields can be found in Bellot Rubio and Orozco Suárez (2019).

The first Zeeman longitudinal magnetograms[8] of the solar surface were recorded by Babcock (1953). His measurements were the first to detect magnetic flux in the quiet Sun. The low spatial resolution of the observations did not allow him to observe structures in the IN. Even network fields were not clearly detected. The first unambiguous observational finding of network and IN fields dates from the 1970s (Livingston and Harvey 1971, 1975; Smithson 1975). Since then, quiet Sun magnetograms have been taken with better and better solar instrumentation. Nowadays, there are instruments that continuously monitor the surface of the full Sun in polarisation, such as the HMI onboard SDO. Other instruments, such as the GREGOR Infrared Spectrograph (GRIS) (Collados et al. 2007) installed at the 1.5 m GREGOR telescope and the CRisp Imaging Spectro-Polarimeter (CRISP) (Scharmer 2006), measure them at very high resolution (better than 100 km), but for limited periods of time.

Figure 6.2 shows a high spatial resolution ($0.2''$), high signal to noise ratio, and high cadence (80 s) longitudinal magnetogram of a quiet Sun region taken with the Narrowband Filter Imager NFI (Ichimoto et al. 2008) onboard the *Hinode* spacecraft (Kosugi et al. 2007) in the Na I 589.6 nm line. The image samples a rectangular area of 67×90 Mm on the solar surface, covering 1-2 supergranular cells. The data speaks by its own: there is a wealth of magnetic signals of both polarities. The stronger ones outline the location of supergranular borders (the network). The rest show mixed polarities all around.

Hinode observations of Zeeman sensitive lines suggest that about 20% of the quiet Sun magnetic flux resides in the internetwork (Wang et al. 1995; Gošić et al. 2014). Seeing-free observations taken from space have made a huge improvement in our understanding of quiet Sun fields. These measurements are not affected by the Earth's atmosphere and hence reach the diffraction limit of the telescope more easily. Long-time series of magnetograms allow following each of the detected magnetic elements during their lifetime and study how they emerge, evolve, interact with other elements, and disappear. It is well known that magnetic fields appear in the internetwork through sudden appearance of single polarity elements or in the form of magnetic dipoles. The amount of magnetic flux injected through these two mechanisms is rather large, about 4×10^{24} Mx day^{-1} over the whole solar IN, enough to replace the entire network flux in about 1 day (Gošić et al. 2016). This poses a serious problem in solar physics. It is not yet known how the flux is fully removed from the surface. So far, there have been two distinct mechanisms: flux cancellation and the sudden disappearance of flux.

Nowadays, full vector magnetograms are also available, i.e., there is access to Stokes Q and U maps as well. For example, HMI takes vector magnetograph observations every 12 min uninterruptedly. The most sensitive vector magnetograms of quiet Sun regions ever taken, with a resolution of about 80 km on the solar surface, have been made by the Imaging Magnetograph eXperiment (IMaX);

[8]Roughly speaking, a longitudinal magnetogram is a map of the line-of-sight component of the magnetic field taken at the Stokes V wing of a magnetic sensitivity spectral line.

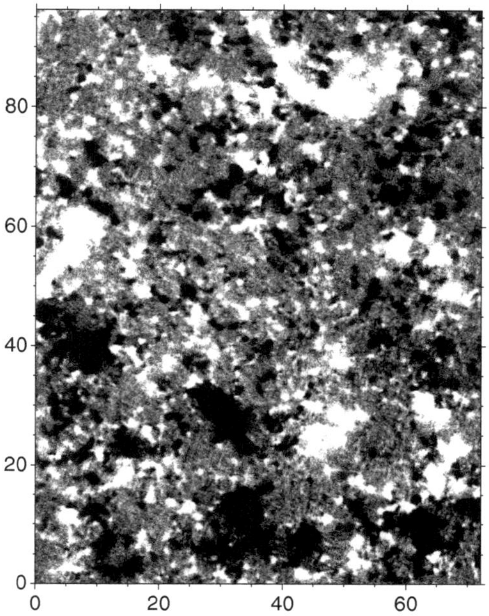

Fig. 6.2 Longitudinal magnetogram taken on 20 January 2010 with the *Hinode* NFI instrument. Data belong to the campaign "Flux replacement in the solar network and internetwork" (*Hinode* Operation Plan 151). Obtained in the two wings of Na I 589.6 nm at ±16 pm from line centre with a cadence of 1 min and of 20 h duration, these are the most sensitive measurement of the quiet Sun ever taken at a resolution of $0''.16$. The total exposure time per image corresponds to 6.4 s. The data have a polarimetric sensitivity of 6 G, attained after a careful calibration to convert the measured circular polarisation signals into longitudinal flux. The dimensions of the field of view are $93'' \times 123''$. The image is clipped at 20 G

(Martínez Pillet et al. 2011) onboard the *SUNRISE* mission (Barthol et al. 2011). IMaX observed the magnetic sensitive Fe I spectral line at 525.02 nm with a cadence of 30 s. As explained before, magnetic elements in the quiet Sun areas are tiny structures. Therefore, the only way to consistently model quiet Sun flux tubes is to use a filling factor. With the *SUNRISE*/IMaX data it has been possible to neglect the contribution of a non-magnetic component, i.e., use filling factors equal one, for the analysis of the vector magnetic field in flux tubes as if they were spatially resolved (Lagg et al. 2010). *SUNRISE*/IMaX has also allowed the direct detection of the magnetic field expansion with height in flux tubes, the study of convective collapse processes in great detail, and the detection of small-scale fields oscillatory behaviour (Martínez González et al. 2011, 2012; Lagg et al. 2010). Moreover, IMaX observations showed that magnetic flux tubes which are seeing as single structures in longitudinal magnetograms consist of multi-cored magnetic structures that share a common identity during their evolution (Requerey et al. 2015).

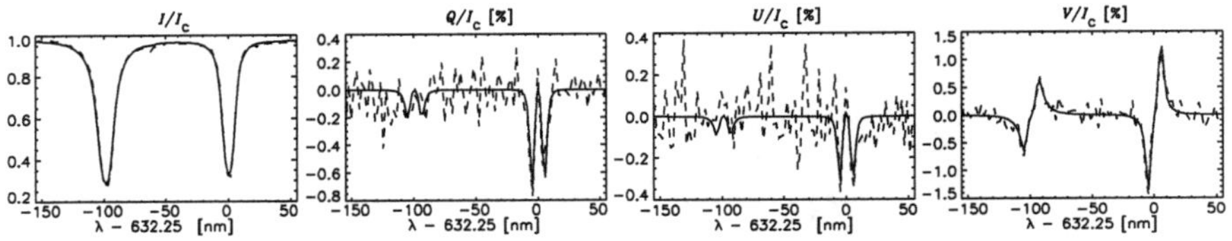

Fig. 6.3 Stokes I, Q U, and V profiles observed with the *Hinode* spectropolarimeter in a quiet Sun internetwork region. The spectral lines belong to the pair of visible Fe I lines at 630 nm. The dashed line represents the observations while the solid ones stand for the best-fit profiles using a simple atmospheric model consisting of a homogeneous magnetic field. The observations can be explained with a magnetic field strength of about 180 gauss inclined 110° with respect to the local vertical. The magnetic field occupies a small fraction of the resolution element (40%) while the rest is field free. Image reproduced with permission from Orozco Suárez et al. (2007a), copyright by AAS

The most fascinating discovery in recent years with high-resolution full vector spectropolarimetry is the detection of ubiquitous linear polarisation signals in the internetwork (Orozco Suárez et al. 2007a,b; Lites et al. 2008; Danilovic et al. 2010). Quiet Sun linear polarisation signals are very weak and often inaccessible to classical magnetographs. Space-based observations as those performed by the Solar Optical Telescope spectropolarimeter (SP) aboard the *Hinode* spacecraft took a qualitative leap forwards and were the first to detect these linear polarisation signals on a regular basis. The *Hinode* SP measures the full Stokes vector of the Fe I 630 nm lines with a resolution of 0.32″. Figure 6.3 shows an example Stokes I, Q U, and V profiles recorded in the solar IN with the *Hinode* SP. The circular polarisation signal reaches 1% but the linear signals, Stokes Q and U, are barely above 0.1%. The analysis of these signals helped settle the controversy between visible and near-IR lines: before *Hinode* existing ground-based spectropolarimetric measurements disagreed about the mean magnetic field strength of internetwork fields. While visible observations pointed towards strong fields (above kilogauss), near-infrared observations suggested weak fields. Thanks to its unprecedented stability and spatial resolution, *Hinode* proved that fields are mostly weak. Furthermore, the observations suggested that there is about five times more magnetic flux contained in the linear polarisation signals (Stokes Q and U) than in the circular ones (Lites et al. 2008). These results were confirmed by, e.g., (Orozco Suárez et al. 2007a), using Stokes inversions[9] which are able to extract the information of the vector magnetic field more accurately.

Researchers have known the existence of linear polarisation signals in quiet Sun regions since the work of Lites and co-workers (Lites et al. 1996). These authors were the first to observe the appearance of small-scale magnetic loops in the quiet Sun, associated with linear signals. They have been the target of many studies using

[9] The reader is referred to del Toro Iniesta and Ruiz Cobo (2016) for detailed review on the inversion of the radiative transfer equation in solar physics.

high-resolution measurements (e.g. Centeno et al. 2007; Ishikawa and Tsuneta 2009; Martínez González et al. 2009; Jin et al. 2009). Nowadays, we know that such structures emerge from subsurface layers in the form of Omega-loops and harbour weak fields. They appear as a patch of linear polarisation followed by two patches of circular polarisation on either side of the linear patch. The circular patches separate from each other with time (Centeno et al. 2007). Through longitudinal flux density measurements around the temperature minimum region taken in the Mg I b 517.3 nm line and intensity narrow-band filtergrams in the Ca II 396.85 nm line, it has been estimated that about 20% of the loops in internetwork regions reach the solar chromosphere (Martínez González et al. 2009). Full vector polarisation measurements in the chromosphere are fundamental to estimate the energy these low lying loops transport into the chromosphere and corona. However, such measurements are still out reach because they require unprecedented polarimetric sensitivity. The situation is different in the solar photosphere. The most sensitive measurements taken with the *Hinode* SP in quiet Sun regions have demonstrated that about 60% of the observed area show linear polarisation signals above 4.5 times the noise level. Interestingly, 100% of the solar surface show circular polarisation signals above the noise (Orozco Suárez and Bellot Rubio 2012; Bellot Rubio and Orozco Suárez 2012).

6.3.2 Active Region Magnetic Fields

Sunspots are the most prominent manifestation of the presence of magnetic fields exceeding 2000 G in the solar surface (Borrero and Ichimoto 2011). Their study is important because they are generated by the solar dynamo operating at the base of the convection zone. The intense fields found in sunspots imprint clear Zeeman signatures in spectral lines.

Figure 6.4 shows an example of a high-spatial resolution observation of a sunspot in the visible Fe I 630.15 nm and 630.25 nm lines obtained with the *Hinode* SP. The effective Landé factors of these spectral lines are 1.66 and 2.5, respectively. The increasing Zeeman splitting of the lines while crossing the umbra of the sunspot is evident. Since the components of the Zeeman splitting are completely resolved, it is relatively easy to measure the strength of the magnetic field by just looking at the peak separation between the two σ components. Note that since the magnetic filling factor is close to one in sunspots, the observed spectral lines are almost 100% polarised. Interestingly, these data recorded the strongest fields ever detected in sunspots through direct measurement of the Zeeman splitting (Okamoto and Sakurai 2018). In particular, fields exceeding 6 kG were recorded. This is not the first time fields as strong as those reported in Okamoto and Sakurai (2018) are found. Some authors (van Noort et al. 2013) claim to have detected fields of up to 7 kG in sunspots penumbrae but this result is still being discussed within the community. It should be noted that the field strength can almost be measured by eye in Fig. 6.4. Nevertheless, the interpretation of polarisation signals in sunspots is

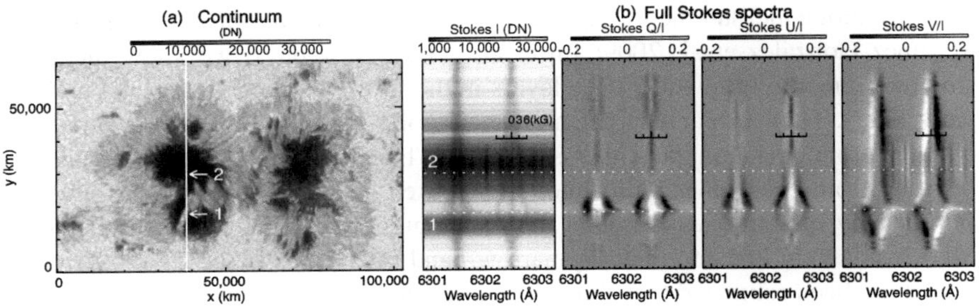

Fig. 6.4 Typical sunspot observations taken with the *Hinode* spectropolarimeter in the Fe I 630 nm pair of lines which carry information from photospheric layers. Left panel (**a**) shows the continuum intensity. The right panels (**b**) show the Stokes spectra recorded along the vertical line depicted in the left panel. The x-axis stands for the wavelength dispersion direction while y-axis represents the slit. The image spatial resolution is about $0.32''$ and the signal-to-noise ratio about 1000 with 4.8 s integration time per slit position. Image reproduced with permission from Okamoto and Sakurai (2018), copyright by AAS

not straightforward when the observations are taken at very high spatial resolution. For instance, the inhomogeneities in the penumbra magnetic field and the flow field produce a rich variety of polarisation signals, usually made up of several components. Their analysis and interpretation require a delicate radiative transfer treatment using several magnetic components along the LoS.

Determining the vector magnetic field in the chromosphere above sunspots is not straightforward. The reason is twofold: firstly, chromospheric lines are far more complicated to model since they are generated in non-local thermodynamic equilibrium; secondly, the polarisation signals in the chromosphere are much weaker and usually strongly affected by plasma flows. The latter makes the interpretation of the signals more difficult because plasma flows not only shift the spectral lines but may also result in emission/absorption features near the chromospheric core. As a result, there are fewer works dealing with the vector magnetic field configuration of sunspots in the chromosphere in comparison with those on the photosphere (Orozco Suárez et al. 2005; Socas-Navarro 2005; Schad et al. 2015; Joshi et al. 2017; Libbrecht et al. 2019).

6.4 Polarimetry Observations in the Chromosphere

The solar chromosphere is probably the most complex layer of the Sun. Chromospheric plasma structures are very dynamic and often associated with optically thin plasmas in non-local thermodynamic equilibrium. They harbour weak magnetic fields which are almost undetectable with current solar instrumentation. Moreover, since the ratio between the plasma pressure to the magnetic pressure is small, fields dominate the dynamics of the chromospheric plasma. The magnetic field is, therefore, the key ingredient for understanding the solar chromosphere. Here I review the most important polarimetric observations of different chromospheric structures: spicules, prominences, and flares.

6.4.1 Solar Spicules

Solar spicules are thin and very dynamic, jet-like chromospheric structures that can be clearly seen at the limb with diameters around 100–2500 km (Beckers 1968; de Pontieu et al. 2007; Pereira et al. 2012). Their rapid evolution and short lifetimes (between 2 and 6 min) make the characterisation of their magnetic properties extremely difficult because of the long integration times needed for recording full Stokes measurements in suitable spectral lines and the required spatial resolution (better than 1 arcsecond). Most properties are not well constrained yet since with present solar instrumentation we can only observe time-averaged bundles of spicules. Polarimetric observations of spicules are, indeed, scarce. The first polarisation measurements date from the twenty-first century. Most measurements have been carried out using line transitions of the He I multiplet line except in Socas-Navarro and Elmore (2005), where the He I 1083.0 nm triplet and the Ca II infrared lines at 849.8 and 854.2 nm were observed simultaneously with the SPINOR[10] instrument.

The first observations were limited to single slit positions and long integrations times in the He I 1083.0 nm triplet (Trujillo Bueno et al. 2005). Those observations showed the joint action of the Zeeman and atomic polarisation and the Hanle effect in this spectral line and were used to constrain the field strength and field orientations in quiet Sun spicules. Observations were improved later (López Ariste and Casini 2005; Ramelli et al. 2006; Centeno et al. 2010). The very first observations able to grasp some spatial information on their properties were carried out by Centeno et al. (2010) and Orozco Suárez et al. (2015). The measurements of Orozco Suárez et al. (2015), depicted in Fig. 6.5, allowed to infer the height dependence of the magnetic field vector in solar spicules. However, the most interesting results have to do with the physics of the polarisation itself, i.e., there are signs of *selective absorption mechanism* at the base of the spicules. As explained in Sect. 6.2, linear polarisation in the blue component can only be produced by *selective absorption* of the lower level $^3S_1$ (Trujillo Bueno et al. 2002b). In the observations of Fig. 6.5, a tiny negative signal around the blue component of Stokes Q and a weak absorption in Stokes I in the red component suggest that the opacity is large enough to make the *selective absorption* process work.

[10] Spectro-Polarimeter for INfrared and Optical Regions, at the Dunn Solar Telescope in Sacramento Peak Observatory.

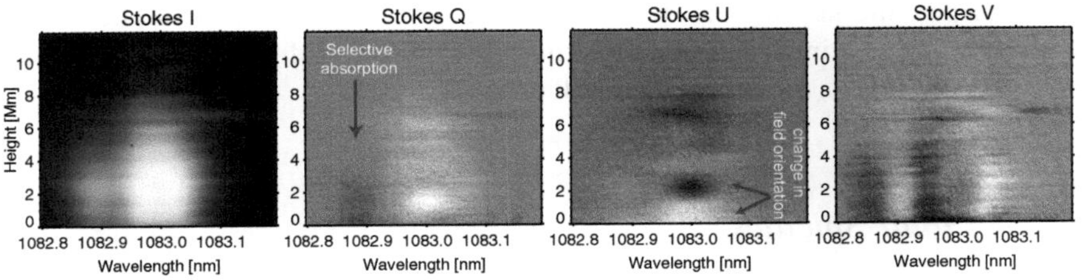

Fig. 6.5 Stokes I, Q, U, and V spectral maps taken with the spectrograph of the Vacuum Solar Telescope of the Observatorio de Izaña (Tenerife) and corresponding to solar spicules. The images have a total integration time of ten seconds and a spatial resolution of $0.7''$. The x-axis represents wavelength and the y-axis the height over the solar limb. The Stokes I panel shows the fine structure of the He I 1083.0 nm triplet (in emission). Notice that in Stokes Q and U, only the red component can be clearly seen above the noise level. The Stokes U sign change represents a change in field orientation with height. A weak blue component signal, due to selective absorption stands above the noise in Stokes Q. Image reproduced with permission from Orozco Suárez et al. (2015), copyright by AAS

6.4.2 *Prominence and Filament Magnetic Fields*

Above the solar chromosphere, within the hot coronal environment, prominences and filaments[11] are highly dynamic structures made of cool neutral plasma ($\sim 5-10$ kK) which is suspended in magnetic dips: concave upwards magnetic field lines that support plasma material against gravity.[12] These fields are thought to be very weak. Although the study of waves, small and large amplitude oscillations and the Kelvin-Helmholtz or/and Rayleigh-Taylor Instabilities taking place in prominences are valid techniques to characterise them, e.g., Arregui et al. (2012), Luna and Karpen (2012), Hillier (2018), most measurements of magnetic fields in solar prominences are based on the interpretation of scattering polarisation signals and the Hanle effect.

Early attempts used longitudinal magnetometry in the H_α, H_β or the He I D3 587.6 nm spectral lines (Zirin and Severney 1961; Rust 1967; Harvey 1969; Tandberg-Hanssen and Anzer 1970). Linear polarisation in prominences was first reported by Hyder in the H_α line in 1965 (Hyder 1965). His measurements (depicted in Fig. 6.6, right panel) did not have enough spatial and spectral resolution. The first full-Stokes spectral profile measurements in the He I D3 line were obtained in 1982 (House and Smartt 1982). These measurements already pointed to the existence of weak fields in prominences (of order a few gauss) which were oriented horizontally with respect to the solar surface. In the modern era, only few spectral lines that present good polarimetric diagnostics of the magnetic field are of interest: Ca II

[11]Filaments are prominences as seen against the brighter solar disc.

[12]It is generally accepted that prominence plasma is supported against gravity by collisions and magnetic tension force. For comprehensive reviews go to, e.g., Parenti (2014).

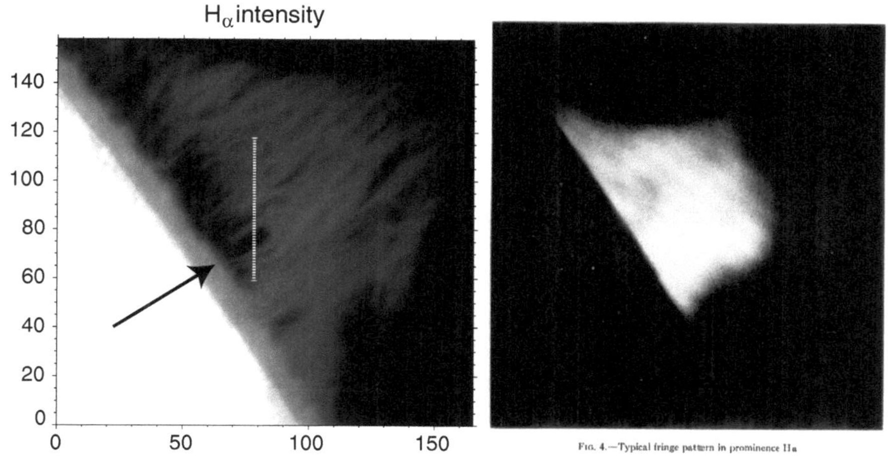

Fig. 6.6 Image of a quiescent solar prominence observed in the 1960s in the H_α spectral line taken from the coronagraph at the Climax observing station of the High Altitude Observatory (right panel). The left panel depicts a similar quiescent prominence but taken in the modern era in the same spectral line with the slit-jaw camera of the Tenerife Infrared Polarimeter (TIP-II) attached to the spectrograph of the German Vacuum Tower Telescope (Tenerife, Spain). The spatial resolution is about $1''$. The vertical small-scale structuring typical of quiescent prominences can be clearly seen at high spatial resolution. The arrow pinpoints the location of a bubble rising through the main body of the prominence. The dotted line represents the spectrograph slit. Images reproduced with permission from Hyder (1965) and Orozco Suárez et al. (2014a), copyright by AAS

854.2 nm, Mg II b lines, Na I D, and the already mentioned He I D3 and He I 1083.0 nm lines. Nowadays, we know that solar prominences harbour magnetic fields in the order of 10–100 G and most of the time they are oriented rather horizontally with respect to the solar surface (e.g. Merenda et al. 2006; Casini et al. 2003).

The challenge is to measure small-scale magnetic fields in solar prominences. However, high spatial resolution measurements of solar prominences are scarce because of the difficulties of carrying out off-limb observations in ground-based telescopes even with the aid of adaptive optics. Recent high spatial resolution observations taken in the He I 1083.0 nm multiplet show little correlation between the small-scale structuring seeing in prominences (see Fig. 6.6) and the inferred vector magnetic field (Orozco Suárez et al. 2014c,a). The reason is the long integrations times needed to record the full Stokes spectra with good signal-to-noise ratios, typically more than 30 s, and/or lack of spatial resolution of the observations, around 1–$1.5''$. Despite these difficulties, it has been possible to measure polarisation signals in fundamental physical processes occurring in prominences. For instance, in Orozco Suárez et al. (2014b) the authors reported the first attempt to determine Doppler shifts and magnetic fields in a prominence Rayleigh Taylor instability (Hillier 2018). No field strength variations during the passage of the instability were found, probably because the detected signals were too weak and inconspicuous above the noise level. Lately, remarkable spectropolarimetric observations of so-called solar tornadoes have allowed reconstructing the three-dimensional configuration of these mysterious coronal structures (Orozco Suárez et al. 2012; Martínez González et al. 2015, 2016).

6.4.3 Polarimetry of Solar Flares

Flares are the most energetic events that can be observed in the solar atmosphere and probably in other stars during their lifetime. They are related to the violent reorganisation of magnetic field lines: *magnetic field reconnection* (Zweibel and Yamada 2009). They occur both in complex active regions and around solar filaments. The latter are responsible for most of the so-called coronal mass ejections. What happens during solar flares is an open question in solar physics. Progress in understanding them requires direct magnetic field measurements in the chromosphere and corona, where most of the reconnection takes place. However, although changes on the vector magnetic field configuration in the photosphere and even magnetic flux fluctuation have been reported (e.g. Wang and Liu 2010), chromospheric magnetic fields do not seen to change much (e.g. Judge et al. 2014). This is so because of the lack of full spectropolarimetric measurements in chromospheric lines during flares. First direct observation of magnetic field changes in the chromosphere driven by reconnection was reported by Kleint (2017) using the Ca II 854.2 nm spectral line. This author observed an X-class flare and studied the behaviour of photospheric and chromospheric fields during the flare event.

Polarisation in solar flares has also been measured in the He I 1083.0 nm triplet lines. Spectropolarimetric observations in this spectral line showed strong emission and variations in the polarisation signals (Penn and Kuhn 1995). These observations led to the very first determination of magnetic fields in a flare kernel. Other observations have followed (Sasso et al. 2011). The first clear indication of a magnetic reconnection scenario in He I 1083.0 nm measurements should be found in the shape of the intensity profiles: Stokes I becomes in emission during the flare and show the superposition of different components associated with plasma moving at different velocities (Li et al. 2007; Judge et al. 2014, 2015; Kuckein et al. 2015). An example of these polarimetric observations can be found in Fig. 6.7. The He I 1083.0 nm triplet line shows a strong emission in intensity and clear circular polarisation signals. Linear signals were barely above the noise level. The results indicate that the plasma associated with the flare was moving downwards. This is not surprising since the formation of the He I 1083.0 nm multiplet is triggered by coronal EUV irradiation.

In the absence of EUV radiation the triplet levels of the He I 1083 nm line are barely populated. He I is ionised under the presence of EUV radiation allowing the He I triplet levels to be populated through recombination cascade, which preferentially increases the population in the metastable level ($2s3S^1$) of the transition (e.g. Centeno et al. 2008). Hence, the emission comes probably from

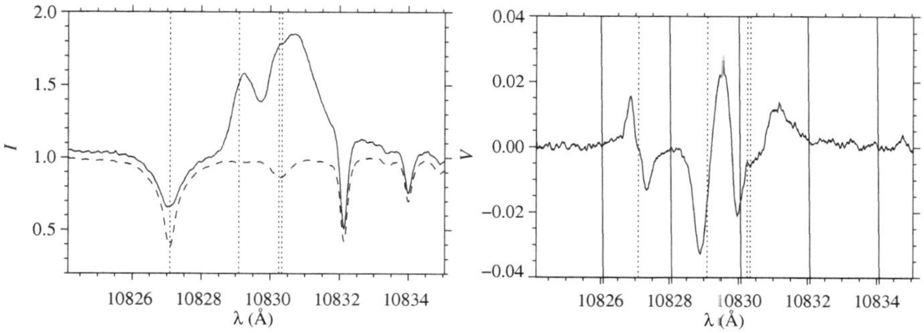

Fig. 6.7 Stokes I and V profiles corresponding to the chromospheric He I 1083.0 nm triplet lines (central wavelengths) and to the photospheric Si I 1082.7 nm line (left spectral line) observed with the Tenerife Infrared Polarimeter (TIP-II) installed at the German Vacuum Tower Telescope (Tenerife, Spain). The profiles were recorded very close to the onset of the flare. The two lines on the right are atmospheric telluric lines. The solid line corresponds to the flare profile while the dashed line represents the average quiet sun profile for comparison purposes. The continuum enhancement in Stokes I is due to the flare. Note that Stokes V is positive for the photospheric line (the blue lobe is positive) indicating that the fields point towards the observer while the chromospheric line triplet shows the opposite behaviour although the field still pointing to the observer. The Stokes V reversal is due to the line going into emission. Image reproduced with permission from Kuckein et al. (2015), copyright by AAS

plasma evaporation in the chromosphere: the measured signals are produced by a violent recombination cascade after and/or during the impulsive phase of the solar flare.

6.5 Polarimetry in the Transition Region and Corona

The plasma emitting radiation from the chromosphere upwards is most of the time optically thin and in non-local thermodynamic equilibrium. Moreover, its structuring and dynamics is dominated by magnetic fields because the plasma $\beta << 1$. It is possible to measure the magnetic field properties in these layers using spectral lines in the ultraviolet and near-infrared spectral bands. For instance, the dipole transition (permitted) Lyman-α lines of hydrogen (121.5 nm) and He I at 30.4 nm and the Mg II h and k lines at $\sim$ 280 nm carry information from the upper chromosphere and transition regions (Trujillo Bueno et al. 2017). Likewise, the forbidden Fe XVI 530.0 nm and Fe XIII 1074.7 nm lines originate in the hot solar corona (see, e.g., Judge 1998). The polarisation signal arising from these lines is fully dominated by atomic level polarisation and the Hanle and Zeeman effects. Unfortunately, the observation of these transitions is not easy. For coronal lines, the main difficulties are the sky brightness and the light scattered within the instrument which contaminates the off-limb polarimetric measurements. In the case of the UV lines the reason is obvious, these line transitions are not accessible from the ground.

Despite these limitations, it has been possible to measure polarisation in particular forbidden coronal spectral lines (House 1977; Casini and Judge 1999; Lin and Casini 2000; Li et al. 2017). Polarimetric observations of the corona concentrate in the Fe XVI 530.0 nm and Fe XIII 1074.7 nm transitions (Mickey 1973; Arnaud and Newkirk 1987; Habbal et al. 2001). The response of the linear polarisation signals in these forbidden lines is rather different from that in other scattering processes in permitted transitions (e.g. the He I 1083 nm multiplet) because the natural line width of forbidden lines is much smaller that the Zeeman splitting, even for very weak magnetic fields. This leads to rather different behaviour of the polarisation signals since the sub-levels in coronal lines barely interfere between each other as discussed in Lin and Casini (2000). Moreover, their linear polarisation signal is not sensitive to the field strength because they are in the Hanle saturation regime for very small field strengths. To infer the full vector from coronal line measurements it is necessary to observe Stokes V as well. Typical Stokes V signals are below 10^{-4} Lin et al. (2000). For this reason Stokes V measurements in coronal lines are scarce (e.g. Harvey 1969; Lin et al. 2004). The first observations of circular polarisation in the coronal Fe XIII 1074.7 nm line were reported by Lin and Casini (2000). In Tomczyk et al. (2007), outstanding coronal observations ($\sim$4.5$''$ resolution) of the four Stokes vector in the Fe XIII 1074.7 nm line have been reported using the Coronal Multi-Channel Polarimeter (CoMP) instrument at NSO's Sacramento Peak Observatory in New Mexico (Tomczyk et al. 2008).

In contrast to the forbidden coronal lines, the Hanle effect in the transition region lines is sensitive to stronger fields. For instance, the sensitivity goes from 10 to 100 G in the case of the Lyman-α line of hydrogen at 121.5 nm (Trujillo Bueno 2014). The papers (Belluzzi and Trujillo Bueno 2012; Belluzzi et al. 2012) show how to interpret the scattering linear polarisation of this Lyman-α line. The discovery of the linear polarisation produced by scattering processes in this line is due to the *Chromospheric Lyman-Alpha Spectro-Polarimeter* (*CLASP*) rocket experiment (Kobayashi et al. 2012; Nakurage et al. 2015, 2017). The observations (Kano et al. 2017), depicted in Fig. 6.8, presented the first measurements of linear polarisation signals due to scattering of anisotropic radiation in the transition region. The data showed linear signals of the order of 0.1% in the Lyman-α line core and up to a few percent in the nearby wings.

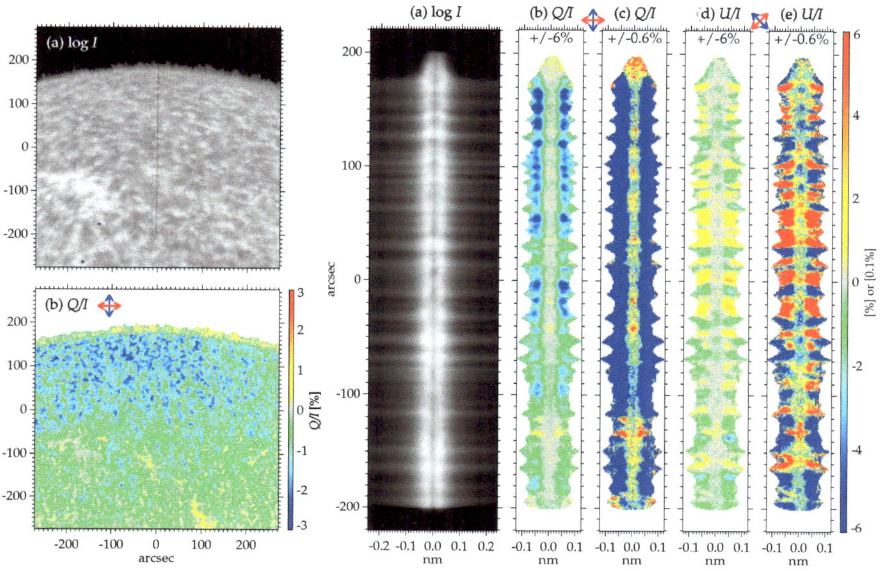

Fig. 6.8 Observations taken with the Chromospheric Lyman-Alpha Spectro-Polarimeter (CLASP) rocket experiment in its September 2015 flight. Left panels show data coming from the instrument broadband slit-jaw camera in the Lyman-α (**a**) intensity and (**b**) Q/I images taken by the CLASP slit-jaw camera. The spectrograph's slit can be seen in the intensity image covering around 400″. Right panels show the variation of the intensity (**a**) and the Stokes Q and U fractional polarisations (**b**,**c**,**d**,**e**) profiles of the hydrogen Lyman-α along the spectrograph slit. Image reproduced with permission from Kano et al. (2017), copyright by AAS

Acknowledgements This work has been supported by Spanish Ministry of Science, Innovation and Universities through the project RTI2018-096886-B-C51. The author acknowledges support from a *Ramón y Cajal* fellowship RYC-2016-20150 by the spanish Ministerio de Economía, Industria y Competitividad. He also thanks J. C. del Toro Iniesta, L. R. Bellot Rubio, and J. Trujillo Bueno for comments that greatly improved the manuscript. IAA-CSIC acknowledges financial support from the Spanish Research Agency (AEI/MCIU) through the "Center of Excellence Severo Ochoa" award for the Instituto de Astrofísica de Andalucía (SEV-2017-0709).

References

Arnaud J, Newkirk G Jr (1987) Astron Astrophys 178:263
Arregui I, Oliver R, Ballester JL (2012) Living Rev Sol Phys 9:2
Asensio Ramos A, Trujillo Bueno J, Landi Degl'Innocenti E (2008) Astrophys J 683:542
Babcock HW (1953) Astrophys J 118:387
Barthol P, Gandorfer A, Solanki SK et al (2011) Sol Phys Rev 268:1
Beckers JM (1968) Sol Phys Rev 3:367
Bellot Rubio LR, Orozco Suárez D (2012) Astrophys J 757:19
Bellot Rubio LR, Orozco Suárez D (2019) Living Rev Sol Phys 16:1

Belluzzi L, Trujillo Bueno J (2012) Astrophys J Lett 750:L11
Belluzzi L, Trujillo Bueno J (2013) Astrophys J Lett 774:L28
Belluzzi L, Trujillo Bueno J, Štěpán J (2012) Astrophys J Lett 755:L2
Belluzzi L, Trujillo Bueno J, Landi Degl'Innocenti E (2015) Astrophys J 814:116
Bogod VM (2011) Astrophys Bull 66:190
Bogod VM, Gelfreikh GB (1980) Sol Phys Rev 67:29
Borrero JM, Ichimoto K (2011) Living Rev Sol Phys 8:4
Brosius JW, White SM (2006) Astrophys J Lett 641:L69
Brosius JW, Landi E, Cook JW et al (2002) Astrophys J 574:453
Casini R, Judge PG (1999) Astrophys J 522:524
Casini R, López Ariste A, Tomczyk S, Lites BW (2003) Astrophys J Lett 598:L67
Casini R, Bevilacqua R, López Ariste A (2005) Astrophys J 622:1265
Casini R, López Ariste A, Paletou F, Léger L (2009) Astrophys J 703:114
Centeno R, Socas-Navarro H, Lites B et al (2007) Astrophys J Lett, 666, L137
Centeno R, Trujillo Bueno J, Uitenbroek H, Collados M (2008) Astrophys J 677:742-750
Centeno R, Trujillo Bueno J, Asensio Ramos A (2010) Astrophys J 708:1579
Collados M, Lagg A, Díaz Garcí A, JJ et al (2007) In: Heinzel P et al (eds) The physics of chromospheric plasmas, vol 368, pp 611–616. ASP, San Francisco, CA
Danilovic S, Beeck B, Pietarila A et al (2010) Astrophys J Lett 723:L149
de Pontieu B, McIntosh S, Hansteen VH et al (2007) Publ Astron Soc Jpn 59:655
del Toro Iniesta JC (1996) Vistas Astron 40:241
del Toro Iniesta JC (2003) Introduction to Spectropolarimetry, pp 244. Cambridge University Press, Cambridge. ISBN 0521818273
del Toro Iniesta JC, Ruiz Cobo B (2016) Living Rev Sol Phys 13:4
Emslie AG, Brown JC (1980) Astrophys J 237:1015
Gandorfer A (2000) The second solar spectrum: a high spectral resolution polarimetric survey of scattering polarization at the solar limb in graphical representation. Volume I: 4625 Å to 6995 Å. VdF, Zurich
Gandorfer A (2002) The second solar spectrum: a high spectral resolution polarimetric survey of scattering polarization at the solar limb in graphical representation. Volume II: 3910 Å to 4630 Å. VdF, Zurich
Gary DE, Hurford GJ (1994) Astrophys J 420:903
Gošić M, Bellot Rubio LR, Orozco Suárez D, Katsukawa Y, del Toro Iniesta JC (2014) Astrophys J 797:49
Gošić M, Bellot Rubio LR, del Toro Iniesta JC, Orozco Suárez D, Katsukawa Y (2016) Astrophys J 820:35
Grebinskij A, Bogod V, Gelfreikh G et al (2000) Astron Astrophys Suppl 144:169
Habbal SR, Woo R, Arnaud J (2001) Astrophys J 558:852
Hale GE (1908) Astrophys J 28:315
Hanle W (1924) Z Phys 30:93
Happer W (1972) Rev Mod Phys 44:169
Harvey JW (1969) PhD thesis, Department of Astro-Geophysics, University of Colorado
Hillier A (2018) Rev Mod Plasma Phys 2:1
House LL (1977) Astrophys J 214:632
House LL, Smartt RN (1982) Sol Phys Rev 80:53
Hyder CL (1965) Astrophys J 141:1374
Ichimoto K, Lites B, Elmore D et al (2008) Sol Phys Rev 249:233
Ishikawa R, Tsuneta S (2009) Astron Astrophys 495:607
Jin C, Wang J, Zhou G (2009) Astrophys J 697:693
Joshi J, Lagg A, Hirzberger J, Solanki SK (2017) Astron Astrophys 604:A98
Judge PG (1998) Astrophys J 500:1009
Judge PG, Kleint L, Donea A, Sainz Dalda A, Fletcher L (2014) Astrophys J 796:85
Judge PG, Kleint L, Sainz Dalda A (2015) Astrophys J 814:100
Kano R, Trujillo Bueno J, Winebarger A et al (2017) Astrophys J Lett 839:L10

Kleint L (2017) Astrophys J 834:26
Kobayashi K, Kano R, Trujillo-Bueno, J et al (2012) Fifth hinode science meeting, vol 456, p 233
Kosugi T, Matsuzaki K, Sakao T et al (2007) Sol Phys Rev 243:3
Kuckein C, Collados M, Manso Sainz R (2015) Astrophys J Lett 799:L25
Lagg A, Solanki SK, Riethmüller TL et al (2010) Astrophys J Lett 723:L164
Landi Degl'Innocenti E (1992) Solar observations: techniques and interpretation, vol 71. Cambridge University Press, Cambridge
Landi degl'Innocenti E (1998) Nature 392:256
Landi Degl'Innocenti E, Landolfi M (2004) Astrophysics and space science library, vol 307. Kluwer, Dordrecht
Li H, You J, Yu X, Du Q (2007) Sol Phys Rev 241:301
Li H, Landi Degl'Innocenti E, Qu Z (2017) Astrophys J 838:69
Libbrecht T, de la Cruz Rodríguez J, Danilovic S, Leenaarts J, Pazira H (2019) Astron Astrophys 621:A35
Lin H, Casini R (2000) Astrophys J 542:528
Lin H, Penn MJ, Tomczyk S (2000) Astrophys J Lett 541:L83
Lin H, Kuhn JR, Coulter R (2004) Astrophys J Lett 613:L177
Lites BW, Rutten RJ, Kalkofen W (1993) Astrophys J 414:345
Lites BW, Leka KD, Skumanich A, Martinez Pillet V, Shimizu T (1996) Astrophys J 460:1019
Lites BW, Kubo M, Socas-Navarro H et al (2008) Astrophys J 672:1237
Livingston W, Harvey J (1971) Sol Magn Fields 43:51
Livingston WC, Harvey J (1975) Bull Am Astron Soc 7:346
López Ariste A, Casini R (2005) Astron Astrophys 436:325
Loukitcheva M, White SM, Solanki SK, Fleishman GD, Carlsson M (2017) Astron Astrophys 601:A43
Luna M, Karpen J (2012) Astrophys J Lett 750:L1
Martin SF (1984) Small-scale dynamical processes in quiet stellar atmospheres. In: Keil SL (ed) Proceedings of the conference held in Sunspot, NM, 25–29 July, 1983. National Solar Observatory, p 30
Martínez González MJ, Bellot Rubio LR (2009) Astrophys J 700:1391
Martínez González MJ, Asensio Ramos A, Manso Sainz R et al (2011) Astrophys J Lett 730:L37
Martínez González MJ, Bellot Rubio LR, Solanki SK et al (2012) Astrophys J Lett 758:L40
Martínez González MJ, Manso Sainz R, Asensio Ramos A et al (2015) Astrophys J 802:3
Martínez González MJ, Asensio Ramos A, Arregui I et al (2016) Astrophys J 825:119
Martínez Pillet V, Del Toro Iniesta JC, Álvarez-Herrero A et al (2011) Sol Phys Rev 268:57
Merenda L, Trujillo Bueno J, Landi Degl'Innocenti E, Collados M (2006) Astrophys J 642:554
Mickey DL (1973) Astrophys J Lett 181:L19
Müller D, Marsden RG, St. Cyr OC, Gilbert HR (2013) Sol Phys Rev 285:25
Nakajima H, Nishio M, Enome S et al (1994) IEEE Proc 82:705
Narukage N, Auchère F, Ishikawa R, et al (2015) Appl Opt 54:2080
Narukage N, Kubo M, Ishikawa R, et al (2017) Sol Phys Rev 292:40
Okamoto TJ, Sakurai T (2018) Astrophys J Lett 852:L16
Orozco Suárez D, Bellot Rubio LR (2012) Astrophys J 751:2
Orozco Suarez D, Lagg A, Solanki SK (2005) Chromosph Coronal Magn Fields 59(1), 596
Orozco Suárez D, Bellot Rubio LR, del Toro Iniesta JC et al (2007a) Astrophys J Lett 670:L61
Orozco Suárez D, Bellot Rubio LR, Del Toro Iniesta JC et al (2007b) Publ Astron Soc Jpn 59:S837
Orozco Suárez D, Asensio Ramos A, Trujillo Bueno J (2012) Astrophys J Lett 761:L25
Orozco Suárez D, Asensio Ramos A, Trujillo Bueno J (2014a) Nat Promin Role Sp Weather 300:112
Orozco Suárez D, Díaz AJ, Asensio Ramos A, Trujillo Bueno J (2014b) Astrophys J Lett 785:L10
Orozco Suárez D, Asensio Ramos A, Trujillo Bueno J (2014c) Astro Astrophys 566:A46
Orozco Suárez D, Asensio Ramos A, Trujillo Bueno J (2015) Astrophys J Lett 803:L18
Parenti S (2014) Living Rev Sol Phys 11:1
Penn MJ, Kuhn JR (1995) Astrophys J Lett 441:L51

Pereira TMD, De Pontieu B, Carlsson M (2012) Astrophys J 759:18
Pesnell WD, Thompson BJ, Chamberlin PC (2012) Sol Phys Rev 275:3
Quémerais E, Lamy P (2002) Astron Astrophys 393:295
Ramelli R, Bianda M, Merenda L, Trujillo Bueno T (2006) Sol Polarisation 4:358, 448
Requerey IS, Del Toro Iniesta JC, Bellot Rubio LR et al (2015) Astrophys J 810:79
Rust DM (1967) Astrophys J 150:313
Sasso C, Lagg A, Solanki SK (2011) Astron Astrophys 526:A42
Schad TA, Penn MJ, Lin H, Tritschler A (2015) Sol Phys Rev 290:1607
Scharmer GB (2006) Astron Astrophys 447:1111
Schou J, Scherrer PH, Bush RI et al (2012) Sol Phys Rev 275:229
Smithson RC (1975) Bull Am Astron Soc 7:346
Socas-Navarro H (2005) Astrophys J Lett 631:L167
Socas-Navarro H, Elmore D (2005) Astrophys J Lett 619:L195
Socas-Navarro H, Trujillo Bueno J, Landi Degl'Innocenti E (2004) Astrophys J 612:1175
Solanki SK, del Toro Iniesta JC, Woch J et al (2015) Polarimetry 305:108
Stenflo JO (1982) Sol Phys Rev 80:209
Stenflo JO (2003). In: Dwivedi BN (ed) Dynamic sun. Cambridge University Press, Cambridge, pp 128–147. ISBN 0-521-81057-4
Stenflo JO (2017) Space Sci Rev 210:5
Stenflo JO, Keller CU (1997) Astron Astrophys 321:927
Stenflo JO, Twerenbold D, Harvey JW (1983a) Astron Astrophys Suppl 52:161
Stenflo JO, Twerenbold D, Harvey JW, Brault JW (1983b) Astron Astrophys Suppl 54:505
Stenflo JO, Keller CU, Gandorfer A (2000) Astron Astrophys 355:789
Suarez-Garcia E, Hajdas W, Wigger C et al (2006) Sol Phys Rev 239:149
Tandberg-Hanssen E, Anzer U (1970) Sol Phys Rev 15:158
Thernisien AFR, Howard RA, Vourlidas A (2006) Astrophys J 652:763
Tomczyk S, McIntosh SW, Keil SL et al (2007) Science 317:1192
Tomczyk S, Card GL, Darnell T et al (2008) Sol Phys Rev 247:411
Trujillo Bueno J (2001) Adv Sol Polarimetry Theor Observ Instrum 236:161
Trujillo Bueno J (2010) Astrophys Space Sci Proc 19:118
Trujillo Bueno J (2014) Solar Polarization 7. In: Nagendra KN, Stenflo JO, Qu ZQ, Sampoorna M (eds) Proceedings of a workshop held at Expo Garden Hotel, Kunming, China, 9–13 September, 2013. ASP conference series, vol 489. Astronomical Society of the Pacific, San Francisco, p 137
Trujillo Bueno J, Landi Degl'Innocenti E (1997) Astrophys J Lett 482:L183
Trujillo Bueno J, Casini R, Landolfi M, Landi Degl'Innocenti E (2002a) Astrophys J Lett 566:L53
Trujillo Bueno J, Landi Degl'Innocenti E, Collados M, Merenda L, Manso Sainz R (2002b) Nature 415:403
Trujillo Bueno J, Shchukina N, Asensio Ramos A (2004) Nature 430:326
Trujillo Bueno J, Merenda L, Centeno R, Collados M, Landi Degl'Innocenti E (2005) Astrophys J Lett 619:L191
Trujillo Bueno J, Landi Degl'Innocenti E, Belluzzi L (2017) Space Sci Rev 210:183
Tsuneta S, Ichimoto K, Katsukawa Y et al (2008) Sol Phys Rev 249:167
van de Hulst HC (1950) Bull Astron Inst Neth 11:135
van Noort M, Lagg A, Tiwari SK, Solanki SK (2013) Astron Astrophys 557:A24
Wang H, Liu C (2010) Astrophys J Lett 716:L195
Wang J, Wang H, Tang F, Lee JW, Zirin H (1995) Sol Phys Rev 160:277
Wedemeyer S, Bastian T, Brajša R et al (2016) Space Sci Rev 200:1
White SM (2001) Magn Fields Hertzsprung-Russell Diagr 248:67
White SM (2005) Chromosph Coronal Magn Fields 10(1):596
White SM, Kundu MR (1997) Sol Phys Rev 174:31
Zirin H, Severny A (1961) Observatory 81:155
Zweibel EG, Yamada M (2009) Annu Rev Astron Astrophys 47:291

Chapter 7
Spectropolarimetry of Stars Across the H-R Diagram

Swetlana Hubrig and Markus Schöller

Abstract The growing sample of magnetic stars shows a remarkable diversity in the properties of their magnetic fields. The overall goal of current studies is to understand the origin, evolution, and structure of stellar magnetic fields in stars of different mass at different evolutionary stages. In this chapter we discuss recent measurements together with the underlying assumptions in the interpretation of data and the requirements, both observational and theoretical, for obtaining a realistic overview of the role of magnetic fields in various types of stars.

7.1 Introduction

The presence of a magnetic field in the Sun was detected by Hale (1908) and the first detection of a magnetic field in a distant star was achieved by Babcock (1947). Since then, a lot of effort has been put into investigations of solar and stellar magnetism to understand how and why so many stars, including our own Sun, are magnetic, and what the implications might be for life on Earth and elsewhere in the universe. The past two decades have seen a significant step forward in our understanding of the occurrence of magnetic fields in stars of different mass and different age. This became possible through the significant progress in spectropolarimetric surveys of surface magnetic fields that made use of the currently available spectropolarimeters ESPaDOnS (Echelle SpectroPolarimetric Device for the Observation of Stars) at the Canada-France-Hawaii telescope, NARVAL at the Bernard Lyot 2 m Cassegrain telescope, the Main Stellar Spectrograph of the Russian 6 m telescope, FORS 1 and FORS 2 (FOcal Reducer low dispersion Spectrographs) at the ESO Very Large

---

S. Hubrig (✉)
Leibniz-Institut für Astrophysik Potsdam (AIP), Potsdam, Germany
e-mail: shubrig@aip.de

M. Schöller
European Southern Observatory, Garching, Germany
e-mail: mschoell@eso.org

R. Mignani et al. (eds.), *Astronomical Polarisation from the Infrared to Gamma Rays*, Astrophysics and Space Science Library 460,
https://doi.org/10.1007/978-3-030-19715-5_7

Telescope (VLT), and HARPSpol (High Accuracy Radial velocity Planet Searcher) at ESO/La Silla.

As of today, magnetic fields are found at all stages of stellar evolution, from brown dwarfs and young T Tauri stars to the end products of stellar evolution: white dwarfs and neutron stars. The incidence of magnetic fields in stars is diverse. The presence of a convective envelope is a necessary condition for significant magnetic activity in stars and magnetic activity is found all the way from late A-type stars, e.g. in Altair (Robrade and Schmitt 2009) with very shallow convective envelopes, down to the coolest fully convective M-type stars. All solar type stars appear to be magnetic, the stronger the more rapidly they rotate (e.g. Pallavicini et al. 1981). This is understood through the $\alpha\Omega$-dynamo, which is thought to operate in the convective envelope of these stars (e.g. Schmitt 1987). In intermediate mass main sequence stars, only about 10% are found to have kG-strength large-scale magnetic fields. Here, the correlation of magnetic fields with stellar rotation is opposite to that of solar type stars: whereas most intermediate mass stars are rapid rotators throughout their main sequence life, the magnetic stars are mostly slow rotators (e.g. Mathys 2017; Hubrig et al. 2018a).

In spite of the fact that recent years have seen a significant increase in observational studies of stellar magnetism, the most important aspects, such as the origin of stellar magnetic fields and the evolution of magnetic field configurations—the large-scale organised magnetic fields observed in massive stars or the complex magnetic topology observed in low-mass stars—are still not understood.

For massive and intermediate mass stars with radiative envelopes, it has been argued that their magnetic fields could be fossil relics of the fields that were present in the interstellar medium from which these stars have formed (e.g. Moss 2003). A search for the presence of magnetic fields in massive stars located in active sites of star formation, in the ρ Ophiuchus star-forming cloud and in the Trifid nebula, led to the detection of magnetic fields of several kG in two early B-type stars, the B2 V star HD 147933 and the B1 V star HD 164492Cb (Hubrig et al. 2018b, 2014a). However, the fossil field hypothesis has several problems as it does not explain the low ($\sim$ 10%) occurrence of magnetic stars and their broad range of field strengths. Moreover, a study of the distribution of magnetic Ap stars in the H-R diagram using accurate Hipparcos parallaxes by Hubrig et al. (2000) showed that these stars are concentrated towards the centre of the main-sequence band and only rarely can be found close to the zero-age main sequence (ZAMS). Similar studies using Gaia DR2 data are on the way, but their results are not published yet. Alternatively, magnetic fields may be generated by strong binary interaction, i.e., in stellar mergers, or during a mass transfer or common envelope evolution (e.g. Tout et al. 2008). The resulting strong differential rotation is considered as a key ingredient for the generation of magnetic fields (Petrovic et al. 2005). For Galactic O-type stars, it was shown that the fraction of very close binaries is so high that about one third of them interacts with a companion during their main sequence evolution through mass transfer or merging (Sana et al. 2012).

In cool, low-mass stars, the key unresolved aspects are the origin of the solar-like activity cycle, the origin and character of differential rotation and its role in the large-scale dynamo, and the origin of high energy coronal activity. Knowledge of the evolution of the magnetic field and its configuration is extremely important for our understanding of the past and future of the Sun and the solar system. Notably, low-mass stars are complex magnetohydrodynamical objects that require state-of-the-art observations and numerical simulations to pin down the physical processes that are responsible for the diversity and dynamics of these stars. The origin of stellar activity in low-mass stars resides most likely in the stellar convection zones. The mechanism involves helical gas motions and large-scale differential rotation. The most straightforward combination of these effects is the $\alpha\Omega$-dynamo, which can reproduce many basic properties of the solar activity cycle. Differential rotation generates large-scale toroidal fields by winding up the poloidal field. Differential rotation is the main field generator in the $\alpha\Omega$-dynamo, which therefore maintains predominantly axisymmetric fields that are dominated by the toroidal component. On the other hand, differential rotation alone is not sufficient for dynamo action, as it cannot prevent the decay of the poloidal field. The α-effect, which is caused by the Coriolis force, is vital to the dynamo, because it produces a poloidal field from the toroidal field, thus completing the field-amplifying dynamo process. In contrast to differential rotation, the α-effect can maintain a magnetic field by itself. This type of dynamo is known as the α^2-dynamo. Unlike the $\alpha\Omega$-dynamo, it prefers non-axisymmetric field configurations that are not dominated by one component. Mean field magnetohydrodynamics provides a very successful theoretical model for stellar differential rotation and the accompanying meridional flows (Kitchatinov and Rüdiger 1999; Küker and Stix 2001).

In the following sections, we discuss magnetic field diagnostics, observations of magnetic fields in upper main-sequence, pre-main-sequence, and low-mass active stars, finishing with post-main-sequence stars and degenerate stellar remnants.

7.2 Magnetic Field Diagnostics

The currently applied observing techniques differ for different types of stars, involving the Zeeman effect, the Paschen-Back effect, and the molecular Zeeman effect. Observations of broad-band linear polarisation, which is caused by different saturation levels of the π and σ components of a spectral line in the presence of a magnetic field, were reported in the past for a few chemically peculiar magnetic Ap stars (e.g. Leroy 1989; Landolfi et al. 1993). This differential effect for Zeeman components is qualitatively similar for all lines, so that in broad-band observations the contributions of all lines add up. Monitoring of broad-band linear polarisation of a few Ap stars over a few years was used to establish the presence of their extremely slow rotation. Studies of broad-band linear polarisation do not represent a widely used technique of stellar magnetic field diagnosis, and nowadays they are superseded by the observations of linear polarisation in individual spectral lines.

The most complete way to study stellar magnetic fields is to observe both linear and circular polarisation, i.e. all four Stokes parameters I, Q, U, and V.

However, such observations are very challenging as in disc-integrated observations the signal can be canceled or significantly decreased due to mixed polarity fields. Furthermore, measurements of linear polarisation are sensitive to instrumental polarisation. Therefore, only large-scale magnetic fields can be studied using the full Stokes vector. Importantly, physically realistic models of the geometric structure of magnetic fields in stellar atmospheres can be achieved only if linear polarisation measurements are implemented.

Observing the spectrum of a star in the two senses of circular polarisation is in almost all cases the most sensitive method of detecting weak magnetic fields on a stellar surface. The measurement of the Stokes V parameter permits to detect the mean longitudinal magnetic field, which is a line intensity weighted average over the visible hemisphere of a star of the magnetic field component directed along the line-of-sight. To measure longitudinal magnetic fields, in particular observing with low-resolution spectropolarimeters, a regression analysis approach is frequently used (e.g. Bagnulo et al. 2001). The mean longitudinal magnetic field measurement $\langle B_\mathrm{z} \rangle$ in this approach is based on the relation

$$\frac{V}{I} = -\frac{g_\mathrm{eff}\, e\, \lambda^2}{4\pi\, m_\mathrm{e}\, c^2}\, \frac{1}{I}\, \frac{\mathrm{d}I}{\mathrm{d}\lambda}\, \langle B_\mathrm{z} \rangle\ , \tag{7.1}$$

where V is the Stokes parameter that measures the circular polarisation, I is the intensity in the unpolarised spectrum, g_eff is the effective Landé factor, e is the electron charge, λ is the wavelength, m_e is the electron mass, c is the speed of light, $\mathrm{d}I/\mathrm{d}\lambda$ is the wavelength derivative of Stokes I, and $\langle B_\mathrm{z} \rangle$ is the mean longitudinal magnetic field. Although the longitudinal magnetic field measurements present the canonical diagnostics of stellar magnetic fields, other observational and analytical methods to diagnose the presence and structure of magnetic fields in stellar atmospheres are frequently applied.

In a number of Ap and Bp stars whose projected rotational velocity is sufficiently small and whose magnetic field is strong enough to exceed the rotational Doppler broadening, certain lines with suitable Zeeman patterns show resolved Zeeman split lines in unpolarised light (Preston 1971; Mathys 1990). They are used to estimate the line-intensity weighted average of the modulus of the magnetic field over the visible stellar hemisphere, the mean magnetic field modulus. Simultaneous knowledge of the mean magnetic field modulus and the longitudinal magnetic field over the stellar rotation period permits to set constraints on the magnetic field geometry.

Since the Zeeman signatures in the spectra of many stars are generally very small, and increasing the signal-to-noise ratio (S/N) by increasing the exposure time is frequently limited by the short rotation period of the star, multi-line approaches as proposed by Semel (1989) are commonly used to increase the S/N. The most widely used approach is the Least Squares Deconvolution (LSD; Donati et al. 1997). The main assumption in LSD is the application of the weak field approximation, that is, the magnetic splitting of spectral lines is assumed to be smaller than their Doppler broadening. Furthermore, it is assumed that the local line profiles are self-similar and can be combined into an average profile. Due to non-linear effects in the summation and the effect of blends, the resulting LSD profiles should not be considered as observed, but rather processed Zeeman signatures. In a number of recent studies, a novel magnetometric technique, the multi-line Singular Value Decomposition (SVD) method was applied. The software package to study weak magnetic fields using the SVD method was introduced by Carroll et al. (2012). The basic idea of SVD is similar to the principal component analysis (PCA) approach, where the similarity of individual Stokes V profiles permits to describe the most coherent and systematic features present in all spectral line profiles as a projection onto a small number of eigenprofiles.

Other popular methods to exploit the information contained in polarised spectral line profiles currently include the moment technique, in particular the determination of the crossover effect and of the mean quadratic field (e.g. Mathys 1993), and Zeeman-Doppler imaging (ZDI). In the moment technique, the shapes of the spectral lines observed in Stokes I and Stokes V profiles can be characterised by their moments of various orders about the line centre and can be interpreted in terms of quantities related to the magnetic field. The analysis is usually based on the consideration of samples of reasonably unblended lines and critically depends on the number of lines that can be employed. ZDI is an inversion technique applied to time-series of two or four Stokes parameters, introduced by Semel (1989). It allows to recover the distribution of the temperature (on the surface of cool stars) or of chemical spots (on the surface of Ap and Bp stars) and of the magnetic field vector over the stellar surface. In most studies, however, only Stokes I and Stokes V observations are used to obtain Zeeman-Doppler stellar images. Obviously, in such cases, the solution is not unique and is constrained by a number of assumptions. In the studies of cool stars with temperature spots it is however possible to remedy this situation by the implementation of full polarised radiative transfer in both atomic and molecular lines. Employing molecular lines is essential to localise magnetic fields inside cool regions on the stellar surface (e.g. Sennhauser et al. 2009).

A novel technique to detect surface magnetic fields in A- and F-type stars, which uses the autocorrelation of spectra in integral light, was recently suggested by Borra and Deschatelets (2015). The original idea of the autocorrelation technique is to measure the width of spectral lines and to disentangle the magnetic broadening from other broadening effects, using the differences in wavelength dependence. Such a technique can be applied to low-resolution spectra, down to stars that are too faint for high-resolution observations even with large telescopes. The availability of this technique allows one to use publicly available data archives from numerous spectrographs all over the world to obtain reliable statistics on the occurrence of magnetic fields in A and F stars in clusters and the field at different evolutionary stages. Using this technique, it will become possible even to search for magnetic fields in a sample of early B-type stars in the Magellanic Clouds using low-resolution spectroscopic observations.

7.3 O- and Early B-Type Stars

Massive stars with initial masses ranging from 8 to 100 $M_{\odot}$ are of particular interest, as they end their evolution with a supernova explosion, producing neutron stars or black holes. Magnetic O-type stars with masses larger than 30 $M_{\odot}$ and their WR descendants have been suggested as progenitors of magnetars (Gaensler et al. 2005). A magnetic mechanism for the collimated explosion of massive stars, relevant for long-duration gamma-ray bursts, X-ray flashes, and asymmetric core collapse supernovae was proposed by, e.g., Uzdensky and MacFadyen (2006).

Recent observations indicate that probably only a small fraction of about 7±3% of O-type stars with masses exceeding 18 $M_{\odot}$ (Grunhut et al. 2017) and about 6±3% of early B-type and O-type stars (Schöller et al. 2017) possess measurable, mostly dipolar magnetic fields. Progress in the studies of massive stars became possible mostly through the MiMeS (The Magnetism in Massive Stars, Wade et al. 2011), MAGORI (MAGnetic field ORIgin, Hubrig et al. 2011), and BOB (B-fields in OB stars, Morel et al. 2015) surveys, which made use of the spectropolarimeters ESPaDOnS, NARVAL, FORS2, and HARPSpol. The vast majority of magnetic massive stars exhibit a smooth, single-wave variation of the longitudinal magnetic field during the stellar rotation cycle. These observations are considered as evidence for a dominant dipolar contribution to the magnetic field topology.

Braithwaite and Spruit (2004) showed that stars with radiative envelopes can host stable large-scale magnetic field configurations, as those observed in the Ap/Bp stars, independent from the star's rotation. The implication is that up to about 30–50% of all massive stars might be magnetic, and that so far only the tip of the iceberg has been seen (Stello et al. 2016). From Kepler and Kepler K2 photometry, Balona (2013, 2016) concluded that about 40% of A-type stars and 30–50% of late O and B-type stars show rotational modulation due to the presence of star spots. Pulsating stars, eclipsing binaries, and ellipsoidal variables of high amplitude were avoided in these studies. Since it is difficult to conceive of starspot formation without the presence of a magnetic field, these results indicate that the occurrence of magnetic fields in A, B, and late O-type stars is greatly underestimated.

No complete volume-limited study of the incidence of magnetism in massive stars has been presented so far. Thus one of the major goals of recent studies is to build trustworthy statistics on the occurrence of magnetic fields in massive stars and their topology at different evolutionary stages. This is critical to answer the principal question of the possible origin of such fields. Previously observed samples greatly suffer from a number of biases in the target selection related to specific spectral classifications. We know already that all five Galactic stars with Of?p classification are magnetic and at least 10% of the early type B stars belong to the magnetic He-rich group, containing stars with overabundances of Si and He. O-type stars with Of?p classification exhibit recurrent, and apparently periodic, spectral variations in Balmer, He I, C III, and Si III lines, sharp emission or P Cygni profiles in He I and the Balmer lines, and strong C III emission around 4650 Å (Walborn 1972). A search for the presence of magnetic fields at different evolutionary stages is also predominantly carried out in bright stars that are slow rotators. Since very young massive stars at the ZAMS are usually faint, this evolutionary stage of massive stars remains completely unexplored. The results recently presented in several papers reflect the lack of observations of young massive stars close to the ZAMS (e.g. Castro et al. 2015).

It is quite possible that a significant fraction of the magnetic massive stars remains undiscovered due to a more complex magnetic field topology related to the presence of a subsurface convection zone. Multi-wavelength observations of massive stars show cyclical line profile structures, the discrete absorption components (DACs). Their presence indicates that stellar prominences can emerge on the stellar surface, probably due to localised magnetic activity, more or less in analogy to what is observed on the solar surface (e.g. Sudnik and Henrichs 2016). Action of a subsurface convection zone (Cantiello et al. 2009) would be the most likely driving mechanism that generates magnetic spots, which are the source of those prominences. This zone may also be the source of a global magnetic field, winding up toroidally with stochastic buoyancy breakouts, causing corotating magnetic bright spots at the surface of the star (Cantiello and Braithwaite 2011).

Atmospheric magnetic fields can impact surface rotation rates via magnetic braking (Weber and Davis 1967; ud-Doula et al. 2008), introduce chemical abundance inhomogeneities and peculiarities (Hunger and Groote 1999), and confine the stellar wind in a magnetosphere (e.g. Friend and MacGregor 1984; ud-Doula and Owocki 2002). The ejected matter streams along the field lines towards the magnetic equator and collides in the vicinity of the equatorial plane, leading to strong shocks and thereby heating plasma to X-ray temperatures (Babel and Montmerle 1997).

The strongest longitudinal magnetic field in an O star of more than 5 kG was detected in the Of?p star NGC 1624-2, suggesting a dipole component of about 20 kG (Wade et al. 2012). Among the early B-type stars, the record holder is the He-rich star CPD -62° 2124 with a dipole component of at least 21 kG (Hubrig et al. 2017).

Noteworthy, the survey of the BOB consortium, aiming to establish a lower boundary of magnetic fields in massive stars, revealed the presence of very weak Zeeman signatures across the line profiles in high-quality HARPSpol spectra of two early B-type stars. This work indicated that also magnetic fields of the order of a few Gauss can be present in massive stars, but possibly escape their detection due to insufficient measurement accuracy (e.g. Fossati et al. 2015).

Wolf-Rayet (WR) stars form an integral part of the late evolution of the most massive stars. While the WR core He-burning phase lasts only $\sim$ 10% of the O-star main sequence phase, WR winds are at least ten times stronger than on the main sequence. With wind-momenta in WR stars reaching up to 30 times the available photon momentum, magnetic fields have been proposed as additional or alternative wind-driving mechanisms to radiation pressure (Cassinelli et al. 1995). The detection of magnetic fields in WR stars is however difficult, chiefly because the line spectrum is formed in the strong stellar wind. This does not only imply a dilution of the field at the place of line formation. The biggest problem is the wind broadening of the emission lines by Doppler shifts with wind velocities of a few 1000 km s^{-1}. de la Chevrotiére et al. (2014) used ESPaDOnS spectra to search for magnetic fields in eleven WR stars. They reported that the data show evidence supporting marginal detections of a magnetic field of the order of a few hundred Gauss and less in three WR stars. A search for magnetic fields in several WR stars using low-resolution FORS 2 spectropolarimetry showed the presence of a magnetic field in the cyclically variable star WR 6 at a significance level of 3.3σ (258$\pm$78 G). In this work, a theoretical approach to estimate the magnetic field strength in stars with strong emission lines was considered for the first time (Hubrig et al. 2016).

7.4 Ap and Bp Stars

Globally ordered magnetic fields are observed in roughly 10% of the intermediate and massive main-sequence stars with spectral types between approximately B2 and F0. These stars, generally called chemically peculiar Ap and Bp stars, exhibit strong overabundances of certain elements, such as iron peak elements and rare earths, and underabundances of He, C, and O, relative to solar abundances, and are characterised observationally by anomalous line strengths. Massive Bp stars usually show overabundances of He and Si. As the star rotates, the magnetic field and surface abundance distribution is observed from various aspects, resulting in variability of the measured magnetic field and spectral line strengths.

The variable magnetic fields are generally diagnosed through mean longitudinal magnetic field, mean magnetic field modulus, and net broad-band linear polarisation measurements. Chemical abundance anomalies are commonly believed to be due to radiatively driven microscopic diffusion in stars rotating sufficiently slowly to allow such a process to be effective (e.g. Michaud 1970). While two thirds of all Galactic O-type stars appear to be members of close binary systems (Sana et al. 2012), the study of Carrier et al. (2002) indicated a dearth of magnetic Ap stars in close binaries. Only for two close binary systems with Ap magnetic components, HD 98088 (e.g. Babcock 1958; Abt et al. 1968) and HD 161701 (Hubrig et al. 2014b), is the structure of the magnetic field known. Remarkably, in both systems carries the surface of the Ap component, which is facing the companion, a positive magnetic field. It is, however, not clear whether tidal forces may play a role during the dynamic process of tidal synchronisation, to align the field geometries in the observed way. Further systematic searches for magnetic fields in binary systems should be conducted to properly evaluate theoretical models of the origin of their magnetic fields.

Currently, 84 stars with magnetically split components are known and their study permits to establish the general properties of their magnetic fields (e.g. Mathys 2017). The star with the strongest magnetic field currently known, "Babcock's star" HD 215441 is a B8 V star with a surface dipole magnetic field strength of 34 kG (Babcock 1960; Preston 1969). Zeeman split lines were recently discovered in a few early Bp stars with low projected rotational velocity and strong, kG order magnetic

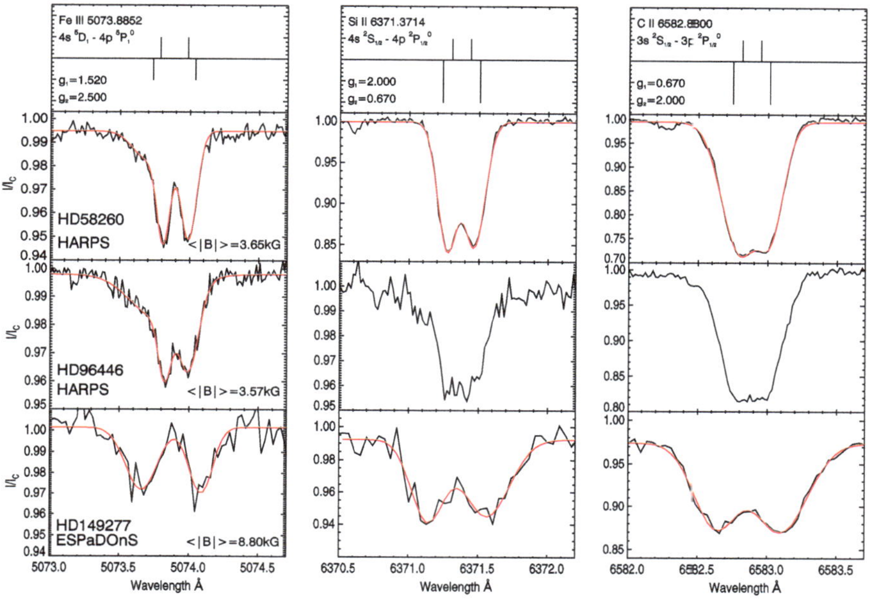

Fig. 7.1 The magnetically split lines Fe III λ5074, Si II λ6371, and C II λ6583 in high-resolution Stokes *I* spectra of the early B-type stars HD 58260, HD 96446, and HD 149277. The red lines denote the fit of a multi-Gaussian to the data. For two lines in the spectrum of HD 96446, the splitting is not sufficient to allow a proper fit

fields. Examples of magnetically split lines in three such stars are presented in Fig. 7.1.

The evolution of magnetic field configurations in stars with masses below 6 $M_\odot$ using a sample of Ap and Bp stars with accurate Hipparcos parallaxes and definitely determined longitudinal magnetic fields was until now considered only in the study by Hubrig et al. (2007). This work showed that stronger magnetic fields are detected in younger stars. No evidence was detected for any loss of angular momentum during the main-sequence life. Also the magnetic flux was found to be constant over the stellar life time on the main sequence. An excess of stars with large obliquities β (the orientation of the magnetic axis with respect to the rotation axis) was detected in both higher and lower mass stars. The evolution of β appeared to be somewhat different for stars with M<3 $M_\odot$ than for stars with M>3 $M_\odot$, for which a strong hint for an increase of β with time spent on the main sequence was discovered.

Using the indirect surface mapping ZDI method is a valuable option to derive magnetic field maps, but was applied only for a few Ap/Bp stars during the last years. A summary of ZDI results is presented in the work of Kochukhov et al. (2019). The authors find that the global magnetic field geometry changes little from one star to another, with nearly all stars showing dominant dipolar fields with a varying degree of distortion.

7.5 Pre-main Sequence Stars

Studies of magnetic fields in stars at early evolutionary stages, before they arrive on the main sequence, are of special interest to get an insight into the magnetic field origin. Moreover, such studies are extremely important because they enable us to improve our insight into how the magnetic fields in these stars are generated and how they interact with their environment, including their impact on the planet formation process and the planet-disc interaction.

It is generally accepted that magnetic fields are important ingredients of the star formation process (e.g. McKee and Ostriker 2007) and are already present in stars in the pre-main sequence (PMS) phase. The PMS low-mass T Tauri stars stand out by their very large emissions in chromospheric and transition-region lines. The accretion of circumstellar disc material onto the surface of these stars is controlled by a strong stellar magnetic field. These stars are usually grouped into two classifications: classical T Tauri stars, which show evidence of a circumstellar disc and mass accretion onto the central star in the form of excess emission in X-rays, the UV, optical, and infrared, and weak-line T Tauri stars, which do not show evidence for significant mass accretion. A model for the generation of magnetic fields in fully convective PMS stars was presented by, e.g., Küker and Rüdiger (1999), Chabrier and Küker (2006), and Küker (2009), who considered a dynamo process of α^2-type.

The accretion of circumstellar disc material onto the surface of T Tauri stars is assumed to be controlled by a strong stellar magnetic field. Furthermore, the magnetic field appears to be critical for explaining their rotational properties (e.g. Herbst et al. 2007) and probably plays a critical role in driving the observed outflows (Mohanty and Shu 2008). Magnetic field measurements of these stars over the last decades show the presence of kG magnetic fields. Basri et al. (1992) were the first to detect a magnetic field on the surface of a T Tauri star using Zeeman broadening of magnetically sensitive lines in Stokes I. Modelling Zeeman-broadened Stokes I profiles often involves template spectra of magnetically active and inactive stars, which are weighted by a filling factor. As this method is sensitive to the magnetic field distribution but has a limited sensitivity to the magnetic field geometry, most recent studies involve circular polarisation observations to characterise surface magnetic fields using the ZDI technique. On the other hand, according to Carroll et al. (2012), a common problem for ZDI is that the real strong magnetic fields that are associated with cool surface regions produce only a fraction of the photon flux compared to the unaffected quiet or even hot surface regions. When studying the correlation between the magnetic field and temperature spots, it is important to be aware that the detectability of the field is weighted by the surface brightness. To prevent such an ill-defined flux-weighting and the mutual effects between the temperature and magnetic field, Carroll et al. have pursued a strategy that performs a simultaneous DI and ZDI to retrieve both the temperature and the magnetic field distribution at the same time. Their reconstruction of the temperature and magnetic field of the weak-line T Tauri star V410 Tau revealed that the majority of the strong fields of 2 kG are located in cool spots, in particular within the large polar spot. The reconstruction clearly showed that both polarities coexist within the large polar spot and that the entire polar-field topology appears to be dominated by a twisted bi-polar structure separated by a relatively sharp neutral line.

To study magnetic field topologies in T Tauri stars, the group of Donati (e.g. Donati et al. 2015) applies a stellar-surface tomographic-imaging package. This imaging code is set up to invert (both automatically and simultaneously) time series of Stokes I and V LSD profiles into brightness and magnetic maps of the stellar surface, where brightness imaging is allowed to reconstruct both cool spots and warm plages, known to contribute to the activity of very active stars. This tool is based on the principles of maximum-entropy image reconstruction and on the assumption that the observed variability is mainly caused by rotational modulation with an added option for differential rotation.

According to Adams and Gregory (2012), in accreting systems, the polar strength of the dipole component is particularly relevant for determining the disc truncation radius and the balance of torques in the star-disc system. Johnstone et al. (2014) found that the dipole strength is correlated with the rotation period: stars with weak dipole components tend to be rotating faster than stars with strong dipole components. Recent observations of magnetic fields in a small sample of T Tauri stars suggest that the magnetic field topology changes as a function of age, i.e. there is a relation between the large-scale magnetic field and the position in the H-R diagram. Gregory et al. (2012) reported that the magnetic field gains complexity with the evolution of the star: the magnetic field is mainly dipolar when the star is completely convective. After the development of the radiative core the dipolar component loses power compared to high order components of the multipole expansion. Based on magnetohydrodynamic simulations, Zaire et al. (2017) suggest that the presence of a tachocline could be an important factor in the development of magnetic fields with higher multipolar modes.

The presence of magnetic fields in higher mass PMS stars, the Herbig Ae/Be stars, has long been suspected, in particular on account of Hα spectropolarimetric observations pointing out the possibility of magnetospheric accretion, similar to that of classical T Tauri stars (e.g. Vink et al. 2002). While models of magnetically driven accretion and outflows successfully reproduce many observational properties of the classical T Tauri stars, the picture is completely unclear for the Herbig Ae/Be stars, mostly due to the poor knowledge of their magnetic field topology. So far, the magnetic field geometry was constrained only for two Herbig Ae/Be stars, V380 Ori (Alecian et al. 2009) and HD 101412 (Hubrig et al. 2011), and only about 20 Herbig stars were reported to host magnetic fields (Hubrig et al. 2015, and references therein).

The two best studied Herbig Ae/Be stars exhibit a single-wave variation in their mean longitudinal magnetic field during the stellar rotation cycle. This behaviour is usually considered as evidence for a dominant dipolar contribution to the magnetic field topology. Presently, the Herbig Ae star HD 101412 possesses the strongest magnetic field ever measured in any Herbig Ae star, with a surface magnetic field $\langle B \rangle$ of up to 3.5 kG. HD 101412 is also the only Herbig Ae/Be star for which the rotational Doppler effect was found to be small in comparison to the magnetic splitting, presenting several spectral lines resolved into magnetically split components observed in unpolarised light at high spectral resolution (Hubrig et al. 2010).

Notably, the task of magnetic field measurements in Herbig stars is very challenging, as the work of Hubrig et al. (2015) demonstrates, in which the authors compiled all magnetic field measurements reported in previous spectropolarimetric studies. This study indicates that the low detection rate of magnetic fields in Herbig Ae stars, about 7% (Alecian et al. 2013), can indeed be explained not only by the limited sensitivity of the published measurements, but also by the weakness of these fields. The obtained density distribution of the rms longitudinal magnetic field values reveals that only a few stars have magnetic fields stronger than 200 G, and half of the sample possesses magnetic fields of about 100 G and less. Consequently, the currently largest spectropolarimetric survey of magnetic fields in several tens of Herbig stars by Alecian et al. (2013) using ESPaDOnS and NARVAL cannot be considered as representative: the measurement accuracy in this study is worse than 200 G for 35% of the measurements, and for 32% of the measurements it is between 100 and 200 G. Clearly, to improve our understanding of the origin of magnetic fields in Herbig Ae/Be stars and their interaction with the protoplanetary disc, it is of utmost importance to study magnetic fields with high accuracy measurements in a representative sample of Herbig Ae/Be stars.

Zeeman signatures in the spectra of Herbig Ae/Be stars are generally very small, and increasing the S/N by increasing the exposure time is frequently limited by the short rotation period of the star. Therefore, multi-line approaches such as LSD and SVD are commonly used to increase the S/N. An example of a very weak Zeeman feature in the HARPSpol spectra of the Herbig Ae star PDS 2 detected using the SVD method (Hubrig et al. 2015) is presented in Fig. 7.2.

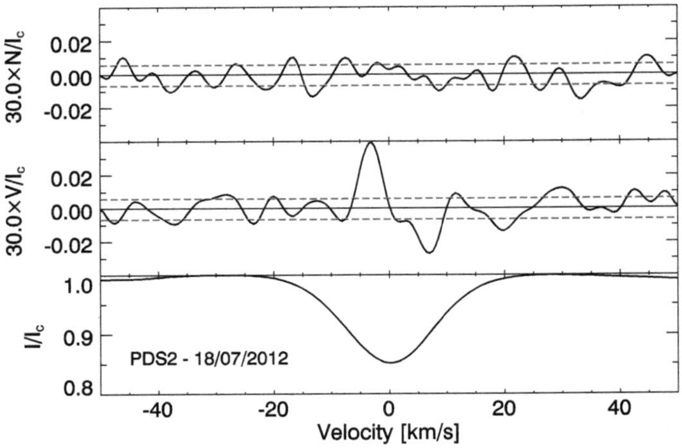

Fig. 7.2 Detection of a mean longitudinal magnetic field in the Herbig Ae star PDS 2. The Stokes I (bottom), Stokes V (middle), and diagnostic null (N; top) SVD profiles were used to determine $\langle B_z \rangle = 33 \pm 5$ G. The Stokes V and N profiles were expanded by a factor of 30. The red dashed lines indicate the standard deviations for the Stokes V and N spectra. This figure corresponds to the lower part of Fig. 2 in Hubrig et al. (2015)

Spectropolarimetric observations of a sample of 21 Herbig Ae/Be stars observed with FORS 1 were used to search for a link between the presence of a magnetic field and other stellar properties by Hubrig et al. (2009). This study did not indicate any correlation of the strength of the longitudinal magnetic field with disc orientation, disc geometry, or the presence of a companion. No simple dependence on the mass-accretion rate was found, but the range of the observed field values qualitatively supported the expectations from magnetospheric accretion models with dipole-like field geometries. Both the magnetic field strength and the X-ray emission showed a decline with age in the range of $\sim 2 - 14$ Myr probed by the sample, supporting a dynamo mechanism that decays with age. In particular, it was found that magnetic fields are strong in very young Herbig stars and become very weak or completely disappear at the end of their PMS life. In line with these results, Hubrig et al. (2000, 2005, 2007) showed that magnetic fields in Ap stars with masses less than $3\,M_{\odot}$ are rarely found close to the ZAMS and that kG magnetic fields appear in A stars already evolved from the ZAMS. Thus, it is very unlikely that Herbig stars are the progenitors of the magnetic Ap stars. Importantly, the Herbig stars seem to obey the universal power-law relation between magnetic flux and X-ray luminosity established for the Sun and main-sequence active dwarf stars (Pevtsov et al. 2003). Future work on stellar properties of magnetic Herbig stars should involve a larger and more representative sample to determine the existing relations at a higher confidence level.

As already mentioned in Sect. 9.4, in agreement with the merging scenario for the origin of magnetic fields in magnetic stars with radiative outer zones, the number of close binary systems with Ap components is very low. Similarly, studies of Herbig Ae stars by Wheelwright et al. (2010) and Duchêne (2015) indicate the lack of close binary systems with $P_{\mathrm{orb}} < 20$ d. Weak magnetic fields were recently detected in two Herbig Ae systems, in AK Sco and in the presumed binary HD 95881. Using high quality HARPSpol observations, Järvinen et al. (2018) obtained $\langle B_z \rangle = -83 \pm 31$ G for the secondary component in the system AK Sco and $\langle B_z \rangle = -93 \pm 25$ G for HD 95881. It is of interest that for AK Sco they observe the magnetic field in the secondary component in the region of the stellar surface facing permanently the primary component, meaning that the magnetic field geometry in the secondary component is likely related to the position of the primary component. A similar magnetic field behaviour, where the field orientation is linked to the companion, was previously detected in HD 98088 and HD 161701, the two close main-sequence binaries with Ap components mentioned above. Obviously, a search for magnetic fields and the determination of their geometries in close binary systems is very important as the knowledge of the presence of a magnetic field and of the alignment of the magnetic axis with respect to the orbital radius vector in Herbig binaries may hint at the mechanism of the magnetic field generation.

The weakness of the observed magnetic fields put into question the current understanding of the magnetospheric accretion process in intermediate-mass pre-main sequence stars. Importantly, Cauley and Johns-Krull (2014) studied the He I λ10830 morphology in a sample of 56 Herbig Ae/Be stars. They suggested that early Herbig Be stars do not accrete material from their inner discs in the same manner as T Tauri stars, while late Herbig Be and Herbig Ae stars show evidence for magnetospheric accretion. Furthermore, they proposed more compact magnetospheres in Herbig Ae/Be stars compared to T Tauri stars. Ababakr et al. (2017) presented Hα linear spectropolarimetry of a sample of 56 Herbig Ae/Be stars. A change in linear polarisation across this line was detected in 42 (75%) objects, indicating that the circumstellar environment around these stars has an asymmetric structure on small spatial scales, which is typically identified with a disc. A second outcome of their research was the confirmation that Herbig Ae stars are similar to T Tauri stars in displaying a line polarisation effect, while depolarisation is more common among Herbig Be stars. Using near-infrared multi-epoch spectroscopic data obtained with the CRIRES and X-shooter spectrographs installed at the VLT, Schöller et al. (2016) examined the magnetospheric accretion in the Herbig Ae star HD 101412. Spectroscopic signatures in He I 10830 and Paγ, two near-infrared lines that are formed in a Herbig star's accretion region, were presenting temporal modulation. The authors showed that this modulation is governed by the rotation period of this star and that the observed spectroscopic variability was explained within the magnetic geometry established earlier from magnetic field measurements by Hubrig et al. (2011).

To summarise, it appears that the as yet small number of magnetic Herbig Ae/Be stars can be due to both the weakness of the magnetic fields and the large measurement errors. According to Alecian (2014), the magnetic properties of Herbig Ae/Be stars must have been shaped at a very early phase of the stellar evolution. Using pre-main-sequence evolutionary tracks calculated with the CESAM code (Morel 1997), she concluded that even stars above three solar masses will undergo a purely convective phase before reaching the birthline. Therefore, it is reasonable to assume that the weak magnetic fields detected in a number of Herbig Ae/Be stars are just leftovers of the fields generated by pre-main-sequence dynamos during the convective phase. If this scenario is valid, we should expect a significantly larger number of Herbig stars possessing weak magnetic fields.

7.6 Low-Mass Stars

Magnetic fields play a key role in the evolution of solar-like low-mass stars as they determine the angular momentum loss, shape the stellar wind, and produce high-energy electromagnetic and particle radiation. A solar mass star spends approximately 10 Gyr on the main sequence. Over the main-sequence life, magnetic activity declines with age and is closely related to a loss of angular momentum (e.g. Skumanich 1972). Long-term monitoring of stellar activity using chromospheric emission measured in the cores of Ca II H & K lines revealed that the chromospheric flux increases with the rotation rate (Baliunas et al. 1995; Hall et al. 2007). However, not all active stars undergo smooth activity cycles and the activity of the most active stars has the tendency to fluctuate erratically, while regular activity cycles are observed in older stars. Egeland et al. (2017) studied the long-term variability of a sample of five solar analog stars using composite chromospheric activity records up to 50 years in length and synoptic visible-band photometry about 20 years long. Their sample covered a large range of stellar ages, which they used to represent the evolution of activity for solar mass stars. The authors found that young, fast rotators have an amplitude of variability many times higher than that of the solar cycle, while old, slow rotators have very little variability.

Typical aims in polarimetric studies of low-mass stars are to determine the topology of the surface magnetic field and to study the relation between magnetic fields and cool spots. Most recent studies apply the ZDI technique using Stokes IV spectropolarimetry and it is not seldom that no correlation between cool spots and a detected magnetic field is found. Rosén et al. (2015) compared ZDI results using Stokes IV and the full set of Stokes $IVQU$ parameters for the evolved binary RS CVn-type variable II Peg. They found significant differences in the magnetic field solutions and underestimation of the total magnetic energy. The retrieval of the meridional field component especially benefited from using linear polarisation.

Naturally, spectropolarimetric observations of low-mass stars provide excellent constraints for theoretical dynamo models. One would expect stars with an internal structure similar to that of the Sun to show the same type of dynamo operation. On the other hand, ZDI studies frequently find large spots at high latitudes and large regions of a near-surface azimuthal field (e.g. Järvinen et al. 2015). Petit et al. (2008) suggested that the magnetic field is mostly poloidal for low rotation rates and that more rapid rotators host a large-scale toroidal component in their surface field. Furthermore, they inferred that a rotation period lower than about 12 d is necessary for the toroidal magnetic energy to dominate over the poloidal component. Metcalfe et al. (2016) proposed that a change in the character of differential rotation is the mechanism that ultimately disrupts the large-scale organisation of magnetic fields in sun-like stars. The process begins at a Rossby number (associated with convective motions and presenting the ratio of the rotation period to the convective turnover time) close to one, where the rotation period becomes comparable to the convective turnover time. The Vaughan-Preston gap (Vaughan and Preston 1980) can then be interpreted as a signature of rapid magnetic evolution triggered by a shift in the character of differential rotation, which is an emergent property of turbulent convection in the presence of Coriolis forces.

The red dwarfs are the coolest objects which include late K and M dwarfs. The majority of red dwarfs exhibit evidence of strong magnetic activity, which is expressed in strong optical flares, enhanced UV, X-ray, and radio emission. The process of generating strong magnetic fields in fully convective objects and those with very deep convection zones is not yet understood. First direct magnetic field measurements in these dwarfs were achieved using the Zeeman split Fe I line at 8468.4 Å (Johns-Krull and Valenti 1996). More recent spectropolarimetric observations demonstrated that early type M dwarfs have magnetic fields with a strong toroidal component, reminiscent of those of active K and G dwarfs, whereas the lowest-mass M dwarfs exhibit magnetic fields that are highly organised and strongly poloidal (e.g. Morin et al. 2008a) and some of them show indications of polar star spot features (Morin et al. 2008b).

Starspots are the primary evidence of magnetic activity harbouring the strongest field and the coldest plasma. Spectropolarimetric surveys of red dwarfs revealed the presence of numerous atomic and molecular signatures formed in star spots and the chromosphere. First detections of polarisation in TiO, CaH, and FeH transitions in starspots on M and K dwarfs were presented by Berdyugina et al. (2006, 2008) using ESPaDOnS. It was the first time that polarimetry of cool atmospheres with the help of molecular physics was used to obtain new insights into the magnetic structures of cool atmospheres. For the red dwarf AU Mic, the authors identified four spots, one of positive and three of negative polarity. These spots are 500–700 K cooler than the photosphere and harbour a maximum magnetic field of 5.3 kG.

Brown dwarfs, often referred to as failed stars as they do not sustain hydrogen fusion in their core, show magnetic activity similar to that of low-mass stars and produce huge flares (Paudel et al. 2018). The first direct detection of a strong, 5 kG magnetic field on the surface of an active brown dwarf was presented in the work of Berdyugina et al. (2017). The surface magnetic field of the brown dwarf with a mass of about 55 M_J and at an age of 22 Myr was detected in circularly polarised signatures in the 819 nm sodium line. According to the authors, the young age of the magnetic brown dwarf implies that it may still maintain a disc, which may facilitate bursts via magnetospheric accretion, like in higher-mass T Tauri stars.

The study of magnetic fields of low-mass stars became particularly important after the discovery of exoplanets around such stars. Multi-wavelength observations with different observing strategies concluded that a giant exoplanet in a short-period orbit can induce activity in the photosphere and upper atmosphere of its parent star. This makes the host star's magnetic activity a probe of the planet's magnetic field. The large-scale magnetic field of the planet-host star HD 189733 with a mainly toroidal surface magnetic field with a strength of 20–40 G has been observed over multiple years. The ZDI analysis of these observations revealed a structural evolution of the field between observations (e.g. Fares et al. 2013).

7.7 Post-main-Sequence Stars

The evolved binary components in RS CVn-type variables show remarkable activity, displaying strong chromospheric plages, coronal X-ray and microwave emission and flares detected in the optical, UV, radio, and X-rays. For these systems, magnetic activity cycles have been suggested to be the cause of the detected period modulations. One of the RS CVn systems exhibiting substantial period variations is HR 1099 with a high level of magnetic activity in all available activity indicators. Studies using photometric observations reported different magnetic activity periods, from 14.1 yr (Muneer et al. 2010) to 19.5 yr (Lanza 2006). The Applegate mechanism proposed by Applegate (1992) has recently been discussed by Völschow et al. (2018) as a potential intrinsic mechanism to explain transit timing variations in various kinds of close binary systems. In the framework of the Applegate model, period modulations can be explained by gravitational coupling of the orbit to variations in the shape of the magnetically active star in the system. The variable deformation of the active star is produced by variations in the distribution of angular momentum as the star goes through its activity cycle. The torque needed to redistribute the angular momentum can be exerted by a mean subsurface magnetic field of several kilogauss.

In former studies, an exciting topic of the magnetic activity of RS CVn-type variables and the FK Comae stellar group was the flip-flop phenomenon. The FK Comae stellar group consists of single G–K giants showing high activity and rapid rotation. In many such active stars, the spots were found to concentrate on two permanent active longitudes, which were half a period apart (see, e.g., Berdyugina et al. 1998). In some of these stars, the dominant part of the spot activity concentrated on one of the active longitudes, and abruptly switched the longitude every few years. This flip-flop phenomenon was discovered in FK Comae (Jetsu et al. 1993). Light curves and Doppler imaging showed evidence of differential rotation of modest strength. However, in his most recent work, Jetsu (2018) showed that the period search methods used in previous studies do not necessarily produce correct rotation period estimates, if there are multiple interfering periodicities. Based on his new two-dimensional period finding method, Jetsu argued that the observed deceptive light curves were the interference of two real constant period light curves of long-lived starspots. The slow motion of these long-lived starspots with respect to each other caused the observed light curve changes.

Hubrig et al. (1994) carried out magnetic field measurements for thirteen bright G–K giants of different X-ray luminosity and concluded that magnetic fields in such objects are indeed observable using spectropolarimetry. Later on, direct detections of magnetic fields in the photospheres of cool, evolved giants advancing on the red giant branch and the asymptotic giant branch were presented by Konstantinova-Antova et al. (2013) and Auriére et al. (2015). In the sample of 48 red giants studied by Aurière et al., magnetic fields were detected via Zeeman signatures in 29 giants. The majority of the magnetic giants are either in the first dredge up phase or at the beginning of core He burning, i.e. in phases when the convective turnover time is at a maximum. Four giants for which the magnetic field was measured to be outstandingly strong, up to 100 G (with respect to that expected from the relation between rotation period and magnetic field or from their evolutionary status), were interpreted as being probable descendants of magnetic Ap stars.

One of the major open questions regarding the formation of planetary nebulae (PNe) concerns the mechanism that is responsible for their non-spherical, often axisymmetric shaping (e.g. Balick and Frank 2002). Both, central star binarity and stellar rotation in combination with magnetic fields, are among the favourite explanations. Basically, the origin of PNe is understood to be a consequence of the interaction of the hot central star with its circumstellar environment through photoionisation and wind–wind collision. In principle, the role of magnetic fields in shaping PNe may be verified or disproved by empirical evidence, as already suggested by Jordan et al. (2005). Using FORS 1 in spectropolarimetric mode, they reported the detection of magnetic fields of the order of kG in the central stars of the PNe NGC 1360 and LSS 1362. A reanalysis of their data, however, did not provide any significant evidence for longitudinal magnetic fields that are stronger than a few hundred Gauss in these stars (Jordan et al. 2012). Their field measurements had typical error bars of 150 to 300 G. Similar results were achieved in the work of Leone et al. (2011), who concluded that the mean longitudinal magnetic fields in NGC 1360 and LSS 1362 are much weaker, less than 600 G, or that the magnetic field has a complex structure. The most recent search for magnetic fields in central stars of planetary nebulae by Leone et al. (2014), based on spectropolarimetric observations of 19 central stars with the Intermediate-dispersion Spectrograph and Imaging System (ISIS) on the William Herschel Telescope (WHT) and VLT/FORS 2, was partly affected by large measurement uncertainties and reported no positive detection either.

Using low-resolution polarimetric spectra obtained with FORS 2, Steffen et al. (2014) carried out a search for magnetic fields in a sample of 12 central stars covering the whole range of morphologies from round to elliptical/axisymmetric, and bipolar PNe, and including both chemically normal and Wolf-Rayet type (hydrogen-poor) central stars. The sample included two round nebulae (NGC 246, Hen 2-108), five elliptical nebulae (IC 418, NGC 1514, NGC 2392, NGC 3132, Hen 2-131), and three bipolar nebulae (NGC 2346, Hen 2-36, Hen 2-113). Two targets have WR-type central stars (NGC 246, Hen 2-113). In addition, the authors included the two (elliptical) targets of Jordan et al. (2005), NGC 1360 and LSS 1362, for which they originally claimed the detection of kG magnetic fields. Six of the 12 central stars were known binaries.

Formal 3σ detections were achieved for IC 418 ($\langle B_z \rangle = -181 \pm 54$ G), for the WR-type central star Hen 2-113 ($\langle B_z \rangle = -58 \pm 18$ G), and the weak emission line star Hen 2-131 ($\langle B_z \rangle = -120 \pm 32$ G). The mean longitudinal magnetic field $\langle B_z \rangle$ of the central stars was derived by linear regression employing two different methods: a) using a weighted linear regression line through the measured data points and b) generating $M = 10^6$ statistical variations of the original dataset by the bootstrapping technique and analysing the resulting distribution $P(\langle B_z \rangle)$ of the M regression results. Mean and standard deviation of this distribution are identified with the most likely mean longitudinal magnetic field and its 1σ error, respectively. The main advantage of this method is that it provides an independent error estimate. The results of both methods for the central star of the young, elliptical planetary nebula IC 418 are presented in Fig. 7.3. A mean longitudinal magnetic field of negative polarity was detected at a 3σ significance level when using the entire spectrum: $\langle B_z \rangle = -181 \pm 54$ G using the first method and $\langle B_z \rangle = -177 \pm 54$ G for the second method. Most stars in the sample of Steffen et al. (2014) were not studied at the achieved accuracy before, permitting to put constraints on the strength of the magnetic fields in the central stars of planetary nebulae.

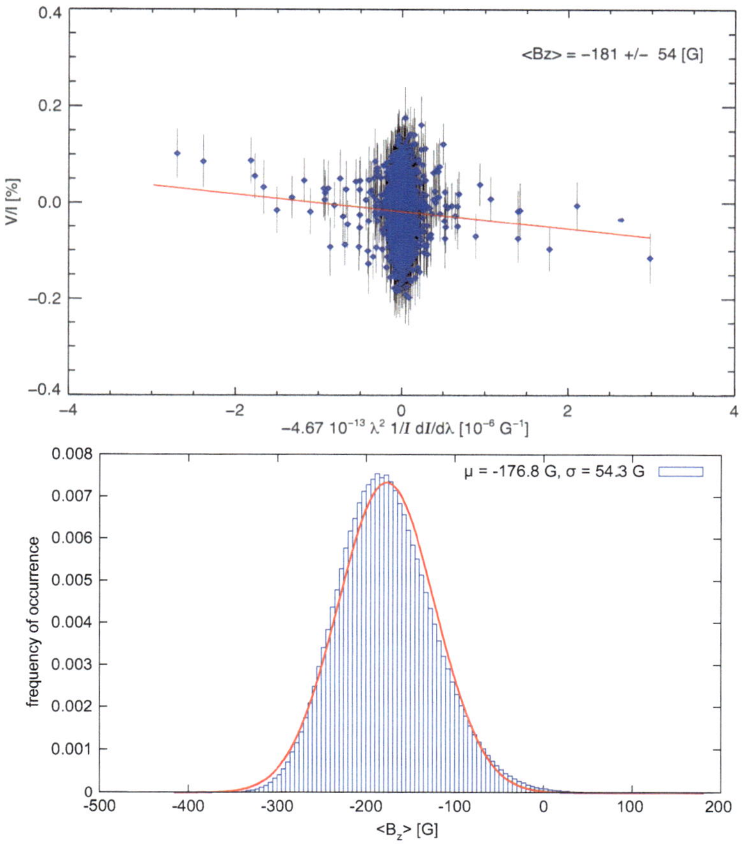

Fig. 7.3 IC 418: Regression detection of a $\langle B_z \rangle = -181 \pm 54\,\mathrm{G}$ mean longitudinal magnetic field, using all lines (top). The corresponding $\langle B_z \rangle$ distribution (bottom) deviates slightly from a Gaussian (red curve) and indicates $\langle B_z \rangle = -177 \pm 54\,\mathrm{G}$. Figure from Steffen et al. (2014, A&A, 570, A88); reproduced with permission, © ESO

7.8 Degenerate Stellar Remnants

White dwarfs represent the final stage of stellar evolution for the majority of stars with masses less than about $8\,M_{\odot}$. The majority of white dwarfs have hydrogen-rich atmospheres (DA). Other types of white dwarfs include stars with atmospheres dominated by helium (DB) and those without a visual helium spectrum (DC). 25% to 30% of white dwarfs show traces of heavy elements, i.e. elements heavier than helium (Zuckerman et al. 2003, 2010), probably due to accreted planetary or asteroidal debris (Jura 2008).

A significant fraction of white dwarfs possess a magnetic field ranging from a few kG up to several hundred MG (e.g. Ferrario et al. 2015). According to Kepler et al. (2013), the incidence of magnetism within the white dwarf population remains uncertain with an incidence of magnetism of about 5% in magnitude limited surveys, or between 10–20% in volume limited surveys (e.g. Kawka et al. 2007). Kawka and Vennes (2014) showed that cool, polluted white dwarfs have a higher incidence of magnetism, up to 40%, than in the general population of white dwarfs. Dufour et al. (2013) reported on a magnetic field incidence of about 70% in the rare class of warm and hot DQ white dwarfs, which present rare carbon-dominated objects known as hot DQ white dwarfs.

To study magnetic fields in white dwarfs, again circular polarisation observations are involved. Similar to Ap and Bp stars, the geometry of the magnetic field in white dwarfs is assumed to be a centred or offset dipole. However, the magnetic field structure in white dwarfs studied over their rotational period shows a large diversity, from a simple offset dipole, over structures with spots, to multipoles (e.g. Kawka et al. 2019). For non-variable magnetic white dwarfs it is assumed that they are either rotating with a very long period or that their magnetic fields are nearly aligned with the rotation axis. A recent statistical analysis of the linear polarisation properties between magnetic and non-magnetic white dwarfs showed no difference in the polarisation degrees (Żejmo et al. 2017; Słowikowska et al. 2018). Such an analysis allowed the authors to select a set of good candidates of faint linear polarimetric standard stars that can be used as stable polarimetric calibration sources.

The mechanism responsible for the presence of magnetic fields in these stars remains unclear, as white dwarfs do not contain any significant convective zones. In the framework of the fossil field theory for the origin of magnetic fields, and assuming magnetic flux conservation, Ap and Bp stars are unlikely progenitors of white dwarfs with weak magnetic fields. Tout et al. (2008) proposed a binary origin, where the magnetic field is formed via a dynamo created during a common envelope phase. Nordhaus et al. (2011) proposed that during a CE phase a low-mass star will be tidally disrupted by its proto-white dwarf companion, forming an accretion disc and generating a dynamo in the disc, which is then transferred to the degenerate core via accretion. García-Berro et al. (2012) suggested that magnetic fields can also be produced by the merger of two white dwarfs. To explain the sharp increase in the incidence of magnetism in cool polluted white dwarfs, Kawka and Vennes (2014) argued that the phenomenon coincides with the likely presence of a perennially crowded circumstellar environment, which may have been caused by the same event that generated the magnetic field. As various classes of white dwarfs show a different incidence of magnetic fields, it is plausible that magnetic white dwarfs are created by several discernible processes.

For a discussion of the magnetic properties of neutron stars see Chap. 12.

References

Ababakr KM, Oudmaijer RD, Vink JS (2017) Mon Not R Astron Soc 472:854
Abt HA, Conti PS, Deutsch AJ, Wallerstein G (1968) Astrophys J 153, 177
Adams FC, Gregory SG (2012) Astrophys J 744:55
Alecian E (2014) In: Mathys G, Griffin ER, Kochukhov O, Monier R, Wahlgren GM (eds) Putting a stars into context: evolution, environment, and related stars, p 84
Alecian E et al (2009) Mon Not R Astron Soc 400:354
Alecian E et al (2013) Astron Astrophys 549:L8
Applegate JH (1992) Astrophys J 385:621
Auriére M et al (2015) Astron Astrophys 574:A90
Babcock HW (1947) Astrophys J 105:105
Babcock HW (1958) Astrophys J 128:228
Babcock HW (1960) Astrophys J 132:521
Babel J, Montmerle T (1997) Astron Astrophys 323:121
Bagnulo S, Szeifert T, Wade GA, Landstreet JD, Mathys G (2001) Messenger 104:32
Balick B, Frank A (2002) Arch Rec 40:439
Baliunas SL et al (1995) Astrophys J 438:269
Balona LA (2013) Mon Not R Astron Soc 431:2240
Balona LA (2016) Mon Not R Astron Soc 457:3724
Basri G, Marcy GW, Valenti JA (1992) Astrophys J 390:622
Berdyugina SV, Jankov S, Ilyin I, Tuominen I, Fekel FC (1998) Astron Astrophys 334:863
Berdyugina SV, Fluri D, Petit P, Fluri DM, Afram N, Arnaud J (2006) In: Casini R, Lites BW (eds) Solar polarization, vol 4. ASP Conference Series, vol 358, p 381
Berdyugina SV et al (2008) In: van Belle G (ed) 14th cambridge workshop on cool stars, stellar systems, and the sun. ASP Conference Series, vol 384, p 175
Berdyugina SV et al (2017) Astrophys J 847:61
Borra EF, Deschatelets D (2015) Astron J 150:146
Braithwaite J, Spruit HC (2004) Nature 431:819
Cantiello M, Braithwaite J (2011) Astron Astrophys 534:A140
Cantiello M et al (2009) Astron Astrophys 499:279
Carrier F, North P, Udry S, Babel J (2002) Astron Astrophys 394:151
Carroll TA, Strassmeier KG, Rice JB, Künstler A (2012) Astron Astrophys 548:A95
Cassinelli JP, Ignace A, Bjorkman JE (1995) Int Astron Union Symp 163:191
Castro N et al (2015) Astron Astrophys 581:A81
Cauley PW, Johns-Krull CM (2014) Astrophys J 797:112
Chabrier G, Küker M (2006) Astron Astrophys 446:1027
de la Chevrotiére A, St-Louis N, Moffat AFJ, MiMeS Collaboration (2014) Astrophys J 781:73
Donati J-F et al (1997) Mon Not R Astron Soc 291:658
Donati J-F et al (2015) Mon Not R Astron Soc 453:3706
Duchêne G (2015) Astrophys Space Sci 355:291
Dufour P, Vornanen T, Bergeron P, Fontaine AB (2013) 18th European white dwarf workshop. ASP Conference Series, vol 469, p 167
Egeland R, Soon W, Baliunas S, Hall JC, Henry GW (2017) Int Astron Union Symp 328:329
Fares R et al (2013) Mon Not R Astron Soc 435:1451
Ferrario L, de Martino D, Gänsicke BT (2015) Space Sci Rev 191:111
Fossati L et al (2015) Astron Astrophys 574:A20
Friend DB, MacGregor KB (1984) Astrophys J 282:591
Gaensler BM et al (2005) Astrophys J 620:L95
García-Berro E et al (2012) Astrophys J 749:25
Gregory SG et al (2012) Astrophys J 755:97
Grunhut JH et al (2017) Mon Not R Astron Soc 465:2432
Hale GE (1908) Astrophys J 28:315

Hall JC, Lockwood GW, Skiff BA (2007) Astron J 133:862
Herbst W, Eislöffel J, Mundt R, Scholz A (2007) In: Reipurth B, Jewitt D, Keil K (eds) Protostars and planets V, p 297. University of Arizona Press, Tucson
Hubrig S, Plachinda SI, Hunsch M, Schröder K-P (1994) Astron Astrophys 291:890
Hubrig S, North P, Mathys G (2000) Astrophys J 539:352
Hubrig S, North P, Schöller M (2005) AIP Conf Proc 784:145
Hubrig S, North P, Schöller M (2007) Astr Nachr 328:475
Hubrig S et al (2009) Astron Astrophys 502:283
Hubrig S et al (2010) Astr Nachr 331:361
Hubrig S et al (2011) Astron Astrophys 525:L4
Hubrig S et al (2014a) Astron Astrophys 564:L10
Hubrig S et al (2014b) Mon Not R Astron Soc 440:L6
Hubrig S, Carroll TA, Schöller M, Ilyin I (2015) Mon Not R Astron Soc 449:L118
Hubrig S et al (2016) Mon Not R Astron Soc 458:3381
Hubrig S et al (2017) Mon Not R Astron Soc 472:400
Hubrig S et al (2018a) Mon Not R Astron Soc 477:3791
Hubrig S, Schöller M, Järvinen SP, Küker M, Kholtygin AF, Steinbrunner P (2018b) Astr Nachr 339:72
Hunger K, Groote D (1999) Astron Astrophys 351:554
Järvinen SP et al (2015) Astron Astrophys 574:A25
Järvinen SP et al (2018) Astrophys J Lett 858:L18
Jetsu L (2018, submitted) Mon Not R Astron Soc, arXiv:1808.02221
Jetsu L, Pelt J, Tuominen I (1993) Astron Astrophys 278:449
Johns-Krull CM, Valenti JA (1996) Astrophys J 459:L95
Johnstone CP, Jardine M, Gregory SG, Donati J-F, Hussain G (2014) Mon Not R Astron Soc 437:3202
Jordan S, Werner K, O'Toole SJ (2005) Astron Astrophys 432:273
Jordan S, Bagnulo S, Werner K, O'Toole SJ (2012) Astron Astrophys 542:A64
Jura M (2008) Astron J 135:1785
Kawka A, Vennes S (2014) Mon Not R Astron Soc 439:L90
Kawka A, Vennes S, Schmidt GD, Wickramasinghe DT, Koch R (2007) Astrophys J 654:499
Kawka A, Vennes S, Ferrario L, Paunzen E (2019) Mon Not R Astron Soc 482:5201
Kepler SO et al (2013) Mon Not R Astron Soc 429:2934
Kitchatinov LL, Rüdiger G (1999) Astron Astrophys 344:911
Kochukhov O, Shultz M, Neiner C (2019) Astron Astrophys 621:A47
Konstantinova-Antova R et al (2013) Bulg Astron J 19:14
Küker M (2009) AIP Conf Proc 1094:77
Küker M, Rüdiger G (1999) Astron Astrophys 346:922
Küker M, Stix M (2001) Astron Astrophys 366:668
Landolfi M, Landi DeglÍnnocenti E, Landi DeglÍnnocenti M, Leroy JL (1993) Astron Astrophys 272:285
Lanza AF (2006) Mon Not R Astron Soc 369:1773
Leone F, Martínez González MJ, Corradi RLM, Privitera G, Manso Sainz R (2011) Astrophys J 731:L33
Leone F, Corradi RLM, Martínez González MJ, Asensio Ramos A, Manso Sainz R (2014) Astron Astrophys 563:A43
Leroy JL (1989) Astron Astrophys 215:360
Mathys G (1990) Astron Astrophys 232:151
Mathys G (1993) Am Soc Prev Cardiol 44:232
Mathys G (2017) Astron Astrophys 601:A14
McKee CF, Ostriker EC (2007) Arch Rec 45:565
Metcalfe TS, Egeland R, van Saders J (2016) Astrophys J Lett 826:L2
Michaud G (1970) Astrophys J 160:641
Mohanty S, Shu FH (2008) Astrophys J 687:1323

Morel P (1997) Astron Astrophys Ser 124:597
Morin J et al (2008a) Mon Not R Astron Soc 390:56
Morin J et al (2008b) Mon Not R Astron Soc 384:77
Morel T et al (2015) Int Astron Union Symp 307:342
Moss D (2003) Astron Astrophys 403:693
Muneer S, Jayakumar K, Rosario MJ, Raveendran AV, Mekkaden MV (2010) Astron Astrophys 521:A36
Nordhaus J, Wellons S, Spiegel DS, Metzger BD, Blackman EG (2011) Proc Natl Acad Sci 108:3135
Pallavicini R et al (1981) Astrophys J 248:279
Paudel RR et al (2018) Astrophys J 861:76
Petit P et al (2008) Mon Not R Astron Soc 388:80
Petrovic J, Langer N, van der Hucht KA (2005) Astron Astrophys 435:1013
Pevtsov AA et al (2003) Astrophys J 598:1387
Preston GW (1969) Astrophys J 156:967
Preston GW (1971) Astrophys J 164:309
Robrade J, Schmitt JHMM (2009) Astron Astrophys 497:511
Rosén L, Kochukhov O, Wade GA (2015) Astrophys J 805:169
Sana H et al (2012) Science 337:444
Schmitt D (1987) Astron Astrophys 174:281
Schöller M et al (2016) Astron Astrophys 592:A50
Schöller M et al (2017) Astron Astrophys 599:A66
Semel M (1989) Astron Astrophys 225:456
Sennhauser C, Berdyugina SV, Fluri DM (2009) Am Soc Prev Cardiol 405:543
Skumanich A (1972) Astrophys J 171:565
Słowikowska A, Krzeszowski K, Żejmo M, Blinov D, Reig P (2018) Mon Not R Astron Soc 479:5312
Steffen M et al (2014) Astron Astrophys 570:A88
Stello D et al (2016) Publ Astron Soc Aust 33:11
Sudnik NP, Henrichs HF (2016) Astron Astrophys 594:A56
Tout CA, Wickramasinghe DT, Liebert J, Ferrario L, Pringle JE (2008) Mon Not R Astron Soc 387:897
ud-Doula A, Owocki SP (2002) Astrophys J 576:413
ud-Doula A, Owocki SP, Townsend RHD (2008) Mon Not R Astron Soc 385:97
Uzdensky DA, MacFadyen AI (2006) Astrophys J 647:1192
Vaughan AH, Preston GW (1980) Publ Astron Soc Pac 92:385
Vink JS, Drew JE, Harries TJ, Oudmaijer RD (2002) Am Soc Prev Cardiol 267:439
Völschow M, Schleicher DRG, Banerjee R, Schmitt JHMM (2018) Astron Astrophys 620:A42
Wade GA et al (2011) Am Soc Prev Cardiol 449:262
Wade GA et al (2012) Mon Not R Astron Soc 425:1278
Walborn NR (1972) Astron J 77:312
Weber EJ, Davis L Jr (1967) Astrophys J 148:217
Wheelwright HE, Oudmaijer RD, Goodwin SP (2010) Mon Not R Astron Soc 401:1199
Zaire B, Guerrero G, Kosovichev AG, Smolarkiewicz PK, Landin NR (2017) Int Astron Union Symp 328:30
Żejmo M, Słowikowska A, Krzeszowski K, Reig P, Blinov D (2017) Mcn Not R Astron Soc 464:1294
Zuckerman B, Koester D, Reid IN, Hünsch M (2003) Astrophys J 596:477
Zuckerman B, Melis C, Klein B, Koester D, Jura M (2010) Astrophys J 722:725

Chapter 8
Dust Polarisation in the Interstellar Medium

Ralf Siebenmorgen and Christian Peest

Abstract The optical to submillimetre continuum radiation of the ISM is polarised by aligned non-spherical dust particles. Modelling of this polarisation spectrum in combination with abundance constraints and simultaneous fit to extinction and emission yields important information on dust grains, including chemistry, mineralogy, particle sizes, and shapes. The pristine nature of dust is better revealed by studying sight-lines that include a single-cloud in the observing beam as compared to attempts at analysing averages of the observing characteristics over multiple-clouds. The propagation of polarised light through dust enshrouded objects is solved by polarisation radiative transfer for which we outline the central methodology.

8.1 Introduction

Observational evidence of cosmic dust has been around since people have looked at the sky. Cosmic dust appears as dark patches against the brighter background of stellar fields. A striking example is the Coalsack Nebula in the southern skies. A detailed characterisation of dust in the interstellar medium (ISM) is of crucial importance to understand the chemical history of the Universe and various fundamental astrophysical processes, such as the formation and evolution of the stars and their planetary systems. Most of the heavy elements are present in the ISM in the form of dust grains. Non-spherical dust grains, aligned along a preferred direction, are most efficient at linearly polarising the stellar background radiation at wavelengths close to the grain size. Therefore, linear polarimetric measurements of stars that are not intrinsically polarised set constraints on the chemical composition,

R. Siebenmorgen (✉)
European Southern Observatory, Garching bei München, Germany
e-mail: Ralf.Siebenmorgen@eso.org

C. Peest
Universität Potsdam, Potsdam, Germany
e-mail: peest@uni-potsdam.de

R. Mignani et al. (eds.), *Astronomical Polarisation from the Infrared to Gamma Rays*, Astrophysics and Space Science Library 460,
https://doi.org/10.1007/978-3-030-19715-5_8

shape and size distribution of the interstellar dust, as well as on the mechanisms responsible for dust grain alignment. Observations of polarised light provide also constraints of the dust density distribution of the ISM as well as its magnetic field structure.

In the following we first review cosmic abundance constraints (Sect. 8.2), dust extinction (Sect. 8.3), emission (Sect. 8.4), dichroic polarisation (Sect. 8.5), polarised emission (Sect. 8.6), and results combining the latter two polarisation data (Sect. 8.7). Then we discuss grain alignment (Sect. 8.8) and present a dust model of the solar neighbourhood that is consistent with the cosmic abundance constraints and simultaneously fits extinction, emission, and polarisation (Sect. 8.9). We also show how the combination of recent *Gaia* distance estimates with available high precision polarisation data can be used to infer the nearby ($R \lesssim 300$ pc) structure of the ISM (Sect. 8.10). The dust model of Sect. 8.9 is then applied to data of the Large Interstellar Polarisation Survey, which in combination with high resolution spectroscopy (Sect. 8.11) is used to discuss sight-lines that includes a single-cloud in the observing beam. We show that for studies of the pristine nature of the dust such single-cloud sight-lines are better suited than analysing averages of the observing characteristics.

We present a scheme and review methods for solving the radiative transfer of polarised light in dust enshrouded media (Sect. 8.12). Finally, we compare results of different treatments for computing cross sections of spheroidal dust particles (Sect. 8.13).

8.2 Dust Element Abundances

The only stable elements created during the phase of primordial nucleosynthesis, roughly 10 s to 20 min after the Big Bang, are hydrogen (75% by mass), helium (25%), and traces of lithium (10^{-7}%). Essentially all other elements have since been produced by stellar nucleosynthesis in stars. In the later stages of their lifetime, evolving and exploding stars expel some of the material they created into the void between the stars, enriching the so-called interstellar medium. These elements can form molecules, and when these cool down, they can form microscopic particles in the solid phase, which are generally called dust particles. Cosmic dust grains in the interstellar or circumstellar medium may clump into pebbles, rocks, planetesimals, and planets.

Spectroscopic measurements of stars allow the determination of the abundances of different gases in cosmic clouds. There is usually a difference between the inferred and the measured cloud compositions. Spectroscopic measurements do not detect atoms locked in solid particles so that some elements are rarer in the gas of the clouds. They are called "depleted from the gas phase" (Jenkins 2014).

Clues to the composition of interstellar dust come from observed elemental depletion in the gas phase. However, dense clouds usually have higher depletion rates than thin clouds. It is generally assumed that the abundances of the chemical elements in the ISM are proportional to those measured in the photosphere of the Sun. The abundances of the elements of the interstellar dust (the condensed phase of matter) are estimated as the difference between the elemental abundances in the solar photosphere (Asplund et al. 2009) and that of the gas phase. Absolute values of the interstellar gas phase abundance of element [X] are given with respect to that of hydrogen [H]. Such [X]/[H] ratios are often derived from the analysis of absorption lines. However, oscillator strengths of some key species, such as CII 2325 Å, are uncertain up to a factor of 2. The most abundant condensable elements in the ISM are C, O, Mg, Si, and Fe. When compared to the values of the Sun, elements such as Mg, Si, and Fe, which form silicates, appear under-abundant in the gas phase. By contrast, oxygen represents a striking exception, because it appears over-abundant towards certain sight-lines (Voshchinnikov and Henning 2010). Another important dust forming element, C, cannot be characterised in detail because it has been analysed only in a restricted number of sight-lines, leading in some cases to inaccurate values of its abundance (Jenkins 2009; Parvathi et al. 2012). It is widely accepted that cosmic dust consists of some combination of silicate and carbon.

8.3 Dust Extinction

The most prominent feature of cosmic dust clouds is obviously that they act as screens and shield the radiation of the stars behind it from our telescopes. The dust scatters or absorbs parts of the stellar radiation that crosses the cloud with probability given by the dust cross section K_{sca} and K_{abs} (cm^2/g-dust). Together these effects are called *extinction* with dust extinction cross section $K_{\mathrm{ext}} = K_{\mathrm{sca}} + K_{\mathrm{abs}}$. The interstellar extinction or reddening curve is obtained from photometry on a reddened and unreddened comparison star of same spectral type, and for sight-lines where dust emission can be neglected. The flux of the reddened star F is reduced by the optical depth $\tau(\lambda) = \int_0^s \rho(s)\, K_{\mathrm{ext}}(\lambda)\, ds$ of the interstellar medium with dust density $\rho(s)$ of the ISM along the sight-line s from the star to the observer:

$$F(\lambda) = F_0(\lambda)\, e^{-\tau(\lambda)} \tag{8.1}$$

The optical depth is related to the dust extinction cross section when normalising to a photometric band such as V. For what is called extinction curve it holds that $\tau(\lambda)/\tau(V) = K_{\mathrm{ext}}(\lambda)/K_{\mathrm{ext}}(V)$. The extinction is generally stronger in the optical and UV part of the spectrum and weaker in the near-infrared (NIR). Extinction due to dust is therefore often called the reddening of stars, because their spectra are less blue than the spectra of similar, but unobscured stars. By comparing photometry or spectra of reddened with unobscured stars we can determine the extinction at different wavelengths.

The observed (dust) extinction curves are not completely smooth and display various spectral bands. The most prominent one is the 2175 Å bump, where graphite and polycyclic aromatic hydrocarbons (PAHs) have strong electronic transitions. At 9.7 µm and 18 µm, features are visible that are assigned to Si-O stretching and O-Si-O bending modes of silicate minerals, respectively. In addition, there are numerous weaker features, such as the 3.4 µm absorption assigned to C-H stretching modes (Mennella et al. 2003) and the diffuse interstellar bands in the optical (Krelowski 2014). The observed 9.7 and 18 µm band profiles can be reproduced better in the laboratory by amorphous silicates than by crystalline structures. Features that are assigned to crystalline silicates, such as olivine (Mg_{2x} Fe_{2-2x} SiO_4 with $x \sim 0.8$), have been detected in AGB and T Tauri stars and in comet Hale-Bopp (see Henning 2010 for a review). However, since these features are not seen in the ISM, the fraction of crystalline silicates in the ISM is estimated to be $\lesssim 2\,\%$ (Min et al. 2007). Dust in the diffuse ISM appears free of ices that are detected in regions shielded by $A_V > 1.6$ mag (Bouwman et al. 2011; Pontoppidan et al. 2003; Siebenmorgen and Gredel 1997; Whittet et al. 1988).

Extinction curves τ/τ_V for about 554 sight-lines are provided by Fitzpatrick and Massa (2007, 2009) and Gordon et al. (2009). The curves are reproduced by a third-order polynomial and a Drude profile, which accounts for the 2175 Å extinction bump. In the range $x = 1/\lambda \geq 3.3\,\mu\text{m}^{-1}$ the extinction curve is fit by

$$\frac{\tau(x)}{\tau_V} = c_1 + c_2\,x + c_3\,D(x, \gamma, x_0) + c_4\,F(x) \tag{8.2}$$

where $F(x)$, which describes the non-linear UV part of the curve, is defined as

$$F(x) = 0.5392(x - 5.9)^2 + 0.05644(x - 5.9)^3 \;:\; x \geq 5.9\,\mu\text{m}^{-1} \tag{8.3}$$

and otherwise $F(x) = 0$ for $x < 5.9\,\mu\text{m}^{-1}$. The Drude profile is given by

$$D(x, \gamma, x_0) = \frac{x^2}{(x^2 - x_0^2)^2 + (x\,\gamma)^2} \tag{8.4}$$

with a damping constant γ and the central wavelength $1/x_0$. At longer wavelengths, specifically for the $UBVJHK$ bands, one may apply the R_V parametrisation presented by Fitzpatrick (2004), where $R_V = A_V/E(B - V)$ is the ratio of *total-to-selective extinction*, a parameter which is difficult to derive. Higher quality optical/NIR extinction curves are expected from the ongoing *HST*/STIS observing programme (Li 2017).

The extinction curve provides strong constraints on dust models and, in particular, on the size distribution of the grains (Weingartner and Draine 2001). Essential work has been published by Mathis et al. (1977), who introduced a power-law scaling of the abundance of dust with particle radii $n(r) \propto r^{-q}dr$, the so-called MRN size distribution, and by Greenberg (1978), who presented the grain core–mantle model. In photometric measurements of stars, one usually assumes that all photons have reached the observer on a direct path and neglects the fraction that has been scattered. However, in observations of distant stars, for instance, when they are in another galaxy, this fraction is not negligible. Krügel (2009), Scicluna and Siebenmorgen (2015), and previous authors investigated how scattered light affects the extinction curve. Depending on the number of scattering events in the beam, the extinction curve is observed to steepen in homogeneous media, and flatten in clumpy media. As a result, clumpy dust distributions are able to reproduce extinction curves with arbitrary total-to-selective extinction R_V. The flattening is also able to "wash out" the 2175 Å bump, and results in a shift of the peak to shorter wavelengths. It is found that in the diffuse ISM at distances below 1 kpc such effects do not impact the extinction curve.

8.4 Dust Emission

The IR and submillimetre spectral range provides an important constraint for dust models. *IRAS* data revealed 12 and 25 μm emission from interstellar clouds that was stronger than expected (Boulanger et al. 1985). At the same time, various PAH emission bands have been detected (Allamandola et al. 1985, 1989; Puget and Leger 1989). Both emission components can only be explained by considering dust particles that are small enough to be stochastically heated. A step forward was taken in the dust models by Desert et al. (1990), Siebenmorgen and Krügel (1992), Dwek et al. (1997), Li and Draine (2001), Draine and Li (2007), Jones et al. (2017). In these models, very small grains and PAHs are treated as an essential grain component along with the (so far) standard carbon and silicate mixture of large grains.

In the mid-IR, there are conspicuous emission bands at 3.3, 6.2, 7.7, 8.6, 11.3, and 12.7 μm, as well as a wealth of weaker bands in the 3–25 μm region. These bands are ascribed to vibrational transitions in PAH molecules, which are planar structures that consist of benzene rings with hydrogen attached. Less perfect structures may be present where H atoms are replaced by OH or CN and some of the C atoms by N atoms (Hudgins et al. 2005). PAHs may be ionised: PAH^+ cations may be created by stellar photons, and PAH^- anions may be created by collisions of neutral PAHs with free e^-. The ionisation degree of PAH has little influence on the central wavelength of the emission bands, but has a large impact on the feature strengths (Allamandola et al. 1999; Galliano et al. 2008). The feature strengths depend on the hardness of the exciting radiation field and on the hydrogenation coverage of the molecules (Siebenmorgen and Heymann 2012).

The *Planck* Collaboration (2011, 2014, 2016) showed that dust in diffuse interstellar clouds emits most of the radiation at around 140 μm (Fig. 8.1). The peak corresponds to a mean dust temperature of ~20 K of the large grains. *Planck* data between 100 and 850 μm (Ysard et al. 2015) are fit by a power-law $I(\nu) \propto \nu^{\beta}\ B_{\nu}(T)$, where $B_{\nu}(T)$ is the *Planck* function at temperature T. Variations of the far-IR (FIR) emission in the diffuse ISM on large scales are revealed by the *Planck* satellite. Bot et al. (2009) find that the flux ratio in cirrus clouds of the ISM at 60 μm/100 μm decreases while the 160 μm/100 μm flux ratio increases. This

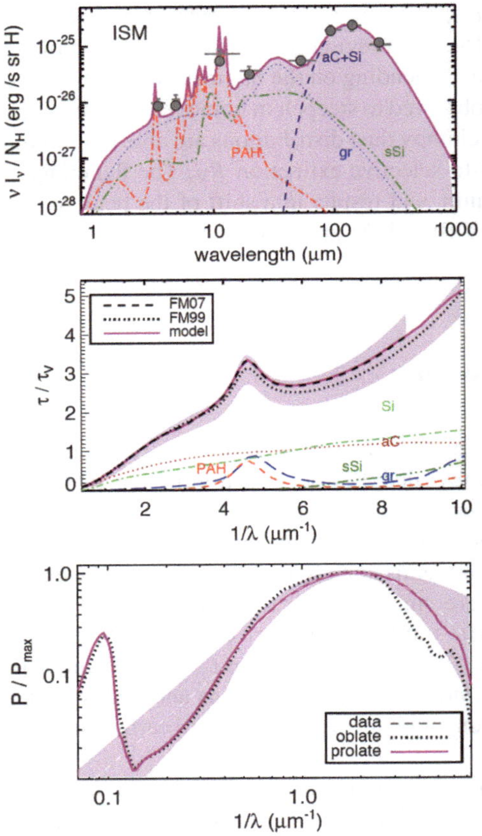

Fig. 8.1 Dust in the solar neighbourhood. **Top:** Dust emission when heated by the ISRF (Mathis et al. 1983). Observations (1σ error bars) at high galactic latitudes from *DIRBE* (Arendt et al. 1998) and FIRAS (Finkbeiner et al. 1999). Model (magenta) convolved with band passes (filled circles). **Middle:** Mean ISM extinction (dashed) by Fitzpatrick (1999), Fitzpatrick and Massa (2007). Contributions of the dust components to the total (magenta) as labeled: large ($r \geq 6\,\mathrm{nm}$) silicates (Si) and amorphous carbon (aC) particles. Small ($r < 6\,\mathrm{nm}$) silicates (sSi), graphite (gr), and PAH. **Bottom:** Linear polarisation normalised to its maximum. Observation (dashed line) as given by Voshchinnikov (2012). Polarisation of silicates is shown for prolate (magenta) and oblate (black dotted) particles. Hatched areas indicate 1σ variation. The model simultaneously fits the observed emission, extinction, and polarisation curves

dependency cannot be explained by changing the interstellar radiation field (ISRF, Mathis et al. 1983) that is heating the dust. Therefore significant changes of dust properties such as grain sizes or emissivities are expected from cloud-to-cloud.

8.5 Dichroic Polarisation

Hiltner (1949) and Hall (1949) published the first measurements of the polarisation of extra solar radiation. They observed distant stars with PD of up to 12%. Hall showed that the PD correlates with reddening along the sight-line towards the star. Additionally, the PA of stars close to each other on the sky was found to be similar, which is not expected if the polarisation is due to effects local to the stars. This suggested that the origin of the polarisation is interstellar matter between the stars and us. Their discovery of polarisation due to extinction of radiation by interstellar dust along the line of sight is called *dichroic* polarisation. It must be distinguished from polarisation that is due to light which scattered on dust particles.

In most dust models large grains are often simplified to be spherical, and the particle shape is generally not discussed further. Spherical particles can at most account for polarisation that is due to scattering but a symmetry argument shows that they cannot cause polarisation due to extinction. Consider a sphere: It is rotationally symmetric and linearly polarised radiation encounters the same sphere no matter what the plane of polarisation is. If the view to the particle is identical, then the extinction is also identical for the different types of linear polarisation. One type of linear polarisation cannot be diminished stronger than another type. Therefore, extinction due to spheres cannot create linear polarisation. The dichroic polarisation must be due to non-spherical particles. The particles must also have a mechanism for at least a partial alignment on large scales. The next simple geometry of a dust particle is obtained by stretching or compressing a sphere along one axis. This results in a spheroidal particle shape. Spheroids are characterised by the ratio a/b between major and minor semi-axes. There are two flavours of spheroids: prolates, such as needles, which are mathematically described by rotation about the major axis of an ellipse, and oblates, such as pancakes, obtained from the rotation of an ellipse about its minor axis. In our notation, the volume of a prolate is the same of a sphere with radius $r = (a \cdot b^2)^{1/3}$, and the volume of an oblate is that of a sphere with radius $r = (a^2 \cdot b)^{1/3}$.

Scattering properties of non-spherical particles are quite different from that of spherical grains. In early models Hong and Greenberg (1980), Kim and Martin (1994), Li and Greenberg (1997), among others, considered infinite cylinders. More realistic particle shapes such as spheroids have been considered more recently. As we will see below, the derivation of cross sections of spheroids is an elaborate task. The influence of the type of spheroidal grain on the polarisation is discussed by Voshchinnikov (2004). Dust models considering spheroidal particles that fit the average Galactic extinction and polarisation curves were presented by Gupta et al. (2005), Draine and Allaf-Akbari (2006), and Draine and Fraisse (2009). The observed interstellar extinction and polarisation curves towards individual sight-lines have been modelled by Voshchinnikov and Das (2008), Das et al. (2010), Draine and Hensley (2013), Voshchinnikov and Hirashita (2014), Siebenmorgen et al. (2014), and for a larger sample more recently by Voshchinnikov et al. (2016) and Siebenmorgen et al. (2018).

The polarisation due to interstellar extinction has been found to be maximal usually around $\lambda_{\rm max} \sim 0.6\,\mu$m. It decreases smoothly to shorter and longer wavelengths. The polarisation curve of the ISM shows an empirical wavelength dependence, first described by Serkowski (1973) and often called Serkowski curve,

$$p(\lambda) = p_{\rm max}\,\exp\left[-k_{\rm p}\,\ln^2\left(\frac{\lambda_{\rm max}}{\lambda}\right)\right]. \tag{8.5}$$

The Serkowski curve includes three free parameters: the maximum polarisation $p_{\rm max}$, the wavelength $\lambda_{\rm max}$ at $p_{\rm max}$, and the width of the spectrum $k_{\rm p}$. The Serkowski curve has been observed for many sight-lines (e.g. Zickgraf and Schulte-Ladbeck 1989; Orsatti et al. 1998; Schultz et al. 2004; Efimov 2009; Bagnulo et al. 2017). The width of the Serkowski curve was initially thought to be $k_{\rm p} = 1.1$, later it is found to scale with $\lambda_{\rm max}$ (Wilking et al. 1982).

In the ISM the maximum polarisation is observed to be $p_{\rm max}/A_{\rm V} \leq 3\,\%$/mag (Whittet 2003), and when assuming a "mean" total-to-selective extinction of $R_{\rm V} = 3.1$ of the ISM, one translate it to $p_{\rm max}/E_{\rm B\text{-}V} \leq 9.3\,\%$/mag. By taking reddening of the stars from Arenou et al. (1992), Drimmel et al. (2003), Lallement et al. (2014), Gontcharov (2017), one finds that a number of stars violate this limit and significant variations from the canonical $R_{\rm V} = 3.1$ value are suggested. In the thermal IR at $\lambda \geq 2.5\,\mu$m polarisation data are fit by a power-law where $p(\lambda) \propto \lambda^{-t}$ with $1.6 \leq t \leq 2$ (Martin et al. 1992; Nishiyama et al. 2006). As can be seen from Fig. 8.1, the Serkowski or power-law fits naturally break down in the 10 μm silicate band. Spectropolarimetry in the 8–13 μm range is presented by Smith et al. (2000) for 55 sources. Typically absorptive polarisation profiles in the mid-IR are similar to the archetypal silicate polarisation found in OMC1 BN and show a mid-IR polarisation of 1–3%/mag. In carbon rich sources the 10 μm polarisation is weaker and ranges between 0.5 and 1%/mag. Carbon-rich sources have nearly featureless mid-IR polarisation that is also ascribed to dichroism in carbon-based grains, providing evidence that such grains can be aligned. Aligned silicates on their own are sufficient to explain the optical/NIR polarisation spectrum.

In Fig. 8.1 we show that prolate particles provide a better fit to the Serkowski curve than oblate dust particles. However, oblate particles provide higher polarisation efficiencies than prolate shapes, therefore in directions with observed high dichroic polarisation a dominance of oblate shapes could be suggested.

8.6 Polarised Emission

Dust not only absorbs but also radiates; polarised emission of aligned non-spherical dust particles is observed in the FIR and sub-millimetre spectral region. Due to atmospheric absorption, these spectral regions are nearly exclusively accessible to satellite and balloon borne missions. Outstanding examples are provided by the *Planck* mission (Planck Collaboration 2015; Fanciullo et al. 2017; Guillet et al. 2018). The intensity of a uniform dust cloud of optical depth τ and temperature T is given by: $I = B(T) \cdot (1 - \exp(-\tau))$, where τ is the optical depth and $B(T)$ the Planck function. The optical depth in the FIR is small. The polarised emission can be taken similar to the case of dichroic polarisation in the optical, where the polarisation in a given direction depends on the optical depths at the position angles of maximum and minimum absorption, respectively, $\tau_{\max}$ and $\tau_{\min}$. The PD can therefore be written as:

$$p_{\mathrm{FIR}} = \frac{\tau_{\max} - \tau_{\min}}{\tau_{\max} + \tau_{\min}} \tag{8.6}$$

The value of p_{FIR} is finite, even for small τ. These optical depths are directly related to the ratio of the longest grain axis to the shortest. Krügel (2003) computes the percentage polarisation of thermal dust emission at wavelengths between 8 and 2000 µm, assuming perfect alignment of the rotational axes of the dust particles. He finds a flat FIR/submm polarisation spectrum. Even a small deviation from a spherical shape of the grain results in significant polarisation that is emitted by the dust. For a ratio of the major-to-minor axis of the grain, a/b = 1.4, the predicted polarisation is in the 20 µm – 2 mm range is as high as 20%. Therefore, even if only a small fraction of the dust particles are aligned, one expects a rather high degree of polarised emission. Indeed, *Planck* observed fractional polarisation of the order of 20–25 % (Guillet et al. 2018).

8.7 Combining Optical and FIR Polarisation

Let us assume that the FIR and optical polarisation of the source is known. Krügel (2003) shows that the polarisation properties at both wavelength regimes are related

$$\frac{p_{\mathrm{V}}}{p_{\mathrm{FIR}}} = \tau_{\mathrm{V}} \frac{\beta_{\mathrm{FIR}}}{\beta_{\mathrm{V}}} , \tag{8.7}$$

where $\beta = \tau_{\max}/\tau_{\min} - 1$. The polarisation fractions p_{V}, p_{FIR}, and optical depth τ_{V} in the visual are determined observationally, leaving the information on grain shape contained within the ratio of β values at the different wavelengths. Because the two wavelength regimes are influenced differently by elongation, this ratio can be used to constrain grain shapes. A correlation of *Planck* data at 353 GHz with observed optical (starlight) polarisation demonstrated that common dependencies such as unknowns of the dust alignment and magnetic field structure can be removed (*Planck* Collaboration 2016; Guillet et al. 2018).

Constraints on the properties of dust particles in the diffuse ISM and their shape are critical for understanding the sub-millimetre emission of dust, which is often

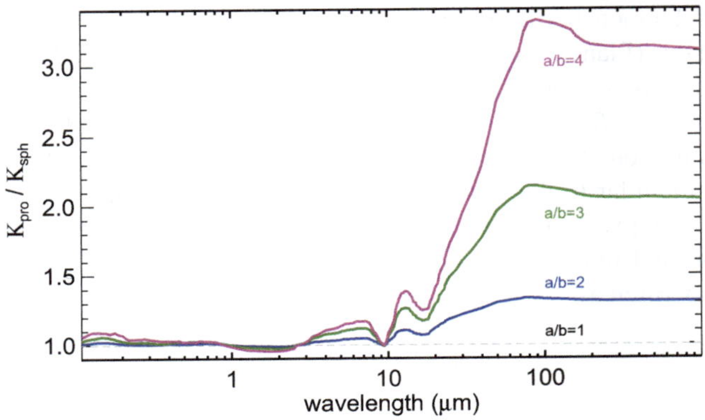

Fig. 8.2 Ratio of the total mass extinction cross section of large prolate and spherical particles with same volume. Prolates, with a/b ratios as labeled, and spheres have same size distribution and composition as derived for the dust in the solar neighbourhood

used to estimate dust masses. As can be seen in Fig. 8.2, estimates based on spherical grain shapes might significantly over-estimate the dust mass. This is of wide interest and might, for example, impact the so-called dust budget crisis of high redshift galaxies (Watson et al. 2015).

8.8 Grain Alignment

The cosmic dust grains are somehow aligned. The mechanisms doing this have to act against the randomisation of motion that happens when grains collide with the gas (mostly hydrogen) around them. Because the interstellar medium is so thin, a collision is expected on average a few times per hour (Davis and Greenstein 1951). The mass of a grain is much higher than the mass of a molecule, therefore randomising the motion of a grain takes tens to hundreds of thousands of years. There are two major theories of how the grains overcome the randomisation: the Davis–Greenstein mechanism and the radiative torque alignment mechanism (Andersson et al. 2015). Both assume that the grains contain metal inclusions. In the Davis–Greenstein relaxation model, the dust grains spin initially randomly. Because of the metal inclusions, they interact with the magnetic field, and spin around the magnetic field lines. Their motion initially contains precession and nutation moments. Over thousands of years the metal inclusions dissipate the energy in these moments and the grains are left with a stable rotation axis (Jones and Spitzer 1967; Purcell 1979). A more recent theory is the radiative torque (RAT) alignment (Dolginov and Mitrofanov 1976). In the RAT model the grains start spinning when they are hit by a photon. Slight irregularities in their shape make them helical, leading to different scattering efficiencies for left- and right-handed circular polarised radiation. Every scattering of a photon will transmit a small amount of torque and finally the grain aligns with the magnetic field (Lazarian and Hoang 2007).

The alignment in both cases is such that the short axis of the spheroid aligns with the magnetic field. Of course in reality there is some spinning and wobbling of the grain (axis of orientation) with the B vector. As oblate grains have only one short axis, they spin like a regular spinning top. Prolate grains will spin like an egg on a counter-top, the magnetic field being vertical. The extinction for radiation polarised parallel to the long axis is higher than for radiation polarised perpendicular to it. A ray of photons crossing a cloud of magnetically aligned grains will therefore exit the cloud polarised parallel to the magnetic field. Radiation emitted by spheroids is polarised preferably parallel to the long axis. Therefore thermal emission from the same cloud will be linearly polarised perpendicular to the magnetic field. This has been observed, e.g., in the Orion molecular cloud and towards the Galactic Centre (Clayton and Mathis 1988; Hildebrand et al. 1984; Werner et al. 1988) and extra-galactic objects (Siebenmorgen et al. 2001). There are additional constraints on the alignment (Lazarian 2007). Firstly, only silicate particles are expected to be aligned, carbonaceous particles might be aligned less efficiently (Mathis 1986). Secondly, aligned silicate particles with radii larger than 0.1 μm are on their own sufficient to account for most of the observations (Draine and Fraisse 2009; Siebenmorgen et al. 2018).

8.9 Dust in the Solar Neighbourhood

Comparing results of the extinction fitting with infrared observations at high galactic latitudes has become a kind of benchmark for the above cited dust models that aim to reproduce dust emission, extinction, and polarisation spectra of the ISM. The average of the extinction curves over many sight-lines is taken to be representative for the diffuse ISM of the Milky Way. It gives a total-to-selective extinction of $R_V = 3.1$. In a similar way one takes an average of available polarisation curves which is then fit by a Serkowski curve with parameters $\lambda_{max} = 554\,$nm and $k_p = 1.66\ \lambda_{max} + 0.01$ (Voshchinnikov 2012). IR/submm observations at high Galactic latitudes ($|b| > 25°$) provide constraints on the typical dust emission spectrum (Arendt et al. 1998; Finkbeiner et al. 1999). The heating of the dust in the solar neighbourhood is approximated by the mean intensity of the interstellar radiation field (Mathis et al. 1983).

Modelling extinction, emission, and polarisation data simultaneously, while respecting cosmic abundance constraints, typically leads to the following set of dust parameters: $[C]/[H]_{tot} = 94 \pm 10\,$ppm of C atoms with respect to H atoms in the ISM, which are in carbon grains, out of which $[C]/[H]_{gr} = 17 \pm 6\,$ppm are in nanometre sized (<6 nm) graphite and $[C]/[H]_{PAH} = 10 \pm 9\,$ppm in PAH. The total silicon abundance in dust is $[Si]/[H]_{tot} = 20 \pm 2\,$ppm, out of which $[Si]/[H]_{sSi} = 5 \pm 2\,$ppm are in small (<6 nm) silicates. We remind that such abundances are relative measures. The exponent of the MRN dust size distribution is $q = 3.0 \pm 0.4$, with upper sizes for silicates of $r_{+}^{\mathrm{Si}} = 303 \pm 50$ nm and carbon of $r_{+}^{\mathrm{aC}} = 316 \pm 157$ nm. In this model one assumes that only silicates are aligned. The upper size limits of carbon particles have a larger uncertainty than that of silicates, which are further constrained by fitting the polarisation curve. The minimum radius of aligned silicates is $r_{\mathrm{pol}}^{-} = 136 \pm 28$ nm. This model by Siebenmorgen et al. (2018) is shown in Fig. 8.1.

Polarimetry imposes additional constraints on the dust properties; in particular, simultaneous modelling of extinction and polarisation results in steeper dust-grain size-distributions and smaller mean grain sizes than when using extinction data only.

8.10 ISM Structure

Polarisation has been measured for thousands of stars at precision down to $\sigma(p) \leq 0.02\,\%$, see catalogs by Heiles (2000), Fosalba et al. (2002), and more recently a compilation by Gontcharov and Mosenkov (2018). The distribution of the PD p, efficiency $p_{\mathrm{max}}/E_{\mathrm{B-V}}$, and position angle in Galactic coordinates are presented by Lallement et al. (2014), and using *Gaia* DR2 distances (Gaia Collaboration et al. 2018a,b,c) of the stars by Gontcharov and Mosenkov (2018). The maps show the three dimensional distribution of dust in the solar neighbourhood. The Gould Belt, that is located within 100 pc from the Sun, becomes visible as an unpolarised region.

8.11 Cloud-to-Cloud Variation of Dust

Interstellar absorption spectra, as evidenced by Fitzpatrick and Massa (2007), diffuse interstellar bands (DIBs, Heger 1922; Hobbs et al. 2008), as well as polarisation curves of the ISM (Efimov 2009) differ from one sight-line to the next. The variations are so strong that one is wondering if blind averages over such curves comprises meaningful physical signatures of the dust. Furthermore, the interpretation of the extinction and polarisation of the interstellar medium along a particular sight-line is also strongly biased by the fact that multiple interstellar clouds may be present within the observing beam. Fortunately one can resolve individual Doppler components in interstellar atomic and/or molecular lines by means of high-resolution spectroscopy ($\lambda/\Delta\lambda \sim 10^5$). Useful lines are CaII H and K, NaI 5889 and 5895 D_2 and D_1, the KI 7699 Å lines, and also features of CH, CH^+, and CaI near 4300, 4232, and 4227 Å, respectively. Sight-lines where the lines show one Doppler component are thought to be dominated by a single absorbing cloud. Of course in some sight-lines there may be more than one component with the same radial velocity although such cases are thought to be rare.

A Large Interstellar Polarisation Survey (LIPS) using spectropolarimetric facilities in both hemispheres has been carried out. Results obtained with the FORS2 instrument of the ESO Very Large Telescope (VLT) are presented by Bagnulo et al. (2017). The spectra cover the wavelength range 380–950 nm at spectral resolving power of about 880. Serkowski-curve parameters, as well as the wavelength gradient of the PA for the interstellar polarisation along 76 different sight-lines are published. These sight-lines have been used to simultaneously fit extinction and polarisation curves applying the dust models of Sect. 8.9, and are complemented by available VLT/UVES high-resolution spectra. This allowed the extraction of the small number of 15 single-cloud sight-lines. It is found for sight-lines that intersect several clouds and hence different cloud compositions that the pristine nature of the extinction and polarisation curve is indeed lost. As shown as open symbols in Fig. 8.3 one loses the possibility to investigate the relation between the physical parameters of the dust and the observing characteristics of the extinction and polarisation. The only possibility to investigate such relations is to inspect sight-lines dominated by a single-cloud (filled symbols in Fig. 8.3).

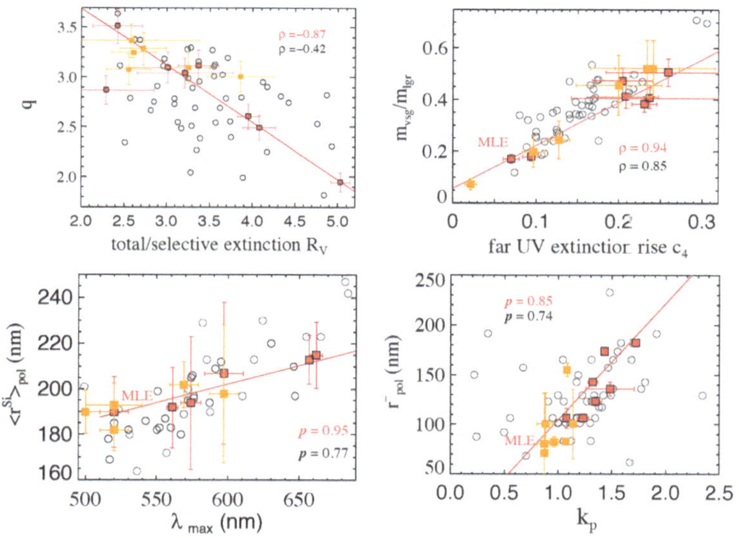

Fig. 8.3 Dust parameters versus observed extinction and polarisation characteristics. Influence of *top-left*: the exponent of the dust size distribution q on the total to selective extinction R_V, *top-right*: the mass ratio of small and large (≥6 nm) grains m_{vsg}/m_{lgr} on the strength of the far UV rise of the extinction curve (c_4), *bottom-left*: the mean size of aligned silicates $< r_{pol}^{Si} >$ versus the wavelength at which the polarisation has its maximum (Serkowski parameter λ_{max}), and *bottom-right*: the minimum size of aligned silicate grains r_{pol}^{-} versus the width of the polarisation curve (Serkowski parameter k_p). **Symbols:** Sight-lines to individual stars that intersect multiple dust clouds are shown as open circles and their Pearson correlation coefficient ρ is given in black. Sight-lines to stars where only a single dust cloud is penetrated are shown as red squares together with their Person coefficient and a straight line maximum-likelihood estimator in red. In the course of further archive research we identify a few additional single-cloud sight-lines that we mark in orange

For single-cloud sight-lines several correlations between the observed extinction and polarisation characteristics and the physical parameters of the dust are confirmed and even previously undetected correlations are detected (Fig. 8.3). One notices that interstellar polarisation from multiple-clouds is smaller than that from single-cloud sight-lines. The interpretation of polarisation data from multiple-clouds sight-lines is complicated. After the radiation is polarised in a dust cloud, the intersection of a second or more clouds leads to de-polarisation (Sect. 8.12). This fact can be understood by cloud-to-cloud variations of the grain properties, dust alignment efficiency, and the direction of the magnetic field.

It was noticed earlier that the mean radius of silicates correlates with $\lambda_{\max}$ (Chini and Krügel 1983) and Mathis (1986) explained the relation between $k_{\rm p}$ and $\lambda_{\max}$ by varying the minimum size of aligned grains. In the single-cloud cases the correlations between the parameters of the Serkowski curve and the silicate dust parameters become more obvious. For example, the wavelength at which the polarisation reaches maximum $\lambda_{\max}$ correlates with the mean radius of aligned silicates $\langle r \rangle^{\rm Si}_{\rm pol}$ much stronger for the single-cloud sight-lines with a Pearson coefficient of $\rho = 0.95$ as compared to $\rho = 0.77$ for the full sample. The same holds for the width of the correlation of the polarisation spectrum $k_{\rm p}$ with the minimum radius of aligned silicates $r^{-}_{\rm pol}$ where the Pearson coefficient of $\rho = 0.85$ is again higher than for the full sample showing $\rho = 0.74$ (Fig. 8.3). No dependencies of the Serkowski parameters on the mean radius of large carbon particles $\langle r \rangle_{\rm aC}$ are found. The data also show large variations of the dust characteristics from cloud-to-cloud. However, when a number of clouds are averaged, a similar "mean" of the dust parameters is retrieved representing of what is called the Milky Way mean extinction curve. Studies of single-cloud sight-lines are most suitable to investigate the pristine nature of interstellar dust (Siebenmorgen et al. 2018).

8.12 Propagation of Polarised Light in Dusty Media

Dust obscured objects cannot be studied directly in the UV/optical, since the dust shields most of the visible light. To derive the UV/optical component or to constrain the morphological structure from available IR/submm observations, a detailed model of the interaction of photons with the dust is required. The problem is to solve the radiative transfer (RT) equation, see, e.g., Krügel (2008) for a comprehensive textbook description. Analytically, the RT problem can only be done in some simple configurations, for example, by assuming spherical or disc symmetry in which either scattering or absorption is neglected and the wavelength dependency of the dust cross section is strongly simplified, e.g. by a gray body. Nature, however, is usually not well approximated by such assumptions and numerical modelling is the only way to solve this problem.

In a dusty medium the intensity $I_\nu(s)$ at position s in the direction of a unit vector k is weakened by extinction and enhanced by thermal dust emission and other photons that are scattered from any directions into the beam k. It holds, dropping frequency dependencies, that

$$\frac{1}{\rho}\frac{\mathrm{d}I}{\mathrm{d}s} = \Gamma - K_{\mathrm{ext}}\, I \,, \tag{8.8}$$

where $\rho(s)$ is the dust density distribution of the object, K_{ext} the extinction cross section, and $\Gamma(s)$ the source term. It includes emission and scattering by dust and is given by

$$\Gamma = K_{\mathrm{abs}} \int B(T)\, P(T)\, \mathrm{d}T + \frac{K_{\mathrm{sca}}}{4\pi} \int \Phi(k', k)\, I(k')\, \mathrm{d}\Omega \,, \tag{8.9}$$

where K_{abs}, K_{sca} are the absorption and scattering cross section. We compute cross sections in units cm^2 per gram of dust. The dust temperature distribution function $P(T)$ is computed following Guhathakurta and Draine (1989), Siebenmorgen et al. (1992). The phase function $\Phi(k', k)$ of photons scattered from direction k' into k is difficult to compute. Numerical treatments such as the rejection sampling method are applied (Baes and Camps 2015; Peest 2018) or it is assumed that $\Phi(k', k)$ can be approximated by some mathematical formulae such as the Henyey–Greenstein phase function (Henyey and Greenstein 1941).

This description of the radiative transfer problem of Eq. (8.8) is incomplete, as it does not consider the polarisation of the radiation. For polarised light the radiation field I is represented by the Stokes vector $\mathbf{S}$,

$$I \Rightarrow \mathbf{S} = (I, Q, U, V)^T \,. \tag{8.10}$$

The first element of the Stokes vector describes the intensity of the radiation, the second Q and third term U the linear polarisation, and V the circular polarisation (Stokes 1852). Changes to the Stokes vector are described by 4×4 Müeller matrices

$$\mathbf{S}_{\mathrm{new}} = \mathbf{M} \cdot \mathbf{S} \,. \tag{8.11}$$

The linear polarisation refers to a reference direction, which can be any direction perpendicular to the propagation direction. In analogy to observations, Peest (2018) introduced a North direction $\mathbf{d}_{\mathrm{N}}$, where North is "up" when looking towards the source. The East direction is then "left". Directions and angles used in this section are defined in the left panel of Fig. 8.4. Concerning handedness there are various conventions in the literature (Hamaker and Bregman 1996; Peest et al. 2017) and we apply the one set by the IAU (Contopoulos and Jappel 1974). The Stokes vector changes depending on the choice of North. It is multiplied with a rotation matrix $\mathbf{R}(\varphi)$ when rotating the North direction by φ about the propagation direction of the

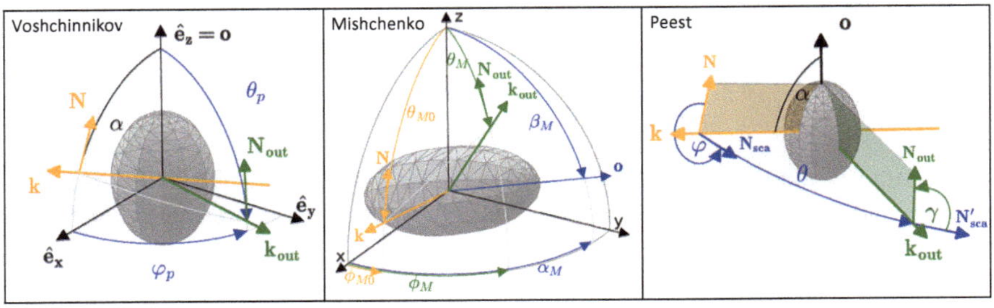

Fig. 8.4 Definition of the scattering angles and North of the incoming k and outgoing (scattered) k_{out} photon direction as defined by Voshchinnikov in the particle frame (*left*), Mishchenko (*middle*), and as used in our Monte Carlo radiative transfer code MCpol (*right*), see Peest (2018) for further details

radiation (Fig. 8.4),

$$\mathbf{S}_{\rm rot} = \mathbf{R}(\varphi)\,\mathbf{S} = \begin{pmatrix} 1 & 0 & 0 & 0 \\ 0 & \cos 2\varphi & \sin 2\varphi & 0 \\ 0 & -\sin 2\varphi & \cos 2\varphi & 0 \\ 0 & 0 & 0 & 1 \end{pmatrix} \mathbf{S}\,. \tag{8.12}$$

For scattering on *spherical* grains: North is usually chosen in the plane of scattering, defined by the propagation direction before scattering $\mathbf{k}$ and the propagation direction after scattering $\mathbf{k}'$.

For scattering on *spheroidal* grains: North is in the plane of incidence, defined by the propagation direction before scattering $\mathbf{k}$ and the grain orientation $\mathbf{o}$ (Fig. 8.4).

Scattering events of polarised light are described by a 4×4 scattering matrix $\mathbf{Z}$. For incoming radiation it describes the probability of scattering towards any direction. Up to seven elements of the $\mathbf{Z}$ matrix are independent (Bohren and Huffman 1998). Because of the rotational symmetry of spheroids, the direction of the incoming radiation can be described by one angle, the angle of incidence α. It is defined by the direction of propagation and the grain symmetry axis (Fig. 8.4).

The probability that a photon is scattered or absorbed by a spheroidal dust particle depends on the polarisation of the incoming radiation. The absorption by spheroids give rise to dichroism and birefringence. *Dichroism* means that the extinction differs for differently polarised radiation. *Birefringence* means that the travel speed through the medium differs for different polarisation or in other words that a wave that passes through a medium of complex refractive index $m = n + \mathrm{i} \cdot k$ is phase-shifted with respect to a wave travelling through vacuum (Krügel 2003).

Extinction of polarised light is represented by a 4×4 extinction matrix $\mathbf{K}$. For spheroids with North in the plane of incidence the extinction matrix has a block diagonal shape

$$\mathbf{K} = \begin{pmatrix} K_{\text{ext}} & K_{\text{pol}} & 0 & 0 \\ K_{\text{pol}} & K_{\text{ext}} & 0 & 0 \\ 0 & 0 & K_{\text{ext}} & K_{\text{cpol}} \\ 0 & 0 & -K_{\text{cpol}}l & K_{\text{ext}} \end{pmatrix} \tag{8.13}$$

By replacing in Eq. (8.8) the intensity by the Stokes vector and the cross sections by the Müeller matrices the RT equation for polarised light can be written in analogy to Eq. (8.8) as:

$$\frac{1}{\rho}\,\frac{\mathrm{d}\mathbf{S}}{\mathrm{d}s} = \tilde{\Gamma} - \mathbf{K}\,\mathbf{S} \tag{8.14}$$

where the source term $\tilde{\Gamma}$ includes the Stokes vector for polarised emission S_{em}, the 4×4 scattering $\mathbf{Z}$ and extinction $\mathbf{K}$ matrices.

$$\tilde{\Gamma}(s) = \mathbf{S}_{\text{em}} \int B(T)\; P(T)\; \mathrm{d}T + \frac{1}{4\pi} \int \mathbf{Z}(k', k)\; \mathbf{S}(k')\; \mathrm{d}\Omega\,. \tag{8.15}$$

The Stokes vector of polarised emission is

$$\mathbf{S}_{\text{em}} = \left(K_{\text{abs}},\, K_{\text{abs}}^{\text{pol}},\, 0, 0\right)^{T}\,. \tag{8.16}$$

The absorption cross sections K_{abs} and $K_{\text{abs}}^{\text{pol}}$ depend on the angle of incidence α (Fig. 8.4). North is in the plane given by the propagation direction of the photon and the symmetry axis of the particle.

Unfortunately solving this equation is complicated and Monte Carlo radiative transfer (MCRT) schemes have to be applied. We exemplify such MCRT treatments for unpolarised radiation as outlined by Krügel (2008): Initially a Cartesian grid with cubes is set to represent the model space. For semi-adaptive mesh refinement each cube may be divided into subcubes. The dust density $\rho(s)$ is assumed to be uniform within each dust cell. The cells have a temperature which is initially zero Kelvin. The flight path of a huge number of photon packages ($\sim 10^9$) through the model is followed. Each package is generated in a cell which represents the source, e.g. a star. The photon packages have a unique luminosity, frequency, and an initial arbitrary propagation direction. The distance Δs along the propagation direction from the entrance point to the edge of the current cell is calculated. This gives an optical depth of the cell of $\Delta\tau = \rho\; K_{\text{ext}}\; \Delta s$. A random roll ξ determines at which optical depth the photon will interact in the cell. If the optical depth to the edge of the cell is lower, $\Delta\tau < -\ln\xi$, the photon propagates to the edge of the cell and enters the next cell. If the photon reaches the edge of the model space, it is detected, its direction and frequency is recorded. If the optical depth is higher than the roll, $\Delta\tau \geq -\ln\xi$, the photon interacts. The length s to the interaction point is calculated, $s = -\ln\xi / nK_{\rm ext}$. The photon is propagated to the interaction point. The *albedo* $\Lambda = K_{\rm sca}/K_{\rm ext}$ of the dust is calculated. A new random roll ξ' decides whether the photon scatters, $\Lambda > \xi'$, or is absorbed, $\Lambda \leq \xi'$, by the dust. Scattering changes the propagation direction of the photon and is based on the phase function $\Phi(k', k)$. In case of absorption the temperature in the cell rises and the photon package is re-emitted from the interaction point with a wavelength based on the current cell temperature. This is the Bjorkman and Wood (2001) method to calculate the radiation field in one iteration. Re-emission is considered isotropic. The photon starts from the interaction point or from the edge of the next cell. It repeats the steps as above starting with calculating Δs until it leaves the model space. The simulation finishes when all photons have left the model. The scheme is well suited for parallel computing (Heymann and Siebenmorgen 2012).

A MCRT scheme for polarised light can be set up in a similar way and few enhancements need to be included. Each photon package stores besides its frequency, its accompanying Stokes vector, which is oriented w.r.t. North. The diminishing of the radiation follows no longer a simple exponential ($\propto \exp(-\tau)$), but becomes the solution of the linear differential matrix equation

$$\frac{d}{ds}\mathbf{S}(\nu, \mathbf{s}, \mathbf{k}) = -\rho(\mathbf{s})\mathbf{K}(\nu, \mathbf{k})\mathbf{S}(\nu, \mathbf{s}, \mathbf{k}) \tag{8.17}$$

The closed-form solution was published by Whitney and Wolf (2002) and Lucas (2003). It includes, for each element of the Stokes vector, mixing terms $\cosh(\rho K_{\rm pol} s)$ and $\sinh(\rho K_{\rm pol} s)$. To determine whether the photon package will interact in a given cell, one needs to know its optical depth. For dichroically extinct polarised radiation and photon path of length s_c through a cell with density ρ the optical depth is

$$\tau = -\ln\left(\frac{I(s_c)}{I(0)}\right) \tag{8.18}$$

$$= \rho K_{\rm ext} s_c - \ln\left(\cosh(\rho K_{\rm pol} s_c) - \sinh(\rho K_{\rm pol} s_c)\frac{Q_0}{I_0}\right) \tag{8.19}$$

The interaction point of a photon packet in a cell depends on the polarisation status of the incoming radiation. The length s of the starting point of the photon in the cell to the interaction point can be approximated (Baes et al. 2019)

$$s \sim s_c \, \frac{\ln(\xi)}{\tau} \, . \tag{8.20}$$

The albedo of polarised radiation also depends on Stokes Q

$$\Lambda = \frac{K_{\rm pol} I + K_{\rm pol} Q}{K_{\rm ext} I + K_{\rm ext} Q} \tag{8.21}$$

We have seen that the propagation of polarised light in a dusty medium is described by the radiative transfer equation. It is numerically solved by Monte Carlo Radiative Transfer (MCRT) codes that are presented with different levels of sophistication, for example, by Voshchinnikov and Karjukin (1994), Bianchi et al. (1996), Harries (2000, TORUS), Watson and Henney (2001, Pinball), Misselt et al. (2001), Whitney and Wolf (2002), Lucas (2003), Pinte et al. (2006, MCFOST), Min et al. (2009, MCMAX), Robitaille (2011, HYPERION), see Steinacker et al. (2013) for a review up to that time, and more recently by Goosmann et al. (2014, STOKES), Kataoka et al. (2015, RADMC-3D), Peest et al. (2017, SKIRT), and Reissl et al. (2016, POLARIS) which calculates the polarised emission, scattering, dichroic extinction, and birefringence by spheroidal dust that are aligned assuming different mechanisms. For a more detailed description of our scheme on the propagation of polarised light through a dusty medium and an MCRT treatment to solve it, see Peest (2018).

8.13 Cross Section of Spheroidal Grains

The interaction of light with dust particles is described by the cross sections of the grains. One distinguishes dust cross sections for absorption, scattering, linear, and circular polarisation (Voshchinnikov 2012). One advantage of assuming that interstellar dust grains are spheroids is that the optical properties of spheroids can be derived for a wide range of grain parameters and wavelengths in an (also not easy but) understandable methodology, and with reasonable computer resources. Because of their regular shape, a special method can be used to solve Maxwell's equations and determine the interaction of radiation with them (Asano and Yamamoto 1975; Voshchinnikov and Farafonov 1993). For slightly more general particles, namely all rotationally symmetric objects, there is the so-called T-matrix method (Mishchenko 1991; Mishchenko et al. 1996, 2016; Mishchenko and Travis 1998). For general non-spherical particles, neither method is applicable. The most general method is to bin the shape into a regular grid of dipoles and calculating the optical properties of this system. This is how, e.g., the code DDSCAT works (Purcell and Pennypacker 1973). It has been refined (Draine 1988; Draine and Flatau 1994) and is being applied to a growing number of problems (Yurkin and Hoekstra 2007; Reissl et al. 2018). One critical issue with the discrete dipole approximation is to find how such dipoles are distributed in a realistic grain and for large particle sizes the computer memory needs are exploding.

Ultimately it is irrelevant which method is used to calculate the optical properties of the dust. All methods are orders of magnitudes too slow to be run for every of the trillions of scattering events in a radiative transfer computation. Instead, results have to be pre-computed and used as input files for the codes. The 16 elements of the scattering matrix depend on frequency and particle size, orientation angle of the grain, and two angles describing the direction of the outgoing radiation (Fig. 8.4). Even for a crude grid of such a multi-dimensional matrix, the memory easily amounts to several dozens of GB. Another issue is an ambiguity in the definition of the Stokes vector. The tables of the grain codes computing the cross sections therefore sometimes need to be converted between conventions as summarised by Peest et al. (2017).

We give one example and compare results of elements of the 4×4 Müller scattering matrix Z of a spheroidal dust particle utilising the Voshchinnikov and Mishchenko codes. We consider a prolate grain with ratio of the major axis $a/b = 1.9$, equal-volume sphere radius of $r = 200$ nm, which is irradiated by light at $\lambda = 160$ nm. We take for the dust particle as optical constant $m = n + \mathrm{i}\,k = 2.1 + 0.26\mathrm{i}$. We vary directions of the incoming k_{in} and scattered k_{out} photon directions. Results are computed for the particle frame (Fig. 8.4 (left)). We define the z-axis to be parallel to the major axis $z \parallel a > b$ of the particle and North direction $\mathbf{d}_{\mathrm{N}}$ so that it is in the plane of incidence and perpendicular to k_{in}. The angle of incidence α is between the z-axis, represented by the unit vector e_{z} of the grain and the propagation direction of the radiation k_{in}. The vector pair $(e_{\mathrm{z}}, k_{\mathrm{in}})$ open the plane of incidence.

We find quantitative very good agreement between both completely different treatments. The slight differences that can be sometimes noticed are mainly caused by different binning of the angles when using one of either code. One example is shown in Fig. 8.5 where we vary $0° \leq \varphi \leq 360°$ for three different constant pairs (α, θ) as indicated. However, such excellent agreement between the codes is only visible when applying two basic changes in the output of the Voshchinnikov code. The angle φ shall be counted anti-clockwise so that we replace φ by $360° - \varphi$. Furthermore, we need to change the definition of the Stokes vector to be consistent with that used by Mishchenko. Therefore, we flipped the sign of the elements of the Müller matrix of Z_{13}, Z_{14}, Z_{23}, Z_{24}, Z_{31}, Z_{32}, Z_{41}, and Z_{42}.

8.14 Conclusion

Several advances on the observational and computational side have enlighten the physical composition of dust particles in the ISM. A large polarisation survey in the optical highlights the dichroic nature of the ISM continuum polarisation, which is explained by partly aligned non-spherical grains. Sight-lines with a single dust cloud in the observing beam are most suitable to investigate the pristine nature of interstellar material and estimate the magnetic field structure of the Galaxy. However, while hundreds of extinction and polarisation curves are observed their number for single-cloud sight-lines is very limited. This is asking for further observing surveys.

Recent advances in computer technology enable the use of efficient Monte Carlo techniques to study the transport of radiation in arbitrary three-dimensional geometry for polarised or unpolarised light through a dusty medium. Progress in this field is at an early stage with a few codes existing and easy accessible benchmark results for validation just published. Exploring the physics of the dust by applying such MC treatments is an opportunity for further studies.

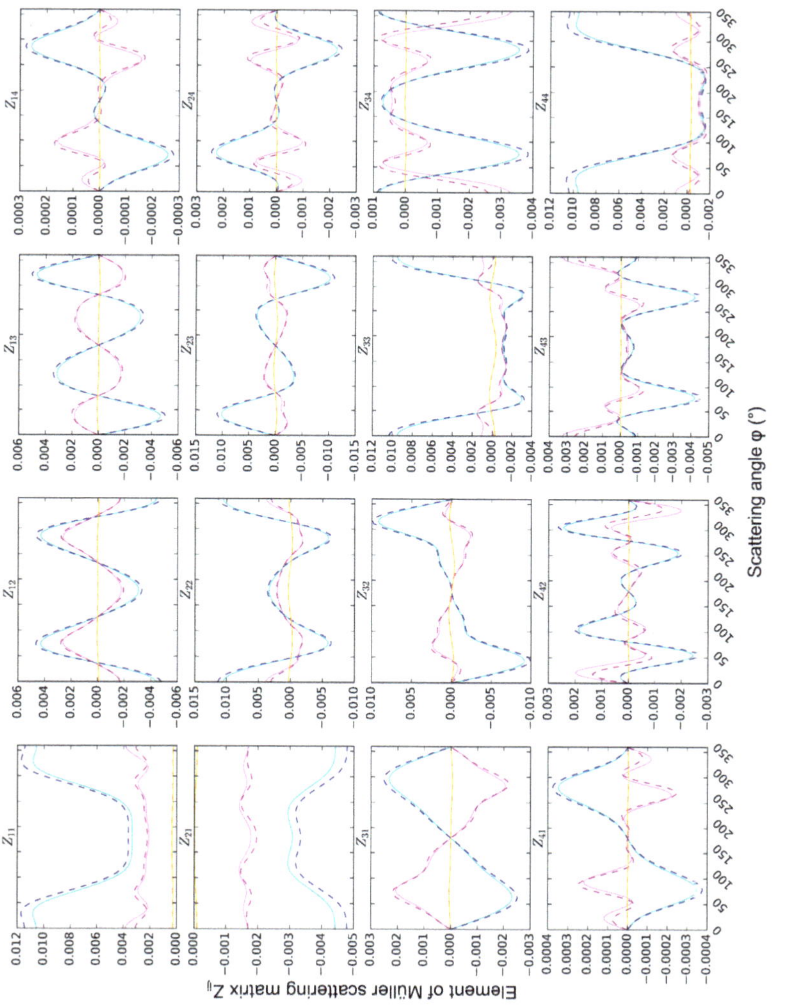

Fig. 8.5 Elements of the scattering matrix Z_{ij} as a function of φ for constant pairs (α, θ) of $(42^\circ, 10^\circ)$ in blue, $(122^\circ, 160^\circ)$ in red, and $(42^\circ, 160^\circ)$ in yellow, respectively. Scattering angle as defined in Fig. 8.4 (left). Results are computed with codes by Voshchinnikov (solid lines) and Mishchenko (dashed lines). Grain properties as described in the text

Acknowledgment We thank M. Baes, E. Krügel, and J. Krelowski for fruitful discussions.

References

Allamandola LJ, Tielens AGGM, Barker JR (1985) Astrophys J 290:L25
Allamandola LJ, Tielens AGGM, Barker JR (1989) Astrophys J Suppl Ser 71:733
Allamandola LJ, Hudgins DM, Bauschlicher CW Jr, Langhoff SR (1999) Astron Astrophys 352:659
Andersson B-G, Lazarian A, Vaillancourt JE (2015) Annu Rev Astron Astrophys 53:501
Arendt RG, Odegard N, Weiland JL et al (1998) Astrophys J 508:74
Arenou F, Grenon M, Gomez A (1992) Astron Astrophys 258:104
Asano S, Yamamoto G (1975) Appl Opt 14:29
Asplund M, Grevesse N, Sauval AJ, Scott P (2009) Annu Rev Astron Astrophys 47:481
Baes M, Camps P (2015) Astron Comput 12:33
Baes M et al (2019) Astron Astrophys, submitted
Bagnulo S, Cox NJL, Cikota A et al (2017) Astron Astrophys 608A:146B
Bianchi S, Ferrara A, Giovanardi C (1996) Astrophys J 465:127
Bjorkman JE, Wood K (2001) Astrophys J 554:615
Bohren CF, Huffman DR (1998) Absorption and scattering of light by small particles. Wiley-Interscience, New York
Bot C, Helou G, Boulanger F, Lagache G, Miville-Deschenes M-A, Draine B, Martin P (2009) Astrophys J 695:469
Boulanger F, Baud B, van Albada GD (1985) Astron Astrophys 144:L9
Bouwman J, Cuppen HM, Steglich M et al (2011) Astron Astrophys 529:46
Chini R, Krügel E (1983) Astron Astrophys 117:289
Clayton GC, Mathis JS (1988) Astrophys J 327:911
Contopoulos G, Jappel A (eds) (1974) Transactions of the IAU, vol 15B. Springer Netherlands, Dordrecht
Das HK, Voshchinnikov NV, Il'in VB (2010) Mon Not R Astron Soc 404:625
Davis L Jr, Greenstein JL (1951) Astrophys J 114:206
Desert FX, Boulanger F, Puget JL (1990) Astron Astrophys 237:215
Dolginov AZ, Mitrofanov IG (1976) Astrophys Space Sci 43:291
Draine BT (1988) Astrophys J 333:848
Draine BT, Allaf-Akbari K (2006) Astrophys J 652:1318
Draine BT, Flatau PJ (1994) J Opt Soc Am A 11:1491
Draine BT, Fraisse AA (2009) Astrophys J 626:1
Draine BT, Hensley B (2013) Astrophys J 765:159
Draine BT, Li A (2007) Astrophys J 657:810
Drimmel R, Cabrera-Lavers A, Lopez-Corredoira M (2003) Astron Astrophys 409:205
Dwek E, Arendt RG, Fixsen DJ, Sodroski TJ, Odegard N et al (1997) Astrophys J 475:565
Efimov YS (2009) Bull Crime Astrophys Obs 105:82
Fanciullo L, Guillet V, Boulanger F et al (2017) Astron Astrophys 602:7
Finkbeiner DP, Davis M, Schlegel DJ (1999) Astrophys J 524:867
Fitzpatrick EL (1999) Publ Astron Soc Pac 111:63
Fitzpatrick EL (2004) In: Witt AN, Clayton, GC, Draine, BT (eds) Astrophysics of dust. Astronomical Society of the Pacific conference series, vol 309. Astronomical Society of the Pacific, San Francisco, p 33
Fitzpatrick EL, Massa D (2007) Astrophys J 663:320
Fitzpatrick EL, Massa D (2009) Astrophys J 699:1209
Fosalba P, Lazarian A, Prunet S, Tauber JA (2002) Astrophys J 564:762
Gaia Collaboration et al (2018a) Astron Astrophys 616:A1

Gaia Collaboration et al (2018b) Astron Astrophys 616:A4
Gaia Collaboration et al (2018c) Astron Astrophys 616:A9
Galliano F, Madden SC, Tielens AGGM et al (2008) Astrophys J 679:310
Gontcharov GA (2017) Astron Lett 43:472
Gontcharov GA, Mosenkov A (2018) Interstellar polarization in the local bubble and the Gould Belt based on Gaia DR2 data. In: Modern star astronomy, vol 1, pp 121. https://doi.org/10.31361/eaas.2018-1.024
Goosmann RW, Gaskell CM, Marin F (2014) Adv Space Res 54:1341
Gordon KD, Cartledge S, Clayton GC (2009) Astrophys J 705:1320
Greenberg JM (1978) In: McDonnell JAM (ed) Cosmic dust. Wiley, New York, p 187
Guhathakurta P, Draine BT (1989) Astrophys J 345:230
Guillet V, Fanciullo L, Verstraete L et al (2018) Astron Astrophys 610A:16
Gupta R, Mukai T, Vaidya DB, Sen AK, Okada Y (2005) Astron Astrophys 441:555
Hall JS (1949) Science 109:166
Hamaker J, Bregman J (1996) Astron Astrophys Suppl Ser 117:161
Harries TJ (2000) Mon Not R Astron Soc 315:722
Heiles C (2000) Astron J 119:923
Heger ML (1922) Lick Obs Bull 10:146
Henning Th (2010) Annu Rev Astron Astrophys 48:21
Henyey LG, Greenstein JL (1941) Astrophys J 93:70
Heymann F, Siebenmorgen R (2012) Astrophys J 751:27
Hildebrand RH, Dragovan M, Novak G (1984) Astrophys J 284:51
Hiltner WA (1949) Science 109:165
Hobbs LM, York DG, Snow TP, Oka T, Thorburn JA, Bishof M, Friedman SD, McCall BJ, Rachford B, Sonnentrucker P (2008) Astrophys J 680:1256
Hong SS, Greenberg JM (1980) Astron Astrophys 88:194
Hudgins DM, Bauschlicher CW Jr, Allamandola LJ (2005) Astrophys J 632:316
Jenkins EB (2009) Astrophys J 700:1299
Jenkins EB (2014), arXiv:1402.4765
Jones RV, Spitzer L Jr (1967) Astrophys J 147:943
Jones AP, Köhler M, Ysard N, Bocchio M, Verstraete L (2017) Astron Astrophys 602:A46
Kataoka A, Muto T, Momose M et al (2015) Astrophys J 809:78
Kim SH, Martin PG (1994) Astrophys J 431:783
Krelowski J (2014) In: Cami J, Cox NLJ (eds) The diffuse interstellar bands. IAU Symposium, vol 297, pp 23–33
Krügel E (2003) The physics of interstellar dust. The Institute of Physics, Bristol. ISBN 0 7503 0861 3
Krügel E (2008) An introduction to the physics of interstellar dust. The Institute of Physics, Bristol. ISBN 978 1 58488 707 2
Krügel E (2009) Astron Astrophys 493:385
Lallement R, Vergely JL, Valette B et al (2014) Astron Astrophys 561:A91
Lazarian A (2007) J Quant Spectrosc Radiat Transf 106:225
Lazarian A, Hoang T (2007) Mon Not R Astron Soc 378:910
Li A (2017) HST Proposal 15037 L. http://adsabs.harvard.edu/abs/2017hst..prop15037L
Li A, Draine BT (2001) Astrophys J 550:L213
Li A, Greenberg JM (1997) Astron Astrophys 323:566
Lucas P (2003) J Quant Spectrosc Radiat Transf 79:921
Martin PG, Adamson AJ, Whittet DCB et al (1992) Astrophys J 392:691
Mathis JS (1986) Astrophys J 308:281
Mathis JS, Rumpl W, Nordsieck KH (1977) Astrophys J 217:425
Mathis JS, Mezger PG, Panagia N (1983) Astron Astrophys 128:212
Mennella V, Baratta G, Esposito A (2003) Astrophys J 587:727
Min M, Waters LBFM, de Koter A et al (2007) Astron Astrophys 462:667
Min M, Dullemond CP, Dominik C, de Koter A, Hovenier JW (2009) Astron Astrophys 497:155

Mishchenko MI (1991) Astrophys J 367:561
Mishchenko M, Travis LD (1998) J Quant Spectrosc Radiat Transf 60:309
Mishchenko MI, Travis LD, Mackowski DW (1996) J Quant Spectrosc Radiat Transf 55:535
Mishchenko MI, Dlugach JM, Yurkin MA et al (2016) Phys Rep 632:1
Misselt KA, Gordon KD, Clayton GC, Wolff MJ (2001) Astrophys J 551:277
Nishiyama S, Nagata T, Kusakabe N et al (2006) Astrophys J 638:839
Orsatti AM, Vega E, Marraco HG (1998) Astron J 116:266
Parvathi VS, Sofia UJ, Murthy J, Babu SRS (2012) Astrophys J 760:36
Peest C (2018) Modeling the polarization of radiation by cosmic dust, PhD thesis, Ghent University
Peest C, Camps P, Stalevski M, Baes M, Siebenmorgen R (2017) Astron Astrophys 601:A92
Pinte C, Menard F, Duchene G, Bastien P (2006) Astron Astrophys 459:797
Planck Collaboration et al (2011) Astron Astrophys 536:A24
Planck Collaboration et al (2014) Astron Astrophys 566:A55
Planck Collaboration et al (2015) Astron Astrophys 571:A11
Planck Collaboration et al (2016) Astron Astrophys 586:A132
Pontoppidan KM, Fraser HJ, Dartois E et al (2003) Astron Astrophys 408:981
Puget JL, Leger A (1989) Annu Rev Astron Astrophys 27:161
Purcell EM (1979) Astrophys J 231:404
Purcell EM, Pennypacker CR (1973) Astrophys J 186:705
Reissl S, Wolf S, Brauer R (2016) Astron Astrophys 593:A87
Reissl S, Stutz AM, Brauer R (2018) Mon Not R Astron Soc 481:2507
Robitaille TP (2011) Astron Astrophys 536:A79
Schultz J, Hakala P, Huovelin J (2004) Baltic Astronomy 13:581
Scicluna P, Siebenmorgen R (2015) Astron Astrophys 584:A108
Serkowski K (1973) In: Greenberg JM, Hayes DS (eds) Interstellar dust and related topics. IAU symposium. Reidel, Dordrecht, vol 52, p 145
Siebenmorgen R, Gredel R (1997) Astrophys J 485:203
Siebenmorgen R, Heymann F (2012) Astron Astrophys 543:25
Siebenmorgen R, Krügel E (1992) Astron Astrophys 259:614
Siebenmorgen R, Krügel E, Mathis JS (1992) Astron Astrophys 266:501
Siebenmorgen R, Krügel E, Laureijs RJ (2001) Astron Astrophys 377:735
Siebenmorgen R, Voshchinnikov NV, Bagnulo S (2014) Astron Astrophys 561:A82
Siebenmorgen R, Voshchinnikov NV, Bagnulo S, Cox NLJ, Cami J, Peest C (2018) Astron Astrophys 611:A5
Smith CH, Wright CM, Aitken DK, Roche PF, Hough JH (2000) Mon Not R Astron Soc 312:327
Steinacker J, Baes M, Gordon KD (2013) Annu Rev Astron Astrophys 51:63
Stokes GG (1852) Trans Camb Philos Soc 9:399
Voshchinnikov NV (2004) Astrophys Space Phys Rev 12:1
Voshchinnikov NV (2012) J Quant Spectrosc Radiat Transf 113:2334
Voshchinnikov NV, Das HK (2008) J Quant Spectrosc Radiat Transf 109:1527
Voshchinnikov NV, Farafonov VG (1993) Astrophys Space Sci 204:19
Voshchinnikov NV, Henning Th (2010) Astron Astrophys 517:A45
Voshchinnikov NV, Hirashita H (2014) Mon Not R Astron Soc 445:301
Voshchinnikov N, Karjukin V (1994) Astron Astrophys 288:883
Voshchinnikov NV, Il'in VB, Das HK (2016) Mon Not R Astron Soc 462:2343
Watson AM, Henney WJ (2001) Rev Mex Astron Astrofis 37:221
Watson D, Christensen L, Knudsen KK et al (2015) Nature 519:327
Weingartner JC, Draine BT (2001) Astrophys J 548:296
Werner MW, Davidson JA, Morris M, Novak G, Platt SR, Hildebrand RH (1988) Astrophys J 333:729
Whitney BA, Wolf MJ (2002) Astrophys J 574:205
Whittet DCB (2003) Dust in the galactic environment. Institute of Physics Publishing, Bristol

Whittet DCB, Bode MF, Longmore AJ et al (1988) Mon Not R Astron Soc 233:321
Wilking BA, Lebofsky MJ, Rieke GH (1982) Astron J 87:695
Ysard N, Köhler M, Jones A et al (2015) Astron Astrophys 577:A110
Yurkin MA, Hoekstra AG (2007) J Quant Spectrosc Radiat Transf 106:558
Zickgraf FJ, Schulte-Ladbeck RE (1989) Astron Astrophys 214: 274

Chapter 9
Optical Polarimetry of Small Solar System Bodies: From Asteroids to Debris Disks

Irina Belskaya, Alberto Cellino, Anny-Chantal Levasseur-Regourd, and Stefano Bagnulo

Abstract We present a review on polarimetric properties of small Solar system bodies. The main results of polarimetric investigations of asteroids, comets, transneptunian objects and Centaurs are discussed focusing on recent findings. We also discuss the polarimetric observations of dust clouds in our Solar system.

9.1 Introduction

Polarimetry is a traditional technique which has been used to study Solar system bodies since the nineteenth century. The solar light scattered at visible wavelengths by surfaces of Solar system bodies or dust particles is in a state of partial linear polarisation. The fraction of linear polarisation P varies as a function of the configuration of the Sun–observer–target system. In particular, there is a dependence of P upon the so-called phase angle α, which is the angle between the directions to the Sun and to the observer as seen from the target (see Fig. 9.1a). Observations of Solar system bodies show that the plane of linear polarisation is usually either normal or parallel to the scattering plane (the Sun–observer–target plane). Thus, the

I. Belskaya (✉)
Institute of Astronomy, V.N. Karazin Kharkiv National University, Kharkiv, Ukraine
e-mail: irina@astron.kharkov.ua

A. Cellino
INAF - Osservatorio Astrofisico di Torino, Pino Torinese, Italy
e-mail: alberto.cellino@inaf.it

A. C. Levasseur-Regourd
LATMOS, Sorbonne Université, IPSL-CNRS, UVSQ, Paris, France
e-mail: aclr@latmos.ipsl.fr

S. Bagnulo
Armagh Observatory and Planetarium, College Hill, Armagh, UK
e-mail: stefano.bagnulo@armagh.ac.uk

R. Mignani et al. (eds.), *Astronomical Polarisation from the Infrared to Gamma Rays*, Astrophysics and Space Science Library 460,
https://doi.org/10.1007/978-3-030-19715-5_9

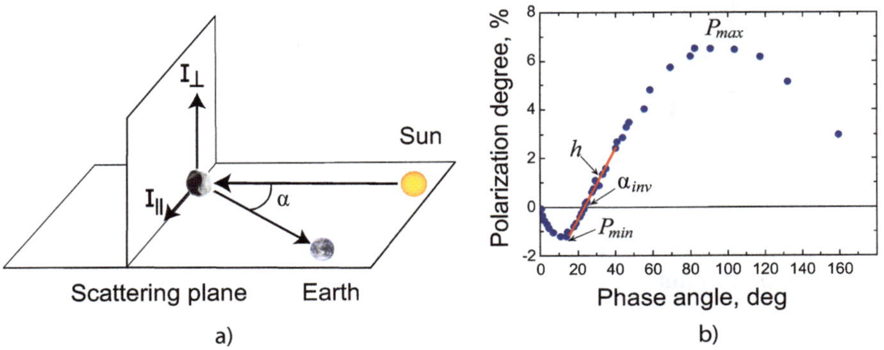

Fig. 9.1 Illustration of geometry of scattering (**a**) and an example of typical phase-angle dependence of polarisation degree for a Solar system body (the data points are from Lyot (1929) for the Moon) (**b**)

degree of linear polarisation P_r can be described in terms of the intensities of the scattered light polarised along the planes perpendicular $I_\perp$ and parallel $I_\parallel$ to the scattering plane:

$$P_r = \frac{(I_\perp - I_\parallel)}{(I_\perp + I_\parallel)}$$

As can be seen, when the maximum value of measured light intensity is in the direction parallel to the scattering plane (and the minimum intensity correspondingly is in the direction normal to the scattering plane) the P_r parameter is negative. The use of P_r is just a useful convention to merge two kinds of information into one single parameter, namely the degree of measured linear polarisation P and the orientation of the polarisation plane. In terms of Stokes parameters, P_r is the reduced second Stokes parameter $P_Q = Q/I$ if we adopt the direction perpendicular to the scattering plane as a reference direction. For symmetry reasons the reduced third Stokes parameter $P_U = U/I$ is expected to be zero.

The degree of linear polarisation varies with the phase angle α and the wavelength λ of the observations, and does not depend on the distances of the scattering surfaces to the Sun and to the observer. The typical polarisation phase-angle curve $P_r(\alpha)$ in visible light measured for Solar system bodies (Fig. 9.1b) is characterised by the presence of a negative polarisation branch at small phase angles, i.e., the case when the polarisation plane is parallel to the scattering plane. Its extreme negative value called polarisation minimum $P_{\min}$ typically occurs at $\alpha \sim 8 - 10°$. Around the polarisation inversion angle α_{inv}, where the orientation of the plane of linear polarisation is changed, the trend is mostly linear and its slope at the inversion angle is generally called polarimetric slope h (see Fig. 9.1b). The positive polarisation at high phase angles around 100° reaches a maximum $P_{\max}$.

The behaviour of polarisation phase curve $P_r(\alpha)$ strongly depends on the physical properties of a scattering layer or cloud, such as the particle size, complex refractive index and thus geometric albedo. The interpretation of polarimetric observations is mainly based on laboratory and numerical modelling of light scattering by particles and particulate media. Thanks to the progress both in instrumentation and in theoretical modelling polarimetric investigations of Solar system bodies have become more intensive during the last two decades. These advances in optical polarimetry were summarised in the books *Polarimetric remote sensing of Solar system objects* by Mishchenko et al. (2010) and *Polarimetry of stars and planetary systems* (Kolokolova et al. 2015). Here we overview the main results of polarimetric investigations of small bodies which include asteroids (Sect. 9.2), transneptunian objects and Centaurs (Sect. 9.3), comets (Sect. 9.4), and dust clouds (Sect. 9.5) focusing on new findings obtained since 2015. Section 9.6 is devoted to summary and conclusions.

9.2 Asteroids

Polarimetric observations of asteroids began in the 1960s. At that time, polarimetry together with photometry, spectrophotometry and radiometry was one of the main techniques to study the physical properties of asteroids. Later the relative contribution of polarimetry became much smaller compared to other techniques. In fact, there are less than 400 asteroids for which polarimetric measurements have been made so far, while photometric and radiometric surveys have covered thousands of asteroids. This is connected first of all with the complexity of polarimetric observations which require (1) the use of larger size telescopes (compared to photometry) in order to measure P_r with uncertainties $\leq 0.1\%$; (2) careful characterisation of instrumental effects; (3) observations of the selected target during several months to cover a wide range of phase angles.

Another problem that has prevented the widespread use of polarisation techniques in asteroid studies is that the interpretation of polarimetric observations in terms of physical characteristics is not straightforward. The interpretation is mainly based on empirical relationships between polarimetric parameters, geometric albedo, particle size and composition. On the other hand, the accumulation of observational data can lead to improvement of theoretical models and interpretation.

Most asteroid observations were generally made using 1–2-m class telescopes equipped with photopolarimeters. It means that only relatively bright asteroids with apparent visible magnitudes brighter than $13–14^m$ were observed. In the last decade imaging polarimeters based on CCD detectors started to be used allowing to gain 2–3 magnitudes in many cases (e.g., Gil-Hutton and García-Migani 2017). Larger size telescopes, like the ESO VLT are occasionally used for fainter asteroids of particular interest (space mission targets, dynamical family members, near-Earth objects) (e.g., Cellino et al. 2018). Polarimetric observations are recorded in the Asteroid Polarimetric Database (APD) at the Small Bodies Node of the Planetary Data

System and annually updated (Lupishko 2017). Some recent reviews on asteroid polarimetry have been given by Cellino et al. (2015a) and Belskaya et al. (2015).

9.2.1 Negative Polarisation

An example of well-measured asteroid phase curves is shown in Fig. 9.2. They demonstrate diversity in the depth and width of the negative polarisation branch. The phase curves are fitted with the following relation typically used for main-belt asteroids (Kaasalainen et al. 2003; Muinonen et al. 2009; Cellino et al. 2015b) which may be observed from Earth in the restricted phase-angle range up to ∼30°:

$$P_r = A\,(e^{\frac{-\alpha}{B}} - 1) + C \cdot \alpha, \tag{9.1}$$

where α represents the phase angle, and A, B, C are parameters characterising the behaviour of each single object, and are determined using least square techniques.

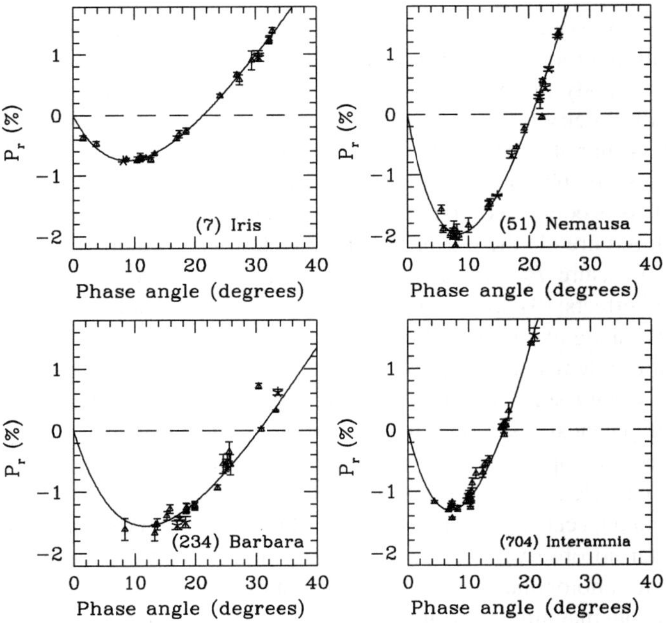

Fig. 9.2 Phase-polarisation curves in the V filter of the asteroids (7) Iris (S-class), (51) Nemausa (C-class), (234) Barbara (L-class) and (704) Interamnia (F-class). Data fit using Eq. (9.1). For the references and meaning of the taxonomic classes, see text

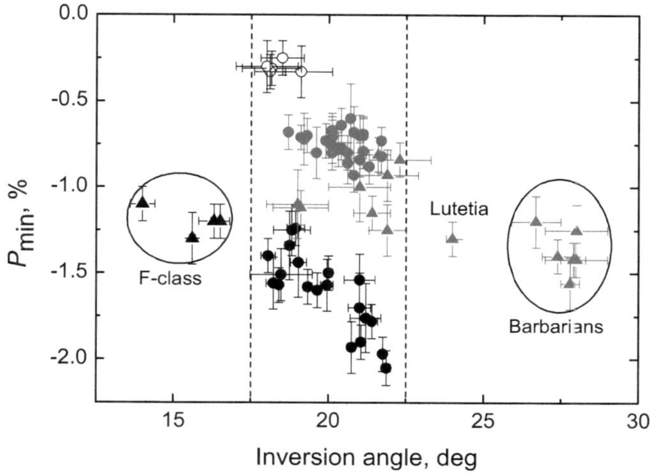

Fig. 9.3 The relationship P_{min} inversion angle α_{inv} for asteroids of different classes. Low-albedo C-class, moderate-albedo S-class and high-albedo E-class asteroids are shown by black, grey and open circles, respectively. Asteroids of other classes are shown by triangles. Specific asteroids with α_{inv} outside of typical values are indicated

For a long time the location of the objects in the plane P_{min} inversion angle α_{inv} has been considered as indicative on the average sizes of regolith particles (see Dollfus et al. 1989). The main asteroid classes, moderate-albedo S (silicate), low-albedo C (carbonaceous) and high-albedo E (enstatite), are well-distinguished in the plot by their P_{min} but have rather similar inversion angles within 18–22° (Fig. 9.3). These values of inversion angles are typical for pulverised meteorites and silicates with grain sizes ~30–300 μm and are between inversion angles measured for bare rocks (<18°) and for lunar-type texture (~21–25°) (see Dollfus et al. 1989). This led to the conclusion that asteroid surfaces are covered by a coarser regolith compared to the lunar surface. Nowadays, these classic results, that received confirmations at the epochs of the first landings on the Moon, are still considered important.

Recent years have seen the discovery of classes of objects, in particular the so-called Barbarians, characterised by unusually high values of inversion angle α_{inv} (Cellino et al. 2006; Gil-Hutton et al. 2008; Masiero and Cellino 2009). They are therefore located in a region of the $P_{\mathrm{min}} - \alpha_{\mathrm{inv}}$ plane where no objects were previously known to exist (Fig. 9.3).

Interestingly, there are reasons to believe that Barbarians are very primitive objects, since they exhibit in their reflectance spectra the signatures of unusually high amounts of refractory, spinel-rich compounds that are common in the so-called calcium magnesium-rich inclusions (CAIs) in primitive meteorites (Devogèle et al. 2018a). The phase-polarisation curve of the prototype of this class, asteroids (234) Barbara, characterised by an inversion angle around 30°, is shown in Fig. 9.2. The lowest inversion angle was found for objects belonging to the low-albedo

F-class (Belskaya et al. 2005), which is not distinguished in recent asteroid taxonomies. An example is given by the phase-polarisation curve of (704) Interamnia (see Fig. 9.2). Interestingly, a low value of the inversion angle characterises also (4015) Wilson-Harrington, an F-class asteroid that after its discovery was found to display cometary activity. The same property has been found also for the comet 2P/Encke observed at large heliocentric distances when a coma was not developed (Boehnhardt et al. 2008) and for the main-belt comet 133P/Elst–Pizarro (Bagnulo et al. 2010). Thus, asteroids of the same taxonomic class tend to group together in the plane $P_{\rm min} - \alpha_{\rm inv}$. This suggests that the location of asteroids in this plot can be strongly related to surface composition rather than to substantial variations in the surface texture from dust free surface to very small-grained regolith. According to Gil-Hutton and García-Migani (2017) the inversion angle of asteroids strongly depends on the index of refraction of their surface, while the phase angle of the polarisation minimum provides information about the distance between scatter particles.

Some studies have shown that a taxonomic classification based only on polarimetric parameters is possible. Penttila et al. (2005) published a preliminary study of the subject. More recently, Belskaya et al. (2017) carried out an extensive analysis of the subject. They found that most asteroids show phase-polarisation curves close to the average curve of the class, defined based on spectral reflectance data, to which they belong. Based on their polarimetric properties, a taxonomic classification for 283 main-belt asteroids was derived by the above authors.

Gil-Hutton et al. (2017) found that small V-type objects show a shallow minimum and a smaller inversion angle in comparison with the largest V-type asteroid (4) Vesta, which could be explained by differences in the regolith properties of small and large size asteroids.

Both the polarimetric slope h and the extreme value of negative polarisation $P_{\rm min}$ have long been known to be diagnostic of the albedo. The relationships between the geometric albedo $p_{\rm V}$ and polarimetric parameters have the following view:

$$\log(p_{\rm V}) = C_1 \log(h) + C_2, \tag{9.2}$$

$$\log(p_{\rm V}) = C_3 \log(P_{\rm min}) + C_4, \tag{9.3}$$

where C_1–C_4 are empirically derived constants.

A recent analysis and reassessment of the situation was published by Cellino et al. (2015b) and Lupishko (2018). They used different sets of data to calibrate the h-albedo and $P_{\rm min}$-albedo relations. Cellino et al. (2015b) used a small list of asteroids for which the albedo is considered to be known with good accuracy. In their analysis, they took into account some known effects of degeneration of the $P_{\rm min}$-albedo relation for low-albedo objects, and suggested different criteria to choose among different albedo polarisation relations, depending upon the quality and quantity of available data for any given object. They also proposed to use a new polarimetric parameter, which they called ψ, defined as the difference of the values of P_r measured at the phase angles of 30° and 10°. Instead, Lupishko

(2018) used all available data on asteroid albedos and proposed a new generalised calibration of the h-albedo and P_{min}-albedo relations by averaging the coefficients of individual calibrations. We note that the uncertainties in the polarimetric albedos due to different calibrations are typically small.

The determination of the albedo based on polarimetric parameters can lead to more accurate results with respect to the other technique traditionally used in asteroid science to determine the sizes and the albedos of asteroids, namely thermal radiometry. The determination of asteroid sizes using thermal IR data is generally very reliable, because the thermal flux mostly depends on the size of the emitting body, and only weakly upon its albedo. In the case of the albedo, conversely, it is necessary in principle to measure simultaneously the thermal flux and the visible flux of scattered sunlight, but this never happens in practice. A first in situ validation of the determination of the albedo by means of polarimetry was given by the Hayabusa space mission, which visited the near-Earth asteroid (25143) Itokawa. According to Ishiguro et al. (2017), the value of albedo obtained by means of polarimetric observations by Cellino et al. (2005), namely 0.24 ± 0.01, was fully confirmed by analysis of data from the Hayabusa on-board camera.

9.2.2 *Positive Polarisation*

The number of small Solar system objects for which some estimate of the maximum of positive polarisation P_{max} has been derived is still extremely small. The reason is that P_{max} is reached only at very high values of phase angle, and the only objects that can be observed in these conditions from the ground are near-Earth objects, and most of them are quite faint. In many cases, even during favourable apparitions, it can happen that the determination of P_{max}, which can occur at phase angles above 100°, can be unsuccessful, because the object moves so rapidly in the sky to become not observable from a given observing site before reaching the phase-angle values corresponding to P_{max}. This happened, for instance, during the observations of (3200) Phaethon by Devogèle et al. (2018b) in the 2017 apparition of this object.

For a long time, the only near-Earth asteroids for which some estimate of P_{max} were available belonged to the S taxonomic class, consisting of objects having moderate albedo, as (1685) Toro, (4179) Toutatis and (23187) 2000 PN_9. The corresponding value of linear polarisation for these objects was around 7–8%. The first successful measurements of a low-albedo asteroid in the vicinity of polarisation maxima were obtained for the near-Earth asteroids (3200) Phaethon at its closest approach to the Earth. Phaethon is believed to be the parent body of the Geminid meteor shower. Several groups of observers carried out polarimetric observations of Phaethon using different instruments in 2016 and 2017 (Devogèle et al. 2018b; Ito et al. 2018; Kiselev et al. 2018; Shinnaka et al. 2018). They found extremely high positive polarisation (40–50% at phase angles around 100°) varying over Phaethon's surface (Borisov et al. 2018). An even larger value of linear polarisation at maxima is expected in the case of another small near-Earth asteroid, (152679) 1998 KU_2, observed by Kuroda et al. (2018).

First polarimetric observations of one more low-albedo asteroid (101955) Bennu also revealed extremely high positive polarisation (Cellino et al. 2018). The observations were made with the FORS2 instrument of the ESO VLT in the phase-angle range from 16° to 57°. The degree of polarisation found at $\alpha = 57°$ is 21% in the R-band which is 5% higher than that for Phaethon at the same phase angle. This assumed the higher value of polarisation maximum for Bennu. Moreover, Bennu's phase dependence is characterised by a small inversion angle, which led Cellino et al. (2018) to propose that this asteroid might be a member of the F taxonomic class, suggesting a possible link with comets. The first OSIRIS-REx results on hydroxyls at the Bennu's surface seem to confirm this conclusion.

Currently 16 near-Earth asteroids were observed at large phase angles. Most observations were made in a single filter, V or R. They are plotted all together in Fig. 9.4. The observations of low-albedo asteroids (Devogèle et al. 2018b; Ito et al. 2018; Kiselev et al. 2018; Shinnaka et al. 2018; Kuroda et al. 2018; Cellino et al. 2018), moderate-albedo asteroids (Fornasier et al. 2015 and the references therein; Ishiguro et al. 2017) and high-albedo E-type asteroids (Kiselev et al. 2002; De Luise et al. 2007) are shown by different symbols.

In the case of near-Earth objects, which can be observed over a much wider interval of phase angles than main-belt asteroids, the following empirical trigonometric approximation, first proposed by Lumme and Muinonen (1993) is typically used:

$$P_r = A \sin^B(\alpha) \cdot \cos^C\left(\frac{\alpha}{2}\right) \sin(\alpha - \alpha_{\text{inv}}), \tag{9.4}$$

where α is the phase angle, α_{inv} is the polarisation inversion angle and A, B, C are free parameters to be derived from data. The main advantage of the function is that

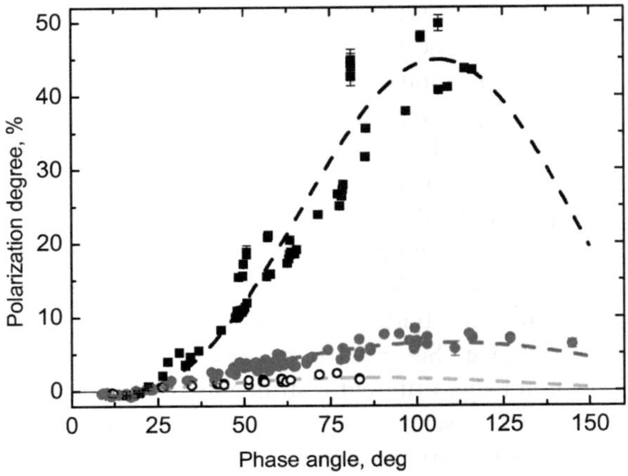

Fig. 9.4 Composite polarisation phase-angle dependences for near-Earth asteroids in the V and R filters (see text for references). Low-albedo, moderate-albedo and high-albedo asteroids are shown by black, grey and open circles, respectively. The fit of each albedo groups by the trigonometrical function (Eq. (9.4)) is shown by dash lines

it gives physically reasonable behaviour of the polarisation-phase dependence in a wide range of phase angles (e.g., Penttila et al. 2005).

The location of polarisation maximum varies from ~80° for high-albedo asteroids to ~120° for low-albedo objects (see Fig. 9.4). According to modelling by Petrov and Kiselev (2017) the position of the positive polarisation maximum depends on the imaginary part of the refractive index in a non-monotonic way. The value of polarisation maxima for measured asteroids varies from 2% to 50% and strongly correlates with albedo. Polarimetric measurements close to polarisation maxima can be very important in albedo determination of potentially hazardous NEAs. It is possible to obtain reliable values of albedos (and thus sizes) even with a single polarimetric measurement due to large differences in the polarisation degree for high-, moderate- and low-albedo asteroids.

9.2.3 Spectropolarimetry

By means of an analysis of all multi-bands polarisation measurements available in the literature, Belskaya et al. (2009) carried out a systematic investigation of the dependence of P_r upon the wavelength. Later, Bagnulo et al. (2015) published a similar analysis for a dozen of asteroids, only, but based on spectropolarimetric data covering full wavelength intervals, mostly from 380 nm to 920 nm.

These investigations have shown an interesting behaviour of the gradient of linear polarisation as a function of wavelength for objects belonging to different taxonomic classes, characterised by different values of the albedo. The main discovery was that the gradient of polarisation changes its sign if the same object is observed at phase angles corresponding to either the negative or the positive branch of polarisation. In particular, low-albedo objects exhibit a positive gradient of polarisation (the linear polarisation increasing for increasing phase angle) in the positive polarisation branch, whereas in the negative polarisation branch the degree of polarisation tends to decrease for increasing wavelength. Interestingly, the behaviour of objects of moderate albedo is just the opposite. This behaviour is not yet fully understood from a theoretical point of view, and constitutes a challenge and an important constraint for the development of detailed models of light-scattering properties of planetary surfaces.

The analysis carried out so far has also shown that objects belonging to taxonomic classes exhibiting similar reflectance spectra (at visible wavelengths) may have quite different polarisation spectra. This suggests that spectropolarimetry can be a very useful tool for the purposes of asteroid taxonomic classification. Bagnulo et al. (2015) also found a few cases in which the so-called Umov law (Umov 1905) seems to be violated, namely cases in which the degree of linear polarisation and the reflectance at different wavelengths are positively correlated.

The spectropolarimetry technique presents intrinsic advantages with respect to separate applications of spectroscopy and polarimetry, because it can provide at the same time both the reflectance spectrum of an object, and its degree of linear polarisation, as well as a detailed measurement of the gradient of polarisation as a function of wavelength. The drawback is that in most cases the success of this technique depends on the availability of observing time at large, and normally oversubscribed, telescopes. On the other hand, a simpler multi-band photo-polarimetric observation can hardly provide the richness of information provided by a full spectropolarimetric observation covering simultaneously and with a very good sampling a large interval of wavelengths, but it can be more feasible in practical terms.

9.2.4 Rotational Variations

The first really unambiguous case of polarisation variations (in the negative branch) synchronous with rotation has been found for asteroid (4) Vesta (for references see Cellino et al. 2016). A priori, this behaviour suggests the presence of some heterogeneity of the asteroid's surface, in terms of albedo, or composition, or regolith properties, or a combination of them. The extensive exploration of Vesta's surface by NASA's Dawn mission made it possible to interpret the data set of polarimetric measurements of Vesta, obtained in the past by means of remote, disk-integrated measurements, in terms of a correspondence between the degree of linear polarisation and the variation of local properties of the surface visible to ground-based observers at different epochs. In particular taking profit of the detailed analysis of Vesta's surface performed by the Dawn mission, Cellino et al. (2016) obtained for the first time a link between disk-integrated polarimetric measurements and the local properties of the visible surface of an asteroid. Their results confirmed the relation between albedo and polarisation being the main responsible of the sinusoidal variation of polarisation of Vesta during its rotation, being due to a region of low albedo covering about 180° in longitude. The authors also found, however, that albedo alone is not sufficient to justify completely the observed behaviour and some surface slope and/or mineralogical composition effect seems to play some non-negligible role. The single polarimetric data obtained from the ground were then analysed by splitting them into subsets corresponding to observations of regions of different albedo on Vesta's surface. In this way, four distinct polarimetric slopes were obtained, which nicely fit the traditional form of the slope albedo relation. In particular, excellent agreement was found with the slope-albedo relation obtained by Cellino et al. (2015b) after excluding from the analysis asteroids having albedos lower than about 0.08.

Recently a variation of the linear polarisation (positive branch) with rotation was found for the NEA (3200) Phaethon (Borisov et al. 2018). The authors concluded that the observed change in the linear polarisation might be related to differences in the thickness of the surface regolith in different areas or local topographic features.

9.3 Transneptunian Objects and Centaurs

Investigation of physical properties of transneptunian objects (TNOs) and Centaurs requires very large telescopes due to their remoteness and thus faintness. Other limitation in application of polarimetric technique to study these distant objects is a very restricted phase-angle range. The ground-based observations of Centaurs can be performed up to phase angle $\alpha \sim 7°$, while observations of TNOs can be made only up to 2°. Nevertheless, the pioneering study of polarisation properties of these bodies started in 2002 (Boehnhardt et al. 2004) revealed very interesting results.

Currently, polarimetric measurements are available for 13 TNOs and Centaurs (Boehnhardt et al. 2004; Rousselot et al. 2005; Bagnulo et al. 2006, 2008; Belskaya et al. 2008, 2010, 2012). All these observations were carried out at the ESO VLT using the FORS instrument in service mode. The description of the instrument and data reduction procedure is given in detail by Bagnulo et al. (2011). There are also several polarimetric measurements of the Pluto–Charon system obtained in 1972–1988 (Kelsey and Fix 1973; Breger and Cochran 1982; Avramchuk et al. 1992).

These observations gave a first idea about the phase-angle behaviour of linear polarisation of TNOs and Centaurs. They demonstrate that largest and smaller size objects have distinct polarimetric properties (Bagnulo et al. 2008). Polarisation behaviours of the measured Centaurs are different from each other and from TNOs (Belskaya et al. 2010). An example of diversity in polarisation-phase dependences is shown in Fig. 9.5. Although the available observations were made in different phase-angle ranges, the differences in polarisation phase-angle behaviours of these objects are evident.

The detailed polarisation-phase dependence covered the phase-angle range from 0.5° to 4.2° was obtained for the Centaur Chiron (Bagnulo et al. 2006, 2010).

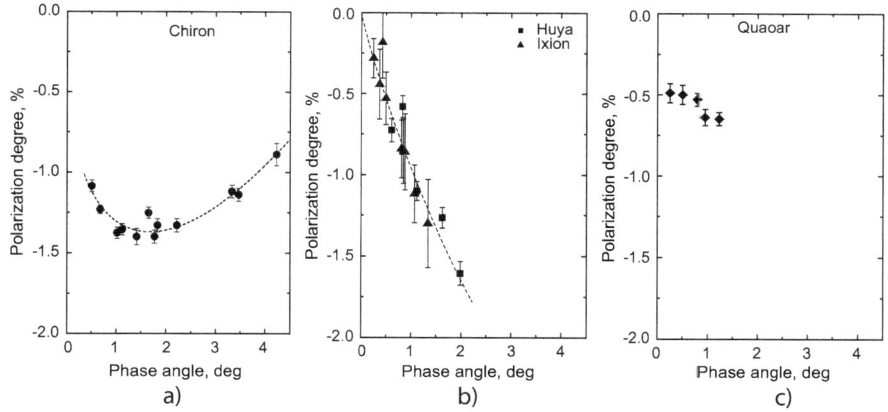

Fig. 9.5 Polarisation-phase dependence in the R filter measured for Centaur Chiron (**a**), resonant TNOs Huya and Ixion (**b**) and classical TNO Quaoar (**c**). See text for references

It shows a pronounced branch of negative polarisation with a minimum of about -1.4% occurring at phase angle as small as 1.5° (Fig. 9.5a). It is the smallest phase angle at which the minimum of negative polarisation occurs as compared to any other Solar system object observed so far. The phase curve of Chiron is highly asymmetric with a predicted inversion angle $\alpha_{\rm inv} \sim 7 - 9°$ (Belskaya et al. 2010). Polarimetric measurements of three other Centaurs, Chariklo, Pholus, and (29981) 1999 TD10 (Belskaya et al. 2010; Rousselot et al. 2005) revealed a great diversity in polarisation properties with the negative polarisation varying from -1% to -2%.

The resonant TNOs Huya and Ixion also show a deep negative branch of polarisation at small phase angles, which reaches $P = -1.6\%$ at phase angle $\alpha \sim 2°$ (Fig. 9.5b). Similar polarisation-phase behaviour was measured for two other TNOs, the resonant object (26375) 1999 DE9 and (20000) Varuna belonging to the classical population (Bagnulo et al. 2008). All these TNOs show very similar polarisation phase-angle behaviours within the errors with a rapid increase (in absolute terms) of the negative polarisation in the measured phase-angle range from near zero to $\alpha \sim 1 - 2°$.

The polarisation-phase dependence of the large classical object (50000) Quaoar demonstrates a different behaviour compared to Centaurs and smaller size TNOs (Fig. 9.5c). The negative polarisation branch is shallower with slow changes of the measured polarisation in the phase-angle range 0.25°–1.2° (Bagnulo et al. 2006). Very slow changes of polarisation with phase angle were also found for the largest transneptunian objects Eris, Pluto and Makemake (Kelsey and Fix 1973; Belskaya et al. 2008, 2012). Two more large TNOs, Haumea and Orcus, were measured at a single phase angle (Bagnulo et al. 2008; Belskaya et al. 2012) and new observations are needed to investigate their polarisation-phase behaviour.

Application of numerical modelling of coherent-backscattering to the observations of TNOs showed that a possible way to explain their polarisation properties is to assume two-component surface media consisting of Rayleigh scatterers with small and large single-scattering albedos. Examples of model fits are given in Boehnhardt et al. (2004), Bagnulo et al. (2006) and Rousselot et al. (2005).

Laboratory measurements of polarisation of particulate surfaces confirmed this conclusion. Dougherty and Geake (1994) measured similar deep negative polarisation branches at small phase angles for a very thin frost layer of submicron ice crystals on the top of a dark surface. Their polarisation-phase behaviours are very similar to those of Centaurs and small TNOs. Poch et al. (2018) found strong dependence of polarisation phase curves on particle size and the degree of sintering of ice surfaces.

The methane-ice rich and water-ice rich large TNOs display distinctly different polarisation properties. The methane-rich TNOs (Eris, Pluto and Makemake) show similar very shallow phase-angle dependences with depths of polarisation of $P \sim -0.3\%$ (see Belskaya et al. 2012). The negative polarisation branches measured for water-ice rich bodies are far deeper as compared to methane-ice rich TNOs. Two water-ice rich objects of different sizes and albedos, Haumea and Quaoar, measured at the same phase angle $\alpha \sim 1°$ show practically identical polarisation equal to $P \sim -0.7\%$ (Bagnulo et al. 2006, 2008). Thus, the measured diversity in polarisation-phase dependences of TNOs and Centaurs can be naturally explained by different abundance and composition of ices at their surfaces.

The surface albedo seems to not play a dominant role in determining the polarisation-phase behaviour of distant Solar system bodies. The correlation between polarisation and albedo of TNOs and Centaurs is rather weak and cannot be used for their albedo estimates as in case for asteroids. This may be partly due to the impossibility to estimate a real value of polarisation minima due to the very restricted phase-angle range of ground-based observations. However, the similar polarisation for objects of different albedos, as in case of Haumea and Quaoar, demonstrates that deviates from the $P_{\rm min}$-albedo relationship are rather large.

The main conclusion from the polarimetric measurements of TNOs and Centaurs is that their surface microstructure should be very different from that of small bodies in the inner part of the Solar system.

9.4 Comets

Comets have long been noticed to develop, when passing relatively close to the Sun on their elongated orbits, comae and tails. J.F. Arago was the first to provide evidence, from observations with his polariscope of bright comet Tralles in 1818, for tiny particles reflecting solar light in what is now called the dust tail (Arago 1858). It was progressively understood that comets are porous small bodies of ices and refractory materials, so-called nuclei, which release gases, together with dust particles, whenever they are heated enough by solar radiation. Gases and dust particles form a thin and bright coma that surrounds the nucleus. Long whitish dust tails and bluish plasma tails expand under the effects of solar radiation pressure and solar wind, respectively; trails of debris with sizes above hundreds of microns (e.g., Reach et al. 2007) are also detectable in the infrared domain. Cometary nuclei formed in the protosolar nebula, and were later kept far away from the Sun in the Kuiper belt or in the Oort cloud. Their dust particles, extensively studied through polarimetry, are thus tracers of our Solar System formation and early evolution, as now established by the Rosetta mission to comet 67P/Churyumov–Gerasimenko (thereafter 67P/C-G). Current progresses in cometary polarimetry are related, not only to the use of large telescopes, but also to the use of narrow interference filters, needed to avoid any contamination from gaseous emissions within the coma. Of major interest are progresses in interpretation of the observational data through numerical simulations (i.e., theoretical modelling) and experimental simulations (in the laboratory or under microgravity conditions). The first (and up to now unique) in situ polarimetric observations of cometary dust took place in March 1986, during the flyby of comet 1P/Halley by the Giotto spacecraft. The OPE instrument provided local measurements along the trajectory, from about 20,000 to 900 km nucleus-distance, at a constant phase angle of 73° and for three visible wavelengths (442, 576 and 718 nm). Two mean features were discovered: (1) jets with an increase in polarisation, i.e., a change in dust properties, (2) a trend to a decrease in polarisation in the inner coma, with possibly a change in polarimetric colour (Levasseur-Regourd et al. 1999, 2005). Numerous remote polarimetric observations of comets have taken place since 1990s. Of major importance were the retrieval of polarimetric phase curves and the development of polarimetric imaging. Recent reviews on cometary polarimetry have been given by Levasseur-Regourd (2010) and Kiselev et al. (2015). We mostly emphasise below new observations, together with their interpretation, in the context of the ground-truth on cometary dust particles provided by the 26-months long Rosetta rendezvous with comet 67P/C-G.

9.4.1 Phase Angle and Wavelength Dependence of the Linear Polarisation

Polarimetric phase curves were retrieved for the first time for 1P/Halley, from Earth-based and Giotto data (Dollfus et al. 1988). Such phase curves, obtained in the 0–120° phase-angle range, may be fitted through trigonometric function (Eq. (9.4)). They are reminiscent of polarimetric curves obtained for regoliths, and typical of scattering by irregular particles with sizes greater than the wavelength. The maximum in polarisation is reached by 90–100° of phase angle. Whenever the spectral domains of observations (mostly through green and red emission-free interference filters) are separated, different types of polarimetric behaviours may be pointed out, between comets with quite a low maximum in polarisation, and comets with a higher maximum in polarisation (Levasseur-Regourd et al. 1996). It should be added that new comet C/1995 O1 Hale-Bopp presented the highest polarisation ever observed up to now (e.g., Kiselev and Velichko 1997; Hadamcik et al. 1997) and that comets with a high maximum in polarisation usually present a significant silicate emission feature in the infrared domain.

9.4.2 Imaging Polarimetry

Polarimetric imaging techniques have been developed with the return of 1P/Halley and the use of CCD detectors. Since a polarimetric image is obtained at given phase angle α and wavelength λ, changes in the polarisation of the coma may provide immediate clues to changes in the properties of the scattering dust particles, within dust jets and possibly in inner comae. Examples of polarimetric images of comets are presented in Fig. 9.6. Evidences may be found for an increase of polarisation in jet-like features and, whenever the resolution is sufficient, for a decrease in polarisation, so-called a halo, in the innermost coma. Polarimetric images of the comae surrounding sub-nuclei after the fragmentation of 73P/Schwassmann–Wachmann 3 provided unique clues to differences between some of the sub-nuclei and to the presence of secondary fragments (Hadamcik and Levasseur-Regourd

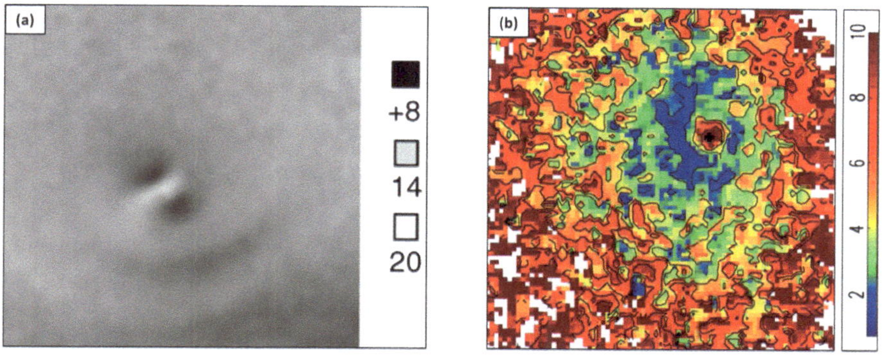

Fig. 9.6 Polarimetric images of cometary comae in R filters, within comparable fields of view of about 75,000 km:
(**a**) C/1995 O1 Hale-Bopp on 9 April 1997 at $\alpha = 44°$, providing evidence for some curved jets with higher linear polarisation ($\approx$20%), ahead of a near-nucleus halo with a lower polarisation ($\approx$8%);
(**b**) 67P/C-G on 8 November 2015 at $\alpha = 33°$, indicating the presence of different dust particles, with a lower polarisation ($\approx$1% to 3%) near the nucleus.
Adapted from: (**a**) Hadamcik and Levasseur-Regourd (2003); (**b**) Rosenbush et al. (2017)

2016). A few polarimetric observations of comet 67P/C-G had taken place before it was the target of the Rosetta mission (Meyers and Nordsieck 1984; Chernova et al. 1993). More observations took place in preparation of the Rosetta mission during the 2008–2009 return of the comet (Hadamcik et al. 2010; Stinson et al. 2016). Coordinated France–India campaigns have confirmed a post-perihelion increase in activity, interpreted as a seasonal effect, and provided evidence, after perihelion, for huge jets with dust properties different from those of dust in the background coma (Hadamcik et al. 2016). The seasonal effect may be accentuated the bi-lobal shape of the nucleus discovered by Rosetta cameras, while numerous outburst events have been noticed to take place after perihelion (e.g., Sierks et al. 2016; Grün et al. 2016).

Polarimetric images of 67P/C-G have been obtained in 2014–2016, i.e., during the *Rosetta* mission, from the *HST* and the BTA, as well as from the VLT, the WHT and the Rohzen Observatory (Snodgrass et al. 2017). *HST* observations were made with ACS/WFC instrument in 2014, 2015 and 2016. In 2014, the comet, still far away from the Sun, was at low galactic latitudes. For a phase angle of 15.5° the polarisation was nominal, of about −2% (Hines and Levasseur-Regourd 2016). In November 2015, 3 months after perihelion, the comet was quite active; for a phase angle of 33°, the polarisation was above average values, in agreement with what had been noticed at its previous apparition (Hadamcik et al. 2016). High-resolution maps of the linear polarisation have been obtained from observations at the 6-m BTA in November–December 2015 and April 2016 (Rosenbush et al. 2017). They provide evidence for the complex structure of the coma in polarisation, likely to correspond to different properties of the dust particles within the coma; a spiral-shaped lower polarisation is noticed in the inner coma, possibly corresponding to a low-polarisation halo, and an enhanced and redder polarisation is noticed in jet-like features (Rosenbush et al. 2017).

9.4.3 Interpretation of the Linear Polarisation of Comets

Simulations, i.e., laboratory experiments and numerical models, are both mandatory to try and interpret the polarimetric observations in terms of properties of the dust particles. Progresses in the simulations rely on technical progresses and on computational advances, such as discrete dipolar approximation (DDA) and T-matrix methods, as well as on realistic hypotheses (see Kiselev et al. 2015 for a review).

Progresses in polarimetric measurements on analogues of cometary dust, that is to say on dust clouds, have been made possible, thanks to the careful choices of well-documented analogues and the development of elaborate instruments. For analogues of porous dust particles, lunar simulants, volcanic ashes, black carbon aggregates or mixtures of dust from minerals and carbonaceous chondrites have been used. Typical examples are given by the PROGRA2 (French acronym for optical properties of astronomical and atmospheric grains) instruments, which measure intensity and linear polarisation in visible or near-infrared domains, either in the laboratory or during parabolic flights campaigns (e.g., Renard et al. 2014); and also by the CODULAB (Cosmic Dust Laboratory), which measures the Stokes parameters at 520 nm in the laboratory (e.g., Muñoz et al. 2017).

Significant progresses in numerical models stem from attempts to fit the data with non-spherical particles and to operate at different wavelengths. It was progressively understood, from such approaches, that comet dust particles had to be agglomerates or aggregates of smaller grains with possibly a mixture of fluffy and more compact particles. Typical examples are those of a model fitting the observations of a given comet in two wavelengths with mixtures of minerals and complex organics building up aggregates and compact particles (Lasue et al. 2009); and also of a model fitting the observations of seven distant comets with fractal aggregates of monomers and mixtures of spheroids and aggregate particles, consisting of water-ice, silicates and organics (Dlugach et al. 2018). The fractal models that had been used have been validated by the ground-truth, provided by Rosetta, of the presence of fractal dust particles in the coma of 67P/C-G (e.g., Mannel et al. 2016; Levasseur-Regourd et al. 2018).

9.5 Dust Clouds in the Solar System

Within our Solar system, dust particles that do not form a regolith are not only present in cometary comae and tails. Dust debris, mostly from comets and asteroids, form a wide flattened cloud in the interplanetary medium, with heterogeneities and possibly much smaller dust clouds within it. Out of the scope of this review, dust clouds are indeed also present in planetary atmospheres and in stellar environments (e.g., debris disks, exo-zodiacal clouds).

9.5.1 Zodiacal Cloud

Polarimetric observations of the zodiacal light, i.e., the solar light scattered by dust particles within the interplanetary dust cloud, have been made since the 1970s, mostly from Earth and also Earth's orbit. By then, one of their main purposes was to provide information on the interplanetary dust in relation with deep-space missions. Nowadays, topics of major interest are related to a better knowledge of their main sources (asteroidal collisions and cometary activity) and to possible models of exo-zodiacal dust clouds. A recent review on zodiacal light polarimetric properties has been given by Lasue et al. (2015).

Observations are delicate since contaminations from the atmospheric airglow, critical for ground-based observations, and from the galactic or extragalactic point sources and diffuse sources need to be avoided (see e.g., Levasseur-Regourd 1992). The retrieval of the local polarisation is even more difficult, since the measured polarisation is integrated along a line of sight, corresponding to changing distances to the Sun and to changing phase angles.

For a line-of-sight tangent to the direction of motion of the observer, a rigorous inversion is feasible at the location of the observer; this approach typically leads, in the near-ecliptic symmetry plane of the zodiacal cloud, to a local linear polarisation of about (30 $\pm$ 3)% for α =90° at 1 au from the Sun (Dumont 1973; Levasseur-Regourd et al. 1990). Inversion is nevertheless possible in the near-ecliptic symmetry plane of the interplanetary dust cloud, assuming some homogeneity and steady state of the cloud over the period of observations (e.g., Levasseur-Regourd et al. 2001). This nodes-of-lesser-uncertainty approach provides a partial phase curve at 1.5 au, from about 0° to 40° and at 90°, as well as the variation of the local linear polarisation at 90° phase angle, for solar distances ranging from 0.1 au to 1.5 au (Levasseur-Regourd et al. 2001; Lasue et al. 2015).

The retrieved phase curve is smooth, with a slightly negative branch, an inversion at α_{inv} = 15 $\pm$ 5°, a slope at inversion h = 0.2 $\pm$ 0.1 (%/deg) and a maximum by 30%. A numerical light-scattering model for such a phase curve, using spheroids and aggregates thereof, has suggested the presence of low-absorbing minerals and absorbing complex organics, with a significant amount of fluffy aggregates (Lasue et al. 2007). This is now found to fairly agree with the properties of cometary dust, as discovered from Rosetta exploration of comet 67P/Churyumov-Gerasimenko (Levasseur-Regourd et al. 2018). Other recent approaches, typically from infrared observations (Rowan-Robinson and May 2016) and dynamical studies (Nesvorný et al. 2010; Carrillo-Sánchez et al. 2016) indeed strongly suggest Jupiter family comets to be the main sources of interplanetary dust along the orbit of the Earth.

The retrieved variation with solar distance of the local polarisation at 90° corresponds to a progressive decrease with decreasing solar distance, from values about 32% at 1.5 au, 30% at 1 au, to 20% at 0.4 au, and presents an abrupt decrease closer to the Sun. Since interplanetary dust particles are very slowly spiralling towards the Sun under Poynting–Robertson effect, this trend has long been suspected to originate in partial sublimation of semi-volatile organics. It is perfectly reproduced, as well as the trend of the phase curve, by measurements made with the PROGRA2 instrument on interplanetary dust analogues (Hadamcik et al. 2019, Fig. 7). Such analogues consist of mixtures of minerals and organics (about 40%, such as carbon black, lignite, crushed carbonaceous chondrites meteorites); the particles have sizes in the 10–200 μm range, and consist of both fluffy aggregates and more compact particles.

9.5.2 *Dust Clouds near Earth–Moon Lagrange Points*

Within the Solar System, numerous objects have been found in the vicinity of stable Lagrange points, for instance, the Trojan asteroids distributed on two regions near the Sun–Jupiter L_4 and L_5 points. It had been suggested, from photographic observations of bright patches near the L_5 point of the Earth–Moon system, that interplanetary dust particles might gather by there (Kordylewski 1961). However, these Kordylewski dust clouds have long remained hypothetical (see Slitz-Balogh et al. 2019 and the references therein).

Recent investigations of a four-body system, consisting of the Sun, the Earth, the Moon and a low-mass dust particle, have allowed Slitz-Balogh et al. (2018), to map the size and shape of a cloud of dust particles that could remain trapped for about 10 years near L_5, and to suggest that the detection of such a cloud, certainly difficult through observation of solar scattered light or of its thermal emission, would be easier by polarimetry. Polarimetric images in three colours have been obtained in August 2017, with a teleobjective equipped with a CCD camera (7.5 × 5° field-of-view). The images seem to provide evidence for an increase in polarisation of about 10%, which might be compared with the zodiacal polarisation in the direction of observations, and could come from this elusive dust cloud (Slitz-Balogh et al. 2019).

9.6 Summary and Conclusions

Thanks to recent observations at large telescopes various classes of Solar system bodies have been probed with polarimetry. In the previous sections we have discussed the main observational features in optical polarisation of small bodies. Similar phase curve features are observed for planetary satellites (see Rosenbush et al. 2015; Bagnulo et al. 2017 for a review). Of course, detailed comparisons of the polarimetric properties displayed by objects located in very different regions of the Solar system are very difficult. The reason is that the fraction of linear polarisation depends strongly upon the phase angle at the epochs of observation, and bodies located at very different heliocentric distances, like near-Earth objects and Transneptunian objects can hardly be compared due to the completely different ranges of phase angle reachable for them. However, we note that the main parameters characterising negative and positive branches of polarisation of various classes of Solar system bodies are within the same range and can be summarised as follows:

1. The depth of negative polarisation branch varies from -0.2% to -2%. Higher values of negative polarisation were measured only in some local places (circumnucleus haloes) of polarimetric images of comets (e.g., Hadamcik and Levasseur-Regourd 2003).
2. The width of negative polarisation branch never exceeds 30°. The extreme values of inversion angle are found for the asteroid with spinel-rich composition (so-called Barbarians).
3. The negative branch is rather symmetrical with the position of minima at $\sim 10^{\circ}$ for most dark and moderate albedo objects and asymmetrical for high-albedo surfaces (see Rosenbush et al. 2015).
4. The exception case is the distant objects (Centaurs, TNOs, Uranian satellites). The location of their polarisation minima is very close to opposition (1–2°) assuming highly asymmetric negative polarisation branch. Rosenbush et al. (2015) pointed out that polarimetric properties of Uranian satellites are different from Saturnian and Jovian satellites of similar albedo.
5. The value of polarisation maxima varies from 2% to 50% and the phase angle of maxima is from $\sim 80^{\circ}$ (high-albedo asteroids) to $\sim 140^{\circ}$ (low-albedo asteroids).

Although recently no space probe visiting a small Solar system body was equipped with a polarimeter, the beginning of the era of in situ space exploration has marked a new era also in the field of interpretation of polarimetric data obtained by means of remote observations. It becomes possible to build a link between ground-based measurements and the local properties (composition, regolith properties and surface morphology at macroscopic scales) of surfaces from space data.

We should also note that in the current era of discovery of so many extrasolar planets and planetary systems around other stars, the role of polarimetric investigations will be probably become increasingly important. An example can be that of the effect on the polarisation state of a central star surrounded by clouds of scattering materials (dust or cometary or asteroidal belts).

In terms of perspectives for the future, we can expect that the importance of polarimetric results obtained in recent years can possibly lead to a better availability of larger telescopes. As the sizes of the largest telescopes increase, instruments that were considered large until a few years ago, namely telescopes in the 4–5 m class, might become increasingly available to polarimetric studies, if properly equipped with state-of-the art polarimetric detectors. Potential applications may include systematic observations of near-Earth objects, taking into account the possibility offered by polarimetric measurements to derive reliable estimates of the albedo.

An ideal development of polarimetric studies of Solar system objects, finally, could come in the future by the development of some dedicated space-based detector. Given the importance of polarisation data for all the fields of modern astrophysics, even at visible wavelengths, this might represent a real change of paradigm in the years to come.

References

Arago JF (1858) Les comètes. In: L'Astronomie populaire. Gide, Paris

Avramchuk VV, Rakhimov VIu, Chernova GP, Shavlovskii VI (1992) Photometry and polarimetry of Pluto near its perihelion. Kinematics Phys Celest Bodies 8:30–37

Bagnulo S, Boehnhardt H, Muinonen K, Kolokolova L, Belskaya I, Barucci MA (2006) Exploring the surface properties of Transneptunian Objects and Centaurs with polarimetric FORS1/VLT observations. Astron Astrophys. https://doi.org/10.1051/0004-6361:20054518

Bagnulo S, Belskaya I, Muinonen K, Tozzi GP, Barucci MA, Kolokolova L, Fornasier S (2008) Discovery of two distinct polarimetric behaviours of trans-Neptunian objects. Astron Astrophys. https://doi.org/10.1051/0004-6361:200810938

Bagnulo S, Tozzi GP, Boehnhardt H, Vincent J-B, Muinonen K (2010) Polarimetry and photometry of the peculiar main-belt object 7968=133P/Elst-Pizarro. Astron Astrophys. https://doi.org/10.1051/0004-6361/200913339

Bagnulo S, Belskaya I, Boehnhardt H, Kolokolova L, Muinonen K, Sterzik M, Tozzi G-P (2011) Polarimetry of small objects of the solar system with large telescope. J Quant Spectrosc Radiat Transf. https://doi.org/10.1016/j.jqsrt.2011.05.004

Bagnulo S, Cellino A, Sterzik MF (2015) Linear spectropolarimetry: a new diagnostic tool for the classification and characterisation of asteroids. Mon Not R Astron Soc. https://doi.org/10.1093/mnrasl/slu154

Bagnulo S, Belskaya I, Cellino A, Kolokolova L (2017) Polarimetry of small bodies and satellites of our Solar System. Europhys Phys J Plus. https://doi.org/10.1140/epjp/i2017-11690-6

Belskaya IN et al (2005) The F-type asteroids with small inversion angles of polarisation. Icarus. https://doi.org/10.1016/j.icarus.2005.04.015

Belskaya I, Bagnulo S, Muinonen K, Barucci MA, Tozzi GP, Fornasier S, Kolokolova L (2008) Polarimetry of the dwarf planet (136199) Eris. Astron Astrophys. https://doi.org/10.1051/0004-6361:20078241

Belskaya IN, Levasseur-Regourd AC, Cellino A, Efimov YS, Shakhovskoy NM, Hadamcik E, Bendjoya Ph (2009) Polarimetry of main belt asteroids: wavelength dependence. Icarus. https://doi.org/10.1016/j.icarus.2008.09.010

Belskaya IN, Bagnulo S, Barucci MA, Muinonen K, Tozzi GP, Fornasier S, Kolokolova L (2010) Polarimetry of Centaurs (2060) Chiron, (5145) Pholus and (10199) Chariklo. Icarus. https://doi.org/10.1016/j.icarus.2010.06.005

Belskaya IN, Bagnulo S, Stinson A, Tozzi GP, Muinonen K, Shkuratov YuG, Barucci MA, Fornasier S (2012) Polarimetry of trans-Neptunian objects (136472) Makemake and (90482) Orcus. Astron Astrophys. https://doi.org/10.1051/0004-6361/201220202

Belskaya I, Cellino A, Gil-Hutton R, Muinonen K, Shkuratov Y (2015) Asteroid polarimetry. In: Michel P et al (eds) Asteroids IV. University of Arizona, Tucson, pp 151–163

Belskaya IN et al (2017) Refining the asteroid taxonomy by polarimetric observations. Icarus. https://doi.org/10.1016/j.icarus.2016.11.003

Boehnhardt H, Bagnulo S, Muinonen K, Barucci MA, Kolokolova L, Dotto E, Tozzi GP (2004) Surface characterisation of 28978 Ixion (2001 KX76). Astron Astrophys. https://doi.org/10.1051/0004-6361:20040005

Boehnhardt H, Tozzi GP, Bagnulo S, Muinonen K, Nathues A, Kolokolova L (2008) Photometry and polarimetry of the nucleus of comet 2P/Encke. Astron Astrophys. https://doi.org/10.1051/0004-6361:200809922

Borisov G et al (2018) Rotational variation of the linear polarisation of the asteroid (3200) Phaethon as evidence for inhomogeneity in its surface properties. Mon Not R Astron Soc. https://doi.org/10.1093/mnrasl/sly140

Breger M, Cochran WD (1982) Polarimetry of Pluto. Icarus. https://doi.org/10.1016/0019-1035(82)90061-6

Carrillo-Sánchez JD, Nesvorny D, Pokorny JD, Plane JMC (2016) Sources of cosmic dust in the Earth's atmosphere. Geophys Res Lett. https://doi.org/10.1002/2016GL071697

Cellino A, Yoshida F, Anderlucci E, Bendjoya P, Di Martino M, Ishiguro M, Nakamura AM, Saito J (2005) A polarimetric study of asteroid 25143 Itokawa. Icarus. https://doi.org/10.1016/j.icarus.2005.08.005

Cellino A, Belskaya IN, Bendjoya Ph, Di Martino M, Gil-Hutton R, Muinonen K, Tedesco EF (2006) The strange polarimetric behavior of asteroid (234) Barbara. Icarus. https://doi.org/10.1016/j.icarus.2005.09.001

Cellino A, Gil-Hutton R, Belskaya I (2015a) Polarimetry of asteroids. In: Kolokolova L, Levasseur-Regourd AC, Hough J (eds) Polarimetry of stars and planetary systems. Cambridge University Press, Cambridge, pp 360–37

Cellino A, Bagnulo S, Gil-Hutton R, Tanga P, Canada-Assandri M, Tedesco EF (2015b) On the calibration of the relation between geometric albedo and polarimetric properties for the asteroids. Mon Not R Astron Soc. https://doi.org/10.1093/mnras/stv1188

Cellino A et al (2016) The Dawn exploration of (4) Vesta as the ground truth to interpret asteroid polarimetry. Mon Not R Astron Soc 456. https://doi.org/10.1093/mnras/stv2683

Cellino A, Bagnulo S, Belskaya IN, Christou AA (2018) Unusual polarimetric properties of (101955) Bennu: similarities with F-class asteroids and cometary bodies. Mon Not R Astron Soc. https://doi.org/10.1093/mnrasl/sly156

Chernova GP, Kiselev NN, Jockers K (1993) Polarimetric characteristics of dust particles as observed in 13 comets: comparison with asteroids. Icarus. https://doi.org/10.1006/icar.1993.1063

De Luise F et al (2007) Physical investigation of the potentially hazardous asteroid (144898) 2004 VD17. Icarus 191. https://doi.org/10.1016/j.icarus.2007.05.018

Devogèle M et al (2018a) New polarimetric and spectroscopic evidence of anomalous enrichment in spinel-bearing calcium-aluminium-rich inclusions among L-type asteroids. Icarus 304. https://doi.org/10.1016/j.icarus.2017.12.026

Devogèle M et al (2018b) The phase-polarisation curve of asteroid (3200) Phaethon. Mon Not R Astron Soc. https://doi.org/10.1093/mnras/sty1587

Dlugach JM, Ivanova OV, Mishchenko MI, Afanasiev VL (2018) Retrieval of microphysical characteristics of particles in atmospheres of distant comets from ground-based polarimetry. J Quant Spectrosc Radiat Transf. https://doi.org/10.1016/j.jqsrt.2017.10.002

Dollfus A, Bastien P, Le Borgne JL, Levasseur-Regourd AC, Mukai T (1988) Optical polarimetry of P/Halley: synthesis of the measurements in the continuum. Astron Astrophys 206:348–356

Dollfus A, Wolff M, Geake JE, Lupishko DF, Dougherty LM (1989) Photo-polarimetry of asteroids. In: Binzel RP, Gehrels T, Matthews MS (eds) Asteroids II. University of Arizona Press, Tucson, pp 594–616

Dougherty IM, Geake JE (1994) Polarisation by frost formed at very low temperatures, as relevant to icy planetary surfaces. Mon Not R Astron Soc. https://doi.org/10.1093/mnras/271.2.343

Dumont R (1973) Phase function and polarisation of interplanetary scatters from zodiacal light photopolarimetry. Planet Space Sci. https://doi.org/10.1016/0032-0633(73)90189-X

Fornasier S, Beskaya IN, Perna D (2015) The potentially hazardous asteroid 2007 PA8: an unweathered L chondrite analogue surface. Icarus. https://doi.org/10.1016/j.icarus.2014.12.015

Gil-Hutton R, García-Migani E (2017) Polarimetric survey of main-belt asteroids. VI. New results from the second epoch of the CASLEO survey. Astron Astrophys. https://doi.org/10.1051/0004-6361/201731388

Gil-Hutton R, Mesa V, Cellino A, Bendjoya Ph, Penaloza L, Lovos F (2008) New cases of unusual polarimetric behavior in asteroids. Astron Astrophys. https://doi.org/10.1051/0004-6361:20078965

Gil-Hutton R, López-Sisterna C, Calandra MF (2017) Polarimetric survey of main-belt asteroids. V. The unusual polarimetric behavior of V-type asteroids. Astron Astrophys. https://doi.org/10.1051/0004-6361/201628485

Grün E et al (2016) The 2016 Feb 19 outburst of comet 67P/CG: an ESA multi-instrument study. Mon Not R Astron Soc. https://doi.org/10.1093/mnras/stw2088

Hadamcik E, Levasseur-Regourd AC (2003) Imaging polarimetry of cometary dust: different comets and phase angles. J Quant Spectrosc Radiat Transf. 79:661–678

Hadamcik E, Levasseur-Regourd AC (2016) Imaging polarimetry of comet 73P/Schwassmann-Wachmann 3 main fragments during its 2006 apparition. Planet Space Sci. https://doi.org/10.1016/j.pss.2015.12.01

Hadamcik E, Levasseur-Regourd AC, Renard JB (1997) CCD imaging polarimetry of comet Hale-Bopp. Earth Moon Planets. https://doi.org/10.1023/A:1006206613959

Hadamcik E, Sen AK, Levasseur-Regourd AC, Gupta R, Lasue J (2010) Polarimetric observations of comet 67P/Churyumov-Gerasimenko during its 2008–2009 apparition. Astron Astrophys. https://doi.org/10.1051/0004-6361/201014167

Hadamcik E, Levasseur-Regourd AC, Hines DC, Sen AK, Lasue J (2016) Photometric and polarimetric observations of 67P: clues to cometary dust properties in comparison with Rosetta results. Mon Not R Astron Soc. https://doi.org/10.1093/mnras/stx030

Hadamcik E, Lasue J, Levasseur-Regourd AC, Renard J-B (2019) Analogues of interplanetary dust particles to interpret the zodiacal light polarisation. Planet Space Sci. https://doi.org/10.1016/j.pss.2018.04.022

Hines DC, Levasseur-Regourd AC (2016) Polarimetry observations of comets: status, questions, future pathways. Planet Space Sci. https://doi.org/10.1016/j.pss.2015.11.016

Ishiguro M et al (2017) Polarimetric study of Near-Earth asteroid (1566) Icarus. Astron J. https://doi.org/10.3847/1538-3881/aa8b1a

Ito T et al (2018) Extremely strong polarisation of an active asteroid (3200) Phaethon. Nature Commun 9:2486

Kaasalainen S, Piironen J, Kaasalainen M, Harris AW, Muinonen K, Cellino A (2003) Asteroid photometric and polarimetric phase curves: empirical interpretation. Icarus. https://doi.org/10.1016/S0019-1035(02)00020-9

Kelsey JD, Fix LA (1973) Polarimetry of Pluto. Astrophys J 184:633–636

Kiselev N, Velichko F (1997) Aperture polarimetry and photometry of comet Hale-Bopp. Earth Moon Planets. https://doi.org/10.1023/A:1006246310324

Kiselev NN, Rosenbush VK, Jockers K, Velichko FP, Shakhovskoy NM, Efimov YS, Lupishko DF, Rumyantsev VV (2002) Polarimetry of near-Earth asteroid 33342 (1998 WT24). Synthetic phase angle dependence of polarisation for the E-type asteroids. In: Warmbein B (ed) Proceedings of asteroids, comets, meteors – ACM 2002, ESA SP-500. Noordwijk, Netherlands, pp 887–890

Kiselev N, Rosenbush V, Levasseur-Regourd AC, Kolokolova L (2015) Comets. In: Kolokolova L, Levasseur-Regourd AC, Hough J (eds) Polarimetry of stars and planetary systems. Cambridge University Press, Cambridge. https://doi.org/10.1017/CBO9781107358249.022

Kiselev N, Afanasiev V, Petrov D, Rosenbush V (2018) Unusual properties of asteroid (3200) Phaethon from polarimetric observations and modelling. In: EPSC Abstracts 12, EPSC2018-504

Kolokolova L, Hough J, Levasseur-Regourd AC (eds) (2015) Polarimetry of stars and planetary systems. Cambridge University Press, Cambridge. https://doi.org/10.1017/CBO9781107358249

Kordylewski K (1961) Photographische Untersuchungen des Libration Punktes im System Erde-Mond. Acta Astron 11:165

Kuroda D et al (2018) Significantly high polarisation degree of the very low-albedo asteroid (152679) 1998 KU2. Astron Astrophys. https://doi.org/10.1051/0004-6361/201732086

Lasue J, Levasseur-Regourd AC, Fray N, Cottin H (2007) Inferring the interplanetary dust properties from remote observations and simulations. Astron Astrophys. https://doi.org/10.1051/0004-6361:20077623

Lasue J, Levasseur-Regourd AC, Hadamcik E, Alcouffe G (2009) Cometary dust properties retrieved from polarisation observations, Application to C/1995 O1 Hale-Bopp and 1P/Halley. Icarus. https://doi.org/10.1016/j.icarus.2008.09.008

Lasue J, Levasseur-Regourd AC, Lazarian A (2015) Interplanetary dust. In: Kolokolova L, Levasseur-Regourd AC, Hough J (eds) Polarimetry of stars and planetary systems. Cambridge University Press, Cambridge. https://doi.org/10.1017/CBO9781107358249.024

Levasseur-Regourd AC (1992) Natural background radiation, the light of the night sky. In: McNally D (ed) The vanishing universe. Cambridge University Press, Cambridge, pp 64–68

Levasseur-Regourd AC (2010) Inferring properties of dust within solar system small bodies through observations and simulations of the linear polarisation of scattered solar-light. In: Mischchenko MI et al (eds) Polarimetric detection, characterisation, and remote sensing, Springer, Dordrecht, pp 295–312. https://doi.org/10.1007/978-94-007-1636-0

Levasseur-Regourd AC, Dumont R, Renard J-B (1990) A comparison between polarimetric properties of cometary dust and interplanetary dust particles. Icarus. https://doi.org/10.1016/0019-1035(90)90215-U

Levasseur-Regourd AC, Hadamcik E, Renard JB (1996) Evidence for two classes of comets from their polarimetric properties at large phase angles. Astron Astrophys 313:327–333

Levasseur-Regourd AC, McBride N, Hadamcik E, Fulle M (1999) Similarities between in situ measurements of local dust scattering and dust flux impact data within the coma of 1P/Halley. Astron Astrophys 348:636–641

Levasseur-Regourd AC, Mann I, Dumont R, Hanner M (2001) Optical and thermal properties of cometary dust. In: Grün E et al (eds) Interplanetary dust. Springer-Verlag, Berlin. https://doi.org/10.1007/978-3-642-56428-4

Levasseur-Regourd AC, Hadamcik E, Lasue J (2005) Light scattering as a clue to cometary dust structure. Highlights Astron 13:498–500. https://doi.org/10.1017/S1539299600016415

Levasseur-Regourd AC, Agarwal J, Cottin H, Engrand C, Flynn G, Fulle M, Gombosi T, Langevin Y, Lasue J, Mannel T, Merouane S, Poch O, Thomas N, Westphal A (2018) Cometary dust. Space Sci Rev. https://doi.org/10.1007/s11214-018-0496-3

Lumme K, Muinonen K (1993) A two-parameter system for linear polarization of some solar system objects. In: IAU symposium 160, asteroids, comets, meteors (Abstract), p 194

Lupishko DF (ed) (2017) Asteroid Polarimetric Database V9.0.EAR-A-3-RDR APDPOLARIMETRY – V9.0. NASA Planetary Data System, Washington

Lupishko DF (2018) Generalized calibration of the polarimetric albedo scale of asteroids. Solar Syst Res. https://doi.org/10.1134/S0038094618010069

Lyot B (1929) Recherches sur la polarisation de la lumière des planètes et de quelques substances terrestres. Ann Obs Meudon 8:161

Mannel T, Bentley MS, Schmied R, Jeszenszky H, Levasseur-Regourd AC, Romstedt J, Torkar K (2016) Fractal cometary dust a window to the early Solar System. Mon Not R Astron Soc. https://doi.org/10.1093/mnras/stw2898

Masiero J, Cellino A (2009) Polarisation of asteroid (387) Aquitania: the newest member of a class of large inversion angle asteroids. Icarus. https://doi.org/10.1016/j.icarus.2008.10.003

Meyers RV, Nordsieck KH (1984) Spectropolarimetry of comets Austin and Churyumov Gerasimenko. Icarus 58:431–439

Mishchenko MI, Rosenbush VK, Kiselev NN, Lupishko DF, Tishkovets VP, Kaydash VG, Belskaya IN, Efimov YS, Shakhovskoy NM (2010) Polarimetric remote sensing of solar system objects. Akademperiodyka, Kyiv. https://doi.org/10.15407/akademperiodyka.134.291

Muinonen K, Penttilla A, Cellino A, Belskaya IN, Delbo M, Levasseur-Regourd AC, Tedesco EF (2009) Asteroid photometric and polarimetric phase curves: joint linear-exponential modelling. Meteorit Planet Sci. https://doi.org/10.1111/j.1945-5100.2009.tb02003.x

Muñoz O, Moreno F, Vargas-Martin F, Guirado D, Escobar-Cerezo J, Min M, Hovenier JW (2017) Experimental phase functions of millimeter-sized cosmic dust grains. Astrophys J. https://doi.org/10.3847/1538-4357/aa7ff2

Nesvorný D, Jenniskens P, Levison HF, Bottke WF, Vokrouhlicky D, Gounelle M (2010) Cometary origin of the zodiacal cloud and carbonaceous micrometeorites. Implications for hot debris disks. Astrophys J. https://doi.org/10.1088/0004-637X/713/2/816

Penttila A, Lumme K, Hadamcik E, Levasseur-Regourd A-C (2005) Statistical analysis of asteroidal and cometary polarisation phase curves. Astron Astrophys. https://doi.org/10.1051/0004-6361:20042133

Petrov DV, Kiselev NN (2017) Positive branch of asteroid polarisation: observational data and computer modelling. Solar Syst Res. https://doi.org/10.1134/S0038094617040049

Poch O, Cerubini R, Pommerol A, Jost B, Thomas N (2018) Polarimetry of water ice particles providing insights on grain size and degree of sintering on icy planetary surfaces. J Geophys Res Planets. https://doi.org/10.1029/2018JE005753

Reach WT, Kelley MS, Sykes MV (2007) A survey of debris trails from short-period comets. Icarus. https://doi.org/10.1016/j.icarus.2007.03.031Phaethon

Renard J-B, Hadamcik E, Couté B, Jeannot M, Levasseur-Regourd AC (2014) Wavelength dependence of linear polarisation in the visible and near infrared domain for large levitating grains (PROGRA2 instruments). J Quant Spectrosc Radiat Transf 146:424–430

Rosenbush V, Kiselev N, Afanasiev V (2015) Icy moons of the outer planets. In: Kolokolova L, Levasseur-Regourd AC, Hough J (eds) Polarimetry of stars and planetary systems. Cambridge University Press, Cambridge, pp 340–359. https://doi.org/10.1017/CBO9781107358249.020

Rosenbush VK, Ivanova OV, Kiselev NN, Kolokolova LO, Afanasiev VL (2017) Spatial variations of brightness, colour and polarisation of dust in comet 67P/Churyumov-Gerasimenko. Mon Not R Astron Soc. https://doi.org/10.1093/mnras/stx2003

Rousselot P, Levasseur-Regourd AC, Muinonen K, Petit J-M (2005) Polarimetric and photometric phase effects observed on transneptunian object (29981) 1999 TD10. Earth Moon Planets. https://doi.org/10.1007/s11038-006-9079-5

Rowan-Robinson M, May B (2016) An improved model for the infrared emission from the zodiacal dust cloud. Mon Not R Astron Soc 429:2894–2902

Shinnaka Y, Kasuga T, Furusho R, Boice DC, Terai T, Noda H, Namiki N, Watanabe J (2018) Inversion angle of phase-polarisation curve of Near-Earth asteroid (3200) Phaethon. Astrophys J Lett. https://doi.org/10.3847/2041-8213/aadb3d

Sierks H et al (2016) On the nucleus structure and activity of comet 67P/Churyumov-Gerasimenko. Science. https://doi.org/10.1126/science.aaa1044

Slitz-Balogh J, Barta A, Horvath G (2018) Celestial mechanics and polarisation optics of the Kordylewski dust cloud in the Earth-Moon Lagrange point L5 Part I. Mon Not R Astron Soc 480:5550–5559

Slitz-Balogh J, Barta A, Horvath G (2019) Celestial mechanics and polarisation optics of the Kordylewski dust cloud in the Earth-Moon Lagrange point L5 Part II. Mon Not R Astron Soc 482:762–770

Snodgrass C et al (2017) The 67P/Churyumov-Gerasimenko observation campaign in support of the Rosetta mission. Philos Trans R Soc. https://doi.org/10.1098/rsta.2016.0249

Stinson A et al (2016) Polarimetry of comets 67P/Churyumov-Gerasimenko, 74P/Smirnova-Chernykh, and 152P/Helin-Lawrence. Astron Astrophys. https://doi.org/10.1051/0004-6361/201527696

Umov N (1905) Chromatische Depolarisation durch Lichtzerstereuung. Phys Z 6:674–676

Chapter 10
Polarimetry of Binary Systems: Polars, Magnetic CVs, XRBs

Tariq Shahbaz

Abstract Polarimetry provides key physical information on the properties of interacting binary systems, sometimes difficult to obtain by any other type of observation. Indeed, radiation processes such as scattering by free electrons in the hot plasma above accretion discs, cyclotron emission by mildly relativistic electrons in the accretion shocks on the surface of highly magnetic white dwarfs and the optically thin synchrotron emission from jets can be observed. In this review, I will illustrate how optical/near-infrared polarimetry allows one to estimate magnetic field strengths and map the accretion zones in magnetic cataclysmic variables as well as determine the location and nature of jets and ejection events in X-ray binaries.

10.1 Cataclysmic Variables

Cataclysmic variables (CVs) are a class of binary system in which a main sequence secondary star transfers matter to its white dwarf (WD) companion, via Roche lobe overflow. It is estimated that ~20% of all CVs host a magnetic WD (mCVs) (Wickramasinghe and Ferrario 2000; Ferrario et al. 2015). In the "normal" accreting mode known as the "high-state" of mCVs, the accreting material leaving the inner Lagrangian L1 point attaches itself to the magnetic field lines of the WD close and is channelled towards its surface. However, most mCVs also show erratic and abrupt "low-states", during which accretion through the L1 point temporarily stops, presumably due to changes in the mass-transfer rate in the system, related to the atmospheric magnetic activity of the secondary star (Livio and Pringle 1994). During the high-state, as the flow is collimated by the magnetic field, electrons

T. Shahbaz (✉)
Instituto de Astrofísica de Canarias (IAC), La Laguna, Tenerife, Spain

Departamento de Astrofísica, Universidad de La Laguna (ULL), La Laguna, Tenerife, Spain
e-mail: tsh@iac.es

R. Mignani et al. (eds.), *Astronomical Polarisation from the Infrared to Gamma Rays*, Astrophysics and Space Science Library 460,
https://doi.org/10.1007/978-3-030-19715-5_10

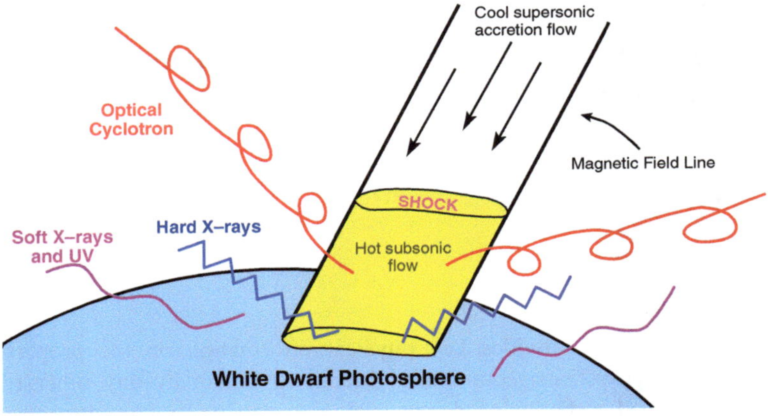

Fig. 10.1 A schematic of the surface of the WD, showing the region of the main emission mechanisms (Potter 2016)

spiral down the magnetic field lines, cooling the plasma flow by cyclotron emission. This radiation is detected as broad overlapping cyclotron humps in the optical and infrared (IR) spectrum and allows to estimate magnetic field strengths within the accretion region. Hard X-rays resulting from a post-shock region just above the accreting pole and softer ultraviolet (UV) and X-rays are emitted from a hot photospheric polar cap. During the low-states, Zeeman spectroscopy can be used to constrain the temperature and magnetic field of the WD (see Fig. 10.1; for a review see Cropper 1990).

MCVs are characterised by strong X-ray emission, high-excitation optical spectra, very stable X-ray and optical periods in their light curves. They are broadly classified into two categories based upon their periodicities, which are in turn assumed to imply different magnetic field strengths. The polars (or AM Her type systems after the prototype system) contain synchronously rotating WDs, whilst in intermediate polars (IPs or DQ Her type systems) the WDs rotate asynchronously with respect to the binary orbit (see Fig. 10.2).

- In polars, the magnetic field strength of the WD (>10 MG) is so strong that the gas leaving the inner Lagrangian point L1 following a ballistic trajectory eventually attaches to the magnetic field of the WD, preventing the formation of an accretion disc as in CVs. Accretion can occur on one or both poles, depending on the orientation of the magnetic dipole of the WD. The material is fully ionised producing highly polarised cyclotron emission due to the presence of the strong magnetic field. Since a large fraction of the optical flux in polars comes from such regions, they have large polarisation values. The radiation is circularly polarised when observed parallel to the magnetic field lines and linearly when observed perpendicular to the field lines. Consequently, polarimetry is a fundamental tool for classifying an object and polar. The magnetic field of the WD extends to

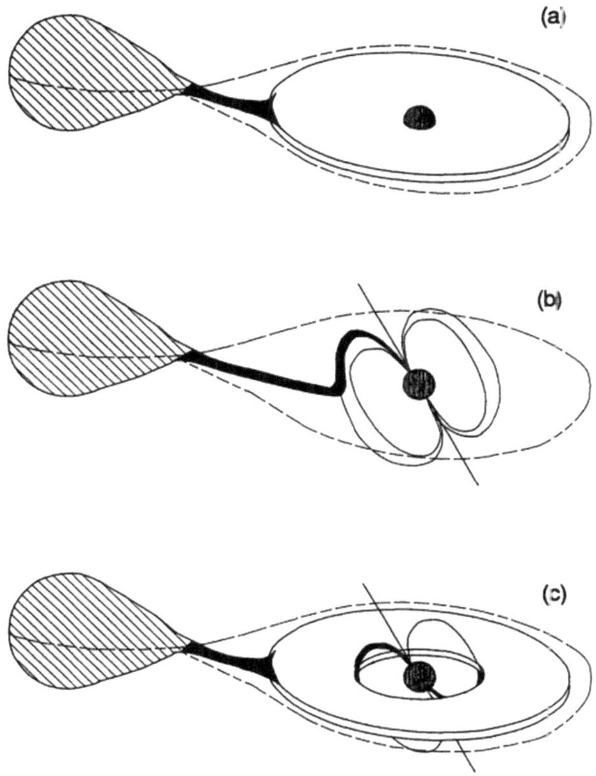

Fig. 10.2 Schematic of the different type of CVs. From top to bottom: a non-magnetic CV with an accretion disc (**a**), a polar (AM Her type systems) in which the accretion disc is prevented from forming (**b**) and an intermediate polar (DQ Her type systems) in which the accretion disc is disrupted (**c**) Norton (1993)

the secondary synchronises the spin and orbital period. Typically the polars are found to have an orbital and spin period of less than 2 h (for a review see Cropper 1990).

- The IPs are asynchronous systems which have a magnetic field strength of <10 MG. The lower magnetic field strength compared to polars implies that matter falls inward towards the WD in much the same way as in non-magnetic CVs. However, the magnetosphere disrupts the inner portion of the accretion disc and the flow then follows the field lines in a curtain-like structure (Patterson 1994). Due to the disc symmetry, the accretion curtains are present for both magnetic poles, resulting in arc-like accretion regions near the WD poles, similar to the auroral ovals on the Earth and Jupiter.

However, there are systems that do not fit into the standard picture of mCVs. AE Aqr was originally classified as an IP (Patterson 1979) but there are a number of unusual observational features that are not naturally explained by this model. It is known to switch between the flaring and quiescent states irregularly, which distinguishes it from other mCVs. Studies of the 33 s WD pulsations in the optical/UV (Eracleous et al. 1994) and X-rays (Reinsch et al. 1995; Choi et al. 1999) show no correlation between their amplitudes and the flaring of the system, which means that the flares are not related to the process of depositing material onto the WD (Eracleous et al. 1994). The Balmer emission lines are single peaked and produce Doppler tomograms that are not consistent with those of an accretion disc (Welsh et al. 1998). Different models were proposed to explain the observational properties of AE Aqr. The first, a "magnetic propeller" model (Wynn et al. 1997) suggests that the accretion rate onto the WD decelerates by means of interaction between its fast rotating magnetosphere and the material inflowing from the secondary. The second "pulsar-like white dwarf" model (Ikhsanov 1998) shows that the observed braking of the white WD can be explained in terms of the canonical pulsar-like spin-down mechanism (Ikhsanov 1998, 2006). Recently, a numerical propeller model of AE Aqr using axisymmetric magneto-hydrodynamic (MHD) simulations has been developed (Blinova et al. 2018). They suggest that an accretion disc forms around the WD that interacts with the magnetosphere of the star in the propeller regime. The model explains the rapid spin-down of the WD through the outflow of angular momentum from the surface of the WD into the wind and predicts that $\sim$85% of the inner disc matter is ejected into the conically shaped winds. The model can also explain the low accretion rate onto the star if the radiative efficiency of accretion is $\sim$7%.

10.1.1 Models

The optical flux and polarisation of polars is highly modulated with orbital phase. These observations are explained by the cooling of the post-shock region through cyclotron emission which depends strongly on the viewing angle and the physical properties of the emitting region. Stokes techniques are consequently a powerful tool to diagnose the physical properties of the accreting region. An increase in circular polarisation is observed when one looks down on the shock (parallel to the magnetic field lines), whereas an increase in linear polarised flux is observed when one views the side of the shock (perpendicular to the magnetic field lines).

The first important step in modelling of polar observations was performed in the 1980s and assumed the emitting region to be a 1-D point source (Chanmugam and Dulk 1981; Meggitt and Wickramasinghe 1982; Wickramasinghe and Meggitt 1985). Wickramasinghe and Meggitt (1985) present a grid of fluxes, and linear and circular polarisation as a function of viewing angle and harmonic, which can be compared to optical data of polars. However, there is observational evidence of extended emitting regions in polars and so the models were improved to 2-D by using a sum of distinct point sources to represent the emitting region (Ferrario and Wickramasinghe 1990; Potter et al. 1998).

More sophisticated models and fitting techniques were then developed referred to "Stokes imaging" (Potter et al. 1998) was the first objective and analytical technique that robustly models the cyclotron emission and maps the accretion zones in magnetic mCVs. This technique allowed objective mapping of the cyclotron-emission regions in terms of location, size and density structure. Stokes imaging uses a cyclotron-emission model and a grid of cyclotron-emission calculations which include cyclotron and free–free opacities (Wickramasinghe and Meggitt 1985). The emission from a single emission spot on the surface of the white dwarf is modelled by first calculating the local magnetic field strength and subsequently the local optical depth parameter. The cyclotron-emission grids are then used to produce variations in the polarised emission as a function of phase. The emission from extended sources is modelled by summing the components of many such emission spots. Projection effects of extended regions are also taken into consideration. Figure 10.3 shows the model fit from Stokes imaging for the IP V347 Pav, where Stokes imaging reproduced the main polarisation morphology (Potter et al. 2000). The main emission region was found to consist with two relatively higher density regions centred on the magnetic equator. The system inclination derived is 68° and the magnetic field strengths are 15 and 20 MG for the upper and lower poles, respectively. Stokes imaging has been advanced by the addition of more realistic stratified accretion shock model in order to calculate cyclotron spectra, and the use of Stokes imaging in combination with Doppler (Marsh and Horne 1988) and Roche tomographic (Watson and Dhillon 2001) techniques, to gain insights into magnetic accretion (Potter et al. 2004).

A more detailed code has also been developed, which adopts a 3-D representation of the accretion column and WD, incorporates the radiative transfer of the cyclotron process in the post-shock region (Costa and Rodrigues 2009). The model has been used to fit the optical near-IR circular polarimetric light curves of V834 Cen (see Fig. 10.4) and the best fit is consistent with previous parameter estimates and is able to reproduce the observed features.

10.1.2 Cyclotron Spectroscopy

Cyclotron spectroscopy is a useful method for extracting on strength of the magnetic fields in polar and IPs. The characteristic feature of polars is the presence of strong linear and circular polarisation at optical and near-IR wavelengths. The polarised emission is due to cyclotron origin and arises from accretion shocks near the surface of the high magnetic field WD. The fundamental cyclotron frequency usually lies in the IR (from ~2 to 10 μm). Indeed, the cyclotron radiation observed in the optical/near-IR consists of the harmonics of the fundamental cyclotron frequency. The spacing between successive cyclotron humps allows one to estimate the magnetic field strength of the cyclotron emission, because cyclotron-emission intensity and polarisation depend on the harmonic n $(=\omega/\omega_c)$ of the fundamental cyclotron frequency ω_c (Wickramasinghe and Meggitt 1985; Ferrario et al. 1993).

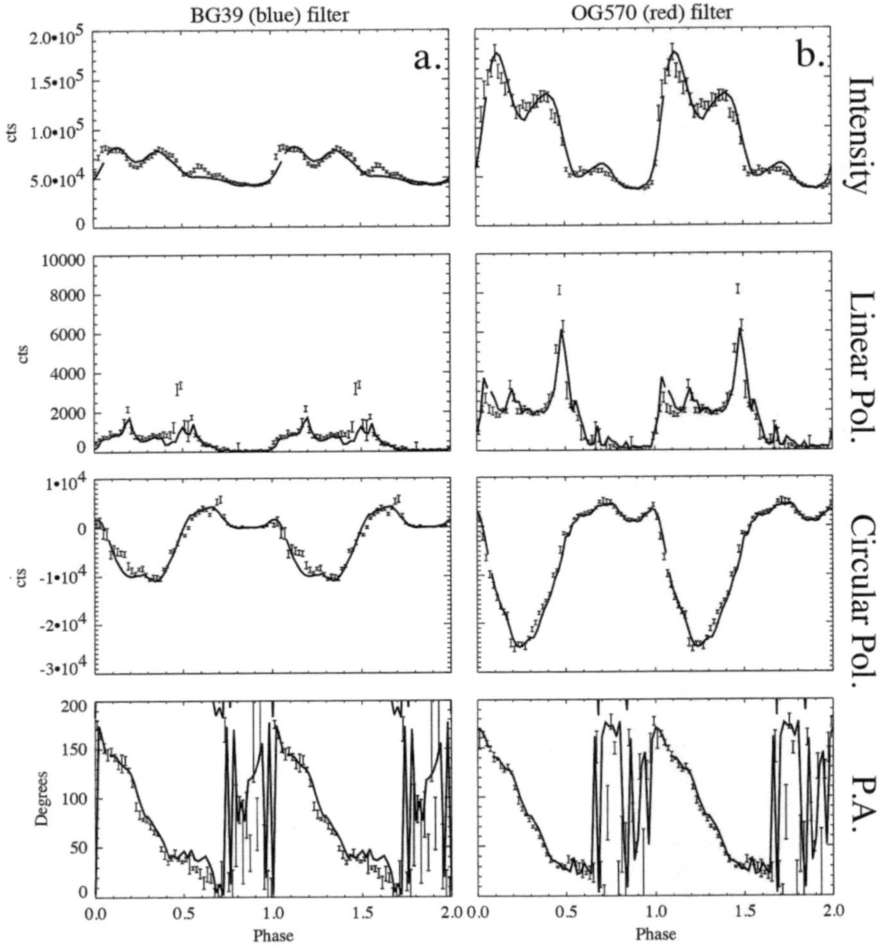

Fig. 10.3 The Stokes imaging optimised solution to the polarimetric observations V347 Pav. Solid curves are the model fit to the blue (**a**) the red (**b**) data. The imaging predicts that the bright-phase emission arises from a large region towards the lower hemisphere, and the faint-phase emission from a single smaller region in the upper hemisphere of the WD (Potter et al. 2000)

The position of the nth harmonic λ_n of the cyclotron emission from a region with magnetic field B and low electron temperatures ($T_e < 10\,\text{keV}$), as a first approximation is given by

$$\lambda_n \sim \frac{10{,}710}{n}\left(\frac{10^8}{B}\right)\,\text{Å} \tag{10.1}$$

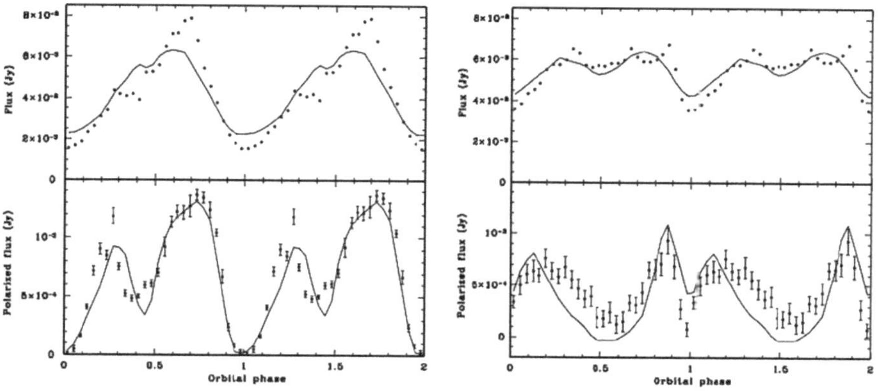

Fig. 10.4 Best fit to the V-band (left) and J-band (right) circular polarisation light curves V834 Cen of using the CYCLOPS (Costa and Rodrigues 2009)

Cyclotron line emission has been observed in a number of polars, where cyclotron harmonics ($n \sim 3$–8) are seen in the optical/near-IR spectra, W Pup (Wickramasinghe et al. 1989), UZ For (Ferrario et al. 1989), MR Ser (Wickramasinghe et al. 1991), DP Leo (Cropper and Wickramasinghe 1993) ST LMi (Ferrario et al. 1993) and EF Eri (Campbell et al. 2008).

The observed circular polarised flux is independent of other sources of non-polarised light (e.g. accretion stream, disc or WD photosphere) and therefore is very useful for comparing with existing cyclotron models. The broadband circular polarised flux spectrum gives estimates to the magnetic field in the cyclotron-emission region, as cyclotron-emission intensity and polarisation are characteristic to each harmonic (Wickramasinghe and Meggitt 1985). An example of the comparison of the observed circular polarised flux spectrum with cyclotron models is shown in Fig. 10.5. IPs are known to be circularly polarised at a level ~1–3% (Ferrario et al. 2015). Magnetic field measurements for IPs which do not show measurable photospheric Zeeman features are difficult, but from the polarised flux spectrum, estimates have been made for PG Gem (8–18 MG, Piirola et al. 1993), PQ Gem (9–21 MG, Vaeth et al. 1996; Potter et al. 1997), V2400 Oph (9–20 MG, Vaeth 1997), RX J2133.7+5107 (>20 MG, Katajainen et al. 2007) and V405 Aur (~30 MG, Piirola et al. 2008). These values are similar to the magnetic field strengths measured in polars, and suggest that some of the polarised IPs may evolve into polars (Chanmugam and Ray 1984).

10.1.3 Time-Resolved Polarimetry of SW Sex Systems

SW Sex stars are a subclass of CVs that possess common but inexplicable features, very different from other CVs (Thorstensen et al. 1991). Their light curves show

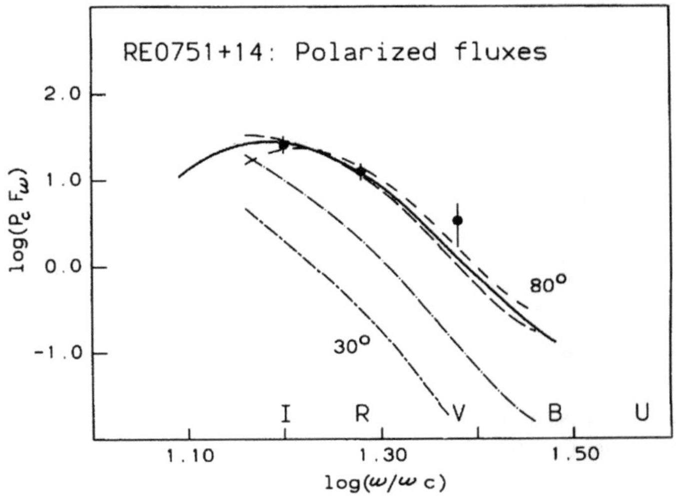

Fig. 10.5 Observed peak circular polarised flux compared with cyclotron models with different magnetic field strengths (Piirola et al. 1993). ©AAS. Reproduced with permission

deep eclipses suggesting a high-inclination accretion disc but their emission lines show single-peaked profiles that do not share the orbital motion of the WD and they exhibit transient absorption features as well as shallow eclipses. This is contrary to the deeply eclipsing, double-peaked profiles one would expect for high-inclination systems (Marsh and Horne 1988). Many models have been proposed to explain the behaviour of the SW Sex systems. Models include gas-stream overflow, magnetic propeller accretion and self-obscuration by the disc rim (for details see Rodríguez-Gil et al. 2007). Dhillon et al. (2013) have proposed that a combination of dominant bright-spot emission and a self-occulting disc can explain the observed properties.

A magnetic origin for the SW Sex systems has been claimed by several authors (e.g. see Martínez-Pais et al. 1999). Although the observation of modulated circular polarisation is an unambiguous proof as it implies the direct detection of a relatively strong magnetic field, the presence of coherent pulsations related to the WD spin period and/or the orbital period also provides strong evidence for the presence of a magnetic WD. Modulated circular polarisation has been observed in the non-eclipsing systems LS Peg (see Fig. 10.6), V795 Her and RXJ1643.7+3402 (Rodríguez-Gil et al. 2001, 2002, 2009). In LS Peg X-ray pulsations associated with the spin period of the WD are observed at the same polarimetric frequency (Baskill et al. 2005), whereas in RXJ1643.7+3402 the circular polarisation period is interpreted as half the beat period between the WD spin period and the orbital period (Rodríguez-Gil et al. 2009). To some extent the observed circular polarisation frequencies can be explained in terms of a magnetic model in which the overflowing gas stream crosses the WD's magnetosphere at approximately the corotation radius (Rodríguez-Gil et al. 2001, 2009).

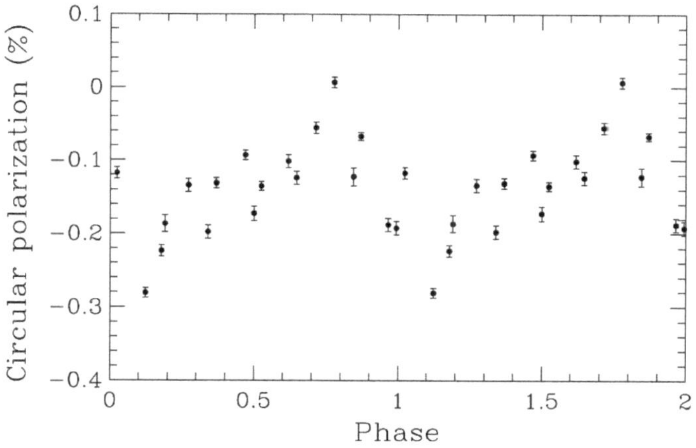

Fig. 10.6 Modulated circular optical polarisation in LS Peg. The period of the modulation is in phase with the spin period of the WD (Rodríguez-Gil et al. 2001). ©AAS. Reproduced with permission

10.1.4 A WD Pulsar

The discovery of pulsar-like behaviour in the close binary system AR Scorpii (hereafter AR Sco) demonstrates that WDs can exhibit many of the same characteristics as neutron star (NS) pulsars. Highly pulsed non-thermal emission is observed from radio to the UV arising from the spinning WD (Marsh et al. 2016).

These pulsations are seen predominantly at the 118 s beat period between the 117 s spin period and the 3.6 h orbital period of the binary. The strong spin modulated emission arising from the WD plus the dominant non-thermal nature of the spectral energy distribution suggests that magnetic interactions power the emission in AR Sco (Marsh et al. 2016; Katz 2017), similar to what is observed in pulsars. Further evidence of pulsar behaviour is given by polarimetric observations (Buckley et al. 2017). Strong linear polarisation (up to 40%) that varies on both the spin period of the WD and the beat period between the spin and orbital period has been detected as well as low-level circular polarisation (~few %). The characteristics of the polarised and non-polarised emission are explained in terms of synchrotron emission from two different regions, one associated with the rotating WD magnetic field and the other as a result of MHD interactions with the M-dwarf companion star. Unlike typical CVs there appears to be little accretion on to the WD from its M-dwarf companion star (Marsh et al. 2016), which means that gravitational potential energy is not the source of the emission. AR Sco is the first example of a WD pulsar. because the pulsed luminosity of AR Sco is powered by the spin-down of the rapidly rotating highly magnetised (<500 MG) WD (Buckley et al. 2017) (Fig. 10.7).

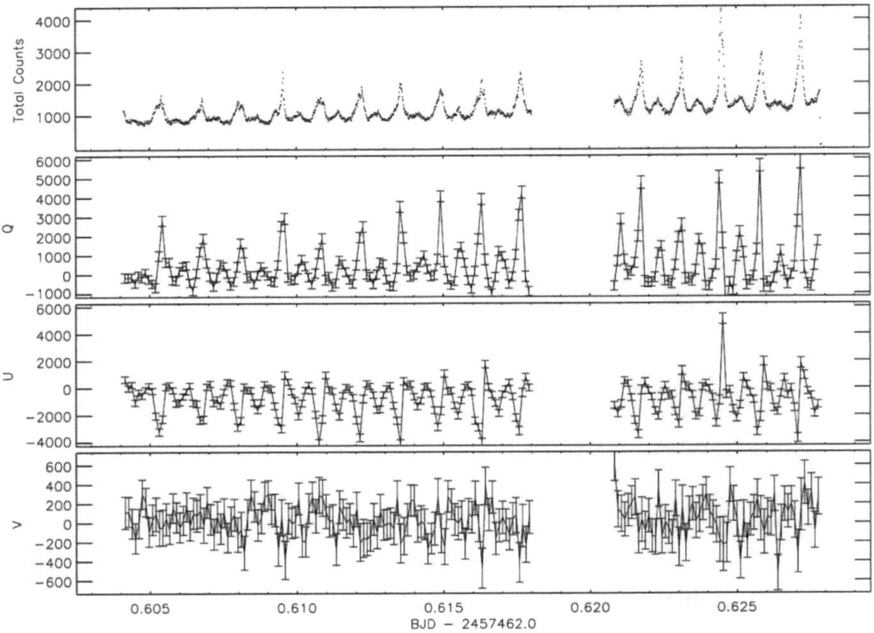

Fig. 10.7 Broad band photopolarimetry of AR Sco. From the top to bottom, the total intensity and the Stokes Q, U and V values (Buckley et al. 2017)

10.2 X-ray Binaries

X-ray binaries (XRBs) are binary systems in which a compact object, a black hole (BH) or a NS, accretes matter from a companion star. During the outburst of a black BH-XRBs, continuously replenished synchrotron-emitting "compact" jets are commonly produced during the hard spectral state (hard-state; Fender and Gallo 2014) (for descriptions of X-ray states see McClintock and Remillard 2006). The hard-state usually occurs at the beginning and end of an outburst, with X-ray luminosities $>10^{33}$ erg s^{-1} and up to ∼1–5% of the Eddington luminosity (e.g. Belloni 2010; Koljonen et al. 2016), with a highly variable hard power-law X-ray spectrum that originates from the inner, radiatively inefficient accretion flow close to the black BH (Narayan and Yi 1994). A classical "flat spectrum" radio jets (similar to those seen in AGN) are commonly observed during the hard-state, when the accretion flow structure allows a large-scale height magnetic field. In the softer X-ray states jets are dramatically quenched at radio frequencies (Fender 2001; Gallo et al. 2003; Russell et al. 2011b) because the geometrically thin disc suppresses any large-scale vertical field (Meier 2001).

Few studies have attempted to uncover the polarimetric signature of the optically thin synchrotron emission from compact jets in XRBs. The polarisation originates close to the base of the jet where the magnetic field should have higher level of ordering over the smaller emission region compared to further out in the jet (Blandford and Königl 1979). Polarimetric measurements of the optically thin power-law synchrotron emission provide a powerful tool to uncover the nature of the magnetic field structure in this region, because it is associated with the start of the particle acceleration in the jet (Polko et al. 2010), which is important for models and simulations of jet production.

10.2.1 The Spectral Energy Distribution

The spectral energy distribution of an XRB can be described by thermal emission from a companion star and accretion disc, synchrotron emission from a jet and a Comptonised corona from the disc. At optical and IR wavelengths the companion star dominates the optical/near-IR emission, whereas the accretion disc dominates at UV to X-ray wavelengths. The radio to IR spectrum of an XRB in the hard-state is due to a ~flat, self-absorbed optically thick synchrotron spectrum with a spectral index of α ~ 0.0 to +0.5, where $F_\nu \propto \nu^\alpha$ (Fender 2001; Fender et al. 2000; Markoff et al. 2001) (see Figs. 10.8, 10.9 and 10.10), which is the principal signature of the presence of a conical, collimated jet in black BH-XRBs. It typically extends at down to at least to millimetre wavelengths and is explained as being due to the superposition of synchrotron-emitting lepton particle distributions at different radii from the compact object (Blandford and Königl 1979; Kaiser 2006; Malzac 2018) (see Fig. 10.9). The higher energy synchrotron emission arises from a small, dense region of the jet, close to the location where the jets are launched, near the compact object (Kaiser 2005). Since α > ~0, the bulk of the radiative power of the jet resides in this higher energy emission and so at some frequency (~IR wavelengths) there is

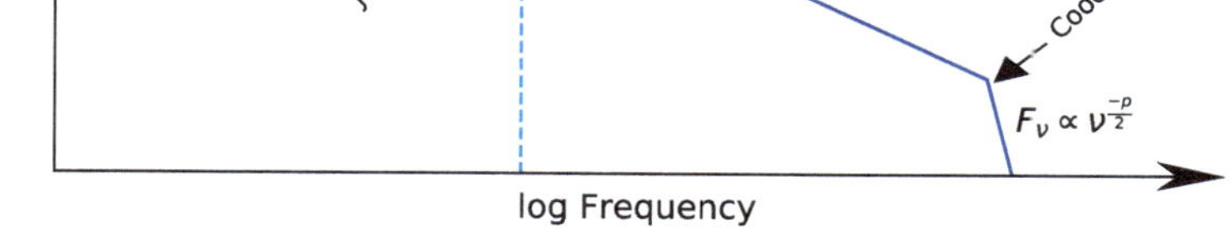

Fig. 10.8 A sketch of spectrum of a steady, hard-state jet. The radio to near-IR spectrum is due to a ~flat, self-absorbed optically thick synchrotron spectrum. The key spectral point is the break from optically thick to optically thin synchrotron emission, between the mid- to near-IR wavelengths

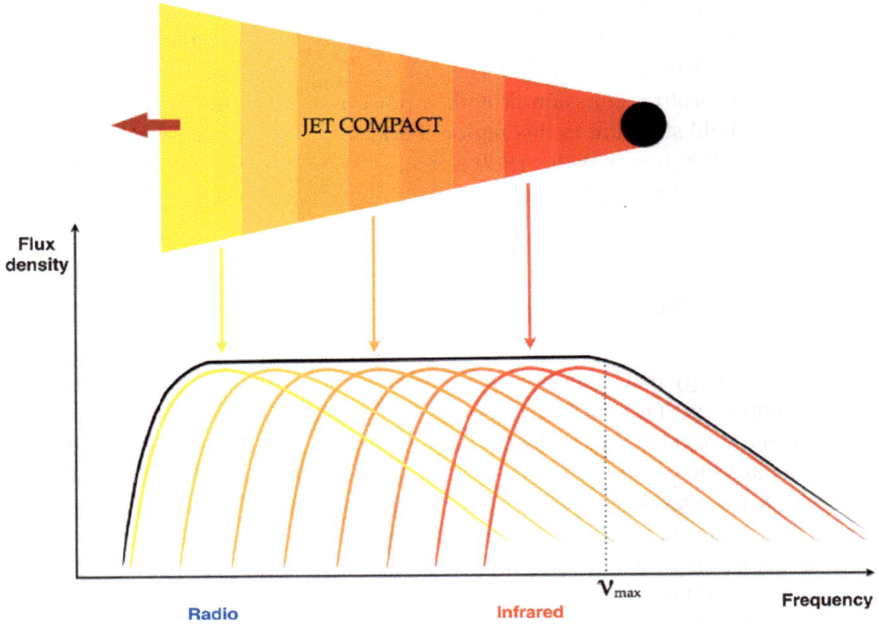

Fig. 10.9 The spectral energy distribution of the jet which is made up of different synchrotron-emitting regions along the length of the jet (Coriat 2010)

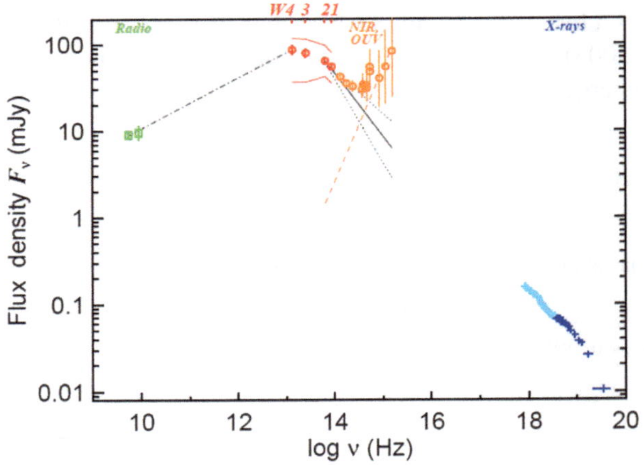

Fig. 10.10 The spectral energy distribution of GX 339-4. The spectrum breaks (jet break) from optically thick to optically thin in the mid-IR (Gandhi et al. 2011). ©AAS. Reproduced with permission

a break in the jet spectrum (e.g. Corbel and Fender 2002; Russell et al. 2006). Above this break frequency, the flat spectral component breaks to an optically thin spectrum corresponding to the point at which the entire jet becomes transparent, close to the jet base in the post-acceleration plasma (see Fig. 10.9). In addition, there is a cut off in the jet spectrum at higher energies, in the X-ray region (Markoff et al. 2001; Maitra et al. 2009) (see Fig. 10.8).

The transition between the optically thick and thin part of the spectrum occurs at the jet break frequency, which usually lies in the mid-IR to near-IR, where the optically thin synchrotron emission is observed, resulting in a power law with index $-1 < \alpha < -0.5$ (Gandhi et al. 2011; Russell et al. 2013b) (see Fig. 10.10). The break frequency has been detected in the BH candidates GX 339-4 (Corbel and Fender 2002; Gandhi et al. 2011; Corbel et al. 2013), XTE J1118+480 (Hynes et al. 2006), XTE J1550-564 (Chaty et al. 2011), V404 Cyg (Russell et al. 2013a; Tetarenko et al. 2019), MAXI J1836-194 (Russell et al. 2013b, 2014), Cyg X-2 (Rahoui et al. 2011), in the quiescent BH-XRBs, Swift J1357.2-933 and A0620-00 (Plotkin et al. 2016; Dinçer et al. 2018; Russell et al. 2018), and in the NS systems 4U 0614+091, 4U 1728-34 and Aql X-1 (Migliari et al. 2010; Díaz Trigo et al. 2017, 2018). The steady jet spectrum evolves with the break frequency (from optically thick to optically thin) moving as a function of time and spectral state as the electron population cools (Russell et al. 2014). During an XRB outburst, as the X-ray spectrum of the source softens, the jet break frequency moves from the near-IR to radio frequencies and then returns to the near-IR at the end of the outburst, when the spectrum gets harder (Russell et al. 2013b; Díaz Trigo et al. 2018). In BH-XRBs, as the source softens during state transitions, the jet break is observed to shift from the IR to millimetre frequencies (Koljonen et al. 2015). In the soft X-ray state, no core jet is detected. The structure of the accretion flow is governed by the presence of large-scale height magnetic fields that determine the position of the break frequency. Since the size of the jet-emitting region scales inversely with the frequency (Blandford and Königl 1979) the break frequency gives information on the size of the base of the jet where the particles are accelerated (Chaty et al. 2011; Ceccobello et al. 2018).

Fast sub-second optical/near-IR variability and correlations have also been observed in various systems (Durant et al. 2008, 2011; Gandhi et al. 2008, 2010). The most likely explanations is a jet origin, because the variability has a spectrum consistent with optically thin synchrotron emission (Hynes et al. 2003, 2006), which is stronger in the near-IR (Casella et al. 2010; Gandhi et al. 2010), maybe due to variable synchrotron emission of the X-ray corona (Veledina et al. 2011).

10.2.2 The Polarisation Spectrum of XRBs

Shahbaz et al. (2008) outlined the expected intrinsic linear polarisation signature of XRBs in the optical/near-IR region. The optical and IR light from XRBs is polarised due to intrinsic polarisation of the emitting photons, or by the (Thompson or possibly Rayleigh) scattering or absorption of unpolarised photons. Hydrogen in the disc is totally ionised and a low linear polarisation is expected in the optical due to Thomson scattering of emitted unpolarised radiation with free electrons in the disc. In the case of the persistent XRBs where the accretion disc dominates the optical light, the linear polarisation has a component that is produced by electron scattering by plasma above the accretion disc. Although a strong interstellar component is present, for Rayleigh and Thompson scattering processes, the level of linear polarisation decreases as a function of increasing wavelength. Electron scattering of the disc radiation by the secondary star is insignificant, because the secondary stars in XRBs have low surface temperatures and therefore low free electron densities. Polarisation of the optical light due to absorption by interstellar dust is inferred by the characteristic constant wavelength-dependent polarisation, given by Serkowski's law (Serkowski 1973; Schultz et al. 2004). There also exists the possibility of a large fractional linear polarisation, or linear polarisation degree (PD_L), from optically thin synchrotron emission. Indeed observations and theoretical results suggest a break between optically thick and thin emission occurs around the near-IR band (e.g. see Corbel and Fender 2002; Russell et al. 2013a and Fig. 10.10).

Optically thick synchrotron emission has a maximum linear polarisation of $\sim$10%, whereas optically thin synchrotron emission can have a PD_L as high as 70% (Blandford et al. 2002). Two polarisation signatures are expected from the jet. For the optically thick part of the spectrum, which results in the flat spectral component observed in the hard X-ray state (Fender 2001), no more than a few % polarisation is expected. At long wavelengths there should be $\sim$1% polarisation from the self-absorbed jet and at short wavelengths a comparable level will be measured due to scattering in the accretion flow. However, in the relatively narrow spectral region in which optically thin synchrotron emission dominates, one expects a strong signature that initially rises to longer wavelengths as the relative jet/disc fraction increases (see Fig. 10.11).

10.2.3 The Theoretical Polarisation Spectrum of the Jet

Relativistic electrons moving in a large-scale and ordered, uniform magnetic field produce synchrotron radiation. If the relativistic electrons produced by the first-order Fermi acceleration have a non-thermal energy distribution, the synchrotron radiation is linearly polarised. The linear polarisation spectrum of optically thin synchrotron emission from a relativistic electron in a uniform magnetic field was first treated by Westfold (1959). Subsequent works have extended these ideas to include power-law electron energy distributions and non-uniform magnetic fields (e.g. Ginzburg and Syrovatskii 1965; Nordsieck 1976; Longair 2011). Indeed, Bjornsson and Blumenthal (1982) derived the more general case for optically thin synchrotron emission from an electron population and magnetic field with an arbitrary distribution of energies and configuration, respectively. They showed that

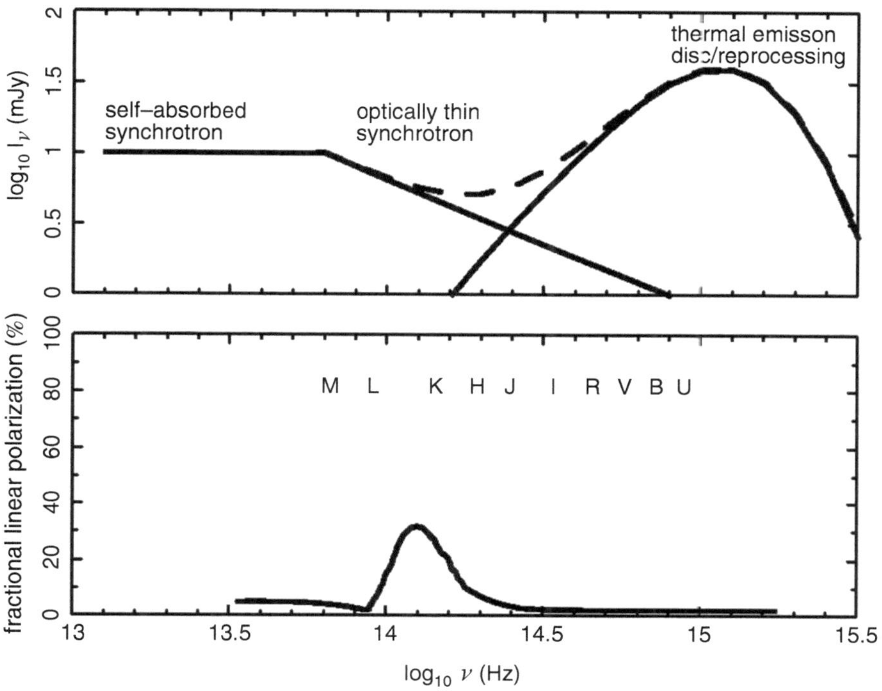

Fig. 10.11 The expected linear polarisation signature in the optical and near-IR spectrum of XRBs. The key spectral points are the break from optically thick (low polarisation <10%) to optically thin (high polarisation <70%) synchrotron emission, and the point at which the thermal (low-polarisation) emission begins to dominate over the jet. These effects should combine to produce a narrow spectral region with relatively high linear polarisation arising from close to the base of the jet (Shahbaz et al. 2008)

the expected PD_L is given by

$$PD_{L,\,thin} = f\frac{p+1}{p+7/3} = f\frac{1-\alpha_{thin}}{5/3-\alpha_{thin}} \tag{10.2}$$

where α_{thin} [=(p-1)/2] is the spectral index of the optically thin synchrotron emission of the form $F_\nu \propto \nu^\alpha$, f represents the ordering of the magnetic field and p is the electron energy distribution ($N(E)dE \propto E^{-p}dE$) (Rybicki and Lightman 1986). The parameter f takes values between 0 and 1, where 0 represents non-uniform, no net field orientation and 1 a perfectly uniform, fully aligned field. If the field is tangled, the differing angles of the polarised light suppress the observed, average polarisation. In the case of a fully ordered field (Rybicki and Lightman 1986; Bjornsson and Blumenthal 1982) the maximum polarisation strength is ~70–80% and is dependent only on the degree of ordering of the field and the energy distribution of the electron population. However, if the spectral index is steeper, the polarisation can be higher. The electric (polarisation) vector is perpendicular to the projection of the magnetic field on the plane of the sky.

For a synchrotron source with a power-law electron energy distribution PD_L is frequency-independent. However, a high energy cutoff in the electron distribution results in frequency-dependent PD_L (Bjornsson 1985). Above the high energy cutoff frequency the polarisation is given by

$$PD_{L,cut} = f\frac{1 - \alpha_{cut}}{5/3 - \alpha_{cut}} \tag{10.3}$$

where α_{cut} is the spectral index defined by power-law emission model. The above situations correspond to a highly idealised situation, assuming optically thin plasma, isotropic distributions of electrons, perfectly ordered homogeneous magnetic fields and the absence of substantial perturbations. A major modification occurs for optically thick plasma where each light ray experiences multiple scattering events. For optically thick, absorbed synchrotron emission ($F_\nu \propto \nu^{5/2}$) the net polarisation spectrum is given by

$$PD_{L,\,thick} = f\frac{3}{6p + 13} \tag{10.4}$$

with a polarisation position angle (PA) that differs by 90° to that of optically thin synchrotron polarisation (Pacholczyk 1970; Rudnick et al. 1978; Blandford et al. 2002; Longair 2011). The polarisation is less polarised than optically thin synchrotron (Blandford et al. 2002). For optically thin synchrotron emission with a typical spectral index of $\alpha_{thin} = -0.7$ ($p = 2.4$), $PD_L = 11\%$. The electric vector in this case is parallel to the projected magnetic field. If the level of ordering of the magnetic field is constant along the jet, with a constant PA, then Eq. (10.4) can be used to describe the polarisation expected from a flat/inverted optically thick jet spectrum.

As outlined in detail in Russell and Shahbaz (2014), Eqs. (10.2)–(10.4) can used to predict PD_L as well as the polarisation PA of the synchrotron-emitting jet spectrum. Given that different components produce emission with different PD_L and PA, it is necessary to calculate the Stokes parameters Q and U for each component at each frequency from the known PD_L and PA values, using the standard equations relating polarisation and PA.

10.2.4 Polarisation in X-ray Persistent XRBs

At optical/near-IR wavelengths, components such as the accretion disc and companion star often dominate the spectrum of BH- and NS-XRBs, suppressing any synchrotron contribution to the polarisation. However, when the synchrotron makes a strong contribution, intrinsic polarisation is detected (Dubus and Chaty 2006; Shahbaz et al. 2008; Russell and Fender 2008; Russell et al. 2011a; Chaty et al. 2011; Baglio et al. 2014). Intrinsic near-IR polarisation has been detected in three BH-XRBs during outburst, GRO J1655-40, XTE J1550-564 and GX 339-4, all during the hard-state when jets are expected to be launched (Russell and Fender 2008; Chaty et al. 2011; Russell et al. 2011a).

In GX 339-4, when the polarisation is strong the polarisation angle is perpendicular to the known jet axis, implying a magnetic field that is parallel to the jet axis (see Fig. 10.14). The low levels of polarisation measured (~1–7%) and variability imply a tangled, rapidly changing magnetic field geometry near the base of the jet (Russell and Fender 2008; Russell et al. 2011a), Mid-IR variability of GX 339-4, also show that either the acceleration region near the base of the jet, or its magnetic field strength, are changing on timescales of minutes (Gandhi et al. 2011), the same timescale as the polarisation variability. During the bright outburst of V404 Cyg in 2015, flares of optical polarisation were also detected and interpreted as synchrotron emission from a variable jet (Lipunov et al. 2016; Shahbaz et al. 2016). Time-resolved optical polarimetry revealed a polarisation flare which decayed fairly smoothly from 4.5 to 3.5% 20 min, with a stable PA that is aligned with the resolved radio jet. This implies that the electric field vector near the base of the jet in V404 Cyg is on average approximately parallel to the jet axis and that the magnetic field is orthogonal to the jet axis. This may be due to the compression of magnetic field lines in shocks in the flow, resulting in a partially ordered transverse field. The polarisation flare also occurs during the initial rise of a major radio flare event that peaks later, and is consistent with a classically evolving synchrotron flare from an ejection event (see Fig. 10.12).

To date, the most detailed polarisation spectrum has been obtained for Cyg X-1. Russell and Shahbaz (2014) modelled the broadband, radio to γ-ray flux spectrum and polarisation spectrum of Cyg X-1 in the hard-state, with a simple phenomenological model consisting of a strongly polarised synchrotron jet (see Sect. 10.2.3) and an unpolarised Comptonised corona and a moderately polarised interstellar dust component (see Fig. 10.13). They were able to show that the origin of the γ-ray, X-ray and some of the IR polarisation is the optically thin synchrotron power law from the inner regions of the jet.

In NS-XRBs, intrinsic near-IR or optical polarisation at similar levels as the BH-XRBs has been detected in Sco X-1, Cyg X-2 and 4U 0614+09 (Russell and Fender 2008; Shahbaz et al. 2008; Russell et al. 2011a; Baglio et al. 2014). Shahbaz et al. (2008) performed near-IR spectropolarimetry of the XRBs Sco X-1 and Cyg X-2 and found them to be intrinsically polarised, with an increasing PD_L at lower frequencies, which is explained by optically thin synchrotron emission from a jet. The PD_L observed is ~1–10%, with evidence for rapid variations in some sources on timescales of seconds–minutes (Russell et al. 2011a). This dependence on wavelength cannot be explained by scattering or interstellar dust absorption and it cannot explain the short timescale variability of the polarisation observed Sco X-1 (Russell et al. 2011a). The PA is usually approximately orthogonal to the axis of the resolved radio jet, which implies the magnetic field is parallel to the jet axis. In general, the

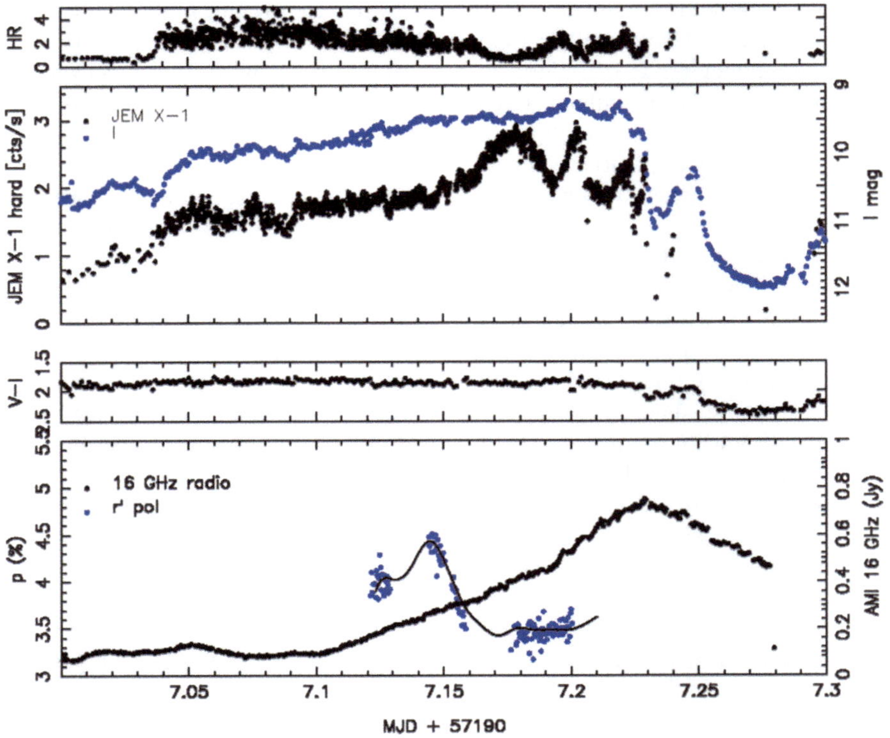

Fig. 10.12 The X-ray, optical and radio light curves of the V404 Cyg during the outburst in 2015 around MJD 57197. From top to bottom: INTEGRAL JEM-X X-ray hardness ratio; INTEGRAL JEM-X X-ray count rate (black dots) with the AAVSO I-band light curve (blue dots); AAVSO optical colour; and the AMI-LA 16 GHz radio data with the r'-band polarisation light curve (blue dots) (Shahbaz et al. 2016)

observations of most XRBs are consistent with a variable, predominantly tangled magnetic field geometry, with field lines preferentially orientated along the jet axis (see Fig. 10.14). However, there are a few exceptions. Cyg X-1 has a highly ordered, stable magnetic field near the base of the jet that is perpendicular to the jet axis (Russell and Shahbaz 2014; Rodriguez et al. 2015). V404 Cyg shows a low-level polarisation flare, with the magnetic field also perpendicular to the jet axis (Shahbaz et al. 2016), which may be due to the compression of tangled magnetic field lines in shocks downstream in the flow or collisions with dense regions of the interstellar medium, resulting in a partially ordered transverse field.

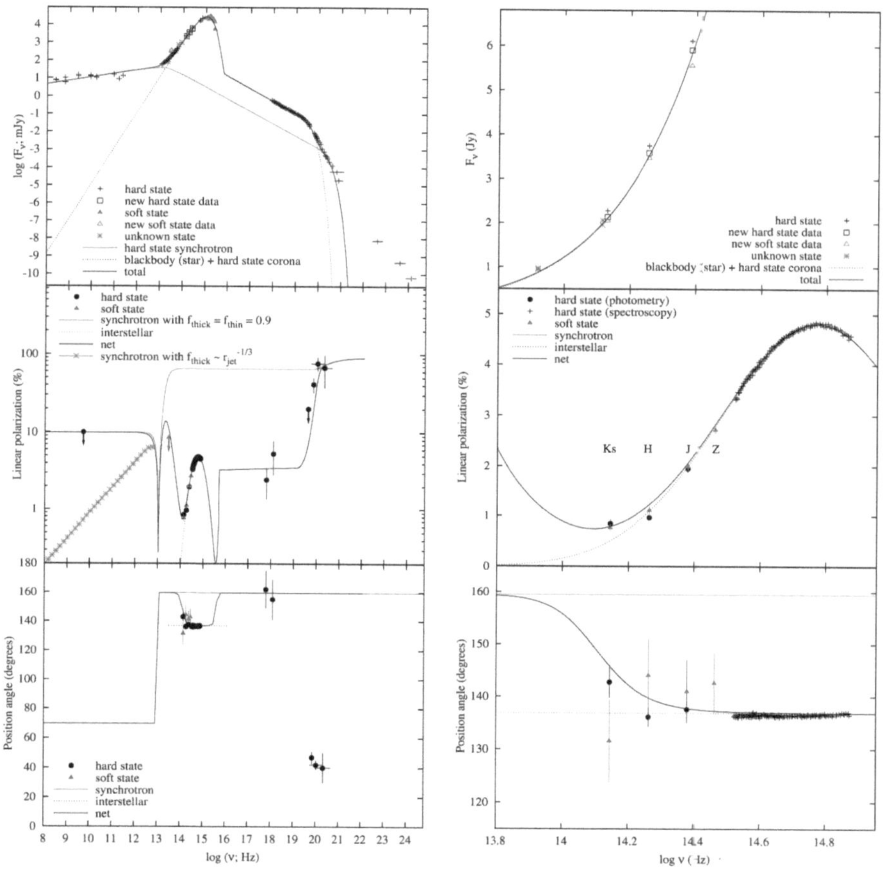

Fig. 10.13 Left: Radio to γ-ray flux density (top), linear polarisation spectrum (middle) and polarisation position angle (bottom) spectrum of Cyg X-1. Right: The same as the left-hand panels but just the near-IR/optical region (Russell and Shahbaz 2014)

10.2.5 Polarisation in X-ray Quiescent XRBs

X-ray transients are a subclass of the XRBs which spend the majority of their time in a state of quiescence, accreting at very low X-ray luminosities $\sim 10^{30} - 10^{34}$ erg s^{-1}, explained by a truncated accretion disc with a radiatively inefficient accretion flow (Narayan and Yi 1994). Although it is not really clear if quiescence is a separate state or simply a low-luminosity version of the hard-state, but observations show that quiescent BH-XRBs have softer X-ray spectra than hard-state systems (e.g. Reynolds et al. 2014) and outflows with weaker particle acceleration (Gallo et al. 2007; Plotkin et al. 2015). There are a few radio detections of XRBs in quiescence, which include V404 Cyg (Gallo et al. 2005), A0620-00 (Gallo et al.

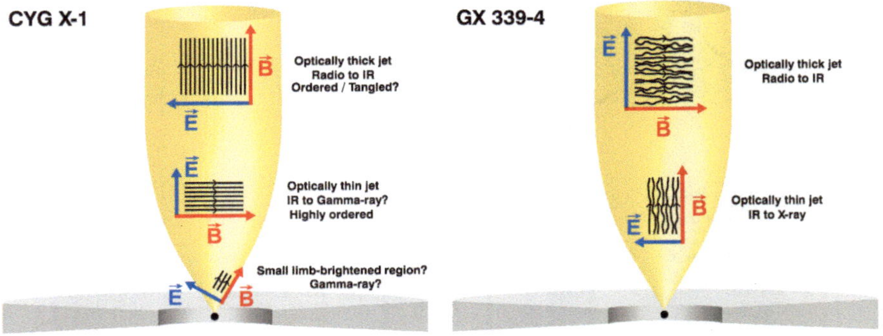

Fig. 10.14 A schematic diagram of the Cyg X-1 jet showing the orientation of the electric (from the near-IR observations) and magnetic field vectors (from observations of the radio jet) (Russell and Shahbaz 2014). For synchrotron, the polarisation PA traces the electric field vector in the emitting region, and the magnetic field vector differs by 90°. In Cyg X-1 (left) a highly ordered magnetic field in the optically thin region near the jet base (as implied by the observations) and an ordered or tangled field in the large-scale jet is shown. In other XRBs, such as GX 339-4 (right), the magnetic fields near the base of BH-XRB jets are moderately tangled, with a preferential orientation of the magnetic field along the jet axis

2006), XTE J1118+480 (Gallo et al. 2014), Swift J1753.5-0127 (Miller-Jones et al. 2011), 47 Tuc X9 (Miller-Jones et al. 2015), M62-VLA1 (Chomiuk et al. 2013), MWC 656 (Dzib et al. 2015) and M15 S2 (Tetarenko et al. 2016); however, there is only one linear polarisation constraint of the quiescent radio emission; <2.11% in V404 Cyg (Rana et al. 2016). A0620-00 and V404 Cyg are the only truly quiescent system with a robust radio detection that seem to follow the same radio/X-ray correlations seen in the hard-state (Gallo et al. 2006).

Jets in quiescent XRBs are weaker than the brighter hard-state jets. The electron may be dominated by partially non-relativistic, perhaps Maxwellian energy distributions, leading to steeper spectral energy distributions that are typical for optically thin synchrotron emission (Shahbaz et al. 2013; Markoff et al. 2015). Evidence for synchrotron emission from the optical/near-IR spectral energy distribution in quiescence has been found in the BH-XRBs A0620-00 (Gallo et al. 2007; Gelino et al. 2010; Froning et al. 2011; Dinçer et al. 2018), Swift J1357.2-0933 (Shahbaz et al. 2013; Plotkin et al. 2016) and the NS-XRB XTE J1814-338 (Baglio et al. 2013). Near-IR polarimetric observations of two quiescent BH-XRBs have revealed the polarimetric signature of synchrotron emission (Russell et al. 2016). In A0620-00, at the longest near-IR bands the PA of the excess 1.3% polarisation changes significantly (see Fig. 10.15). The PA of this weak synchrotron component, most likely from a jet, implies that the orientation of the magnetic field is approximately parallel to the radio jet axis (Kuulkers et al. 1999). In Swift J1357.2-0933 the near-IR flux variability imply a continuously launched, highly variable jet, which has a moderately tangled magnetic field close to the base of the jet. These results imply

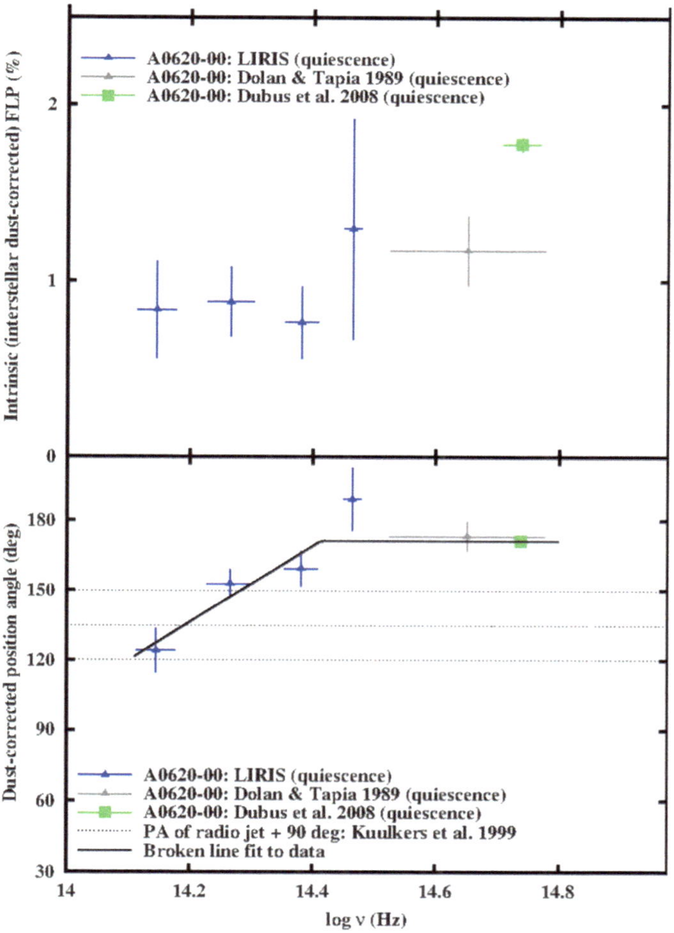

Fig. 10.15 The intrinsic (dust-corrected) near-IR linear polarisation and position angle spectra of A0620-00 in quiescence (Russell et al. 2016). The dust-corrected PD_L of A0620-00 is ~1–2% in the optical and ~0.5–1.0% at near-IR wavelengths. The optical PD_L is most likely due to Thomson scattering within the binary system (Brown et al. 1978), which is observed to vary with orbital phase (Dolan and Tapia 1989a). Note the different PA in the H- and K_s-bands compared to in the optical

that weak jets in low-luminosity accreting systems have magnetic fields which possess similar orientations and tangled fields compared to the more luminous, XRBs at higher accretion rates (see Sect. 10.2.4).

10.2.6 Time-Resolved Polarimetry

Phase-resolved linear polarimetry provides a unique probe of XRBs as one can obtain information on geometrical properties of the binary system from the light scattered by circumstellar material. In principle the orbital inclination (i) can be measured, which is essential in determining the components' masses, the fundamental parameters that determine a star's initial structure and subsequent evolution. Brown et al. (1978) and Rudy and Kemp (1978) have investigated the polarisation produced by Thomson scattering in circumstellar envelopes in binary systems. They showed that the linear polarisation has a characteristic variability around the binary orbit, associated with scattering in the stellar envelope that is distorted by the gravity of the compact object. One assumes a circular binary orbit, scattering due to electrons and that the material is optically thin and co-rotates with the binary. The observed phase-resolved Stokes parameters Q and U light curves can be fitted with fit a truncated (first and second order) Fourier series,

$$
\begin{aligned}
Q &= q_0 + q_1 \cos\lambda + q_2 \sin\lambda + q_3 \cos 2\lambda + q_4 \sin 2\lambda \\
U &= u_0 + u_1 \cos\lambda + u_2 \sin\lambda + u_3 \cos 2\lambda + u_4 \sin 2\lambda
\end{aligned}
\tag{10.5}
$$

where $\lambda = 2\pi\phi$ is and ϕ is the orbital phase. The variations are usually dominated by the second harmonics, which is commonly seen for binary systems showing polarisation modulation. In the $Q - U$ plane the orbital phase-resolved polarisation traces an ellipse twice per orbit. One can derive the orbital parameters either by studying the geometry of the $Q-U$ plane or by considering the Fourier coefficients. A least-square fits can be performed on the observed Q and U light curves to determine the coefficients and q_j and u_j ($j = 0$–4), which can then be compared with theoretical expectations

$$
\begin{aligned}
Q &= \tau_0 [\, (1 - 3\gamma_0) \sin^2 i + \sin 2i (\gamma_1 \cos\lambda - \gamma_2 \sin\lambda) \\
&\qquad - (1 + \cos^2 i)(\gamma_3 \cos 2\lambda - \gamma_4 \sin 2\lambda)] \\
U &= 2\tau_0 [\, \sin i (\gamma_1 \sin\lambda + \gamma_2 \cos\lambda) - \cos i (\gamma_3 \sin 2\lambda + \gamma_4 \cos 2\lambda)]
\end{aligned}
\tag{10.6}
$$

where γ_i are moments of the scatterers' distribution (Brown et al. 1978; Drissen et al. 1986). The binary inclination is then given by

$$
\begin{aligned}
\left[\frac{1 - \cos i}{1 + \cos i}\right]^2 &= \frac{(u_1 + q_2)^2 + (u_2 - q_1)^2}{(u_2 + q_1)^2 + (u_1 - q_2)^2} \\
\left[\frac{1 - \cos i}{1 + \cos i}\right]^4 &= \frac{(u_3 + q_4)^2 + (u_4 - q_3)^2}{(u_4 + q_3)^2 + (u_3 - q_4)^2}
\end{aligned}
\tag{10.7}
$$

Alternatively, in the $Q - U$ plane the orbital phase-resolved polarisation traces an ellipse twice per orbit, and the eccentricity e of this ellipse is related to the binary

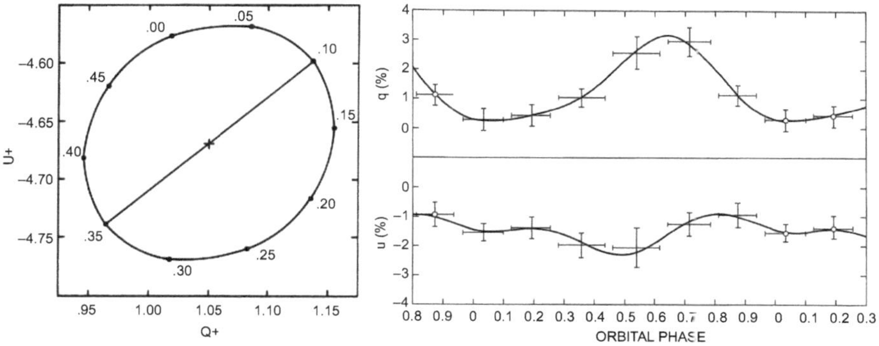

Fig. 10.16 Left: The observed Stokes optical linear polarisation of Cyg X-1 in the Q-U plane (Dolan and Tapia 1989b). The values trace an ellipse, the eccentricity of which can be used to determine the inclination angle. Right: The observed Q and U time-resolved linear V-band polarimetric light curve of A0620-00 (Dolan and Tapia 1989a). The solid line represents the second-order Fourier fit to the data, which is then used to determine the inclination angle. ©AAS. Reproduced with permission

inclination by Rudy and Kemp (1978)

$$\cos i = \sqrt{\frac{1-e}{1+e}} \tag{10.8}$$

Hence, by determining the eccentricity of the observed $Q - U$ ellipse, one can determine the binary inclination, independent of size, shape and position of the circumstellar region (see Fig. 10.16). An extension of the above model to an elliptical binary orbit can be found in Brown et al. (1982) and Simmons and Boyle (1984). Detailed accounts on polarimetric inclination determinations and their uncertainties are given by Aspin et al. (1981), Simmons et al. (1982) and Wolinski and Dolan (1994).

Time-resolved optical polarimetric studies of two XRBs have allowed one to estimate the inclination angle; Cyg X-1 (Kemp et al. 1979; Dolan and Tapia 1989b; Nagae et al. 2009) and A0620-00 (Dolan and Tapia 1989a). In Cyg X-1 a $\sim$5% optical linear polarisation with an amplitude of $\sim$0.3% is observed, whereas in A0620-00 a 3% linear polarisation with a $\sim$2% modulation is observed on the orbital period (see Fig. 10.16). Using the method outlined above, the inclination angle has been estimated to be $i = 62(+5, -37)^\circ$ degrees and $i = 57(+20, -40)^\circ$ for Cyg X-1 and A0620-00, respectively (Dolan and Tapia 1989a,b). The uncertainties in the values are large mainly because of the poor phase-sampling and precision of the Stokes light curves.

Acknowledgements I would like to thank Dave Russell for the many discussions on the works presented in this review.

References

Aspin C, Simmons JFL, Brown JC (1981) Polarimetric accuracy required for the determination of binary inclinations. Mon Not R Astron Soc 194:283

Baglio MC, D'Avanzo P, Muñoz-Darias T, Breton RP, Campana S (2013) The long-term evolution of the X-ray pulsar XTE J1814-338: a receding jet contribution to the quiescent optical emission? Astron Astrophys 559:A42

Baglio MC, Mainetti D, D'Avanzo P, Campana S, Covino S, Russell DM, Shahbaz T (2014) Polarimetric and spectroscopic optical observations of the ultra-compact X-ray binary 4U 0614+091. Astron Astrophys 572:A99

Baskill DS, Wheatley PJ, Osborne JP (2005) The complete set of ASCA X-ray observations of non-magnetic cataclysmic variables. Mon Not R Astron Soc 357:626–644

Belloni TM (2010) States and transitions in black hole binaries. In: Belloni T (ed) Lecture notes in physics, vol 794. Springer Verlag, Berlin, p 53

Bjornsson C-I (1985) Frequency-dependent polarization in blazars. Mon Not R Astron Soc 216:241–253

Bjornsson C-I, Blumenthal GR (1982) Spectral polarization properties of optically thin synchrotron radiation. Astrophys J 259:805–819

Blandford RD, Königl A (1979) Relativistic jets as compact radio sources. Astrophys J 232:34–48

Blandford R, Agol E, Broderick A, Heyl J, Koopmans L, Lee H-W (2002) Compact objects and accretion disks. In: Trujillo-Bueno J, Moreno-Insertis F, Sánchez F (eds) Astrophysical spectropolarimetry. Cambridge University Press, Cambridge, pp 177–223

Blinova AA, Romanova MM, Ustyugova GV, Koldoba AV, Lovelace RVE (2018) Comparisons of MHD propeller model with observations of cataclysmic variable AE Aqr. arXiv e-prints

Brown JC, McLean IS, Emslie AG (1978) Polarisation by Thomson scattering in optically thin stellar envelopes. II – Binary and multiple star envelopes and the determination of binary inclinations. Astron Astrophys 68:415–427

Brown JC, Aspin C, Simmons JFL, McLean IS (1982) The effect of orbital eccentricity on polarimetric binary diagnostics. Mon Not R Astron Soc 198:787–794

Buckley DAH, Meintjes PJ, Potter SB, Marsh TR, Gänsicke BT (2017) Polarimetric evidence of a white dwarf pulsar in the binary system AR Scorpii. Nature Astron 1:0029

Campbell RK, Harrison TE, Schwope AD, Howell SB (2008) Cyclotron modeling phase-resolved infrared spectroscopy of polars. I. EF Eridani. Astrophys J 672:531–539

Casella P, Maccarone TJ, O'Brien K, Fender RP, Russell DM, van der Klis M, Pe'Er A, Maitra D, Altamirano D, Belloni T, Kanbach G, Klein-Wolt M, Mason E, Soleri P, Stefanescu A, Wiersema K, and Wijnands R (2010) Fast infrared variability from a relativistic jet in GX 339-4. Mon Not R Astron Soc 404:L21–L25

Ceccobello C, Cavecchi Y, Heemskerk MHM, Markoff S, Polko P, and Meier D (2018) A new method for extending solutions to the self-similar relativistic magnetohydrodynamic equations for black hole outflows. Mon Not R Astron Soc 473:4417–4435

Chanmugam G, Dulk GA (1981) Polarized radiation from hot plasmas and applications to AM Herculis binaries. Astrophys J 244:569–578

Chanmugam G, Ray A (1984) The rotational and orbital evolution of cataclysmic binaries containing magnetic white dwarfs. Astrophys J 285:252–257

Chaty S, Dubus G, Raichoor A (2011) Near-infrared jet emission in the microquasar XTE J1550-564. Astron Astrophys 529:A3

Choi C-S, Dotani T, Agrawal PC (1999) A study of the X-ray emission of the magnetic cataclysmic variable AE Aquarii. Astrophys J 525:399–406

Chomiuk L, Strader J, Maccarone TJ, Miller-Jones JCA, Heinke C, Noyola E, Seth AC, Ransom S (2013) A radio-selected black hole X-ray binary candidate in the milky way globular cluster M62. Astrophys J 777:69

Corbel S, Fender RP (2002) Near-infrared synchrotron emission from the compact jet of GX 339-4. Astrophys J 573:L35–L39

Corbel S, Aussel H, Broderick JW, Chanial P, Coriat M, Maury AJ, Buxton MM, Tomsick JA, Tzioumis AK, Markoff S, Rodriguez J, Bailyn CD, Brocksopp C, Fender RP, Petrucci PO, Cadolle-Bel M, Calvelo D, Harvey-Smith L (2013) Formation of the compact jets in the black hole GX 339-4. Mon Not R Astron Soc 431:L107–L111
Coriat M (2010) Jets relativistes des trous noirs accrétants. PhD thesis, Paris
Costa JER, Rodrigues CV (2009) Stokes imaging of AM Her systems using 3D inhomogeneous models – I. Description of the code and an application to V834 Cen. Mon Not R Astron Soc 398:240–248
Cropper M (1990) The polars. Space Sci Rev 54:195–295
Cropper M, Wickramasinghe DT (1993) Cyclotron humps in AM HER systems. V – Two poles in DP Leo. Mon Not R Astron Soc 260:696–704
Dhillon VS, Smith DA, Marsh TR (2013) The SW Sex enigma. Mon Not R Astron Soc 428:3559–3568
Díaz Trigo M, Migliari S, Miller-Jones JCA, Rahoui F, Russell DM, Tudor V (2017) ALMA observations of 4U 1728-34 and 4U 1820-30: first detection of neutron star X-ray binaries at 300 GHz. Astron Astrophys 600:A8
Díaz Trigo M, Altamirano D, Dinçer T, Miller-Jones JCA, Russell DM, Sanna A, Bailyn C, Lewis F, Migliari S, Rahoui F (2018) The evolving jet spectrum of the neutron star X-ray binary Aql X-1 in transitional states during its 2016 outburst. Astron Astrophys 616:A23
Dinçer T, Bailyn CD, Miller-Jones JCA, Buxton M, MacDonald RKD (2018) Multiwavength observations of the black hole X-ray binary A0620-00 in quiescence. Astrophys J 852:4
Dolan JF, Tapia S (1989a) The orbital inclination of A0620-00 measured polarimetrically. Publ Astron Soc Pac 101:1135–1139
Dolan JF, Tapia S (1989b) The orbital inclination of Cygnus XR-1 measured polarimetrically. Astrophys J 344:830–834
Drissen L, Lamontagne R, Moffat AFJ, Bastien P, Seguin M (1986) Spectroscopic and polarimetric parameters of the runaway WN7 binary system HD 197406 – is the secondary an X-ray-quiet black hole? Astrophys J 304:188–200
Dubus G, Chaty S (2006) Infrared polarimetry of the microquasars H1743-322, XTE J1550-564 and GRO J1655-40. Astron Astrophys 458:591–595
Durant M, Gandhi P, Shahbaz T, Fabian AP, Miller J, Dhillon VS, Marsh TR (2008) SWIFT J1753.5-0127: a surprising optical/X-ray cross-correlation function. Astrophys J 682:L45
Durant M, Shahbaz T, Gandhi P, Cornelisse R, Muñoz-Darias T, Casares J, Dhillon V, Marsh T, Spruit H, O'Brien K, Steeghs D, Hynes R (2011) High time resolution optical/X-ray cross-correlations for X-ray binaries: anticorrelations and rapid variability. Mon Not R Astron Soc 410:2329–2338
Dzib SA, Massi M, Jaron F (2015) Radio emission from the Be/black hole binary MWC 656. Astron Astrophys 580:L6
Eracleous M, Horne K, Robinson EL, Zhang E-H, Marsh TR, Wood JH (1994) The ultraviolet pulsations of the cataclysmic variable AE Aquarii as observed with the Hubble Space Telescope. Astrophys J 433:313–331
Fender RP (2001) Powerful jets from black hole X-ray binaries in low/hard X-ray states. Mon Not R Astron Soc 322:31–42
Fender R, Gallo E (2014) An overview of jets and outflows in stellar mass black holes. Space Sci Rev 183:323–337
Fender RP, Pooley GG, Durouchoux P, Tilanus RPJ, Brocksopp C (2000) The very flat radio-millimetre spectrum of Cygnus X-1. Mon Not R Astron Soc 312:853–858
Ferrario L, Wickramasinghe DT (1990) Arc-shaped cyclotron emission regions in AM Herculis systems. Astrophys J 357:582–590
Ferrario L, Wickramasinghe DT, Bailey J, Tuohy IR, Hough JH (1989) EXO 033319-2554.2: an eclipsing AM Herculis system showing cyclotron emission features. Astrophys J 337:832
Ferrario L, Bailey J, Wickramasinghe DT (1993) Detection of cyclotron emission features in the infrared spectrum of ST LMi. Mon Not R Astron Soc 262:285–288

Ferrario L, de Martino D, Gänsicke BT (2015) Magnetic White Dwarfs. Space Sci Rev 191:111–169

Froning CS, Cantrell AG, Maccarone TJ, France K, Khargharia J, Winter LM, Robinson EL, Hynes RI, Broderick JW, Markoff S, Torres MAP, Garcia M, Bailyn CD, Prochaska JX, Werk J, Thom C, Béland S, Danforth CW, Keeney B, and Green JC (2011) Multiwavelength observations of A0620-00 in quiescence. Astrophys J 743:26

Gallo E, Fender RP, Pooley GG (2003) A universal radio-X-ray correlation in low/hard state black hole binaries. Mon Not R Astron Soc 344:60–72

Gallo E, Fender RP, Hynes RI (2005) The radio spectrum of a quiescent stellar mass black hole. Mon Not R Astron Soc 356:1017–1021

Gallo E, Fender RP, Miller-Jones JCA, Merloni A, Jonker PG, Heinz S, Maccarone TJ, van der Klis M (2006) A radio-emitting outflow in the quiescent state of A0620-00: implications for modelling low-luminosity black hole binaries. Mon Not R Astron Soc 370:1351–1360

Gallo E, Migliari S, Markoff S, Tomsick JA, Bailyn CD, Berta S, Fender R, Miller-Jones JCA (2007) The spectral energy distribution of quiescent black hole X-ray binaries: new constraints from Spitzer. Astrophys J 670:600–609

Gallo E, Miller-Jones JCA, Russell DM, Jonker PG, Homan J, Plotkin RM, Markoff S, Miller BP, Corbel S, Fender RP (2014) The radio/X-ray domain of black hole X-ray binaries at the lowest radio luminosities. Mon Not R Astron Soc 445:290–300

Gandhi P, Makishima K, Durant M, Fabian AC, Dhillon VS, Marsh TR, Miller JM, Shahbaz T, Spruit HC (2008) Rapid optical and X-ray timing observations of GX 339-4: flux correlations at the onset of a low/hard state. Mon Not R Astron Soc 390:L29–L33

Gandhi P, Dhillon VS, Durant M, Fabian AC, Kubota A, Makishima K, Malzac J, Marsh TR, Miller JM, Shahbaz T, Spruit HC, Casella P (2010) Rapid optical and X-ray timing observations of GX339-4: multicomponent optical variability in the low/hard state. Mon Not R Astron Soc 407:2166–2192

Gandhi P, Blain AW, Russell DM, Casella P, Malzac J, Corbel S, D'Avanzo P, Lewis FW, Markoff S, Cadolle Bel M, Goldoni P, Wachter S, Khangulyan D, Mainzer A (2011) A variable mid-infrared synchrotron break associated with the compact jet in GX 339-4. Astrophys J 740:L13

Gelino DM, Gelino CR, Harrison TE (2010) Spitzer observations of black hole low-mass X-ray binaries: assessing the non-stellar infrared component. Astrophys J 718:1–10

Ginzburg VL, Syrovatskii SI (1965) Cosmic Magnetobremsstrahlung (synchrotron radiation). Annu Rev Astron Astrophys 3:297

Hynes RI, Haswell CA, Cui W, Shrader CR, O'Brien K, Chaty S, Skillman DR, Patterson J, Horne K (2003) The remarkable rapid X-ray, ultraviolet, optical and infrared variability in the black hole XTE J1118+480. Mon Not R Astron Soc 345:292–310

Hynes RI, Robinson EL, Pearson KJ, Gelino DM, Cui W, Xue YQ, Wood MA, Watson TK, Winget DE, Silver IM (2006) Further evidence for variable synchrotron emission in XTE J1118+480 in outburst. Astrophys J 651:401–407

Ikhsanov NR (1998) The pulsar-like white dwarf in AE Aquarii. Astron Astrophys 338:521–526

Ikhsanov NR (2006) On a site of X-Ray emission in AE Aquarii. Astrophys J 640:L59–L62

Kaiser CR (2005) Radio synchrotron spectra for a leaky box approximation. Mon Not R Astron Soc 360:176–184

Kaiser CR (2006) The flat synchrotron spectra of partially self-absorbed jets revisited. Mon Not R Astron Soc 367:1083–1094

Katajainen S, Butters OW, Norton AJ, Lehto HJ, Piirola V (2007) Discovery of polarized emission from the long period intermediate polar RX J2133.7+5107. Astron Astrophys 475:1011–1018

Katz JI (2017) AR Sco: a precessing white dwarf synchronar? Astrophys J 835:150

Kemp JC, Barbour MS, Parker TE, Herman LC (1979) Cygnus X-1 – Polarization evidence for an extended secondary envelope with an eclipsing region. Astrophys J 228:L23–L27

Koljonen KII, Russell DM, Fernández-Ontiveros JA, Markoff S, Russell TD, Miller-Jones JCA, van der Horst AJ, Bernardini F, Casella P, Curran PA, Gandhi P, Soria R (2015) A connection between plasma conditions near black hole event horizons and outflow properties. Astrophys J 814:139

Koljonen KII, Russell DM, Corral-Santana JM, Armas Padilla M, Muñoz-Darias T, Lewis F, Coriat M, Bauer FE (2016) A 'high-hard' outburst of the black hole X-ray binary GS 1354-64. Mon Not R Astron Soc 460:942–955
Kuulkers E, Fender RP, Spencer RE, Davis RJ, Morison I (1999) Multiple ejections during the 1975 outburst of A0620-00. Mon Not R Astron Soc 306:919–925
Lipunov VM, Gorbovskoy E, Kornilov V, Krushinskiy V, Vlasenko D, Tiurina N, Balanutsa P, Kuznetsov A, Budnev N, Gress O, Tlatov A, Rebolo Lopez R, Serra-Ricart M, Buckley DAH, Israelyan G, Lodieu N, Ivanov K, Yazev S, Sergienko Y, Gabovich A, Yurkov V, Levato H, Saffe C, Podesta R, Mallamaci C, Lopez C (2016) MASTER optical polarization variability detection in the microquasar V404 Cyg/GS 2023+33. Astrophys J 833:198
Livio M, Pringle JE (1994) Star spots and the period gap in cataclysmic variables. Astrophys J 427:956–960
Longair MS (2011) High energy astrophysics. Cambridge University Press, Cambridge
Maitra D, Markoff S, Brocksopp C, Noble M, Nowak M, Wilms J (2009) Constraining jet/disc geometry and radiative processes in stellar black holes XTE J1118+480 and GX 339-4. Mon Not R Astron Soc 398:1638–1650
Malzac J (2018) Radiation processes and models. ArXiv e-prints
Markoff S, Falcke H, Fender R (2001) A jet model for the broadband spectrum of XTE J1118+480. Synchrotron emission from radio to X-rays in the Low/Hard spectral state. Astron Astrophys 372:L25–L28
Markoff S, Nowak MA, Gallo E, Hynes R, Wilms J, Plotkin RM, Maitra D, Silva CV, Drappeau S (2015) As above, so below: exploiting mass scaling in black hole accretion to break degeneracies in spectral interpretation. Astrophys J 812:L25
Marsh TR, Horne K (1988) Images of accretion discs. II – Doppler tomography. Mon Not R Astron Soc 235:269–286
Marsh TR, Gänsicke BT, Hümmerich S, Hambsch F.-J, Bernhard K, Lloyd C, Breedt E, Stanway ER, Steeghs DT, Parsons SG, Toloza O, Schreiber MR, Jonker PG, van Roestel J, Kupfer T, Pala AF, Dhillon VS, Hardy LK, Littlefair SP, Aungwerojwit A, Arjyotha S, Koester D, Bochinski JJ, Haswell CA, Frank P, Wheatley PJ (2016) A radio-pulsing white dwarf binary star. Nature 537:374–377
Martínez-Pais IG, Rodríguez-Gil P, Casares J (1999) S193: another non-eclipsing SW Sex star. Mon Not R Astron Soc 305:661–670
McClintock JE, Remillard RA (2006) Black hole binaries. Cambridge University Press, Cambridge, pp 157–213
Meggitt SMA, Wickramasinghe DT (1982) The polarization properties of magnetic accretion columns. Mon Not R Astron Soc 198:71–82
Meier DL (2001) The association of jet production with geometrically thick accretion flows and black hole rotation. Astrophys J 548:L9–L12
Migliari S, Tomsick JA, Miller-Jones JCA, Heinz S, Hynes RI, Fender RP, Gallo E, Jonker PG, Maccarone TJ (2010) The complete spectrum of the neutron star X-ray binary 4U 0614+091. Astrophys J 710:117–124
Miller-Jones JCA, Jonker PG, Maccarone TJ, Nelemans G, Calvelo DE (2011) A deep radio survey of hard state and quiescent black hole X-ray binaries. Astrophys J 739:L18
Miller-Jones JCA, Strader J, Heinke CO, Maccarone TJ, van den Berg M, Knigge C, Chomiuk L, Noyola E, Russell TD, Seth AC, Sivakoff GR (2015) Deep radio imaging of 47 Tuc identifies the peculiar X-ray source X9 as a new black hole candidate. Mon Not R Astron Soc 453:3918–3931
Nagae O, Kawabata KS, Fukazawa Y, Okazaki A, Isogai M, Yamashita T (2009) Multiepoch optical spectropolarimetry of three microquasars, Cyg X-1, LS 5039, and LS I +61deg 303. Astron J 137:3509–3519
Narayan R, Yi I (1994) Advection-dominated accretion: a self-similar solution. Astrophys J 428:L13–L16
Nordsieck KH (1976) Optical polarization models of quasi-stellar objects and BL Lacertae objects. Astrophys J 209:653–662

Norton AJ (1993) Simulation of the X-ray light curves of intermediate polars. Mon Not R Astron Soc 265:316

Pacholczyk AG (1970) Radio astrophysics. Nonthermal processes in galactic and extragalactic sources. Freeman, San Francisco

Patterson J (1979) Rapid oscillations in cataclysmic variables. III – an oblique rotator in an AE Aquarii. Astrophys J 234:978–992

Patterson J (1994) The DQ Herculis stars. Publ Astron Soc Pac 106:209–238

Piirola V, Hakala P, Coyne GV (1993) The discovery of variable polarization over the 13.9 minute spin period of the intermediate polar RE 0751 + 14. Astrophys J 410:L107–L110

Piirola V, Vornanen T, Berdyugin A, GV Coyne SJ (2008) V405 Aurigae: a high magnetic field intermediate polar. Astrophys J 684:558–568

Plotkin RM, Gallo E, Markoff S, Homan J, Jonker PG, Miller-Jones JCA, Russell DM, Drappeau S (2015) Constraints on relativistic jets in quiescent black hole X-ray binaries from broad-band spectral modelling. Mon Not R Astron Soc 446:4098–4111

Plotkin RM, Gallo E, Jonker PG, Miller-Jones JCA, Homan J, Muñoz-Darias T, Markoff S, Armas Padilla M, Fender R, Rushton AP, Russell DM, Torres MAP (2016) A clean sightline to quiescence: multiwavelength observations of the high Galactic latitude black hole X-ray binary Swift J1357.2-0933. Mon Not R Astron Soc 456:2707–2716

Polko P, Meier DL, Markoff S (2010) Determining the optimal locations for shock acceleration in magnetohydrodynamical jets. Astrophys J 723:1343–1350

Potter SB (2016) Stokes imaging: mapping the accretion region(s) in magnetic cataclysmic variables. In: Boffin HMJ, Hussain G, Berger J-P, Schmidtobreick L (eds) Astronomy at high angular resolution. Astrophysics and space science library, vol 439. Springer, Cham, p 179

Potter SB, Cropper M, Mason KO, Hough JH, Bailey JA (1997) Simultaneous UBVRIJK photometric and polarimetric observations of PQ GEM. Mon Not R Astron Soc 285:82–94

Potter SB, Hakala PJ, Cropper M (1998) 'Stokes imaging' of the accretion region in magnetic cataclysmic variables – I. Conception and realization. Mon Not R Astron Soc 297:1261–1268

Potter SB, Cropper M, Hakala PJ (2000) Stokes imaging of the accretion region in magnetic cataclysmic variables – II. V347 Pav. Mon Not R Astron Soc 315:423–430

Potter SB, Romero-Colmenero E, Watson CA, Buckley DAH, Phillips A (2004) Stokes imaging, Doppler mapping and Roche tomography of the AM Herculis system V834 Cen. Mon Not R Astron Soc 348:316–324

Rahoui F, Lee JC, Heinz S, Hines DC, Pottschmidt K, Wilms J, Grinberg V (2011) A multiwavelength study of Cygnus X-1: the first mid-infrared spectroscopic detection of compact jets. Astrophys J 736:63

Rana V, Loh A, Corbel S, Tomsick JA, Chakrabarty D, Walton DJ, Barret D, Boggs SE, Christensen FE, Craig W, Fuerst F, Gandhi P, Grefenstette BW, Hailey C, Harrison FA, Madsen KK, Rahoui F, Stern D, Tendulkar S, Zhang WW (2016) Characterizing X-ray and radio emission in the black hole X-ray binary V404 Cygni during quiescence. Astrophys J 821:103

Reinsch K, Beuermann K, Hanusch H, Thomas H-C (1995) ROSAT observations of intermediate polars: AE AQR and YY DRA. In: Buckley DAH, Warner B (eds) Magnetic cataclysmic variables. Astronomical Society of the Pacific conference series, vol 85. Astronomical Society of the Pacific, San Francisco, p 115

Reynolds MT, Reis RC, Miller JM, Cackett EM, Degenaar N (2014) The quiescent X-ray spectrum of accreting black holes. Mon Not R Astron Soc 441:3656–3665

Rodriguez J, Cadolle Bel M, Alfonso-Garzón J, Siegert T, Zhang X-L, Grinberg V, Savchenko V, Tomsick JA, Chenevez J, Clavel M, Corbel S, Diehl R, Domingo A, Gouiffès C, Greiner J, Krause MGH, Laurent P, Loh A, Markoff S, Mas-Hesse JM, Miller-Jones JCA, Russell DM, Wilms J (2015) Correlated optical, X-ray, and γ-ray flaring activity seen with INTEGRAL during the 2015 outburst of V404 Cygni. Astron Astrophys 581:L9

Rodríguez-Gil P, Casares J, Martínez-Pais IG, Hakala P, Steeghs D (2001) Evidence of magnetic accretion in an SW Sextantis star: discovery of variable circular polarization in LS Pegasi. Astrophys J 548:L49–L52

Rodríguez-Gil P, Casares J, Martínez-Pais IG, Hakala PJ (2002) Detection of variable circular polarization in the SW Sex star V795 Herculis. In: Gänsicke BT, Beuermann K, Reinsch K (eds) The physics of cataclysmic variables and related objects. Astronomical Society of the Pacific conference series, vol 261. Astronomical Society of the Pacific, San Francisco, p 533

Rodríguez-Gil P, Gänsicke BT, Hagen H-J, Araujo-Betancor S, Aungwerojwit A, Allende Prieto C, Boyd D, Casares J, Engels D, Giannakis O, Harlaftis ET, Kube J, Lehto H, Martínez-Pais IG, Schwarz R, Skidmore W, Staude A, Torres MAP (2007) SWSextantis stars: the dominant population of cataclysmic variables with orbital periods between 3 and 4h. Mon Not R Astron Soc 377:1747–1762

Rodríguez-Gil P, Martínez-Pais IG, de la Cruz Rodríguez J (2009) The magnetic SW Sextantis star RXJ1643.7+3402. Mon Not R Astron Soc 395:973–978

Rudnick L, Owen FN, Jones TW, Puschell JJ, Stein WA (1978) Coordinated centimeter, millimeter, infrared, and visual polarimetry of compact nonthermal sources. Astrophys J 225:L5–L9

Rudy RJ, Kemp JC (1978) A polarimetric determination of binary inclinations – results for five systems. Astrophys J 221:200–210

Russell DM, Fender RP (2008) Polarized infrared emission from X-ray binary jets. Mon Not R Astron Soc 387:713–723

Russell DM, Shahbaz T (2014) The multiwavelength polarization of Cygnus X-1. Mon Not R Astron Soc 438:2083–2096

Russell DM, Fender RP, Hynes RI, Brocksopp C, Homan J, Jonker PG, Buxton MM (2006) Global optical/infrared-X-ray correlations in X-ray binaries: quantifying disc and jet contributions. Mon Not R Astron Soc 371:1334–1350

Russell DM, Casella P, Fender R, Soleri P, Pretorius ML, Lewis F, van der Klis M (2011a) Rapid variations of polarization in low-mass X-ray binaries. ArXiv e-prints

Russell DM, Miller-Jones JCA, Maccarone TJ, Yang YJ, Fender RP, Lewis F (2011b) Testing the jet quenching paradigm with an ultradeep observation of a steadily soft state black hole. Astrophys J 739:L19

Russell DM, Markoff S, Casella P, Cantrell AG, Chatterjee R, Fender RP, Gallo E, Gandhi P, Homan J, Maitra D, Miller-Jones JCA, O'Brien K, Shahbaz T (2013a) Jet spectral breaks in black hole X-ray binaries. Mon Not R Astron Soc 429:815–832

Russell DM, Russell TD, Miller-Jones JCA, O'Brien K, Soria R, Sivakoff GR, Slaven-Blair T, Lewis F, Markoff S, Homan J, Altamirano D, Curran PA, Rupen MP, Belloni TM, Cadolle Bel M, Casella P, Corbel S, Dhawan V, Fender RP, Gallo E, Gandhi P, Heinz S, Körding EG, Krimm HA, Maitra D, Migliari S, Remillard RA, Sarazin CL, Shahbaz T, Tudose V (2013b) An evolving compact jet in the black hole X-ray Binary MAXI J1836-194. Astrophys J 768:L35

Russell TD, Soria R, Miller-Jones JCA, Curran PA, Markoff S, Russell DM, Sivakoff GR (2014) The accretion-ejection coupling in the black hole candidate X-ray binary MAXI J1836-194. Mon Not R Astron Soc 439:1390–1402

Russell DM, Shahbaz T, Lewis F, Gallo E (2016) Polarized synchrotron emission in quiescent black hole X-ray transients. Mon Not R Astron Soc 463:2680–2689

Russell DM, Qasim AA, Bernardini F, Plotkin RM, Lewis F, Koljonen KII, Yang Y-J (2018) Optical precursors to black hole X-ray binary outbursts: an evolving synchrotron jet spectrum in Swift J1357.2-0933. Astrophys J 852:90

Rybicki GB, Lightman AP (1986) Radiative processes in astrophysics. Wiley-Interscience, New York

Schultz J, Hakala P, Huovelin J (2004) Polarimetric survey of low-mass X-ray binaries. Baltic Astron 13:581–595

Serkowski K (1973) Interstellar polarization (review). In: Greenberg JM, van de Hulst HC (eds) Interstellar dust and related topics. IAU symposium, vol 52, p 145

Shahbaz T, Fender RP, Watson CA, O'Brien K (2008) The first polarimetric signatures of infrared jets in X-ray binaries. Astrophys J 672:510–515

Shahbaz T, Russell DM, Zurita C, Casares J, Corral-Santana JM, Dhillon VS, Marsh TR (2013) Evidence for quiescent synchrotron emission in the black hole X-ray transient Swift J1357.2-0933. Mon Not R Astron Soc 434:2696–2706

Shahbaz T, Russell DM, Covino S, Mooley K, Fender RP, Rumsey C (2016) Evidence for magnetic field compression in shocks within the jet of V404 Cyg. Mon Not R Astron Soc 463:1822–1830
Simmons JFL, Boyle CB (1984) Interpretation of the variable linear polarization of X-ray transient AO 538–66. Astron Astrophys 134:368–372
Simmons JFL, Aspin C, Brown JC (1982) Bias of polarimetric estimators for binary star inclinations. Mon Not R Astron Soc 198:45–57
Tetarenko BE, Bahramian A, Arnason RM, Miller-Jones JCA, Repetto S, Heinke CO, Maccarone TJ, Chomiuk L, Sivakoff GR, Strader J, Kirsten F, Vlemmings W (2016) The first low-mass black hole X-ray binary identified in quiescence outside of a globular cluster. Astrophys J 825:10
Tetarenko AJ, Sivakoff GR, Miller-Jones JCA, Bremer M, Mooley KP, Fender RP, Rumsey C, Bahramian A, Altamirano D, Heinz S, Maitra D, Markoff SB, Migliari S, Rupen MP, Russell DM, Russell TD, Sarazin CL (2019) Tracking the variable jets of V404 Cygni during its 2015 outburst. Mon Not R Astron Soc 482:2950–2972
Thorstensen JR, Ringwald FA, Wade RA, Schmidt GD, Norsworthy JE (1991) PG0027 + 260 – an example of a class of cataclysmic binaries with mysterious, but consistent, behavior. Astron J 102:272–283
Vaeth H (1997) Cyclotron versus free-free emission from the intermediate polar RX J1712.6-2414. Astron Astrophys 317:476–486
Vaeth H, Chanmugam G, Frank J (1996) The magnetic field of the intermediate polar RE 0751+14. Astrophys J 457:407
Veledina A, Poutanen J, Vurm I (2011) A synchrotron self-Compton-disk reprocessing model for optical/X-ray correlation in black hole X-ray binaries. Astrophys J 737:L17
Watson CA, Dhillon VS (2001) Roche tomography of cataclysmic variables – I. Artefacts and techniques. Mon Not R Astron Soc 326:67–77
Welsh WF, Horne K, Gomer R (1998) Doppler signatures of Hα flares in AE Aquarii. Mon Not R Astron Soc 298:285–302
Westfold KC (1959) The polarization of synchrotron radiation. Astrophys J 130:241
Wickramasinghe DT, Ferrario L (2000) Magnetism in isolated and binary white dwarfs. Publ Astron Soc Pac 112:873–924
Wickramasinghe DT, Meggitt SMA (1985) The polarization properties of magnetic accretion columns. III – A grid of uniform temperature and shock front models. Mon Not R Astron Soc 214:605–618
Wickramasinghe DT, Ferrario L, Bailey J (1989) A 56 MG field at the second pole in VV Puppis. Astrophys J 342:L35–L38
Wickramasinghe DT, Cropper M, Mason KO, Garlick M (1991) Cyclotron humps in AM HER systems. IV – Variations around the orbit in MR SER. Mon Not R Astron Soc 250:692–700
Wolinski KG, Dolan JF (1994) Confidence intervals for orbital parameters determined polarimetrically. Mon Not R Astron Soc 267:5
Wynn GA, King AR, Horne K (1997) A magnetic propeller in the cataclysmic variable AE Aquarii. Mon Not R Astron Soc 286:436–446

Chapter 11
Multi-Wavelength Polarimetry of Isolated Pulsars

Alice K. Harding

Abstract Isolated pulsars are ideal sources for polarisation studies since the radiation from pulsars generally covers a wide range of wavelengths from radio to very-high-energy γ-rays and the pulsed emission is closely tied to the local magnetic field direction. Although radio polarisation data exist for many pulsars, polarisation measurements at higher energies exist for only six sources. All six of these have optical polarisation measurements while only the Crab pulsar has measured polarisation in the X-ray band. One member of a small group of pulsars only visible at optical and X-ray wavelengths, thought to be powered by thermal emission from the cooling neutron star surface, has detected optical polarisation. This chapter will review the current multi-wavelength polarisation observations and the theoretical modelling by which we can probe the pulsar magnetosphere structure and emission mechanisms.

11.1 Introduction

The radiation from rotation-powered pulsars (RPP) has been detected over a very wide frequency band, from radio to very-high energy γ-rays. Both the phase-averaged and phase-resolved properties of this radiation have provided a wealth of information on the mechanisms responsible, the location of the emission and the energies of accelerated particles. In particular, the very high brightness temperatures ($T \sim 10^{30}$ K) of the observed radio emission require a coherent emission mechanism. However, after 50 years of pulsar observations there is still no consensus on the actual mechanism(s) that produce this emission. The emission observed at optical to γ-ray energies is primarily non-thermal and requires particles with a broad spectrum of Lorentz factors up to $\sim 5 \times 10^7$. However, the location and mechanisms of the emission in the different energy ranges is debated. Observations

A. K. Harding (✉)
NASA Goddard Space Flight Center, Greenbelt, MD, USA
e-mail: Alice.K.Harding@nasa.gov

R. Mignani et al. (eds.), *Astronomical Polarisation from the Infrared to Gamma Rays*, Astrophysics and Space Science Library 460,
https://doi.org/10.1007/978-3-030-19715-5_11

of the polarisation of this radiation can provide much of the missing information we need to answer major questions about pulsar emission. Polarisation in fact can reveal particularly well the properties of pulsars since their strong magnetic fields govern the trajectories of the high-energy particles and thus the photon emission directions as the pulsar rotates. The variation of polarisation degree and position angle as a function of pulse phase map out the magnetic field structure and thus provide clues to the location of the radiation within the magnetosphere. Polarisation observations alone though give a tomographic picture. Radiation modelling is needed as well to interpret the observed data and give them physical meaning. This review will therefore discuss both the current state of polarisation observations and theoretical models and predicted polarisation signatures of isolated pulsars.

11.2 Polarisation Observations of Pulsars and Their Nebulae

Over 2500 RPP have been detected in the radio band and phase-resolved polarisation measurements exist for many of these. Typically, pulsar radio emission is very highly polarised and the position angle changes across the pulse period often with an S-shaped variation that is characteristic of the Rotating Vector Model (RVM) (Radhakrishnan and Cooke 1969). The RVM model (see Sect. 11.3.2) assumes that the emission originates near the magnetic poles and that the electric vector is either parallel or perpendicular to the local magnetic field direction. For some pulsars though, the position angle variation with rotational phase is not this simple, and jumps in position angle by $\sim 90°$, thought to be caused by switching between orthogonal modes, is common. For others, notably the Crab pulsar, the phase-resolved polarisation pattern does not resemble the RVM pattern at all (see Sect. 11.2.1). Pulsars also can show circular polarisation that changes sign at the radio pulse peak.

At the present time, polarisation measurements at frequencies above the radio band exist for only six pulsars, three very young pulsars: Crab, B0540-69, B1509-58, two middle-aged pulsars, Vela and B0656+14, and one isolated neutron star RX J1856.5-3754. Table 11.1 lists the current polarisation measurements for these sources (for a comprehensive review of the polarisation observations see Mignani (2018)). In general, the polarisation degree (PD) of radio emission is much higher than that of the higher energy emission. At optical wavelengths and above, polarisation measurements are very dependent of detector field-of-view (FOV). For large FOV, the measured PD and PA are a blend of emission from the pulsar and its nebula, but it is possible to obtain better PD and PA values for just the pulsed emission by either narrowing the FOV, as is possible at optical wavelengths, or by subtracting the off-pulse phases of the pulsar. When this is done, the PD of the pulsed emission at optical and X-ray wavelengths is quite low, in the range $\sim$5–15%, much lower than either the pulsed radio emission or the nebular emission. Another trend is that the PA of the pulsed emission is often roughly aligned with the pulsar proper motion direction, which tends to be roughly aligned with the

Table 11.1 Polarisation observations of isolated pulsars

Pulsar	P (s)	$\dot{E}$ (10^{36} erg s^{-1})	Phase	PD (%)	PA (degrees)	Ref.	
Crab (B0531+21)	0.033	4340			124 ± 0.1	1	Spin axis
					$110 \pm 2 \pm 9$	2	Proper motion
Radio (1.38 GHz)			MP	23.98 ± 0.30	120	3,4	Pulsar
			IP	21.3 ± 1.0	120	3,4	Pulsar
			LFC	98.2 ± 6.7	160.8 ± 1.5	3	Pulsar
Optical			Averaged	5.2 ± 0.3	105.1 ± 1.6	5	Pulsar
			Averaged	5.5 ± 0.1	96.4 ± 0.1	6	Pulsar
X-ray (2.6 keV)			Averaged	15.7 ± 1.5	161.1 ± 2.8	7	Pulsar/Nebula
			Off pulse	19.2 ± 1.0	156.4 ± 1.4	8	Nebula
X-ray (20–120 keV)			Averaged	20.9 ± 5.0	131.3 ± 6.8	9	Pulsar/Nebula
			IP	$33.5^{+18.6}_{-22.3}$	86 ± 18	9	OP subtracted
X-ray (100–380 keV)			Averaged	32.7 ± 5.8	143.5 ± 2.8	10	Pulsar/Nebula
X-ray (0.2–0.8 MeV)			Averaged	47^{+19}_{-13}	100 ± 11	11	
			MP+IP	42^{+30}_{-16}	70 ± 20	11	
B0540–69	0.05	150					
Optical			Averaged	5.6 ± 1.0	79 ± 5	12,13	Pulsar/Nebula
			Averaged	16 ± 4	22 ± 12	14	Pulsar
B1509-58	0.151	17.3					
Radio (660 MHz)			Pulse peak	97 ± 7	150	16	
Optical			Averaged	10.4		12	Pulsar
Vela (B0833-45)	0.089	69			310 ± 1.5	17	Jets
					301 ± 1.8	18	Proper motion
Radio (2.29 GHz)				90	64 ± 1.5	19, 20	

(continued)

Table 11.1 (continued)

Pulsar	P (s)	$\dot{E}$ (10^{36} erg s^{-1})	Phase	PD (%)	PA (degrees)	Ref.	
Optical			Averaged	8.5 ± 0.8		12	Pulsar
			Averaged	8.1 ± 0.7	146.3 ± 2.4	21	Pulsar
				9.4 ± 4	146 ± 11	22	Pulsar
B0656+14	0.384	0.036			93.12 ± 0.38	23	Proper motion
Radio (3.1 GHz)				$80 - 90$	-86 ± 2	24	Pulsar
Optical			Averaged	11.9 ± 5.5	125.8 ± 13.2	25	Pulsar
RX J1856.5-3754	7.06	3.3E-6			100.2 ± 0.2	27, 28	Proper motion
Optical			Averaged	16.43 ± 5.26	145.39 ± 9.44	26	Pulsar

References: (1) Ng and Romani (2004), (2) Kaplan et al. (2008), (3) Slowikowska et al. (2015), (4) Moffet and Hankins (1999), (5) Moran et al. (2013), (6) Slowikowska et al. (2012), (7) Weisskopf et al. (1976), (8) Weisskopf et al. (1978), (9) Chauvin et al. (2017), (10) Vadawale et al. (2018), (11) Forot et al. (2008), (12) Wagner and Seifert (1999), (13) Chanan and Helfand (1990), (14) Mignani (2010), (15) Gaensler et al. (1999), (16) Crawford et al. (2001), (17) Helfand et al. (2001), (18) Dodson et al. (2003) (19) Krishnamohan and Downs (1983), (20) Hamilton et al. (1977), (21) Moran et al. (2014), (22) Mignani et al. (2007), (23) Brisken et al. (2003), (24) Johnston et al. (2007), (25) Mignani et al. (2015), (26) Mignani et al. (2017), (27) Tiengo and Mereghetti (2007), (28) Kaplan et al. (2002)

pulsar spin axis as estimated from the resolved nebular structure at optical and/or X-ray wavelengths. This nebular structure often shows a torus, thought to lie in the spin equator, and jets, thought to lie along the pulsar spin axis. Table 11.1 includes measurements of the direction of proper motion, the spin axis and jets that have appeared in the literature.

11.2.1 *Crab*

Being the most studied source in the sky, the Crab has the most extensive polarisation measurements of any pulsar. It is bright at nearly all wavelengths, making it the perfect source for studying multi-wavelength polarisation characteristics. But that being said, one needs to be cautious about generalising the Crab measurements since it is not always typical of other pulsars. For example, measurement of its radio polarisation has been difficult since the strong flux and polarisation of the nebula is a significant background combined with the apparently unusually low PD of the pulsed radio emission. As will be discussed in Sect. 11.3, this seeming anomaly is likely due to the radio emission from the Crab having its origin in the outer magnetosphere rather than near the polar caps, where relativistic effects cause depolarisation of the pulses. The most recent phase-resolved radio polarisation measurements (Moffet and Hankins 1999; Slowikowska et al. 2015) show very little change in PD or PA through the pulse peaks (see Fig. 11.1). The main pulse (MP) and interpulse (IP) have very similar values of PD ($\sim$21–23%) and PA (nearly aligned with the pulsar spin axis). However, the precursor shows much higher linear PD of $\sim$ 98% and circular PD of $\sim$ 20%. The precursor therefore seems to be a "normal" radio pulse originating near the polar cap.

It is at optical wavelengths that the Crab polarisation has been exquisitely measured. The most recent optical observations of the pulsar and nebula are by OPTIMA (Slowikowska et al. 2009) and *HST* (Moran et al. 2013). Figure 11.2 shows several *HST* fields around the pulsar with increasingly smaller FOVs (Moran et al. 2013). The left field measures primarily the nebular polarisation and has the largest FOV as well as measured PD and PA that are very similar to that of the soft X-ray measurement of the nebula (Weisskopf et al. 1978). The middle field zooms in to the pulsar and its surrounding structures, such as the "inner knot" and the jet, both of which have a higher PD. The PA of the jet is aligned with that of the pulsar spin axis. However, the pulsar PA is closer to the proper motion direction, as shown in the right field. The phase-averaged pulsar PD of 5.2% is in close agreement with that measured by OPTIMA, once the DC component (possibly associated with the knot) is subtracted, but the PA is a bit higher. OPTIMA also measured the phase-resolved polarisation of the Crab (Slowikowska et al. 2009), which shows rapid swings of PA through both the MP and IP, with the MP changing by $\sim$ 140° (see Fig. 11.1). The PD dips at each peak to a very low $\sim$ 3%, which explains the low phase-averaged PD. The optical phase-resolved behaviour is thus very different from that of the radio.

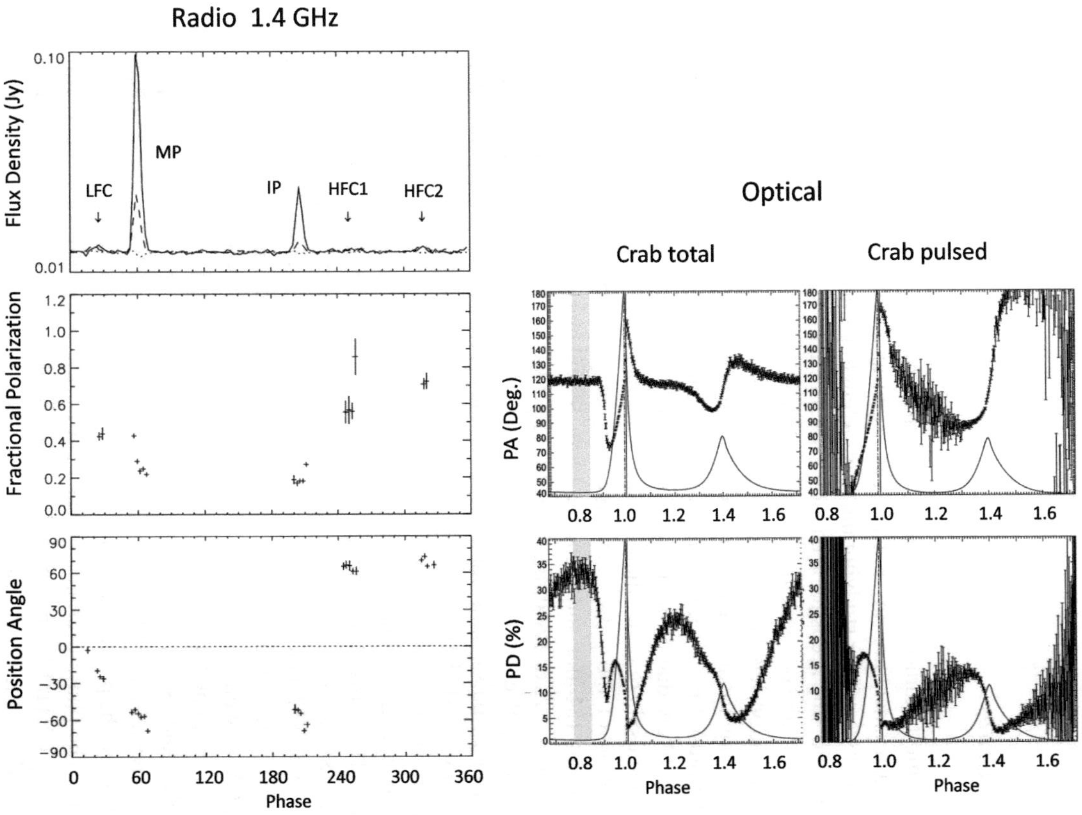

Fig. 11.1 Left: Radio light curve, polarisation fraction and position angle at 1.4 GHz, from Moffet and Hankins (1999), Right: Optical polarisation PA and PD with the light curve superposed, from Slowikowska et al. (2009)

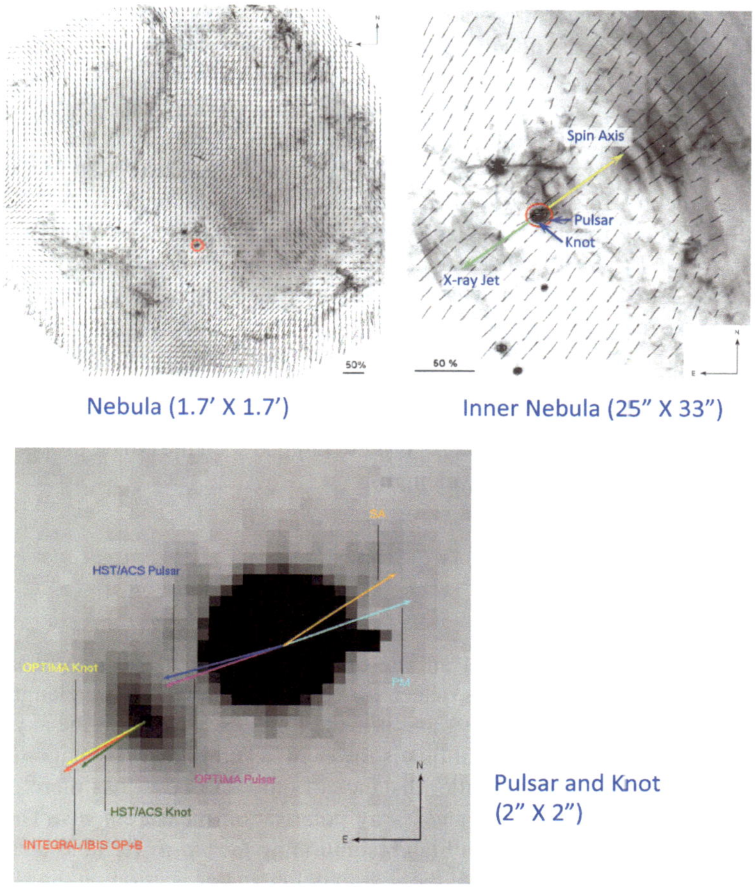

Fig. 11.2 Polarisation of the Crab nebula observed with *HST* at three increasingly smaller FOV (in parentheses), from Moran et al. (2013). The small red circle in the left two panels denotes the pulsar position, and the PA of the pulsar spin axis (SA), proper motion (PM) and X-ray jet. The vectors included are as follows: spin-axis vector (SA) ($124° \pm 0.°1$) (Ng and Romani 2004), proper motion vector (PM) ($110° \pm 2 \pm 9°$) (Kaplan et al. 2008) and the polarisation PA of the pulsar ($105° \pm 1.°6$) and synchrotron knot ($124.°7 \pm 1.°0$) from the *HST*/ACS data. Also included are the phase-averaged OPTIMA measurements of the polarisation PA of the synchrotron knot ($119.°8 \pm 0.°1$) and pulsar ($109.°5 \pm 0.°2$) (Slowikowska et al. 2009), and the phase-averaged *INTEGRAL*/IBIS measurement of the polarisation PA during the off-pulse and bridge emission (OP+B) phases ($122.°7 \pm 7.°7$) (Forot et al. 2008)

There have been a growing number of Crab polarisation measurements at X-ray wavelengths, beginning with the early *OSO-8* observations in soft X-rays (Weisskopf et al. 1976, 1978), who obtained the PD and PA of the nebula plus pulsar and then the nebula alone by taking photons only from the off-pulse phases of the pulsar light curve. The relatively high PD $= 19\%$ of the nebula is expected if the emission is synchrotron radiation. Because of the large FOV, these studies were able to measure the nebular polarisation, but not that of the pulsar alone. This was unfortunately the most recent soft X-ray polarimeter flown, but the approved mission *IXPE* (*Imaging X-ray Polarimetry Explorer*, Weisskopf et al. 2016) has an anticipated launch in 2021. At somewhat higher X-ray energies 20–120 keV *PoGO+*, a balloon-borne Compton scattering polarimeter, measured the polarisation of the total nebula plus pulsar (Chauvin et al. 2017) which agrees with the *OSO-8* result. Their phase-resolved analysis obtained only an upper limit on the PD $< 70\%$ for the MP and a poorly constrained measurement for the IP. At 100–380 keV, *ASTROSAT* measured a somewhat higher PD for the pulsar plus nebula (Vadawale et al. 2018). Finally, the highest-energy (0.2–0.8 MeV) polarisation measurement was obtained by *INTEGRAL/IBIS* (Forot et al. 2008), who found phase-averaged PD $= 47\%^{+19\%}_{-13\%}$ and PD $= 42^{\circ}{}^{+30^{\circ}}_{-16^{\circ}}$ for the MP plus IP. The measurements of the PA of the nebula at X-ray wavelengths are aligned more with the pulsar spin axis, while the PA of the pulse components is more aligned with the pulsar proper motion, similar to the trend at optical wavelengths.

11.2.2 Vela

Vela is a middle-aged pulsar, at an age of $\sim 10^4$ year, but still one of the most energetic. Its pulsed emission is detected over a wide range of wavelengths from radio to Very-High-Energy γ-rays and it is the brightest sources in the sky at GeV energies. Vela's X-ray and optical flux is lower by several orders of magnitude than that of the Crab, despite its smaller distance of only 250 pc. It has therefore been more difficult to obtain optical and X-ray polarisation measurements. The image of the Vela nebula observed with high resolution by *Chandra* (Helfand et al. 2001) shows clear jets structures that are thought to define the projection on the sky of the pulsar spin axis and the pulsar proper motion direction is within a few degrees. The radio linear polarisation is very high, but it is not aligned with the spin or proper motion directions. However, the optical PA measured by *HST* (Moran et al. 2014) and by the VLT (Mignani et al. 2007) is found to lie along the spin and proper motion directions. The pulsar optical PD is measured to be around 8–9%. These characteristics of the optical polarisation, the PA alignment with the proper motion and the relatively low PA, are very similar to that of the Crab pulsar.

11.2.3 B0540-69

First discovered as an X-ray RPP (Seward et al. 1984), PSR B0540-69 lies in the Large Magellanic Cloud in the supernova remnant N47 and is one of the intrinsically brightest RPPs in X-rays and the most luminous γ-ray pulsar (Ackermann et al. 2015). It is often referred to as the "Crab Twin" since the two sources share a number of uncommon characteristics, including phase-aligned light curves at radio to γ-ray wavelengths and dominant, non-thermal X-ray and optical emission. Given its large distance of 55 kpc though, it has had relatively few polarisation measurements. It is very faint at radio wavelengths and no radio polarisation results have been published. There are a few polarisation measurements in the optical band. Chanan and Helfand (1990), using the CTIO 4 m telescope, measured the polarisation integrated over the whole nebula, finding a PD $= 5.6 \pm 1.0\%$ and PA $= 79° \pm 5°$. The PD value was confirmed by Wagner and Seifert (1999), observing with the VLT. Mignani (2010), using *HST*/WFPC2, obtained results for the pulsar PD = $16 \pm 4\%$ and PA $=22° \pm 12°$. They find that the PA direction is aligned with the semi-major axis of the nebula, but no proper motion has been measured for the pulsar. The measured pulsar PD is significantly larger than that of the pulsar plus nebula measured by Chanan and Helfand (1990), which is opposite to the Crab measurements where the phase-averaged pulsar PD is much smaller than that of the nebula plus pulsar or the nebula alone. The pulsar PD $\sim$ 16% is also larger than that measured for other RPPs that lie more in the $\sim$ 5–10% range. It is possible, given the different light curve shape, that this difference is due to the pulsar geometry and will be discussed in Sect. 11.3.

11.2.4 B1509-58

PSR B1509-58 is a very young and energetic Crab-like pulsar in the supernova remnant MSH 15-52 and was also first discovered as a soft X-ray RPP (Seward and Harnden 1982). It was later found to be pulsed at radio, hard X-ray and γ-ray wavelengths. The light curve is a broad, single pulse at all wavelengths, although the radio and high-energy pulses are not aligned, as for the Crab and PSR B0540-69. The radio pulse is very highly polarised and the PA is more or less aligned with the pulsar proper motion direction (which has not actually mean measured but inferred from a tentative association with a runaway star). Wagner and Seifert (1999) report an optical PD = 10.4%, in the range of other pulsars, but there is no reported PA measurement.

11.2.5 B0656+14

PSR B0656+14 is a middle-aged RPP, with characteristics similar to those of Vela. Its pulsations are seen at radio, optical, X-ray and γ-ray wavelengths. Similar to other RPPs, its radio pulse is very highly polarised with a phase-averaged PA that is aligned with the pulsar proper motion direction modulo 180°. Mignani et al. (2015), using the VLT, reported phase-averaged optical PD = $11.9 \pm 5.5\%$ and PA = $125.8° \pm 13.2°$, with the PA direction close to the pulsar proper motion direction. Kern et al. (2003), using the Palomar 5 m telescope, presented phase-resolved polarisation in the optical band. They show a significant variation of the PD and PA across the pulse. The PD is very low (unmeasurable) in the pulse peaks but nearly 100% in the bridge region between the two peaks that are separated by a phase of about 0.5. The PA increases through an angle of nearly 90° in the bridge region, whereas the PA is not measured in the pulse peaks since there is no linear polarised flux there.

11.2.6 RX J1856.5-3754

RX J1856.5-3754 is a member of a source class known as Isolated Neutron Stars (INS) that are radio-quiet pulsars first discovered at X-ray wavelengths. This group of sources, known as "The Magnificent Seven", have periods between 3 s and 11 s and purely thermal X-ray spectra. Several have measured pulse period derivatives that imply high surface magnetic fields $B \sim 10^{13} - 10^{14}$ G. The presence of broad absorption features at a few hundred eV in the spectrum also suggests a very high surface field, if the features come from proton cyclotron scattering. The X-ray luminosities of INS are larger than their spin-down luminosities, so it is thought that thermal cooling of the neutron star rather than rotation powers their emission. A PD = 16% was measured for RX J1856.5-3754 in the optical band by Mignani et al. (2017), using the VLT. The measured PA is not aligned with the pulsar proper motion, unlike that of RPPs. In comparing their results with theoretical predictions of polarisation of radiation from a highly magnetised atmosphere, Mignani et al. (2017) conclude that the measured PD level cannot be achieved without including the effects of the QED vacuum polarisation. Since the model PD is sensitive to the magnetic inclination and viewing angles, they were also able to constrain the pulsar geometry.

11.3 Pulsar Polarisation Models

The polarisation properties of pulsar emission are highly sensitive to both the location of the emitting regions and the magnetic field geometry (see Fig. 11.3). Although the radio emission mechanism has not been resolved, it is generally accepted that it originates near the magnetic poles about 50–100 km above the stellar surface, that it must be coherent and that a dense electron-positron pair plasma is required. Evidence to support this picture comes not only from observed properties of the radio pulses, such as the fact that the pulse widths agree with polar cap opening angles (Rankin 1990), but also from the RVM-expected PA sweeps through the pulse profile seen in many pulsars (Sect. 11.3.2).

As models for the pulsed high-energy (optical and higher) radiation have evolved over the half-century since the discovery of pulsars, both the location of the emitting region and the global magnetic field geometry have evolved. Very early models placed the site of the high-energy emission near the polar caps, which is an expected

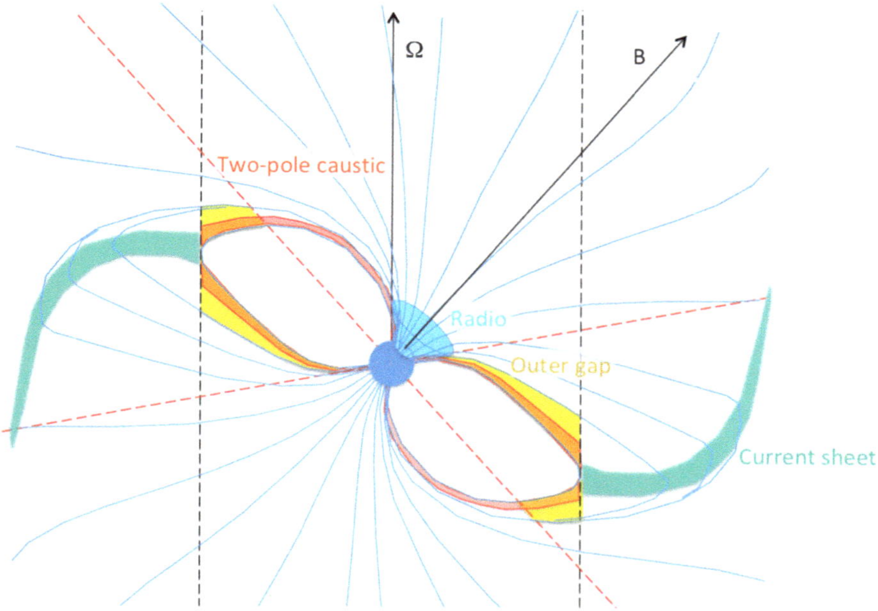

Fig. 11.3 Location of radio (blue) and high-energy emission in the outer gap (yellow), two-pole caustic (red) and current sheet (green) models in the meridional plane containing the spin and magnetic axes of a force-free magnetosphere. The dashed black lines denote the light cylinder and the dotted red lines show projections of the null-charge surface

site of particle acceleration. The accelerated particles radiate γ-rays through curvature and inverse-Compton radiation, initiating electromagnetic cascades through one-photon pair production in the strong magnetic field (Daugherty and Harding 1982; Harding and Muslimov 2002). In this model, the high-energy pulses would appear near the phase of the radio pulses, would be very highly polarised and exhibit the "S-shaped" PA sweep of the RVM. Phase-resolved high-energy polarisation measurements were not available until fairly recently, so this property of the polar cap models could not be tested. However the *Fermi Gamma-Ray Space Telescope*, launched in 2008, discovered many γ-ray pulsars whose pulses appeared at phases significantly different from those of the radio pulses (Abdo et al. 2013). In fact, the variety of light curves agree more with predictions of models where the high-energy emission comes from the outer magnetosphere near or beyond the light cylinder, $R_{\rm LC} = c/\Omega$, where Ω is the pulsar rotation rate. Such outer magnetosphere models, such as outer gap (Cheng et al. 1986; Romani 1996), slot gap and current sheet models had been proposed earlier (see Fig. 11.3). Naturally, polarisation predictions also changed substantially from those of polar cap models and, as will be discussed in the following sections, can better explain the properties of the observed polarisation.

11.3.1 Intrinsic Polarisation of Emission

The proposed pulsar radiation mechanisms, synchrotron radiation (SR), curvature radiation (CR) and inverse-Compton scattering (ICS) all have different intrinsic polarisation properties. Both SR and CR are highly polarised and the polarisation fraction as a function of photon energy ω is

$$P(\omega) = \frac{K_{2/3}(\omega/\omega_c)}{\int_{\omega/\omega_c}^{\infty} K_{5/3}(x)dx} \tag{11.1}$$

where $K_{2/3}(x)$ and $K_{5/3}(x)$ are modified Bessel functions and ω_c are the critical energies for CR

$$\omega_{CR} = \frac{3}{2}\frac{\lambda_c\,\gamma^3}{\rho_c} \tag{11.2}$$

and for SR

$$\omega_{SR} = \frac{3}{2}\gamma^2\,B'\,\sin\psi \tag{11.3}$$

(Westfold 1959) where γ is the particle Lorentz factor, ρ_c the particle trajectory radius of curvature, B' the local magnetic field strength in units of the critical field, $B_{\rm cr} = 4.4 \times 10^{13}$ G and ψ is the particle pitch angle. Asymptotic values of $P(\omega)$ are 0.5 for $\omega \ll \omega_c$ and 1.0 for $\omega \gg \omega_c$. At $\omega = \omega_c$, $P \simeq 0.75$. Thus, the polarisation can reach nearly 100% above the cutoff energy but only for a distribution of nearly mono-energetic particles. Typically though, the radiating particles have a distribution of Lorentz factors and the polarisation fraction depends on the power-law index, p, as

$$P(p) = \frac{p+1}{p+7/3} \tag{11.4}$$

so that $P(p) = 60 - 75\%$ for $p = 1 - 3$. ICS emission in the weakly magnetised limit has a lower polarisation degree than SR or CR (Begelman and Sikora 1987). Krawczynski (2012) presents analytic and numerical results of ICS polarisation in both the Thompson and Klein-Nishina regimes. The predicted PD for a power-law distribution with index p in the non-magnetic Thompson regime is

$$P(p) \sim \frac{(1+p)(3+p)}{11+4p+p^2}. \tag{11.5}$$

In the highly magnetised environment of pulsars, ICS emission is strongly influenced by the magnetic field when the photon energy is near or below the cyclotron resonance. The polarisation of ICS emission in both the Thomson and high field

(QED) regimes is investigated by Wadiasingh et al. (2018), showing that emission near the spectral cutoff is highly polarised. The direction of polarisation for each process is determined by the orientation of the photon electric vector which is parallel to the direction of particle acceleration (Blaskiewicz et al. 1991; Hibschman and Arons 2001). For CR, the particle acceleration and electric vector is along the radius of curvature and therefore in the plane of the field line curvature. For SR, the electric vector is perpendicular to both the magnetic field and photon momentum vectors and is generally orthogonal to that of CR.

11.3.2 Polar Cap: Rotating Vector Model

Since the PA of emission from relativistic particles emitting CR or SR is oriented perpendicular to the magnetic field direction, the change in PA as an observer views different directions with respect to the magnetic pole can constrain the pulsar geometry. Thus the basic tenet of the RVM is that the PA, ψ, of the radiation is determined by Radhakrishnan and Cooke (1969)

$$\tan[\psi(\phi) - \psi_0] = \frac{\sin\alpha \sin(\phi - \phi_0)}{\cos\alpha \sin\zeta - \sin\alpha \cos\zeta \cos(\phi - \phi_0)} \tag{11.6}$$

where α and ζ are the magnetic inclination and observer angles relative to the pulsar spin axis, ϕ is the rotation phase and ψ_0 is a constant offset and ϕ_0 is the phase of maximum PA change, given by

$$\left(\frac{d\psi}{d\phi}\right)_{\psi_0} = \frac{\sin\alpha}{\sin\beta} \tag{11.7}$$

where $\beta = \zeta - \alpha$. RVM fits to radio pulsar PA curves generally constrain the value of the viewing impact angle β, but the values of α and ζ are only constrained within banana-shaped contours that also depend on the pulse width (see Fig. 11.4). Although the pulsed radio emission is mostly linearly polarised, many pulsars also show some circular polarisation, which typically appears for small β where the line of sight is near the magnetic pole and therefore along the field. As seen in Fig. 11.4, the sign of the circular polarisation flips at the phase of magnetic pole crossing.

The RVM is only accurate for emission at altitudes small compared to the light cylinder, where relativistic effects such as aberration and light travel delays are negligible. Thus, it is most applicable to older pulsars with longer periods and thus larger R_{LC}. For short period pulsars, relativistic corotation velocity can have significant effects on the PA pattern. Blaskiewicz et al. (1991) (BCW), presented modifications to the RVM for emission at moderate altitudes (still well within the light cylinder). They find that the PA ϕ_0 is shifted from the pulse profile centroid to a later phase (see Fig. 11.4) by around $4r_{em}/R_{LC}$, where r_{em} is the emission radius (see also Dyks 2008). There is evidence that this occurs in a number of

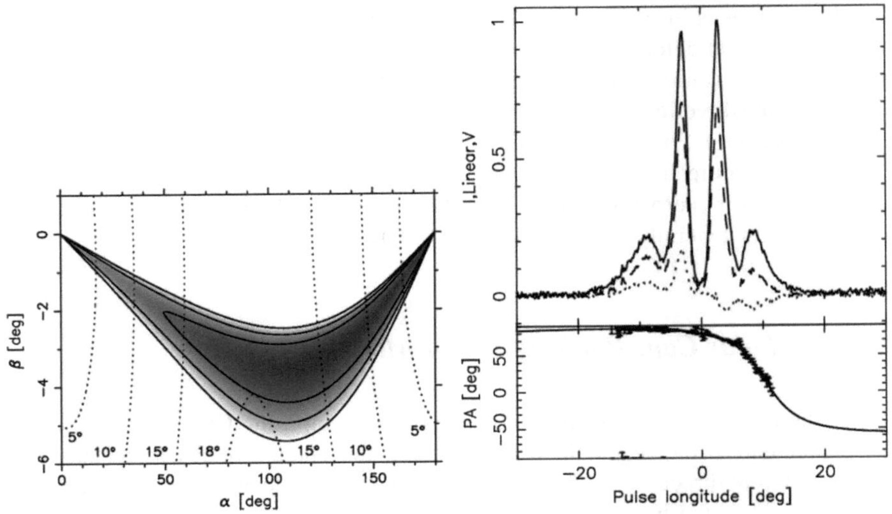

Fig. 11.4 Left: Average radio pulse profile of PSR J0631+1036, as well as the degree of linear polarisation (dashed) and circular polarisation (dotted). The bottom panel is the position angle vs. phase. Right: "Banana"-shaped contours of impact angle β vs. inclination angle α from an RVM fit. Dotted lines are contours of equal pulse width. From Weltevrede et al. (2010)

pulsars and they argue that this effect may be used to estimate the emission radius. Dyks and Harding (2004) propose further corrections to the RVM caused by the sweepback of magnetic field lines near the light cylinder in the vacuum retarded dipole geometry (VRD Deutsch 1955) that produces a shift in the polar cap towards the trailing edge. The correction to the BCW formula for the ϕ_0 introduces a term of order $0.2(r_{\rm em}/R_{\rm LC})^{1/2}$, which at low altitudes can reduce the BCW shift as much as a half. Craig and Romani (2012), also using the VRD magnetic field, introduce further relativistic corrections to the BCW formula for the larger emission radii of young pulsars valid up to $\sim 0.3 r_{\rm em}/R_{\rm LC}$, that can be a significant fraction of $R_{\rm LC}$.

An off-centre magnetic dipole can also affect the interpretation of PA curves of radio pulsars. Burnett and Melatos (2014) modelled the Stokes Parameters for the RVM with an offset magnetic dipole. They show that this model can explain distortions of the PA curves seen in many pulsars. Pétri (2017), using a vacuum static dipole, also presented a study of the change in PA curves due to an off-centred dipole field, giving an exact analytic expression for the phase-resolved PA and Kundu and Petri (2017) presented results for a rotating offset dipole.

11.3.3 Outer Gap and Two-Pole Caustic Models

In outer gap (OG) models, particle acceleration and pair cascades occur in the outer magnetosphere in vacuum gaps that form along the last open magnetic field lines above the null charge surface, $\mathbf{\Omega}\cdot\mathbf{B} = 0$, where the corotation change changes sign, but inside the LC. In a charge-separated magnetosphere, vacuum gaps can form because charges of one sign can escape along the open field lines through the LC, but cannot be replaced since only charges of opposite sign are available on those field lines. The high-energy emission is produced by CR of primary accelerated particles in the gap as well as SR from secondary electron–positron pairs created through photon–photon pair production by interaction of CR photons with thermal X-rays from the polar cap. The two-pole-caustic (TPC) model (Dyks and Rudak 2003) also assumes that the high-energy emission occurs along the last open field lines but starting from the neutron star surface and extending to the LC. Since both OG and TPC models assume the VRD global magnetic field, calculation of particle trajectories and radiation cannot extend beyond the LC where particle velocities will become superluminal. On the other hand, particle velocities stay subluminal in global models, such as force-free magnetospheres, that allow for the influence of currents and/or particles on field structure (see Sect. 11.3.4). Although the original TPC model was a geometrical model, a possible physical equivalent is the slot gap (Arons 1983; Muslimov and Harding 2004).

The pulsations in the OG and TPC models are caustics, resulting from aberration and light travel time delays of emission over a range of altitudes. On trailing field lines the relativistic delays cause phase changes that cancel the phase changes due to the curved dipole field lines, so that all the emission arrives at the same phase to a non-rotating observer. On leading field lines, the relativistic phase changes add to the phases along the curved field and the emission phases are spread out. In the OG and TPC model, the caustics show up as bright patterns of emission on the sky in the observers frame (skymaps). The light curves are formed by cutting through the skymaps of observer angle, ζ, vs. phase, ϕ, at a constant ζ and can show one or two peaks of varying separation depending on magnetic inclination angle α. The OG skymap caustics in each hemisphere are cut off at $\zeta = 90°$, at the null charge surface, while the TPC caustics are not. Observers therefore can only see OG caustics from one hemisphere, and in double-peaked light curves only the second peak is a trailing-edge caustic while the first peak is due to overlapping swept-back field lines near the LC. The peaks in TPC double-peaked light curves are trailing-edge caustics from both poles.

The geometry of the caustics governs the polarisation characteristics of the OG and TPC models. With the peaks formed by emission over ranges of altitude that span field lines of changing direction, rapid PA sweeps and depolarisation are emblematic features. Romani and Yadigaroglu (1995) modelled the PA swing in the OG model and were able to produce PA sweeps through the peaks similar to what is seen in the Crab pulsar optical emission. Dyks et al. (2004) explored the polarisation characteristics of the OG and TPC models through geometric emission models. Taking the emission electric vector parallel to the particle acceleration vector (BCW), they found that the TPC model produces rapid PA swings through both peaks of double-peaked light curves and dips in PD at each peak (Fig. 11.5). However they found that although a PA swing occurs in the second peak of the OG model light curve, the PA swing at the first peak is very weak and requires emission

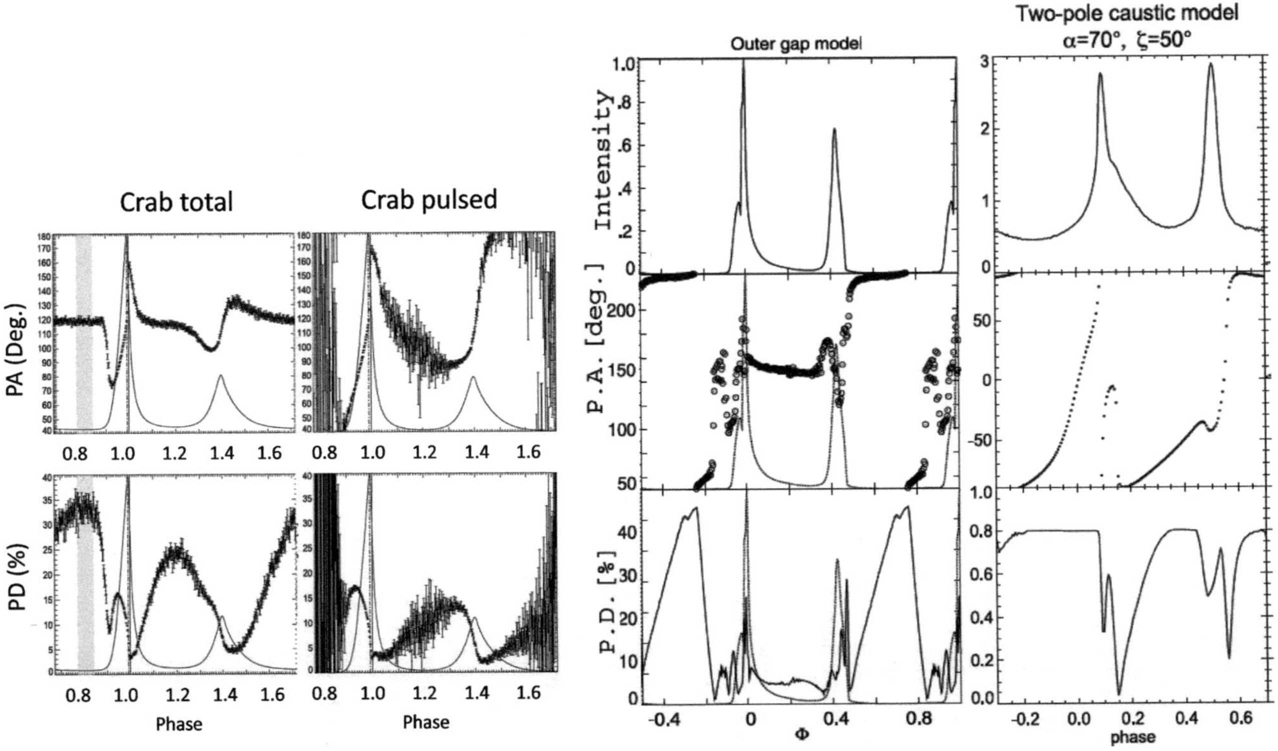

Fig. 11.5 Observed Crab phase-resolved polarisation (Slowikowska et al. 2009), with modelled light curves and polarisation of outer gap (Takata et al. 2007) and two-pole caustic (Dyks et al. 2004) models

very near the LC. Takata and Chang (2007) and Takata et al. (2007) computed the polarisation characteristics of radiation from an OG model. Cognisant of the lack of a strong PA swing at the first peak of the traditional OG model, they extended the emission below the null surface to near the stellar surface and obtained PA swings through both peaks. They modelled the optical and X-ray emission from the Crab as SR, taking into account the fact that the gyromotion produces a changing particle acceleration direction. This effect as well as the caustic emission causes depolarisation, not only at the peaks but between the peaks, where the PD drops below 10% (Fig. 11.4). They modelled the GeV emission as CR and showed that the PD of this emission is expected to be much higher, around 60% in the depolarisation dips at the peaks.

Although the radio emission from the vast majority of pulsars probably originates near the polar caps, there is evidence that the radio emission from a small subset of pulsars may come from high altitudes near or beyond the LC. The Crab, some Crab-like pulsars like PSR B0540-69 and a handful of MSPs have radio peaks that are aligned with peaks at higher energies. Venter et al. (2012) modelled light curves from three MSPs, PSR B1937+21, J0034-0534 and B1957+20, with aligned radio and γ-ray peaks using geometric OG and TPC caustics for both radio and γ-rays, finding that radio emission that is extended along field lines at high altitudes can produce good fits. They noted one prediction of this model is that the radio peaks should be very weakly polarised, unlike the radio peaks from near the polar cap. In fact, these MSPs have weak polarisation and J0034-0534 has no detectable polarisation. The MSP B1821-24 and several other MSPs seem to have radio peaks both from near the polar caps and from high altitude caustics, displaying radio pulses both aligned with the X-ray and γ-ray peaks and aligned with the magnetic poles. Detailed polarisation measurements of B1821-24 show rapid PA swings through the peaks aligned with the γ-ray peaks, as well as dips in PD and an RVM-like swing with high PD through the pole-aligned peak (Bilous et al. 2015). The lower-than-average PD of the Crab radio peaks (Table 11.1) may be consistent with this picture; however, the lack of observed radio PA (Slowikowska et al. 2015) sweep is not. The high PD of the pre-cursor as well as its appearance at the expected phase of the magnetic pole is consistent with a polar cap radio pulse.

11.3.4 Current Sheet Models

It has been only in the last 20 years that global models of the pulsar magnetosphere were developed, due to increases in computing power. Aligned (Contopoulos et al. 1999) and oblique (Spitkovsky 2006) force-free (ideal MHD) magnetosphere models were developed and describe field lines and currents that extend smoothly through the LC, and, importantly, current sheets are seen to form outside the LC. Dissipative global MHD models (Kalapotharakos et al. 2012; Li et al. 2012) that do not enforce the $\mathbf{E} \cdot \mathbf{B} = 0$ condition of force-free magnetospheres in favour of a macroscopic conductivity σ showed for the first time that the sites of the highest $E_{\parallel}$ are near the polar caps and near the current sheet. This is perhaps not surprising since the accelerating electric field is directly proportional to the current, $J = \sigma E_{\parallel}$, and the largest currents are in the current sheet. Recent particle-in-cell (PIC) simulations of global magnetospheres (Chen and Beloborodov 2014; Philippov and Spitkovsky 2014; Brambilla et al. 2018) that compute the self-consistent particle and field distributions with electron–positron pairs screening the $E_{\parallel}$ over most of the magnetosphere confirmed that the current sheet is indeed the site of most particle acceleration in an otherwise near-force-free magnetosphere. The high-energy emission from particles accelerated in the current sheet in PIC models (Kalapotharakos et al. 2018; Cerutti et al. 2016) and dissipative MHD models (Kalapotharakos et al. 2014) produce high-energy light curves that match those of *Fermi* γ-ray pulsars. Interestingly, the light curves in current sheet models have common characteristics with those of OG and TPC models. However, the caustics that form the light curves from the current sheet are different. Outside the LC, particle trajectories become radial in the inertial observer's frame since the drift velocity causes the particles to move backward along the sweptback field lines to keep their velocities subluminal. The photon emission direction is along the particle velocity, so that all the emission from the current sheet arrives in phase. An observer will see a pulse every time the line-of-sight crosses the current sheet, which is once or twice every period although some viewing angles will not have a current sheet crossing.

Polarisation of emission from the striped wind in pulsar current sheets was modelled by Pétri and Kirk (2003). They used an analytic model of the field in the current sheet to compute the light curves and polarisation of the radiation starting from different radii, r_0, from the LC up to $50R_{\rm LC}$, assuming that the emission is SR. Exploring the phase-resolved PD and PA for different α, ζ, r_0 and particle Lorentz factor combinations, they find that the phase-averaged PA is aligned with the pulsar spin axis and the phase-resolved PA makes a sharp sweep, with PD dips at the peaks. The PD between the peaks reaches as high as 28% and at the peak dips to near 0%. The size of the PD dips increase with starting radius so that the best match with the optical PD of the Crab is for $r_0 = 50R_{\rm LC}$. Contopoulos and Kalapotharakos (2010) computed polarisation of CR from a force-free current sheet, obtaining very similar PA sweep and PD dips at the peaks of the light curve. Cerutti et al. (2016) present polarisation predictions for SR emission from the current sheet of PIC simulations that extend to $3R_{\rm LC}$. They find somewhat higher PD that reaches $\sim 30\%$ between the peaks and dips to $\sim 15\%$ at the peaks, with PA sweeps at the peaks. The dips in PD are caused by the formation of caustics in emission from the current sheet while the PA swings are a result of the magnetic field polarity as the line-of-sight passes through the current sheet. Harding and Kalapotharakos (2017) compute the multi-wavelength polarisation properties of emission from near the LC and current sheet, using a force-free magnetic field structure and an emission model where the optical to hard X-ray emission is produced by SR from electron–positron pairs and the GeV emission is CR from accelerated particles (see Fig. 11.6). They find that at optical and X-ray wavelengths where SR dominates, depolarisation results from both the caustics and the gyromotion of the particles (as in the OG models of Takata

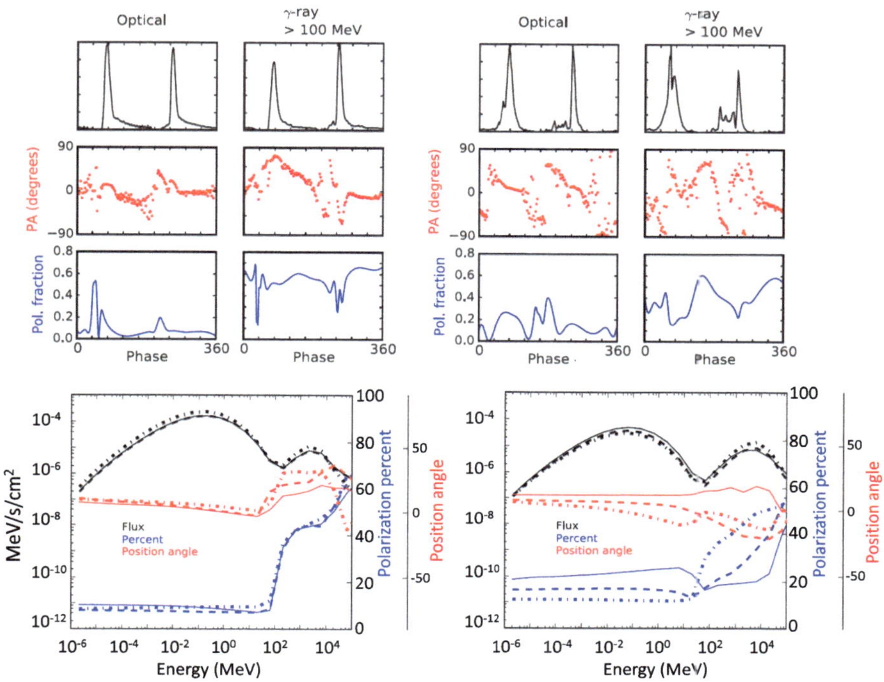

Fig. 11.6 Light curves, position angle and polarisation degree from a model of emission near the light cylinder and current sheet, for phase-resolved (top panels) and phase-averaged (bottom panels) emission for two different radius ranges: $0.7 - 1.3\ R_{LC}$ (left panels) and $1.3 - 2.0\ R_{LC}$ (right panels), for inclination angle of 60° and viewing angle 70° (from Harding and Kalapotharakos 2017)

et al. 2007). The phase-resolved PD reaches near 40% between the peaks and dips to $\sim$ 5% at the peaks, with dependence on both emission radius range and magnetic inclination angle. The GeV is more highly polarised, with PD reaching $\sim$ 60% between the peaks and dipping to ~30–40% between the peaks. In this model, the phase-averaged polarisation shows dramatic energy dependence at the transition between SR and CR, with the PD $\sim$ 10% in the optical through hard X-ray range, jumping to 40–50% above 100 MeV, while the PA lies along the pulsar spin axis in the optical and X-ray bands, but rotating by ~30–40% in the > 100 MeV band. The changes in PD and PA at the SR/CR transition are much stronger if the emission starts at altitudes near or below the LC. Such changes in polarisation from X-ray to γ-ray bands can thus signal a change in emission mechanism and distinguish between SR and CR in the GeV band.

11.4 Conclusion

The small number of pulsars with high-energy polarisation measurements is partly because very few pulsars are bright at optical wavelengths and partly because there have been few X- and γ-ray polarimeters in orbit. However, a number of planned X-ray polarimeters offer improved sensitivity in soft X-ray to γ-ray bands. In the soft X-ray band (2–10 keV), *IXPE* (*Imaging X-ray Polarimetry Explorer*) (Weisskopf et al. 2016) has been selected by NASA as a SMEX mission and will be able to reach a Minimum Detectable Polarisation (MDP) of 5% for bright sources. At higher energies, *AMEGO* (*All-Sky Medium Energy Gamma-Ray Observatory*) is a proposed NASA probe mission that will have sensitivity to polarisation in the Compton scattering (0.2–10 MeV) regime, using silicon strip trackers and CZT and CSI calorimeters. The gas Time Projection Chamber (TPC) technique has also been proposed for high-energy pair production polarimeters. *AdEPT* is a proposed γ-ray polarimeter with a projected MDP of 5% at 1–100 MeV (Hunter et al. 2014), a band that is particularly important for observing the transition between X-ray and γ-ray emission mechanisms. *HARPO* is another proposed TPC polarimeter that would be sensitive in the MeV–GeV energy range (Bernard et al. 2014). It is also possible that the *Fermi Large Area Telescope*, although not designed for polarimetry, will be sensitive to high degrees of polarisation (30–50%) above 200 MeV after 10 years of observations of bright pulsars such as Crab and Vela (Giomi et al. 2016). Such a detection would strongly suggest that CR is at least partly responsible for the observed γ-ray emission.

References

Abdo AA et al (2013) The second fermi large area telescope catalog of gamma-ray pulsars. Astrophys J Suppl 208:17

Ackermann M et al (2015) An extremely bright gamma-ray pulsar in the large magellanic cloud. Science 350:801

Arons J (1983) Pair creation above pulsar polar caps - geometrical structure and energetics of slot gaps. Astrophys J 266:215

Begelman MC, Sikora M, (1987) Inverse compton scattering of ambient radiation by a cold relativistic jet - a source of beamed, polarised conti nuum in blazars? Astrophys J 322:650

Bernard D et al (2014) HARPO: a TPC as a gamma-ray telescope and polarimeter. Proc SPIE 9144:1

Bilous AV, Pennucci TT, Demorest P, Ransom SM (2015) A broadband radio study of the average profile and giant pulses from PSR B1821 -24A. Astrophys J 803:83

Blaskiewicz M, Cordes JM, Wasserman I (1991) A relativistic model of pulsar polarisation. Astrophys J 370:643

Brambilla G et al (2018) Electron-positron pair flow and current composition in the pulsar magnetosphere. Astrophys J 858:81

Brisken WF, Fruchter AS, Goss WM, Herrnstein RM, Thorsett SE (2003) Proper-motion measurements with the VLA. II. Observations of 28 pulsars. Astron J 126:3090

Burnett CR, Melatos A (2014) Stokes tomography of a radio pulsar with an offset magnetic dipole. Mon Not R Astron Soc 440:2519

Cerutti B, Mortier JY, Philippov AA (2016) Polarised synchrotron emission from the equatorial current sheet in gamma-ray pulsars. Mon Not R Astron Soc 463:89
Chanan GA, Helfand DJ (1990) Optical polarisation of the Crab-like supernova remnant 0540-693 in the large magellanic cloud. Astrophys J 352:167–171
Chauvin M et al (2017) Shedding new light on the Crab with polarised X-rays. Sci Rep 7:7816
Chen AY, Beloborodov AM (2014) Electrodynamics of axisymmetric pulsar magnetosphere with electron-positron discharge: a numerical experiment. Astrophys J 795:22
Cheng KS, Ho C, Ruderman M (1986) Energetic radiation from rapidly spinning pulsars. I - outer magnetosphere gaps. II - Vela and Crab. Astrophys J 300:500
Contopoulos I, Kalapotharakos C (2010) The pulsar synchrotron in 3D: curvature radiation. Mon Not R Astron Soc 404:767
Contopoulos I, Kazanas D, Fendt C (1999) The axisymmetric pulsar magnetosphere. Astrophys J 511:351
Craig HA, Romani RW (2012) Altitude limits for rotating vector model fitting of pulsar polarisation. Astrophys J 755:137
Crawford F, Manchester RN, Kaspi VM (2001) Polarisation properties of nine southern radio pulsars. Astrophys J 122:2001
Daugherty JK, Harding AK (1982) Electromagnetic cascades in pulsars. Astrophys J 252:337
Deutsch AJ (1955) The electromagnetic field of an idealized star in rigid rotation in vacuo. Ann Astrophys 18:1
Dodson R, Legge D, Reynolds JE, McCulloch PM (2003) The vela pulsar's proper motion and parallax derived from VLBI observations. Astrophys J 596:1137
Dyks J (2008) Altitude-dependent polarisation in radio pulsars. Mon Not R Astron Soc 391:859
Dyks J, Harding AK (2004) Rotational sweepback of magnetic field lines in geometric models of pulsar radio emission. Astrophys J 614:869
Dyks J, Rudak B (2003) Two-pole caustic model for high-energy light curves of pulsars. Astrophys J 598:1201
Dyks J, Harding AK, Rudak B (2004) Relativistic effects and polarisation in three high-energy pulsar models. Astrophys J 606:1125
Forot M, Laurent P, Grenier IA, Gouiffés C, Lebrun F (2008) Polarisation of the Crab pulsar and Nebula as observed by the INTEGRAL/IBIS telescope. Astrophys J 688:L29
Gaensler BM, Brazier KTS, Manchester RN, Johnston S, Green AJ (1999) SNR G320.4-01.2 and PSR B1509-58: new radio observations of a complex interacting system Mon Not R Astron Soc 305:724
Giomi M et al (2016) In: Aharonian FA, Hofmann W, Rieger FM (ed) High-energy gamma-ay astronomy, AIP Conference Proceedings, vol 1792, p 70022
Hamilton PA, McCulloch PM, Manchester RN, Ables JG, Komesaroff MM (1977) Detection of change in rotation measure of the VELA pulsar. Nature 265:224
Harding AK, Kalapotharakos C (2017) Multiwavelength polarisation of rotation-powered pulsars. Astrophys J 840:80
Harding AK, Muslimov AG (2002) Pulsar polar cap heating and surface thermal X-ray emission. II. Inverse compton radiation pair fronts. Astrophys J 568:862
Helfand DJ, Gotthelf EV, Halpern JP (2001) Vela pulsar and its synchrotron Nebula. Astrophys J 556:380
Hibschman JA, Arons J (2001) Polarisation sweeps in rotation-powered pulsars. Astrophys J 546:382
Hunter S et al (2014) A pair production telescope for medium-energy gamma-ray polarimetry. Astropart Phys 59:18
Johnston S, Kramer M, Karastergiou A, Hobbs G, Ord S, Wallman J (2007) Evidence for alignment of the rotation and velocity vectors in pulsars II. Further data and emission heights. Mon Not R Astron Soc 381:1625
Kalapotharakos C, Kazanas D, Harding AK, Contopoulos I (2012) Toward a realistic pulsar magnetosphere. Astrophys J 749:2

Kalapotharakos C, Harding AK, Kazanas D (2014) Gamma-ray emission in dissipative pulsar magnetospheres: from theory to Fermi observations. Astrophys J 793:97
Kalapotharakos C et al (2018) Three-dimensional kinetic pulsar magnetosphere models: connecting to gamma-ray observations. Astrophys J 857:44
Kaplan DL, van Kerkwijk MH, Anderson J (2002) The parallax and proper motion of RX J1856.5-3754 revisited. Astrophys J 571:447
Kaplan DL, Chaterjee S, Gaensler BM, Anderson J (2008) A precise proper motion for the Crab pulsar, and the difficulty of testing spin-kick alignment for toung neutron stars. Astrophys J 677:1201
Kern B, Martin C, Mazin B, Halpern JP (2003) Optical pulse-phased photopolarimetry of PSR B0656+14. Astrophys J 597:1049
Krawczynski H (2012) The polarisation properties of inverse compton emission and implications for blazar observations with the GEMS X-ray polarimeter. Astrophys J 744:30
Krishnamohan S, Downs GS (1983) Intensity dependence of the pulse profile and polarisation of the Vela pulsar. Astrophys J 265:372
Kundu A, Petri J (2017) Pulsed emission from a rotating off-centred magnetic dipole in vacuum. Mon Not R Astron Soc 471:3359
Li J, Spitkovsky A, Tchekhovskoy A (2012) Resistive solutions for pulsar magnetospheres. Astrophys J 746:60
Mignani RP (2010) HST/WFPC2 observations of the LMC pulsar PSR B0540-69. A & A 515:A110
Mignani RP (2018) Multi-wavelength polarimetry of isolated neutron stars. Galaxies 6:36
Mignani RP et al (2007) VLT observations of the central compact object in the Vela Jr. supernova remnant. A & A 473:883
Mignani RP et al (2015) VLT polarimetry observations of the middle-aged pulsar PSR B0656+14. A & A 583:105
Mignani RP et al (2017) Evidence for vacuum birefringence from the first optical polarimetry measurement of the isolated neutron star RX J1856.5-3754. Mon Not R Astron Soc 465:492
Moffet DA, Hankins TH (1999) Polarimetric properties of the Crab pulsar between 1.4 and 8.4 GHZ. Astrophys J 522:1046
Moran P et al (2013) Optical polarimetry of the inner Crab Nebula and Pulsar. Mon Not R Astron Soc 433:2564
Moran P, Mignani RP, Shearer A (2014) HST optical polarimetry of the Vela pulsar and nebula. Mon Not R Astron Soc 445:835
Muslimov AG, Harding AK (2004) High-altitude particle acceleration and radiation in pulsar slot gaps. Astrophys J 606:1143
Ng C.-Y., Romani RW (2004) Fitting pulsar wind tori. Astrophys J 601:479
Pétri J (2017) Polarised emission from an off-centred dipole. Mon Not R Astron Soc 466:73b
Pétri J, Kirk JG (2003) The polarization of high-energy pulsar radiation in the striped wind model. Astrophys J 627:L37
Philippov A, Spitkovsky A (2014) Ab initio pulsar magnetosphere: three-dimensional particle-in-cell simulations of axisymmetric pulsars. Astrophys J Lett 785:L33
Radhakrishnan V, Cooke DJ (1969) Magnetic poles and the polarisation structure of pulsar radiation. Astrophys J Lett 3:225
Rankin JM (1990) Toward an empirical theory of pulsar emission. IV - Geometry of the core emission region. Astrophys J 352:247
Romani RW (1996) Gamma-ray pulsars: radiation processes in the outer magnetosphere. Astrophys J 470:469
Romani RW, Yadigaroglu I-A (1995) Gamma-ray pulsars: emission zones and viewing geometries. Astrophys J 438:314
Seward FD, Harnden FR Jr (1982) A new, fast X-ray pulsar in the supernova remnant MSH 15–52. Astrophys J 256:L45
Seward FD, Harnden FR Jr, Helfand DJ (1984) Discovery of a 50 millisecond pulsar in the large magellanic cloud. Astrophys J 287:L19

Slowikowska A, Kanbach G, Kramer M, Stefanescu A (2009) Optical polarisation of the Crab pulsar: precision measurements and comparison to the radio emission. Mon Not R Astron Soc 397:103

Slowikowska A, Mignani R, Kanbach G, Krzeszowski K (2012) Decomposition of the optical polarization components of the Crab pulsar and its Nebula. In: Electromagnetic radiation from Pulsars and Magnetars. Proceedings of a Conference held at University of Zielona Gora, Zielona Gora, 24–27 April 2012. Astronomical Society of the Pacific, San Francisco, p 37

Slowikowska A et al (2015) High-time-resolution measurements of the polarisation of the Crab pulsar at 1.38 GHz. Astrophys J 799:70

Spitkovsky A (2006) Time-dependent force-free pulsar magnetospheres: axisymmetric and oblique rotators. Astrophys J 648:L51

Takata J, Chang H-K (2007) Pulse profiles, spectra, and polarization characteristics of nonthermal emissions from the Crab-like pulsars. Astrophys J 670:677

Takata J, Chang H-K, Cheng KS (2007) Polarization of high-energy emission from the Crab pulsar. Astrophys J 656:1044

Tiengo A, Mereghetti S (2007) XMM-Newton discovery of 7 s pulsations in the isolated Neutron star RX J1856.5-3754. Astrophys J 657:L101

Vadawale SV et al (2018) Phase-resolved X-ray polarimetry of the Crab pulsar with the AstroSat CZT imager. Nat Astron 2:50

Venter C, Johnson TJ, Harding AK (2012) Modeling phase-aligned gamma-ray and radio millisecond pulsar light curves. Astrophys J 744:34

Wadiasingh Z, Baring MG, Gonthier PRL, Harding AK (2018) Resonant inverse compton scattering spectra from highly magnetised Neutron stars. Astrophys J 854:98

Wagner SJ, Seifert W (1999) Optical polarisation measurements of pulsars. In: Kramer M, Wex N, Wielebinski N (eds) Pulsar astronomy - 2000 and beyond, ASP Conference Series, vol 202, Proceedings of the 177th Colloquium of the IAU held in Bonn. ASP, San Francisco, p 315

Weisskopf M et al (1976) Measurement of the X-ray polarization of the Crab Nebula. Astrophys J 208:L125

Weisskopf M et al (1978) A precision measurement of the X-ray polarization of the Crab Nebula without pulsar contamination. Astrophys J 220:L117

Weisskopf M et al (2016) The imaging X-ray polarimetry explorer (IXPE). Proc SPIE 9905:990517-1

Weltevrede P et al (2010) Gamma-ray and radio properties of six pulsars detected by the fermi large area telescope. Astrophys J 708:1426

Westfold KC (1959) The polarisation of synchrotron radiation. Astrophys J 130:241

Chapter 12
Polarimetry of Magnetars and Isolated Neutron Stars

Ilaria Caiazzo, Jeremy Heyl, and Roberto Turolla

Abstract Polarised radiation from isolated neutron stars provides key diagnostics on the structure of the neutron-star magnetosphere and the properties of its atmosphere. Furthermore, the detection of a large degree of polarisation is a strong evidence for the presence of vacuum birefringence, which implies that photons of different polarisations travel at different speeds through the magnetised vacuum. We outline how polarisation is generated in neutron-star atmospheres and magnetospheres, how the polarisation is preserved as the radiation travels to us, what the current observations tell us and what are the prospects for future observations.

12.1 Introduction

When a massive star runs out of fuel in its core, nuclear energy cannot provide the pressure needed for it to stand on its own self-gravitation anymore. The collapse of the star gives birth to the spectacular and extremely energetic event called a supernova. A supernova, however, does not destroy the star completely; it leaves over what remains of the star's compact core: either a black hole or a newly born neutron star. The mass, composition and other factors determine in a complicated fashion whether the death of a massive star results in a neutron star or a black hole. Being born in such a dramatic event, a neutron star presents peculiar characteristics, which makes it a uniquely interesting object: it is the densest, most compact object in the universe that we know of, having a mass that is about the mass of the Sun,

I. Caiazzo · J. Heyl (✉)
Department of Physics and Astronomy, University of British Columbia, Vancouver, BC, Canada
e-mail: ilariacaiazzo@phas.ubc.ca; heyl@phas.ubc.ca

R. Turolla
Department of Physics, University of Padova, Padova, Italy

Mullard Space Science Laboratory, UCL, Dorking, UK
e-mail: turolla@pd.infn.it

R. Mignani et al. (eds.), *Astronomical Polarisation from the Infrared to Gamma Rays*, Astrophysics and Space Science Library 460,
https://doi.org/10.1007/978-3-030-19715-5_12

but a radius of only about 10 km; it rotates at rates as fast as 700 times per second; it has incredibly strong magnetic fields, up to 10^{15} G. All these extremes, of gravity, density and magnetic field, cannot ever be achieved in any laboratory on Earth, and therefore, neutron stars represent exquisite laboratories to test fundamental physics.

Observations of neutron stars are abundant in the radio band, as the easiest way to find a neutron star is to look for its radio emission. Indeed, to date, more than 2300 radio pulsars have been found. Even though there is not a universally accepted model for the pulsar emission mechanism, the pulsating nature of the emission is generally interpreted as a beamed radiation, generated in the proximity of the pulsar's magnetic poles, sweeping across the line of sight of the observer as the neutron star rotates. Neutron stars have been observed to emit at higher energies as well, from infrared to gamma rays. Neutron star spectra are in general featureless, as few absorption lines have been observed to date. Some observations in the X-rays have yielded tantalising hints of absorption lines in accreting and isolated neutron stars.

The polarisation signal, on the other hand, is rich in features and carries a wealth of information about the structure of the star and of the magnetic field. Our focus is on isolated neutron stars, so we will examine the theoretical questions of how is polarised radiation generated and propagated through neutron-star atmospheres and magnetospheres. We will not discuss the (accreting) X-ray and gamma-ray pulsars where the emission is unlikely to originate from the surface. We encourage the reader to consult (Mignani 2018) for an up-to-date review of the observational landscape of the polarisation from isolated neutron stars.

12.2 Generation (and Destruction) of Polarised Radiation

The polarised emission of isolated neutron stars comes from their surface, either from an atmosphere (Sect. 12.2.1) or from a condensed surface (Sect. 12.2.2), and thus we will focus on this region first. Afterwards (Sect. 12.2.3), we will see how scattering in the magnetosphere can diminish the extent of polarisation. In section (Sect. 12.3), we will consider how polarisation can change as radiation propagates from the surface (or past the scattering region) to the observer.

12.2.1 Neutron-Star Atmospheres

The emission processes in a neutron star's atmosphere are strongly influenced by the magnetic field. Isolated neutron stars have magnetic fields that range from 10^{11} to 10^{15} G; such strong magnetic fields can constrain the motion of particles in the atmosphere and the geometry of emission.

In the atmosphere of a typical neutron star, the temperature is much less than the electron cyclotron energy,

$$kT = 0.086\,\text{keV}\frac{T}{10^6\text{ K}} \ll \hbar\omega_c = \hbar\frac{eB}{m_e c} = 11.6\,\text{keV}\frac{B}{10^{12}\text{ G}}, \tag{12.1}$$

and much higher than the proton cyclotron energy ($\sim 6\,\text{eV}$ at 10^{12} G). This means that the typical photon energy is not sufficient to excite motion across the magnetic field lines and the scattering and absorption cross sections depend strongly on the polarisation state of the photon and its direction of motion (Canuto et al. 1971). Furthermore, as the typical electron energy is also much smaller than the cyclotron energy, the electrons are found in the ground Landau level and are restricted to move along the field lines.

The cyclotron energy is also much larger than the typical energy of electrons in atoms; therefore, we expect that the structure and binding energies of atoms, if atoms indeed exist at the surfaces of neutron stars, to be dramatically different (Ruderman 1971; Thirumalai and Heyl 2014), so even small atoms such as hydrogen may have a significant neutral fraction in the high temperatures of the neutron-star atmosphere. The composition of the surface of isolated neutron stars is uncertain, and therefore current atmospheres models span a wide range of compositions: hydrogen (Suleimanov et al. 2009), helium (Medin et al. 2008), carbon (Suleimanov et al. 2014), mid-Z elements (Mori and Ho 2007) and iron (Rajagopal et al. 1997). For simplicity, we will consider fully ionised hydrogen atmospheres in the discussion that follows, but the general polarisation properties of emerging radiation depend on the geometry of the polarisation states and on how they interact with free and bound electrons, so the results for hydrogen are illustrative of other compositions.

In the atmosphere, if a photon propagates in a direction that is perpendicular to the field, and its energy is far from the cyclotron energies, its polarisation modes will remain nearly linear within the atmosphere, so that the transverse component of the electric field of the wave is either within the plane containing the magnetic field direction and the wave vector (parallel or ordinary mode) or perpendicular to that plane (perpendicular or extraordinary mode). Also, if the radiation is in the extraordinary mode, there is a small longitudinal component (E_L) along the direction of propagation: the ratio to the transverse field (E_T) is given by Mészáros (1992)

$$\left|\frac{E_L}{E_T}\right| = \left|\frac{\omega_p^2\omega_c}{\omega\left(\omega^2 - \omega_c^2 - \omega_p^2\right)}\right| \approx \frac{\omega_p^2}{\omega\omega_c} \text{ for } \omega \ll \omega_c, \tag{12.2}$$

where $\omega_p = (4\pi n_e e^2/m_e)^{1/2}$ is the plasma frequency, and n_e is the number density of electrons. This longitudinal electric field is typically smaller by a large factor relative to the transverse field. On the other hand, if the photon is propagating along the field, there is no longitudinal component of the photon's electric field, and the transverse electric field in both modes is perpendicular to the magnetic field, and cannot accelerate the electrons unless the energy is close to the cyclotron energy.

If we focus on photons travelling nearly perpendicular to the magnetic field direction (the angle between the direction of propagation of the photon and the magnetic field θ is about $\pi/2$), the non-relativistic scattering cross sections for the two polarisation modes, ordinary (1) and extraordinary (2), become (Canuto et al. 1971; Hamada and Kanno 1974; Kanno and Hamada 1975; Hamada 1975, 1980; Herold 1979; Stoneham 1980; Kostenko and Thompson 2018):

$$\sigma_1 \approx \sigma_T \sin^2\theta \tag{12.3}$$

$$\sigma_2 \approx \sigma_T \left(\frac{\omega^2}{(\omega_c - \omega)^2} + \cos^2\theta \right). \tag{12.4}$$

where in the ordinary mode, the cross section tends to the Thomson cross section (σ_T), while in the extraordinary mode the transverse electric field can only excite the electrons close to the cyclotron resonance. For radiation that propagates approximately along the magnetic field (θ about zero), the cross section for both modes is reduced dramatically, as in both cases the electric field is perpendicular to the magnetic field:

$$\sigma_1 \approx \sigma_T \left(\frac{\omega^2}{(\omega_c + \omega)^2} + \frac{1}{2}\sin^2\theta \right) \tag{12.5}$$

$$\sigma_2 \approx \sigma_T \left(\frac{\omega^2}{(\omega_c - \omega)^2} + \frac{1}{2}\sin^2\theta \right). \tag{12.6}$$

A similar result holds for intermediate angles. As ω gets closer to ω_c, the extraordinary mode's cross section increases, until it becomes larger than the ordinary mode's. Very close to ω_c however, the energy transfer from photons heats up the electrons and Eqs. (12.3)–(12.6) are no longer valid, as damping effects become important (Mészáros 1992).

Because the reduction in the cross section results from the restriction of the electron motion along the field lines and the geometry of the polarisation modes, the cross sections for other processes such as free–free, bound-free and atomic transitions also depend on the polarisation state and the direction of propagation (Pavlov and Kaminker 1975; Daugherty and Ventura 1978; Cadez and Javornik 1981; Nagel and Ventura 1983; Lieu 1983; Potekhin and Lai 2007). The properties of the emission from the atmosphere of neutron stars will depend sensitively on the strength of the magnetic field and its direction relative to the vertical. Within the neutron-star atmosphere both scattering and absorption play a role.

To illustrate the various effects on the generation of polarisation, we will consider a simple plane-parallel atmosphere consisting of magnetised, fully ionised hydrogen from Lloyd (2003). The neutron-star atmosphere is incredibly thin compared to the radius of the star (centimetres vs. kilometres), so the plane-parallel approximation is appropriate; however, across the surface of the star, the magnetic field will vary in magnitude and direction, so the flux emergent through the surface will also depend on the location. For a comprehensive review of neutron-star atmospheres in general, we direct the reader to Potekhin (2014). To calculate the emission from the entire surface, a set of neutron-star atmosphere models must be computed accounting also for the surface temperature distribution. How to add the contribution from the different surface patches is subtle as far as polarisation is concerned and we present results from such a summation in Sect. 12.3.4 and Fig. 12.9.

We will first focus on the situation where the magnetic field is perpendicular to the surface. Figure 12.1 illustrates the various processes at play. The photosphere here is intended as the layer in the atmosphere where the total optical depth is unity. The ordinary mode (indicated as ∥, in orange) is less strongly affected by the magnetic field, so we will start by addressing its behaviour. For small photon energies, the opacity is dominated by free–free absorption, which decreases with photon energy as E^{-2}, so higher energy photons decouple deeper within the atmosphere, as long as their energy is still below the limit at which electron-scattering opacity starts to dominate over free–free opacity. Above this limit (at about $10^{0.5}$ keV in Fig. 12.1), the constant electron-scattering opacity dominates, and the density of decoupling approaches a constant value.

For the extraordinary mode (⊥, in blue) the trends are somewhat more complicated. We will discuss the trends in Fig. 12.1 in conjunction with the polarisation fraction depicted in Fig. 12.2, because the polarisation of the emergent radiation clearly reflects the relative location of the two photospheres. For energies below

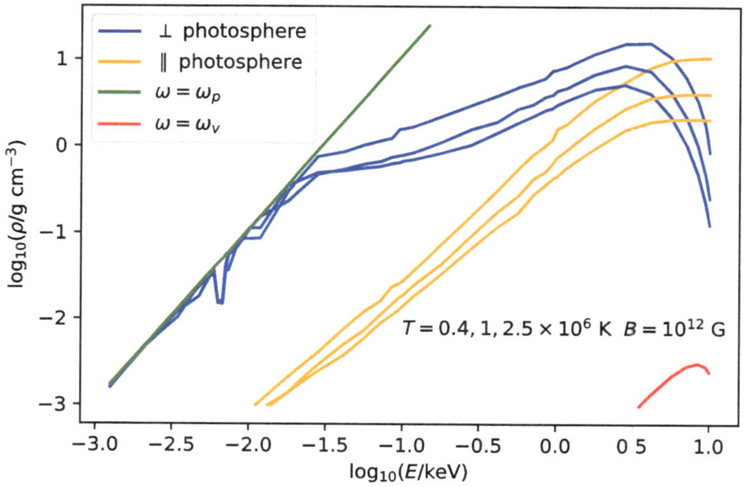

Fig. 12.1 An illustration of the locations of the parallel and perpendicular model photospheres for $B = 10^{12}$ G and $T = 0.5, 1, 2.5, \times 10^6$ K using the models of Lloyd (2003). The radiation in higher temperature models decouples deeper within the star at larger densities

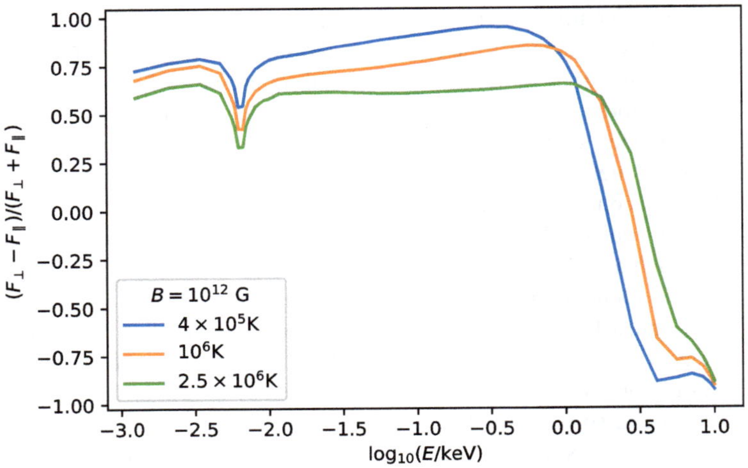

Fig. 12.2 The polarisation fraction as a function of energy for $B = 10^{12}$ G and $T = 0.5, 1, 2.5, \times 10^6$ K using the models of Lloyd (2003)

the proton cyclotron resonance, the opacity is so small that the photosphere lies at the plasma frequency, quite deep in the star compared to the ordinary mode's photosphere, as the collective oscillations of the plasma dominate the generation of the extraordinary photons. The polarisation fraction at these energies is quite high. At the proton cyclotron resonance, the cross sections for scattering and absorption increase, drawing the photosphere to shallower depths and we can see it both in Fig. 12.1 and in Fig. 12.2 as a dip at about 6 eV. The ordinary mode does not interact with the cyclotron resonances, so it is not affected at this energy. Above the proton cyclotron resonance, the photosphere follows the plasma frequency until free–free absorption takes over. Because the electric-field geometry of the extraordinary mode depends on the photon energy, the dependence of the free–free opacity with energy is shallower in this mode, so the depth of the photosphere does not increase as quickly as for the ordinary mode. The two photospheres approach each other. This reduces the extent of polarisation in the total emission. Finally, as the photon energy approaches the electron cyclotron resonance at 11.6 keV, the opacity for the extraordinary mode increases dramatically because of the resonant cross section, and the photosphere of the extraordinary mode lies above that of the ordinary mode and the direction of the net polarisation switches to the ordinary mode (the fraction becomes negative in Fig. 12.2).

We now consider the general case, where the direction of the magnetic field can vary. We can see from the angular dependence of the scattering cross sections in Eq. (12.3) through (12.6) that the cross section is dramatically decreased when the photon is travelling along the direction of the magnetic field. If we examine radiation travelling along the magnetic field direction it should decouple at a larger depth and at a higher density than radiation travelling in other directions. Furthermore, away

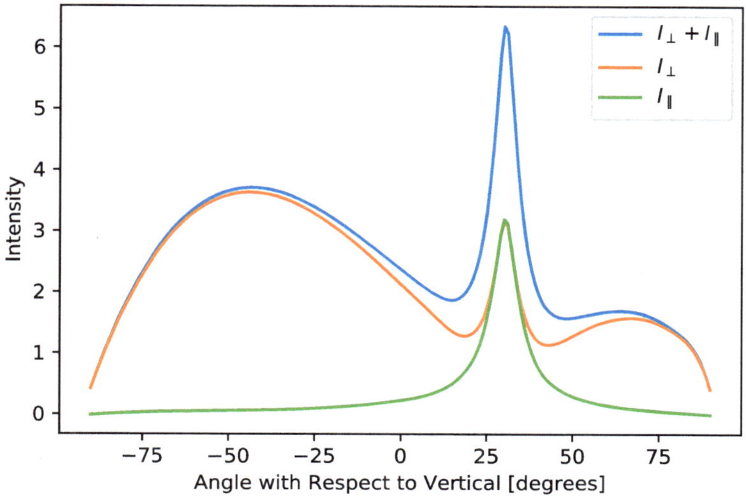

Fig. 12.3 The polarised intensity at $E = 0.32\,\text{keV}$, near the spectral peak, along a slice through sky containing the magnetic field direction. The magnetic field is directed at 30° from the vertical. The atmosphere is calculated for $B = 10^{12}\,\text{G}$ and $T = 10^6\,\text{K}$ using the models of Lloyd (2003). The units of intensity are 10^{19} erg/cm^2/s/sr/keV

from the cyclotron line, the contribution by the two polarisation states of photons travelling along the magnetic field should be nearly equal as their cross sections are also nearly equal. Figure 12.3 depicts the specific intensity for radiation near the peak of the spectrum as a function of the angle between the propagation direction and the vertical for the case in which a magnetic field of 10^{12} G is directed at 30° away from the vertical. The radiation is nearly fully polarised in the extraordinary mode direction and approximately isotropic except for very shallow angles where the intensity is diminished (limb-darkening) and within about 10° of the direction of the magnetic field, where the intensity is much larger and the radiation is not polarised. In this general case, the intensity depends not only on the zenith angle but also on the azimuthal angle relative to the local magnetic field direction. This dramatically increases the numerical effort in calculating a spectral model both relative to the unmagnetised case and to the situation where the magnetic field points in the vertical direction. It is this latter, more restrictive situation that is most often treated in the literature (Ho et al. 2003; Suleimanov et al. 2009, 2012; González-Caniulef et al. 2019), even thought many works, including the model employed in this section, consider a varying inclination of B (Pavlov et al. 1994; Lloyd 2003; Zane and Turolla 2006; González Caniulef et al. 2016).

12.2.2 Condensed Neutron-Star Surfaces

The properties of matter that form the surface regions of neutron stars are strongly affected by the strong magnetic fields (Ruderman 1971). Neutron stars may have a solid surface (Brinkmann 1980; Lai and Salpeter 1997; Lai 2001; Burwitz et al. 2003; Turolla et al. 2004; van Adelsberg et al. 2005; Medin and Lai 2006, 2007; Potekhin 2014); in this case, the emission will essentially depend on the reflectivity of the metallic surface. If one focuses on the interface between the vacuum outside and the condensed surface, one can argue by detailed balance that the intensity emerging from the surface is given by

$$I_{\nu,\mathrm{X/O}}(\theta_k) = \left(1 - R_{\nu,\mathrm{X/O}}(\theta_k)\right) B_\nu(T), \tag{12.7}$$

where $B_\nu(T)$ is the intensity of blackbody radiation at the temperature of the surface and θ_k is the angle with respect to the surface normal. The typical density of the condensed surface is $\approx 10^2$–10^3 g/cm^3, significantly larger than terrestrial metals, and the plasma frequency within the surface is about 1 keV, so even at X-ray energies we would expect the surface to be highly reflective and the emissivity $(1 - R)$ to be small. However, there are some additional complications as the metal is highly magnetised and the ions can damp the radiation within the metal.

To address these complications we can relate the reflectivity to the electromagnetic modes within the condensed surface which is essentially a magnetised plasma (see Sect. 12.3.2). Using the Fresnel equations which establish the boundary conditions across the surface, we find the reflectance of the p-Polarisation is

$$R_\mathrm{p} = \left|\frac{\cos\theta_t - n\cos\theta_k}{\cos\theta_t + n\cos\theta_k}\right|^2, \tag{12.8}$$

and for the s-Polarisation is

$$R_\mathrm{s} = \left|\frac{\cos\theta_k - n\cos\theta_t}{\cos\theta_k + n\cos\theta_t}\right|^2, \tag{12.9}$$

where $n\sin\theta_t = \sin\theta$ (Snell's Law) and n is the index of refraction within the condensed material that forms the surface. We have neglected the index of refraction above the surface and the magnetic permeability of the material. The Fresnel equations are defined in terms of the polarisation states relative to the interface. The labels s and p refer to whether the radiation is polarised with its electric field in the plane containing the incoming ray and the normal to the surface (p) or perpendicular to it (s). The s-Polarisation is parallel to the interface itself. Both above the surface and within the condensed material the propagation modes are determined relative to the magnetic field direction. Both regions are birefringent, so at the interface there are two reflected waves and two transmitted waves, so the complete picture is composed of a reflection coefficient for the X-mode to reflect

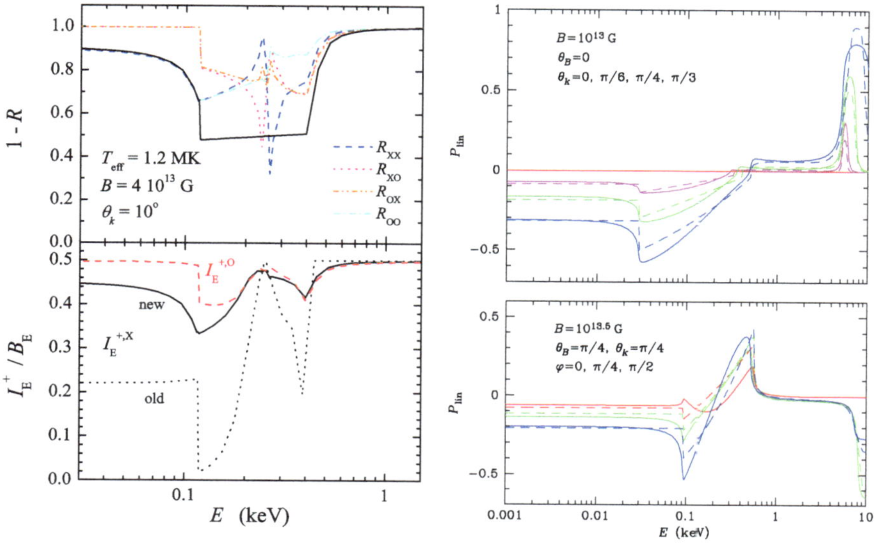

Fig. 12.4 Left: The four surface reflectivities as a function of energy (upper) and the resulting emissivities in the X and O modes (lower). Right: Degree of linear polarisation as a function of photon energy E for condensed Fe surface from Fig. 7 of Potekhin et al. 2012. The upper panel depicts the case where the magnetic field is normal to the surface $\theta_B = 0$ and several directions of the photon relative to the normal direction θ_k: 0° (red), 30° (magenta), 45° (green) and 60° (blue). The lower panel holds the magnetic field and photon direction at 45° relative to the normal and examines the emission as a function of the azimuthal direction: 0° (red), 45° (green) and 90° (blue). The solid lines depict the numerical results, and the dashed lines show a fit

into the X-mode, for the X-mode to reflect into the O and the other possibilities as well as the corresponding transmission coefficients that can be obtained by expanding the propagation modes in terms of the modes defined at the surface.

The left panel of Fig. 12.4 depicts the reflectivities and the emissivity for a condensed iron surface with a thin hydrogen atmosphere above it from Potekhin et al. (2012). The feature in the reflectivities at 0.25 keV results from the proton cyclotron line that affects the polarisation states in the region above the condensed surface. The increase in the reflectivities at about 0.125 keV corresponds to the ion cyclotron frequency within the surface. The feature at 0.4 keV is given by

$$E_C = E_{c,i} + \frac{E_{p,e}^2}{E_{c,e}} \tag{12.10}$$

within the condensed surface. These two frequencies feature strongly in the observed polarisation signature from the surface as local extrema in the polarisation fraction. The right panel depicts the results also from Potekhin et al. (2012) for a condensed iron surface without an atmosphere above it. We see that the extent of polarisation is typically much smaller than for an atmosphere. Furthermore the direction of the polarisation depends on the energy of the photon, and the photon energy where the polarisation switches from the ordinary to the extraordinary mode depends on the strength of the magnetic field through the electron and ion cyclotron frequencies and on the density of the surface layers through the plasma frequency (van Adelsberg et al. 2005). Although the total polarisation is somewhat lower than for the case of the neutron-star atmosphere, the condensed surface leaves many exciting signatures on the polarisation, most importantly that the surface is indeed condensed.

12.2.3 Neutron-Star Magnetospheres

Although the typical density of the plasma in the magnetosphere is too low for many photons to be produced there, the cross section for scattering of photons from the surface may be large within the cyclotron resonance. The observed spectra of isolated neutron stars are characterised by one or two thermal components (below about 1 keV) (Burwitz et al. 2003; Sartore et al. 2012) and possibly a power-law component declining towards higher energies (above 1 keV) especially in magnetars (Marsden and White 2001; Kaspi and Boydstun 2010). The spectra of magnetars often have an additional power-law component that becomes important from 10 keV to 100 keV (Kuiper et al. 2004; Mereghetti et al. 2005; Götz et al. 2006; Beloborodov and Thompson 2007). The source of the low-energy power-law component is believed to be resonant inverse Compton scattering (RCS) onto mildly relativistic electrons/positrons flowing in the twisted magnetosphere (Thompson et al. 2002; Lyutikov and Gavriil 2006). The origin of the harder power-law component is less clear, although it might still be related to RCS, possibly onto a different charge population(s) (Rea et al. 2007, 2008; Baring and Harding 2007; Fernández and Thompson 2007; Beloborodov 2013; Baring et al. 2017; Wadiasingh et al. 2018).

12.2.3.1 RCS in Magnetars

In magnetars the expected particle density due to charges flowing along the magnetic field lines, required to sustain the non-potential field, is too low to build a sizeable Thomson scattering depth (Thompson et al. 2002). On the other hand, this could be easily achieved in the cyclotron resonance (Thompson et al. 2002; Rea et al. 2007, 2008; Baring and Harding 2007; Fernández and Thompson 2007; Nobili et al. 2008a,b; Zane et al. 2009, 2011; Beloborodov 2013; Baring et al. 2017; Wadiasingh et al. 2018). Given the typical energy of the photons from the surface, the cyclotron energy at scattering should be about 1 keV, much less than the rest-mass energy of the electron, so the scattering region must lie at several stellar radii, where the magnetic field is about $10^{-2}-10^{-1}B_{\mathrm{QED}}$. In the frame of the electron, the scattering is non-relativistic and the incoming photon is travelling nearly along the magnetic field lines, so we can use the cross sections from Eqs. (12.5) and (12.6) to understand how the Compton scattering affects the two polarisation states. In particular, for photons travelling along and across the field we see that only the extraordinary or perpendicular mode is resonantly scattered. Therefore, the incoming radiation from the atmosphere, which is mostly polarised perpendicular to the magnetic field direction, remains polarised perpendicular to the field direction after the resonant scattering, and then we expect the high-energy power-law component to be strongly polarised perpendicular to the local magnetic field direction as well. From a more detailed treatment (Taverna et al. 2014), which includes geometric considerations, one find that the resonant scattering can switch the polarisation states. In fact, the cross sections are related in the following way (Herold 1979; Nobili et al. 2008b):

$$\sigma_{1-1} = \frac{1}{3}\sigma_{1-2}, \quad \sigma_{2-2} = 3\sigma_{2-1}, \tag{12.11}$$

where again we use the convention that (1) indicates the ordinary or parallel mode and (2) indicates the extraordinary or perpendicular mode. The resulting emission is highly polarised but less than fully polarised. If one assumes that the initial photons are completely in the extraordinary mode, after a single scattering the polarisation fraction is reduced by 50% and it decreases with subsequent scatterings (Fernández and Thompson 2007). One can conclude that the scattering process typically destroys the polarisation and we shall see this in Sect. 12.4.2.

12.3 Propagation of Polarised Radiation

After a photon is emitted in the atmosphere of an isolated neutron star, its polarisation can change as it propagates through the magnetosphere, even if there are no scattering events. In this section, we will analyse two effects that can rotate a photon's polarisation: the QED effect of vacuum birefringence and plasma birefringence. Furthermore, the competition between these two effects can flip the polarisation direction of a photon if its energy is close to the vacuum resonance.

Throughout this section, in order to describe the evolution of the polarisation of a photon, we will employ the Poincaré sphere formalism: the polarisation of fully polarised light is described by a unit vector

$$\mathbf{s} = \frac{1}{S_0}(S_1, S_2, S_3) = \frac{1}{I}(Q, U, V), \tag{12.12}$$

where S_0, S_1, S_2, S_3 (or I, Q, U, V) are the Stokes parameters and the polarisation states are mapped on the surface of a sphere. In this formalism, Kubo and Nagata derived the equation of motion of the polarisation direction on the Poincaré sphere as light travels through an inhomogeneous birefringent and dichroic medium (Kubo and Nagata 1981, 1983). They obtained the following expression for the evolution of the polarisation of a wave:

$$\frac{\partial \mathbf{s}}{\partial x_3} = \hat{\boldsymbol{\Omega}} \times \mathbf{s} + (\hat{\mathbf{T}} \times \mathbf{s}) \times \mathbf{s}, \tag{12.13}$$

where $\hat{\boldsymbol{\Omega}}$ is the birefringent vector, $\hat{\mathbf{T}}$ is the dichroic vector and x_3 is the length of the photon path through space.

12.3.1 Vacuum Birefringence

Quantum electrodynamics or QED is the relativistic quantum field theory of electrodynamics. It is usually thought to apply only to the realm of the very small; however, its effects can be important on macroscopic scales in extreme environments, like the ones attained around neutron stars. In classical electrodynamics, the polarisation of a photon travelling in vacuum is not affected by the presence of a uniform magnetic field, as Maxwell's equations are linear in the fields and, therefore, light does not interact with light. In QED, the vacuum is filled with virtual particles, electron and positron pairs, and therefore, the Lagrangian presents an additional component which is more than quadratic in the fields. Still, a photon's polarisation does not change as it travels in vacuum unless a strong magnetic field is present: exciting virtual electrons in a direction parallel to the magnetic field is easier than in the perpendicular direction and for this reason, if the direction of a photon's electric field has a component parallel to the local magnetic field, the photon propagates slower than a photon with the opposite polarisation. In other words, in the presence of a strong magnetic field, the vacuum acquires an index of refraction that is different depending on the polarisation direction of the photon: the vacuum becomes birefringent.

This effect was first predicted independently by Heisenberg and Euler (1936) and by Weisskopf (1936), and subsequently derived in full QED by Schwinger (1951). The full Lagrangian derived by Schwinger is given by

$$\mathcal{L} = -F - \frac{1}{8\pi}\int_0^{\infty} dss^{-3}\exp\left(-m^2 s\right)\left\{(es)^2 G\frac{\Re[\cosh(esX)]}{\Im[\cosh(esX)]} - 1 - \frac{2}{3}(es)^2 F\right\}, \tag{12.14}$$

where we can see the usual Lagrangian for electrodynamics, $-F = -\frac{1}{4}F_{\mu\nu}F^{\mu\nu} = -\frac{1}{2}(\boldsymbol{B}-\boldsymbol{E})^2$, and the additional term due to QED, in which $G = \frac{1}{4}F_{\mu\nu}F^{*}_{\mu\nu} = \boldsymbol{E}\cdot\boldsymbol{B}$ and $X^2 = 2(F + iG) = (\boldsymbol{B} + i\boldsymbol{E})^2$. In these expressions, e and m are the charge and mass of the electron, and $F_{\mu\nu}$ and $F^{*}_{\mu\nu}$ are the electromagnetic field tensor and its dual.

The limit for weak magnetic fields ($B \ll B_{\rm QED}$) of Eq. (12.14) corresponds to the effective Lagrangian found by Heisenberg, Euler and Weisskopf:

$$\mathcal{L}_{\rm eff} \simeq \frac{1}{2}(\mathbf{E}^2 - \mathbf{B}^2) + \frac{2\alpha_{\rm QED}^2}{45}\frac{(\hbar/mc)^3}{mc^2}[(\mathbf{E}^2 - \mathbf{B}^2)^2 + 7(\mathbf{E}\cdot\mathbf{B})^2], \tag{12.15}$$

where $\alpha_{\rm QED} = e^2/(\hbar c)$ is the fine-structure constant. By expanding the action for a uniform magnetic field plus a small photon field, Heyl and Hernquist (1997) found the difference between the indices of refraction to be, for $B \ll B_{\rm QED}$ and $E = 0$:

$$n_{\parallel} - n_{\perp} = \frac{\alpha_{\rm QED}}{30\pi}\left(\frac{B}{B_{\rm QED}}\right)^2 \sin^2\theta, \tag{12.16}$$

where in this case, $\parallel$ indicates the polarisation mode in which the photon's electric field has a component parallel to the local magnetic field and vice versa. The value of $B_{\rm QED} = m^2c^3/(e\hbar)$ is about 4.4×10^{13} G.

In the case of the QED vacuum with an external magnetic field to one loop, weak electric field and $\hbar\omega < m_e c^2$, the dichroic vector $\hat{\mathbf{T}} = 0$ (there is no real pair production) and the amplitude of the birefringence vector $\hat{\boldsymbol{\Omega}}$ is proportional to the difference between the indices of refraction for the two polarisation states and to the unperturbed wave number of the photon $k_0 = 2\pi\nu/c$:

$$\hat{\Omega} = k_0(n_{\parallel} - n_{\perp}) = k_0\frac{\alpha_{\rm QED}}{30\pi}\left(\frac{B}{B_{\rm QED}}\right)^2 \sin^2\theta. \tag{12.17}$$

As photons travel in a strong magnetic field, their polarisation can change along the path due to vacuum birefringence. Numerical integrations of Eq. (12.13) show that, for strong enough magnetic fields, the photon polarisation modes are decoupled, and the direction of polarisation follows the direction of the magnetic field (Cheng and Ruderman 1979; Heyl and Shaviv 2000; Fernández and Davis 2011; Taverna et al. 2014, 2015). This effect becomes important when

$$\left|\hat{\boldsymbol{\Omega}}\left(\frac{1}{|\hat{\boldsymbol{\Omega}}|}\left|\frac{\partial\hat{\boldsymbol{\Omega}}}{\partial x_3}\right|\right)^{-1}\right| \gtrsim 0.5. \tag{12.18}$$

From the same expression, we can find the distance from the source at which the polarisation stops following the magnetic field and becomes frozen with respect to the radius of curvature. This distance is called the polarisation-limiting radius or the adiabatic radius.

12.3.2 Plasma Birefringence

At low photon energies or high plasma densities, the plasma may play an important role in the propagation of polarised radiation through the atmospheres and magnetospheres of neutron stars. The typical energy where plasma and vacuum trade-off is in the infrared if we assume that the density of the plasma is approximately the Goldreich and Julian (1969) density, but it will increase with the density of the plasma. If the magnetosphere carries substantial currents as in magnetars (e.g. Thompson et al. 2002), the effect of the plasma will be larger and more important. The combination of plasma and vacuum birefringence results in an eigenvalue equation for the complex amplitudes of the electric field (Mészáros 1992),

$$\begin{bmatrix} \eta_{xx} - n^2 & \eta_{xy} \\ \eta_{yx} & \eta_{yy} - n^2\rho \end{bmatrix} \begin{bmatrix} E_x \\ E_y \end{bmatrix} = 0, \tag{12.19}$$

where η_{ij} are components of the dielectric tensor and we have assumed that the photon propagates along the z-axis and that the magnetic field lies in the $x - z$-plane. The parameter ρ,

$$\rho = 1 - \frac{4\alpha_{\text{QED}}}{45\pi}\left(\frac{B}{B_{\text{QED}}}\right)^2 \sin^2\theta, \tag{12.20}$$

accounts for the magnetisation of the vacuum and the tensor η characterises the dielectric response of the plasma and vacuum. Since we are considering photon frequencies well above the plasma frequency ($\omega \gg \omega_p = \sqrt{4\pi e^2 n/m}$) and far below the cyclotron resonance ($\omega_c = eB/(mc)$), the eigenvalues of the matrix, n_1^2 and n_2^2, are of the order unity. These conditions hold in the neutron star magnetosphere for the photon energies from the visual to the X-rays but do not necessarily hold in the atmosphere.

We can translate this into the formalism of Kubo and Nagata (1983) by noting that the magnitude of $\hat{\Omega}$ is related to the two eigenvalues of the matrix ($n_{1,2}^2$), in particular $|\hat{\boldsymbol{\Omega}}| = k_0 \left|\Delta n_{\parallel\perp}\right|$, where

$$\Delta n_{\parallel\perp} = n_\parallel - n_\perp = \frac{n_\parallel^2 - n_\perp^2}{n_\parallel + n_\perp} \approx \frac{n_\parallel^2 - n_\perp^2}{2} \tag{12.21}$$

$$\approx \frac{1}{2}\left[\left(\eta_{xx} - \rho^{-1}\eta_{yy}\right)^2 + 4\rho^{-1}|\eta_{xy}|^2\right]^{1/2} \tag{12.22}$$

$$\approx \sin^2\theta\left[\frac{\alpha_{\text{QED}}}{30\pi}\left(\frac{B}{B_{\text{QED}}}\right)^2 - \frac{1}{2}\frac{\omega_p^2}{\omega^2}\frac{\omega_c^2}{\omega_c^2 - \omega^2}\right] \tag{12.23}$$

$$\approx \sin^2\theta\frac{\alpha_{\text{QED}}}{30\pi}\left(\frac{B}{B_{\text{QED}}}\right)^2\left[1 - \frac{\omega_p^2}{\omega^2}\frac{1}{v_r}\right], \tag{12.24}$$

where

$$v_r = \frac{\alpha_{\text{QED}}}{15\pi}\left(\frac{B}{B_{\text{QED}}}\right)^2 \frac{\omega_c^2 - \omega^2}{\omega_c^2} \tag{12.25}$$

and where we have assumed $n_1, n_2 \approx 1$.

If $\omega^2 v_r \approx \omega_p^2$, the value of $\Delta n_{\parallel\perp}$ vanishes; in this regime (the vacuum resonance), the modes of the plasma plus vacuum are approximately circular and in the resonance, the difference in index of refraction between the two helicities is given by

$$\Delta n_{\pm,\text{res}} = [n_+ - n_-]_{\text{res}} = -\cos\theta \frac{\omega\omega_p^2}{\omega_c\left(\omega^2 - \omega_{c,i}^2\right)}, \tag{12.26}$$

where $\omega_{c,i} = Z_i eB/(m_i c)$ is the ion cyclotron frequency. In general, the helicity of the photon is an adiabatic invariant for the polarisation states and we have

$$(\Delta n_\pm)^2 = \left(\Delta n_{\pm,\text{res}}\right)^2 + \left(\Delta n_{\parallel\perp}\right)^2 . \tag{12.27}$$

The direction of $\hat{\Omega}$ is given by the direction of the eigenvector with the larger eigenvalue on the Poincaré sphere. The eigenvectors of Eq. (12.19) yield the polarisation vectors

$$\mathbf{e}_+ = [i\cos\theta_m, \sin\theta_m, 0]\,, \ \mathbf{e}_- = [-i\sin\theta_m, \cos\theta_m, 0]\,, \tag{12.28}$$

where we have labelled the states by their helicity rather than parallel and perpendicular. The angle θ_m lies between 0 and $\pi/2$ and characterises the mixing between the parallel and perpendicular polarisation states. We see that for $\theta_m = 0$ in the + state the electric field is parallel to the x-axis and for $\theta_m = \pi/2$ it is parallel to the y-axis. We can determine the value of θ_m from the ratio of the linear to the circular portion of the birefringence

$$\tan 2\theta_m = \frac{\Delta n_{\pm,\text{res}}}{\Delta n_{\parallel\perp}}. \tag{12.29}$$

We map $\mathbf{e}_1$ onto the Poincarè sphere using the definitions

$$I = S_0 = |\mathbf{e}_1 \cdot \hat{x}|^2 + |\mathbf{e}_1 \cdot \hat{y}|^2 = 1 \tag{12.30}$$

$$Q = S_1 = |\mathbf{e}_1 \cdot \hat{x}|^2 - |\mathbf{e}_1 \cdot \hat{y}|^2 = \cos 2\theta_m \tag{12.31}$$

$$U = S_2 = \left|\mathbf{e}_1 \cdot \frac{1}{\sqrt{2}}\left(\hat{x} + \hat{y}\right)\right|^2 - \left|\mathbf{e}_1 \cdot \frac{1}{\sqrt{2}}\left(\hat{x} - \hat{y}\right)\right|^2 = 0 \tag{12.32}$$

$$V = S_3 = \left|\mathbf{e}_1 \cdot \frac{1}{\sqrt{2}}\left(\hat{x} + i\hat{y}\right)\right|^2 - \left|\mathbf{e}_1 \cdot \frac{1}{\sqrt{2}}\left(\hat{x} - i\hat{y}\right)\right|^2 = \sin 2\theta_m, \tag{12.33}$$

yielding the direction of $\hat{\boldsymbol{\Omega}}$ on the Poincarè sphere. Finally we have

$$\hat{\boldsymbol{\Omega}} = |\Delta n_\pm| \frac{\omega}{c} \begin{bmatrix} \cos 2\theta_m \\ 0 \\ \sin 2\theta_m \end{bmatrix} . \tag{12.34}$$

We find that $|\Delta n_{\parallel\perp}| \gg \Delta n_{\pm,\mathrm{res}}$ so the modes of the plasma and vacuum are linear except in the vicinity of the cyclotron resonance and of the two vacuum resonance frequencies where $\Delta n_{\parallel\perp} \approx 0$.

In the neutron star magnetosphere, we expect the net charge density of the plasma to be at least the Goldreich and Julian (1969) density $\rho = -\mathbf{\Omega} \cdot \mathbf{B}/(2\pi c)$. The total charge density could be some multiple ζ (the multiplicity) of this, yielding an estimate of the local plasma frequency of

$$\omega_p = (2\zeta\Omega\omega_c|\cos\beta|)^{1/2} = 14.88\ \mathrm{GHz}(\zeta\cos\beta)^{1/2}\left(\frac{P}{1\ \mathrm{s}}\right)^{-1/2}\left(\frac{B}{10^{12}\ \mathrm{G}}\right)^{1/2}, \tag{12.35}$$

where β is the angle between the spin axis and the magnetic field direction locally. The first vacuum resonance energy in the magnetosphere is typically at

$$\hbar\omega_{v,1} = 0.068\ \mathrm{eV}(\zeta\cos\beta)^{1/2}\left(\frac{P}{1\ \mathrm{s}}\right)^{-1/2}\left(\frac{B}{10^{12}\ \mathrm{G}}\right)^{-1/2}. \tag{12.36}$$

The vacuum resonance energy increases as $r^{3/2}$ further into the magnetosphere if the density follows the Goldreich–Julian expression and the magnetic field is dipolar. Although the plasma density may be sufficiently large for this energy to reach into the visual range ($\zeta \sim 10^3$), we do not expect the magnetospheric density to be large enough ($\zeta \sim 10^8$) for this energy to reach the X-ray regime; in this case the magnetosphere would become somewhat opaque even outside of the resonant energies (see Sect. 12.3.3 for the role of plasma birefringence in the atmosphere).

Below the vacuum resonance energy, the plasma will dominate the birefringence and we can calculate the magnitude of birefringent vector

$$|\hat{\mathbf{\Omega}}| = \zeta|\cos\beta|\frac{\Omega\omega_c}{\omega c}\sin^2\theta. \tag{12.37}$$

The polarisation-limiting radius for low photon energies will be determined by the plasma birefringence. On the other hand, for radiation in the visual and bluer for pair multiplicities less than 10^3, the vacuum birefringence dominates the evolution of the polarisation in the magnetosphere, and the magnitude of the birefringence vector can be calculated from Eq. (12.17).

This section considered just the weak-field limit. The strong-field limit is somewhat more complicated. For a detailed treatment of the propagation of radiation through the combined plasma and vacuum for arbitrary fields consult, (Lai and Ho 2002; Ho and Lai 2003).

12.3.3 *The Vacuum Resonance in the Neutron-Star Atmosphere*

As the density of plasma in the atmosphere is much larger than in the magnetosphere, the energy of the vacuum resonance is also much larger and typically in the range of X-ray energies. Therefore, it may be important for the propagation of X-ray radiation above the photosphere. At the resonance we have

$$\hbar\omega_v = \hbar\omega_p v_r^{-1/2} = 2.0\left(\frac{n}{10^{22}\mathrm{cm}^{-3}}\right)^{1/2}\frac{10^{12}\,\mathrm{G}}{B}\ \mathrm{keV}. \tag{12.38}$$

Deep within the atmosphere, the density is large and the birefringence is dominated by the plasma: we have $\Delta n_{\parallel\perp} < 0$, $|\Delta n_{\parallel\perp}| \gg \Delta n_{\pm}$ and $2\theta_m \approx \pi$. We can see from Eqs. (12.30) to (12.33) that, in this regime, the polarisation modes are linear: $S_1 = -1$ and $S_3 = 0$; also, the $+$ mode is polarised along the y-axis (perpendicular to the magnetic field). As the radiation in the $+$ mode propagates upward and through the vacuum resonance, it will remain in the $+$ mode if the condition of Eq. (12.18) holds. We call this the adiabatic criterion. At low densities the vacuum will dominate so $\Delta n_{\parallel\perp} > 0$, $2\theta_m \approx 0$ and therefore, for the $+$ mode we have $S_1 = 1$ and $S_3 = 0$. As the photon crosses the resonance, the polarisation is transformed from perpendicular to parallel.

We can calculate whether the adiabatic criterion holds as the radiation passes through the vacuum resonance. The index of refraction difference reaches a minimum value of $\Delta n_{\pm,\mathrm{res}}$ at the resonance precisely. In the resonance, the change in the value of $\hat{\boldsymbol{\Omega}}$ is entirely in its direction at a rate of $2\theta_m' = (\Delta n_{\parallel\perp})'/\Delta n_{\pm}$ so we have

$$\left|\hat{\boldsymbol{\Omega}}\left(\frac{1}{|\hat{\boldsymbol{\Omega}}|}\left|\frac{\partial\hat{\boldsymbol{\Omega}}}{\partial x_3}\right|\right)^{-1}\right| = \left|\frac{\omega}{c}\frac{\left(\Delta n_{\pm,\mathrm{res}}\right)^2}{\left(\Delta n_{\parallel\perp}\right)'}\right| \tag{12.39}$$

$$= \frac{\omega}{c}H_\rho\left[\cos\theta\frac{\omega\omega_p^2}{\omega_c\left(\omega^2-\omega_{ci}^2\right)}\right]^2\left[\sin^2\theta\frac{\alpha_{\mathrm{QED}}}{30\pi}\left(\frac{B}{B_{\mathrm{QED}}}\right)^2\right]^{-1} \tag{12.40}$$

$$= \frac{\omega^3}{c}H_\rho\left[\frac{\cot\theta}{\omega_c\left(1-\frac{\omega_{c,i}^2}{\omega^2}\right)}\right]^2\frac{2\alpha_{\mathrm{QED}}}{15\pi}\left(\frac{B}{B_{\mathrm{QED}}}\right)^2 = \left(\frac{E_\gamma}{E_{\mathrm{ad}}}\right)^3. \tag{12.41}$$

This yields the adiabatic energy (Lai and Ho 2002; Ho and Lai 2003)

$$E_{\rm ad}^3 = \frac{15\pi}{2\alpha_{\rm QED}} \left(1 - \frac{\omega_{c,i}^2}{\omega^2}\right)^2 \tan^2\theta \frac{\hbar c}{H_\rho} \left(mc^2\right)^2 \approx (2.6\,{\rm keV})^3 \left(1 - \frac{\omega_{c,i}^2}{\omega^2}\right)^2 \tan^2\theta \frac{1\,{\rm cm}}{H_\rho}, \tag{12.42}$$

where H_ρ is the density scale height along the ray (in the $\hat{k}$ direction), typically $kT/(Am_u\hat{k}\cdot\mathbf{g}) \approx 8\,\text{mm}$ for a temperature $T = 10^6\,\text{K}$, surface gravity $\hat{k}\cdot\mathbf{g} = 10^{14}\,\text{cm s}^{-2}$ and for hydrogen atmosphere $A = 0.5$.

If one considers blackbody emission, for effective temperatures greater than about 5×10^6 K, the energy of the typical photon is greater than the adiabatic energy, so much of the radiation will pass through the resonance adiabatically. Typically, light element atmospheres peak at higher photon energies than $3kT_{\rm eff}$, so the effect would be even more pronounced. If the vacuum resonance occurs in the neutron star above both the photosphere for parallel polarisation and the photosphere for perpendicular polarisation (which we found in Sect. 12.2.1), the main effect is to switch the polarisation coming from the surface from mostly perpendicular to the magnetic field to mostly parallel, if the energy of the photon is greater than $E_{\rm ad}$. Below this energy, the polarisation would remain perpendicular.

For sufficiently strongly magnetised neutron stars (magnetars), the vacuum resonance lies above the photosphere for radiation polarised perpendicular to the magnetic field and below the photosphere for parallel photons. In this case, the effective photosphere for perpendicularly polarised photons with energies greater than the adiabatic energy will lie at the vacuum resonance itself which is at a lower temperature than the photosphere for low-energy perpendicularly polarised radiation. Even in this case, the bulk of the radiation will emerge in the perpendicular polarisation, and the emission of parallel photons will be diminished for photon energies above the vacuum resonance energy at the photosphere. We can estimate the number density of electrons at the photosphere for the parallel polarisation to be $n \approx (\sigma_T H_\rho)^{-1}$ so

$$\hbar\omega_{v,\parallel} \approx 4\,{\rm keV} \left(\frac{B}{B_{\rm QED}}\right)^{-1} \left(\frac{g}{10^{14}{\rm cm\,s^{-2}}}\right)^{1/2} \left(\frac{T}{10^6\,{\rm K}}\right)^{-1/2}. \tag{12.43}$$

Although this equation is not accurate for magnetic fields approaching or exceeding $B_{\rm QED}$, we see that the role of the vacuum resonance in the formation of the spectrum will be crucial for the magnetars where the magnetic field exceeds $B_{\rm QED}$ and the temperatures exceed several million degrees. The number density at the photosphere for the perpendicular polarisation is larger by a factor of ω^2/ω_c^2, giving the energy of photons in the vacuum resonance at this surface of

$$\hbar\omega_{v,\perp} \approx 45\,{\rm keV} \left(\frac{g}{10^{14}{\rm cm\,s^{-2}}}\right)^{1/4} \left(\frac{T}{10^6\,{\rm K}}\right)^{-1/4} \tag{12.44}$$

or approximately the electron cyclotron energy, whichever is smaller.

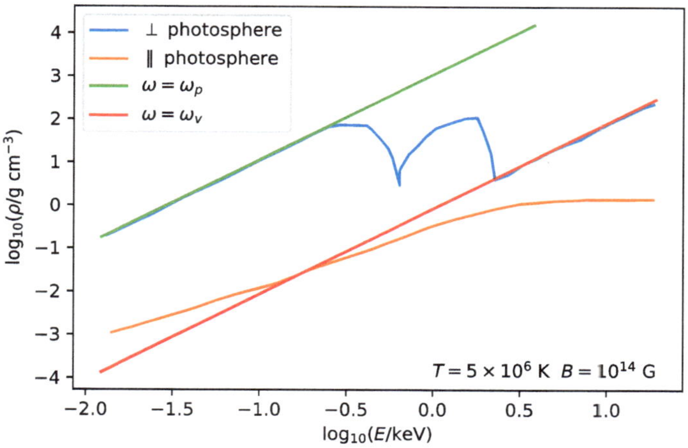

Fig. 12.5 An illustration of the locations of the parallel and perpendicular model photospheres for $B = 10^{14}$ G and $T = 5 \times 10^6$ K using the models of Lloyd (2003). This is similar to Fig. 12.1, but for the strong field case. The proton cyclotron line has moved into the X-rays; the key new thing is that the ⊥-mode photosphere lies along the vacuum resonance line at high energies, because the ⊥-mode photons have a significant parallel component at the vacuum resonance so the opacity for them is really high there. The vacuum resonance is effectively their photosphere. In fact a significant amount of energy is deposited in this layer so it forms a local maximum in the temperature of the atmosphere

Figure 12.5 illustrates how the vacuum resonance affects the location of the photosphere for the perpendicular mode photons. At low energies, the photosphere lies at a density where the plasma frequency equals the frequency of the photon, then it runs at nearly constant density punctuated by a dramatic drop in the density at the proton cyclotron line and then above the adiabatic energy it follows the density at which photon frequency equals the vacuum resonance energy (this is a constant multiple of the plasma frequency that depends on the strength of the magnetic field). For photons above the adiabatic energy, the structure of the atmosphere is rather complicated. At high densities, where the birefringence is plasma dominated, but not so high that the plasma is opaque to photons in the perpendicular mode, the bulk of the energy flux is carried by photons in the perpendicular mode, while those in the parallel mode are trapped. As the radiation approaches the density of the vacuum resonance, both the photons in the parallel mode and those in the perpendicular mode become circularly polarised and both couple strongly to the plasma. At the resonance density, the flux carried in the perpendicular mode is dumped back into the plasma. Just below the resonance density the modes become mainly linear again, so the perpendicular mode is no longer well coupled to the plasma and again travels freely, and the photosphere for the perpendicular mode follows the vacuum resonance density above the adiabatic energy. For stronger magnetic fields, the proton cyclotron line can lie above the adiabatic energy, so the structure of the line, even without polarisation information, will be affected by the resonance (Ho and Lai 2003).

This picture that above the adiabatic energy, all of the radiation behaves adiabatically and below this energy it does not is a gross approximation. In detail, the behaviour near the vacuum resonance will also depend on the imaginary portion of the index of refraction (Pavlov and Shibanov 1979; Mészáros 1992), and the entire mode description may collapse. How to treat photons passing through the vacuum resonance within the atmosphere is still uncertain, and it is often treated as such in the calculations (Zane et al. 2001; Lai and Ho 2002; Ho and Lai 2003; González-Caniulef et al. 2019). However, it is clear that when the radiation passes through the vacuum resonance outside the atmosphere, as in more weakly magnetised stars (see Fig. 12.2), the resonance can switch the final polarisation state of the radiation depending on whether the energy lies above or below the adiabatic energy, leaving an imprint of the local density scale height on the outgoing radiation, which could be a powerful diagnostic of the surface gravity in the polarisation of the outgoing radiation.

12.3.4 Polarisation-Limiting Radii of Neutron Stars

The effect of vacuum birefringence decouples the polarisation modes, and as photons travel through the neutron star magnetosphere, the parallel mode stays parallel and the perpendicular mode stays perpendicular. The effect wears off at a distance from the star where the magnetic field becomes too weak. If we consider the magnetic field around the neutron star to be a dipole field with $B \approx \mu r^{-3}$, where μ is the magnetic dipole moment of the star and r is the distance from the centre of the star, from Eq. (12.18) we can find a condition for vacuum birefringence to be strong enough to affect the polarisation of photons travelling in the magnetosphere of the star:

$$\left| \frac{\alpha}{15} \frac{\nu}{c} \frac{\mu^2 \sin^2\theta}{r^6 B_{\text{QED}}^2} \frac{r}{6} \right| \geq 0.5, \tag{12.45}$$

where θ is the angle between the local magnetic field and the line of sight. We can define the polarisation-limiting radius to be the distance from the star at which the equality holds:

$$r_{\text{PL, V}} = \left(\frac{\alpha}{45}\frac{\nu}{c}\right)^{1/5} \left(\frac{\mu}{B_{\text{QED}}} \sin\theta\right)^{2/5} \tag{12.46}$$

$$\approx 1.2 \times 10^7 \text{cm} \left(\frac{\mu}{10^{30}\ \text{G cm}^3}\right)^{2/5} \left(\frac{\nu}{10^{17}\ \text{Hz}}\right)^{1/5} (\sin\theta)^{2/5}. \tag{12.47}$$

As photons are emitted from the surface of the neutron star, the polarisation modes are decoupled, and their polarisation follows the magnetic field lines out to

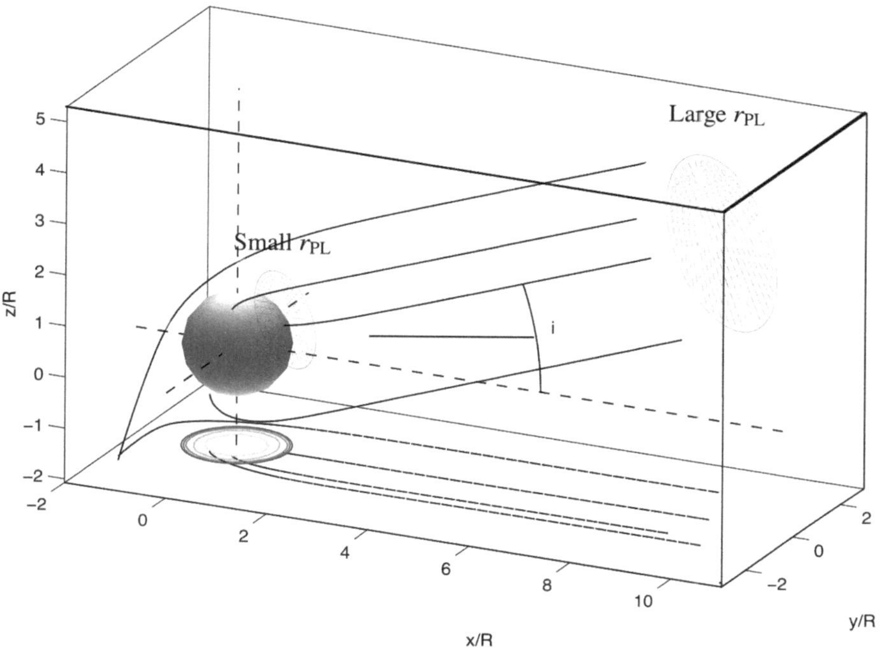

Fig. 12.6 Radiation leaving the surface of a neutron star follows geodesics so that the bundle of rays that reaches the distant observer is approximately cylindrical. The three-dimensional coordinates (x, y, z) are given in terms of the radius of the neutron star (R). If the polarisation-limiting radius is small, the final polarisation will reflect the magnetic field structure near to the star where the bundle covers a large fraction of hemisphere so the field structure varies a lot over the bundle at this point, and the final polarisation will also vary a lot over the image. On the other hand, if the polarisation-limiting radius is large, the field structure over the ray bundle is simpler, and the polarisation direction will not vary much over the image. Since the magnetic field is assumed to be that of a dipole (aligned with the z-axis), it has axial symmetry and different images will be distinguishable by the observer's magnetic inclination angle i. Adapted from Shaviv et al. (1999)

the polarisation-limiting radius. Afterwards, the modes re-couple and the photon polarisation stays constant all the way to the observer. Figure 12.6 depicts this behaviour for two stars with different magnetic dipole moments. In the case of a highly magnetised neutron star, $B \sim 10^{12}$ G, the polarisation-limiting radius is several neutron star radii away, according to Eq. (12.47), and therefore, the observed polarisation reflects the magnetic field structure at a large distance from the star and not at the surface. For a weakly magnetised star, the polarisation-limiting radius is comparable to the radius of the neutron star and therefore the polarisation direction of the observed photons reflects the magnetic field structure close to the star.

Figure 12.7 depicts the final polarisation states across the image of the neutron star surface assuming that the radiation is initially in the extraordinary mode, that is, the electric field is parallel to the local magnetic field. The left panel shows the case where the vacuum birefringence is neglected, and the right panel shows the case where the surface field is about 10^{12} G and the frequency is 10^{17} Hz or an energy of

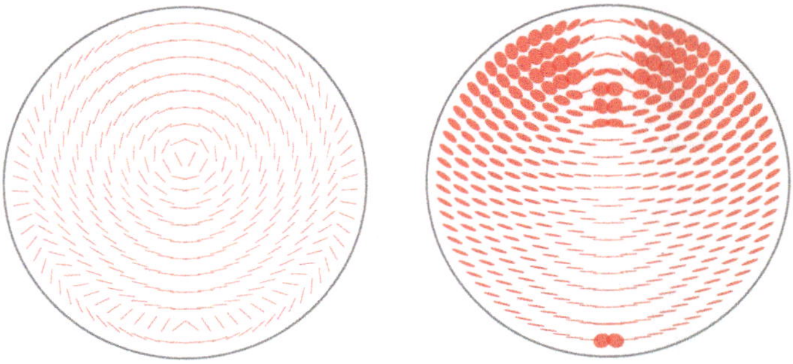

Fig. 12.7 The polarised emission map of a neutron star overlaid on the apparent image of the NS. The left panel depicts the observed map of polarisation directions if one assumes that the surface emits only in the extraordinary mode (perpendicular to the local field direction) and neglects the vacuum birefringence induced by QED. The right panel shows the polarisation map including birefringence for a frequency of $\nu = (\mu/(10^{30}\,\mathrm{G\,cm}^3))^{-2}10^{17}$ Hz. The ellipses and short lines describe the polarisation of a light ray originating from the surface element beneath them. The lines and the major axes of the ellipses point towards the direction of the linear component of the polarisation direction. The minor to major axis ratio provides the amount of circular polarisation (s_3). The observer's line of sight makes an angle of 30° with the dipole axis. For comparison, if one assumes that the entire surface is emitting fully polarised radiation, the net linear polarisation on the left is 13%, while it is 70% on the right

about 0.4 keV. This is appropriate for a thermally emitting neutron star such as one of the X-ray dim neutron stars (XDINS). The effect of the vacuum polarisation is to comb the polarisation direction to be aligned perpendicular to the direction of the magnetic axis of the star and dramatically increase the observed total polarisation from about 13% to about 70%. For more strongly magnetised neutron stars the effect is more dramatic.

The studies of the effects of QED on the net polarisation from astrophysical sources are more mature for magnetars and XDINS. For the magnetars and XDINS, the polarisation-limiting radius is much larger than the radius of the star. Because we expect the emission in these sources to come from a larger region of the stellar surface or magnetosphere (in the case of the non-thermal emission from magnetars, Taverna et al. 2014), we expect a large increase in the observed polarisation fraction due to QED. Although the ratio of the polarisation-limiting radius to the stellar radius is also large for the X-ray pulsars (XRP), the effect for these objects is more subtle. The QED effects for more weakly magnetised stars such as millisecond XRPs (ms XRPs) have not yet been explored.

At low photon energies, the plasma dominates the birefringence, yielding an alternative estimate for the polarisation-limiting radius (Cheng and Ruderman 1979) of

$$
\begin{aligned}
r_{\mathrm{PL,P}} &= \left[\frac{2}{6\pi} \frac{1}{\nu} \frac{\mu e}{mc} \frac{\Omega}{c} \zeta \, |\cos\beta| \sin^2\theta \right]^{1/2} \\
&\approx 2.0 \times 10^6 \mathrm{cm} \left(\frac{P}{1\ \mathrm{s}} \right)^{-1/2} \left(\frac{\mu}{10^{30}\ \mathrm{G\ cm}^3} \right)^{1/2} \left(\frac{\nu}{10^{14}\ \mathrm{Hz}} \right)^{-1/2} \\
&\quad \times \left(\zeta \, |\cos\beta| \sin^2\theta \right)^{1/2} .
\end{aligned}
\tag{12.48}
$$

Figure 12.8 depicts both polarisation-limiting radii as a function of frequency for a neutron star with a period of 1 s and a surface field of 10^{12} G, assuming that the plasma density is the Goldreich and Julian (1969) value. In the general picture, if the radiation is in one of the two normal modes of the combined plasma and vacuum when it leaves the surface, then it will remain in that mode out to the maximal polarisation-limiting radius as determined by the plasma or vacuum. If this radius is much larger than the star, the birefringence can dramatically increase the extent of the observed polarisation (Heyl and Shaviv 2002). Furthermore, if the polarisation-limiting radius is a large fraction of the light cylinder, the direction of the polarisation can be dragged along with the rotation of the star and significant circular polarisation can be generated (Heyl and Shaviv 2000). Linearly polarised

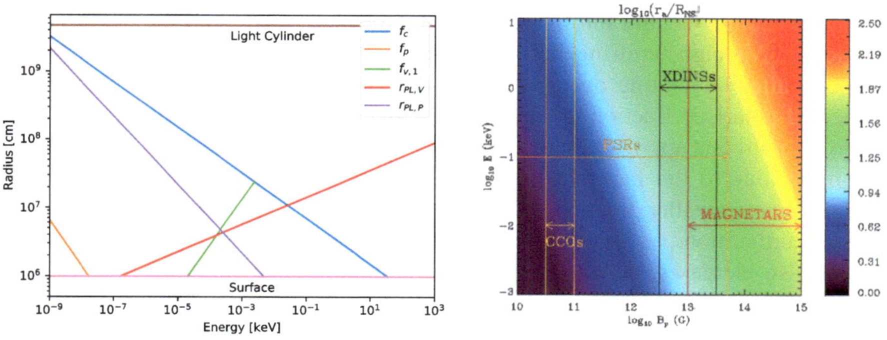

Fig. 12.8 Left: The polarisation-limiting radii for the plasma ($r_{\mathrm{PL,P}}$) and vacuum ($r_{\mathrm{PL,V}}$) as a function of frequency, and the plasma (f_{p}), vacuum resonance ($f_{\mathrm{v,1}}$) and cyclotron (f_{c}) frequency as a function of radius for a neutron star with a surface field of 10^{12} G and a spin period of 1 s. Right: The ratio of the polarisation-limiting radius (vacuum) to the radius of the neutron star for various magnetic field strengths and energies assuming a dipole field geometry. In the figure the polarisation-limiting radius is denoted by r_a (the adiabatic radius). Figure 1 right from Taverna et al. (2015)

radiation can also be converted to circularly polarised radiation, if the radiation happens to pass through the polarisation-limiting radius when it is propagating approximately tangential to the field (Wang and Lai 2009; see near the polar cap in the right panel of Fig. 12.7).

12.4 Observational Consequences and Results

Our focus has been the emission from the atmosphere and the magnetosphere of isolated neutron stars in cooling and magnetically powered neutron stars; therefore, we will focus on observations in the visual and near infrared, where we can now measure the polarisation of radiation from these objects, and in the X-rays, where we will soon have the capability. We will not discuss the expected polarisation from rotation-powered pulsars or polarisation in the gamma rays.

Figure 12.9 depicts the broadband polarisation from the entire visible surface of a neutron star with a temperature and field strength at the magnetic pole of $10^{6.5}$ K and 2×10^{12} G, respectively. The thermal flux through the surface varies as $B^{0.4}\cos^2\zeta$, where ζ is the angle of the local magnetic field with respect to the normal. We plot the polarised fraction in terms of the total flux polarised perpendicular and parallel to the projection of the magnetic moment of the star onto the sky. A value of 1 indicates radiation fully polarised perpendicular to the moment, and -1 indicates radiation fully polarised parallel to the moment. In the upper panel the magnetic moment makes an angle of 30° with respect to the line of sight, and in the lower panel the angle is 60°. In each panel, the upper set of curves traces the result including vacuum birefringence in the magnetosphere and the lower curves neglect it. Vacuum birefringence dramatically increases the expected polarisation fraction after integrating over the stellar surface. With vacuum birefringence, the expected polarised fraction is larger for smaller neutron stars and larger at higher energies until one approaches the cyclotron resonance. The trend with stellar radius is simply due to the fact that the bundle of rays for smaller stars is smaller so it subtends a smaller fraction of the magnetosphere at the polarisation-limiting radius; furthermore, as the energy of the photons increases, the polarisation-limiting radius also increases, increasing the expected polarised fraction. Without vacuum birefringence, both of these trends are reversed. Furthermore, the polarised fraction is larger when the magnetic field makes a larger angle with the line of sight. If one assumes that the flux is largest when the magnetic field is closest to the line of sight, one would expect the polarised fraction and the flux to be somewhat anti-correlated. In particular, the off-pulse radiation is more polarised than on pulse, so polarised emission off-pulse may come from the neutron star itself rather than the background.

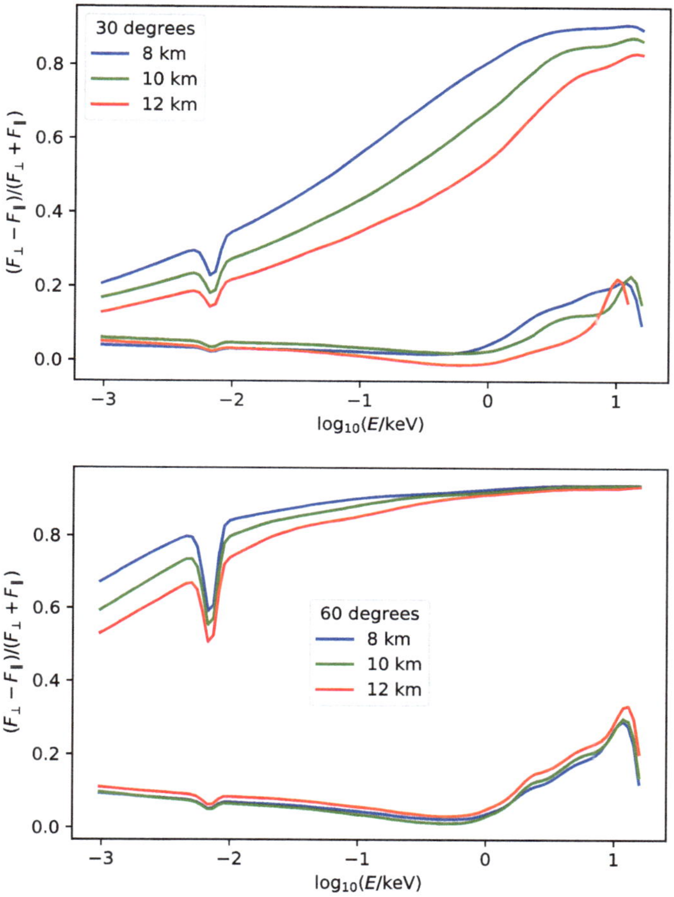

Fig. 12.9 The extent of the polarisation averaged over the stellar surface as a function of energy, angle and stellar radius using the fully ionised hydrogen atmospheres discussed in Sect. 12.2.1. Here the perpendicular direction is defined to be perpendicular to the projection of the magnetic moment of the star into the sky. In the upper panel, the magnetic pole makes an angle of 30° with the line of sight. In the lower panel, the angle is 60°. The lower set of curves trace the results without vacuum polarisation, and the results for the upper curves include it. The mass of the star is 1.4 $M_\odot$. The effective temperature at the magnetic pole is $10^{6.5}$ K with a magnetic moment of $\mu = 10^{30}\,\mathrm{G\,cm}^3$. Locally the thermal flux emerging through the surface is proportional to $B^{0.4}\cos^2\zeta$, where ζ is the angle of the local magnetic field with respect to the normal

12.4.1 Optical Polarisation

Measuring the polarisation of the surface emission from neutron stars in the infrared, optical and ultraviolet is especially tantalising not only because the instruments exist today, but also because observations in these bands could provide diagnostics on the size of the neutron star, as shown in Fig. 12.9, and on the density of the magnetospheric plasma (from Fig. 12.8 and also Shannon and Heyl 2006).

Table 12.1 summarises the detections of isolated neutron stars in the infrared, optical and ultraviolet. The first set of stars are rotation-powered pulsars. We expect the emission in these bands to be dominated by the pulsar wind nebulae and the magnetosphere for the younger pulsars. The final few old objects are millisecond pulsars and in principle we are seeing emission from the surfaces. Unfortunately, these objects are extremely faint.

To see the emission from the neutron star itself or its immediate environment we have to focus on the second two classes of sources: the XDINS and the magnetars. Both are typically brighter than the millisecond pulsars. The XDINS are usually quite close and suffer little absorption due to dust and intervening gas, whereas the magnetars are inherently brighter but typically much further away and suffer many magnitudes of extinction in the visual, so studies often focus on the infrared. The brightest of the XDINS is RXJ J1856.5−3754, and it was discovered to have 16% polarised flux in the V-band averaged over phase (Fig. 12.10 and Mignani et al. 2017). The magnetic field, somewhat stronger than 10^{12} G, tends to increase the expected polarisation, thanks to the vacuum birefringence (González Caniulef et al. 2016). The key result is that the phase-resolved polarised fraction in this band is only about 10% without vacuum birefringence and is 40–60% with the effects of vacuum birefringence included.

It is essentially impossible to account for the observed polarisation of the emission in the V-band without birefringence, if the emission is coming from a large portion of the surface (Mignani et al. 2017). Pulsation in the energy range of 0.15–1.2 keV amounts to only $\sim$ 1.2% (Tiengo and Mereghetti 2007). This low level of pulsation constrains the magnetic axis to lie within a few degrees of the spin axis and the spin axis to lie within 20–40° of the line of sight (Ho 2007). The pulsation signal alone does not change when these two angles are swapped, so an alternative is that line of sight lies within a few degrees of the spin axis and the magnetic axis lies within 20–40° of the spin axis. If the magnetic axis and the spin axis are nearly aligned (the first possibility), we would expect the phase-averaged polarisation to be nearly as large as the phase-resolved value, so we can exclude this possibility, as the phase-resolved polarisation fraction would be significantly larger than that found by (Mignani et al. 2017).

The interpretation that the large observed polarisation from RX J1856.5−3754 results from vacuum polarisation hinges on the assumption that the plasma birefringence is unimportant in this regime. From Fig. 12.8 we see that if the plasma indeed has the Goldreich and Julian (1969) density, this would be the case. Unlike magnetars, XDINS have no twists, so a Goldreich and Julian (1969) density would

Table 12.1 UVOIR identifications of INSs: RPPs (first group), XDINSs (second) and magnetars (third)

Name	Age	Mag	d(kpc)	A_V	IP	S	P	T
Crab	3.10	16.6	1.73	1.6	nUV, O, nIR, mIR	Y	Y	Y
B1509−58	3.19	25.7^R	4.18	5.2	O, nIR		Y	
B0540−69	3.22	22.0	49.4	0.6	fUV, nUV, O, nIR	Y	Y	Y
J0205+6449	3.73	27.4^g	3.2	2.5	O			
Vela	4.05	23.6	0.23	0.2	nUV, O, nIR	Y	Y	Y
Geminga	5.53	25.5	0.07	0.07	nUV, O, nIR	Y		Y
B0656+14	5.05	25.0	0.29	0.09	nUV, O ,nIR	Y	Y	Y
J1741−2054	5.58	25.3	0.38	0.68	O			
B1055−52	5.73	24.9^U	0.72	0.22	nUV, O			
B1929+10	6.49	25.6^U	0.33	0.15	nUV			
B1133+16	6.69	28	0.35	0.12	O			
B0950+08	7.24	27.1	0.26	0.03	nUV, O			
J0108−1431	8.3	26.4^U	0.2	0.03	O			
J0437−4715	9.20		0.14	0.11		Y		
J2124−3358	10.0	27.8^B	0.41	< 0.24	fUV, nUV, O			
J1308.6+2127	6.11	28.6^R	<1	0.14	O			
J0720.4−3125	6.27	26.7	0.30	0.3	nUV, O			
J214303.7+065419	6.56	27.2^B	0.34	0.18	O			
J1856.5−3754	6.60	25.7	0.14	0.12	nUV, O	Y	Y	
J1605.3+3249		26.8^R	<1	0.06	O			
J0420.0−5022		27.5^B	0.35	0.07	O			
1E 1547.0−5408	3.14	18.5	9	17	nIR			
SGR 1806−20	3.14	20.1	15.1	29	nIR			
1E 1048.1−5937	3.63	21.3	3.0	6.10	O, nIR		Y	Y
XTE J1810−197	3.75	20.8	4.0	5.1	nIR			
SGR 0501+451	4.10	19.1	5	5	O, nIR			
4U 0142+61	4.84	20.1	1.73	1.62	O, nIR, mIR			Y
1E 2259+586	5.34	21.7	3.0	5.7	nIR, mIR			

Each group is sorted according to the inferred spin-down age. Column 1 gives the INS name, while columns 2–5 give the spin-down age (in logarithmic units), magnitude, distance and interstellar extinction A_V. For RPPs and XDINSs, magnitudes refer to the V band, unless otherwise indicated by the superscript, while for magnetars they refer to the K_s band. For the latter, only magnitudes in quiescence are listed, with the variability range typically spanning ≈1–2 magnitudes. Columns 6–8 give the broadband imaging photometry (IP) coverage in the near-ultraviolet (nUV), optical (O) and near/mid-infrared (nIR/mIR) spectral bands, and a flag (Y) for spectroscopy (S), polarisation (P) and timing (T) measurements. Updated from Mignani (2011) with data from Moran et al. (2013), Mignani et al. (2012, 2016, 2017, 2018, 2019) and Rangelov et al. (2017)

be realistic. However, if the plasma density exceeds a few hundred times the Goldreich–Julian density, the birefringence of the plasma will dominate in the V-band (Shannon and Heyl 2006). In fact, Lyutikov and Thompson (2005) argue that the eclipses in PSR J0737-3039 indicate a pair multiplicity of nearly 10^5 in the mag-

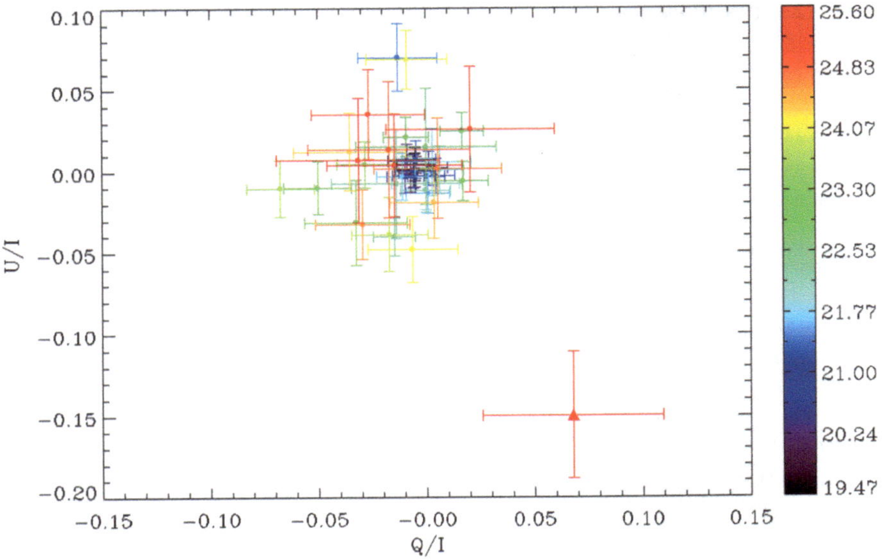

Fig. 12.10 The phase-averaged observed Stokes parameters of RX J1856.5−3754 (red triangle, Fig. 2 of Mignani et al. 2017) in the V-band (lower-right) and of other stars in the field. The colours denote the apparent magnitude of the various stars according to the scale at the right

netosphere of PSR J0737-3039B. Of course, PSR J0737-3039B has had a unique history, and its magnetic field is ten times smaller than RX J1856.5−3754. However, the spin-down luminosity of RX J1856.5−3754 is twice that of PSR J0737-3039B, so a large pair multiplicity is indeed a possibility for RX J1856.5−3754. The role of vacuum versus plasma birefringence is straightforward to determine by measuring the polarisation fraction at several energies in the UVOIR range. If the vacuum dominates, the polarisation fraction will increase with frequency; if plasma dominates, it will increase with wavelength.

The path forward in the UVOIR region of the spectrum is obvious: press forward by measuring the polarisation of more and fainter XDINS in the optical and near-ultraviolet and examine the magnetars in the infrared. In this regime of the spectrum, polarisation will yield key diagnostics on the state of the plasma and the structure of the magnetic field. With the advent of ever larger ground-based telescopes and JWST, all optimised for high-sensitivity observations in the near infrared, the path appears inviting as well. However, none of these future instruments will have a polarimetric capability at first light, and the ultrasensitive JWST will never have it.

12.4.2 X-ray Polarisation

A vast theoretical literature has examined the role of polarisation for magnetars in particular. Although not all of the magnetars and XDINS have been detected in the UVOIR bands, Table 12.1 gives a summary of the landscape of these objects. Magnetars are typically young and hot, so their emission matches well with the capability of the current generation of X-ray polarimeters, *IXPE* (Soffitta 2017) and *eXTP* (Zhang et al. 2016, 2019) in the 2–8 keV energy range. The XDINS are generally too cool and faint to be measured by *IXPE* and *eXTP*. For this reason, although their thermal emission will be highly polarised, we cannot measure it until soft X-ray polarimeters are launched, such as *REDSoX* (Günther et al. 2017), *LAMPP* (She et al. 2015) or *XPP*. Consequently, our focus shall be on magnetars. The conclusions for the thermal emission from XDINS will be similar as for magnetars but at lower photon energies.

To demonstrate the importance of QED vacuum birefringence on the X-ray polarisation from neutron stars, we focus on the magnetar 4U 0142+61. 4U 0142+61 exhibits a X-ray spectrum from 2 to 8 keV (the range of sensitivity of *IXPE* and *eXTP*) that is dominated by thermal emission (Tendulkar et al. 2015). As argued by several authors (Heyl et al. 2003; Ho et al. 2003; Heyl et al. 2005; van Adelsberg and Perna 2009; Potekhin et al. 2016), this thermal emission is expected to be nearly fully polarised as it is emitted. In Fig. 12.11 we show a simulation for 4U 0142+61: we have used the phase-resolved spectral fits of Tendulkar et al. (2015) to simulate the polarised spectra in XIMPOL (Baldini et al. 2016) for a 50.0 ks observation with *eXTP*. Because we expect, when QED is included, that the final polarisation

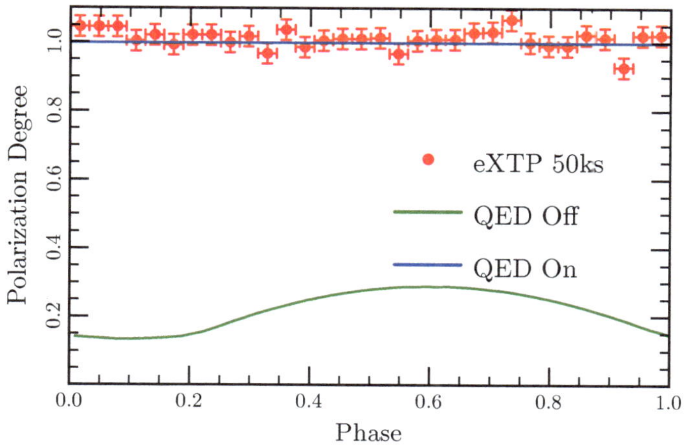

Fig. 12.11 The expected degree of polarisation from the magnetar 4U 0142+61 from the thermal emission, assuming a two black-body model in which the radiation is initially fully polarised in the perpendicular mode. The crosses give the results for a short simulated observation with *eXTP*. The results for *IXPE* are similar with a 150 ks observation

will follow the magnetic field direction at a large distance from the star, the final polarisation fraction with QED will be nearly 100% (Heyl et al. 2003). On the other hand, without QED, the expected final polarisation will be lower, as apparent from the left panel of Fig. 12.7, and will depend on where the emission originates. Using the fits of Tendulkar et al. (2015) as a guide, we assumed that the hot blackbody component originates from a hot-spot of ten-degree radius and that the cool blackbody originates from the entire stellar surface. We assumed that the hard power-law component does not contribute in the 2–8 keV range. Even without QED birefringence, we expect the first component to remain highly polarised because it originates from a small region near the polar cap, where the projected field is well aligned (see the left panel of Fig. 12.7), but the second component will not, because it comes from the rest of the surface. Figure 12.11 presents the results of this calculation along with a short simulated observation with *eXTP*. Even a relatively short observation of this magnetar can detect the effect of QED vacuum birefringence dramatically.

Taverna et al. (2015) examine in detail how the geometry of the magnetic field will affect the observed extent and direction of polarisation assuming that the emission from the surface follows the blackbody distribution and a modified dipole pattern such that the minimum effective temperature is 100 eV, and the effective temperature at the pole is 150 eV. Rather than integrate the Stokes parameters, they also make the additional assumption that the final emergent polarisation is perpendicular to the direction of the magnetic field along the ray at the polarisation-limiting radius (or the adiabatic radius, Heyl and Shaviv 2000). They find that the polarisation ranges from 10–25% in the visual up to nearly 100% in the X-ray band. Furthermore, they argue that the observed polarisation evolution with phase can probe in detail the structure of the magnetic field, in this case the twist of the field lines between the two magnetic hemispheres of the star (see Fig. 12.12).

For the other magnetars, much of the emission in the band of *eXTP* and *IXPE* has been scattered by currents in the magnetosphere. If we use the twisted magnetosphere model (Thompson et al. 2002; Nobili et al. 2008a; Albano et al. 2010; Taverna et al. 2014), we can obtain an estimate of the expected emission including the polarisation in the 2–8 keV energy range. Although the radiation at emission is nearly fully polarised perpendicular to the local magnetic field direction as we argued in Sect. 12.2.3, unless the polarisation changes as it propagates to us, we expect the net polarisation over the entire region to be small, because the radiation comes from a large portion of the neutron-star magnetosphere. Fernández and Davis (2011) argued that vacuum polarisation is again important in this case because the currents that upscatter the photons typically lie within polarisation-limiting radius.

Taverna et al. (2014) calculate the polarised emission from the RCS emission as a function of the viewing geometry, twist of the magnetic field and the velocity of the currents. They have assumed that the seed photons originate fully polarised in the extraordinary mode and follow a blackbody distribution of uniform temperature ($kT = 0.47$ keV) over the surface of the neutron star. Figure 12.13 depicts the extent of the total polarisation that remains after passing through the magnetosphere for

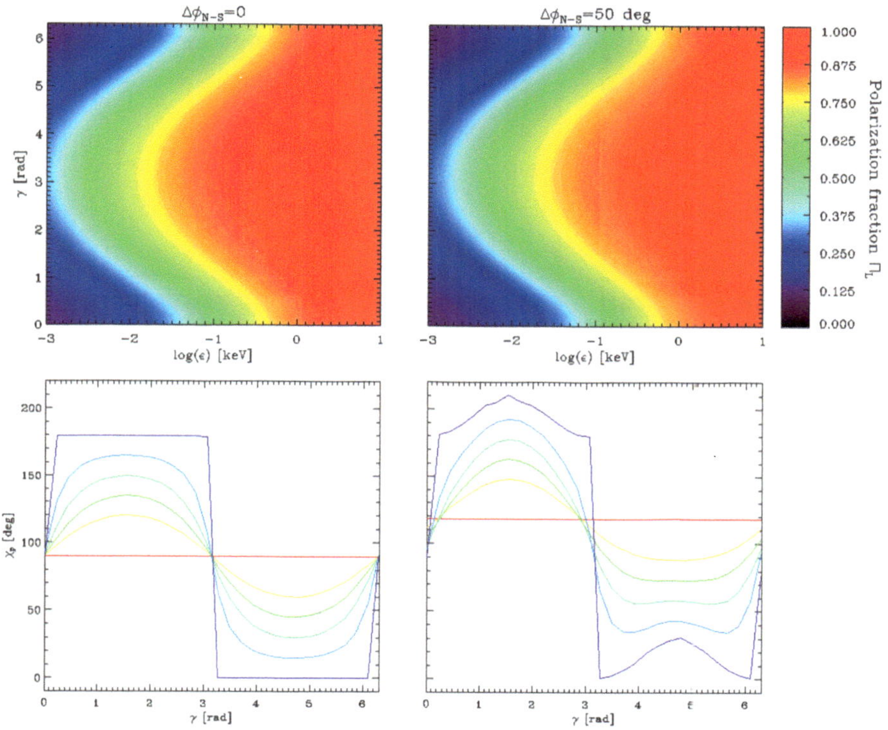

Fig. 12.12 Polarisation observables for the cases of a pure dipolar magnetic field (left-hand column) and a globally twisted dipole with twist angle of 50°. Top row: Polarisation fraction in the energy-phase plane for the inclination of the magnetic axis with respect to the spin axis is fixed to 5°, and the angle between the spin axis and the line of sight is 15°. Bottom row: Polarisation angle as a function of the rotational phase for a fixed photon energy ($E = 0.02$ keV), line of sight inclination of 90° and magnetic inclination of 0° (red), 30° (orange), 45° (green), 60° (light blue), 75° (blue) and 90° (violet). All the plots are obtained for seed photons 100 per cent polarised in the X-mode and values of $R = 10$ km, $M = 1.4\mathrm{M}_{\odot}$ and $B_p = 10^{13}$ G. Figure 9 from Taverna et al. (2015)

different viewing angles with respect to the magnetic poles (θ is the angle between the line of sight and the north magnetic pole), energy and configuration of the magnetic field. Let us first focus on the right-most panel which exhibits a small twist and a large velocity. The total current increases with the twist, and the current is the product of the number density of charges and the velocity, so this model has the fewest charges, and the smallest typical optical depth to scattering. It also has the least high-energy emission (see Fig. 2 of Taverna et al. 2014). For positive values of $\cos\theta$ and low energies, the emergent radiation is nearly fully polarised, because little of it has been scattered. In the crucial 2–8 keV range, the polarised fraction is about 50% corresponding to a single scattering (see Fig. 8 of Fernández and Davis 2011). For the radiation to achieve higher energies, multiple scatterings

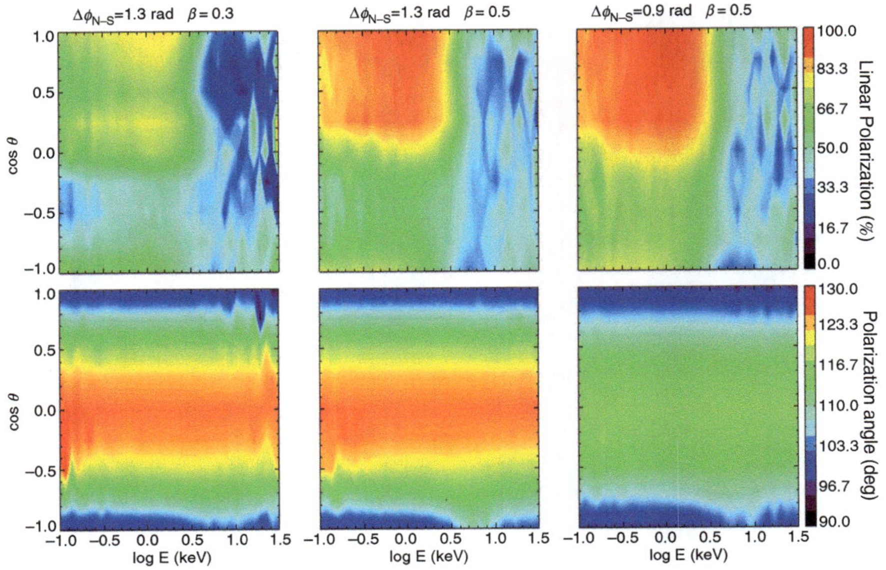

Fig. 12.13 Contour plots for the polarisation fraction and angle as a function of viewing geometry, twist angle and current velocity for $B_p = 5 \times 10^{14}$ G. Figure 1 from Taverna et al. (2014)

are required, and the resulting polarisation fraction is diminished. Moving to the middle panel, the twist is increased while the velocity of the charges remains the same, so the increase in current results from an increase in the density of the charges. In this case the extent of nearly fully polarised radiation is diminished because scattering is more important. Furthermore, the total high-energy emission is larger in this case. The increase in the twist angle has a dramatic effect on the observed direction of the polarisation from a comparison of the bottom panels. Finally, the left-most panels depict the same amount of twist (total current) as the middle panels but with decrease in the velocity of the charges (an increase in the density and scattering optical depth). Here nearly all of the radiation from the surface suffers at least one scattering event and the emergent polarised fraction is only 50% even at low energies. Furthermore, radiation in the 2–8 keV range is now generated through multiple scatterings so the extent of polarisation is diminished in this regime. The combination of the extent of polarisation and the polarisation angle in the 2–8 keV range provides key diagnostics on the total magnetosphere current and the velocity of the charge carriers.

Figure 12.14 depicts the results of this model to simulate an observation with eXTP from Santangelo et al. (2019). In particular, the red dotted curves (Model OFF) trace the polarisation degree and direction as a function of rotational phase for the magnetar 1RXS J170849.0−400910, neglecting the effects of vacuum birefringence. In this case, the expected extent of polarisation is low, typically about 10%. On the other hand, if one includes the effects of vacuum birefringence

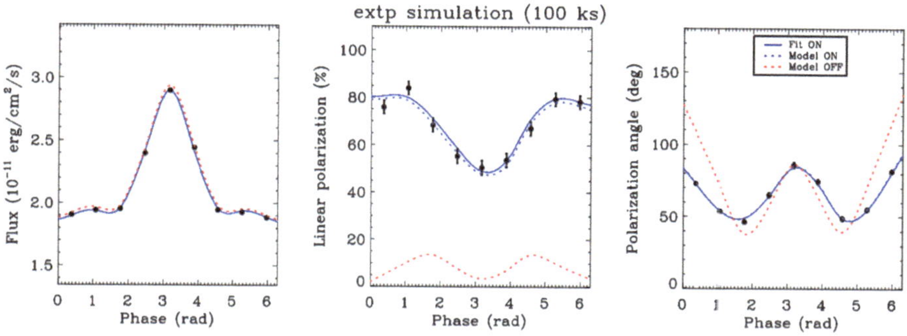

Fig. 12.14 Degree and angle of polarisation as computed according to the 'twisted magnetosphere' model with $\Delta\phi_{N-S} = 0.5$ rad, $\beta = 0.34$, $\chi = 90°$ and $\xi = 60°$ for a source with properties similar to those of the AXP 1RXS J170849.0−400910 (adapted from Fig. 19 of Santangelo et al. 2019). The data points with error bars show the results from a simulated 100 ks observation with *eXTP* including vacuum birefringence. The results for *IXPE* are similar with a 300 ks observation. The light curves for the QED-on and QED-off models are identical by design

(blue curve, Model ON), the expected polarisation fraction ranges from 50 to 80%. Even a relatively short 100.0 ks observation with *eXTP* (data points) will distinguish between these two situations and also provide key diagnostics on the twisted magnetosphere model for the hard X-ray emission from magnetars.

12.5 Future Prospects

With the X-ray polarimetry missions *IXPE* and *eXTP* both at advanced stages of preparation, the future for the study of the polarisation from neutron stars looks bright. In the X-rays, it will be straightforward to determine whether the magnetised vacuum is indeed birefringent, verifying one of the first predictions of QED. Furthermore, the direction and extent of the observed polarisation as a function of energy and phase will provide crucial information for the geometry of the neutron star magnetic field, rotation and emission regions and probe the neutron-star atmosphere. For more weakly magnetised objects or measurements of lower energy emission, the extent of linear polarisation could provide constraints on the radii of the neutron stars as well, especially in concert with observations of the polarisation in the X-rays. Broadband, time-resolved polarimetry is crucial to understand these objects in the greatest detail. Observations in the far and near infrared will provide diagnostics of the plasma surrounding these objects. Consequently, high time-resolution polarimeters on the next generation of UVOIR telescopes whether the extremely large ground-based telescopes such as GMT, TMT and ELT or space-based observatories such as LUVOIR will be excellent probes of neutron-star physics and high-energy astrophysics in general.

References

Albano A, Turolla R, Israel GL, Zane S, Nobili L, Stella L (2010) Astrophys J 722:788–802
Baldini L, Muleri F, Soffitta P, Omodei N, Pesce-Rollins M, Sgro C, Latronico L, Spada F, Manfreda A, Di Lalla N (2016) XIMPOL: a new X-ray polarimetry observation-simulation and analysis framework. In: 41st COSPAR scientific assembly, COSPAR meeting, vol 41
Baring MG, Harding AK (2007), Astrophys Space Sci 308:109–118
Baring MG, Wadiasingh Z, Gonthier PL, Harding AK (2017) Hard X-ray quiescent emission in magnetars via resonant Compton upscattering. J Phys Conf Ser 932:012021
Beloborodov AM (2013) Astrophys J 762:13
Beloborodov AM, Thompson C (2007) Astrophys J 657:967–993
Brinkmann W (1980) Astron Astrophys 82:352–361
Burwitz V, Haberl F, Neuhäuser R, Predehl P, Trümper J, Zavlin VE (2003) Astron Astrophys 399:1109–1114
Cadez A, Javornik M (1981) Astrophys Space Sci 77:299–318
Canuto V, Lodenquai J, Ruderman M (1971), Phys Rev D 3:2303–2308
Cheng AF, Ruderman MA (1979) Astrophys J 229:348–360
Daugherty JK, Ventura J (1978) Phys Rev D 18:1053–1067
Fernández R, Davis SW (2011) Astrophys J 730:131
Fernández R, Thompson C (2007) Astrophys J 660:615–640
Goldreich P, Julian WH (1969) Astrophys J 157:869
González Caniulef D, Zane S, Taverna R, Turolla R, Wu K (2016) Mon Not R Astron Soc 459:3585–3595
González-Caniulef D, Zane S, Turolla R, Wu K (2019) Mon Not R Astron Soc 483:599–613
Götz D, Mereghetti S, Tiengo A, Esposito P (2006) Astron Astrophys 449:L31–L34
Günther HM, Egan M, Heilmann RK, Heine SNT, Hellickson T, Frost J, Marshall HL, Schulz NS, Theriault-Shay A (2017) REDSoX: Monte-Carlo ray-tracing for a soft x-ray spectroscopy polarimeter. In: Society of photo-optical instrumentation engineers (SPIE) conference series, vol 10399, p 1039917
Hamada T (1975) Publ Astron Soc Jpn 27:275–286
Hamada T (1980) Publ Astron Soc Jpn 32:117
Hamada T, Kanno S (1974) Publ Astron Soc Jpn 26:421
Heisenberg W, Euler H (1936) Z Phys 98:714–732
Herold H (1979) Phys Rev D 19:2868–2875
Heyl JS, Hernquist L (1997) J Phys A Math Gen 30:6485–6492
Heyl JS, Shaviv NJ (2000) Mon Not R Astron Soc 311:555–564
Heyl JS, Shaviv NJ (2002) Phys Rev D 66(2):023002
Heyl JS, Shaviv NJ, Lloyd D (2003) Mon Not R Astron Soc 342:134–144
Heyl JS, Lloyd D, Shaviv NJ (2005) Astrophysics, ArXiv e-prints, astro-ph/0502351
Ho WCG (2007) Mon Not R Astron Soc 380:71–77
Ho WCG, Lai D (2003) Mon Not R Astron Soc 338:233–252
Ho WCG, Lai D, Potekhin AY, Chabrier G (2003) Astrophys J 599:1293–1301
Kanno S, Hamada T (1975) Publ Astron Soc Jpn 27:545–552
Kaspi VM, Boydstun K (2010) Astrophys J 710:L115–L120
Kostenko A, Thompson C (2018) Astrophys J 869:44
Kubo H, Nagata R (1981) J Opt Soc Am 71(3):327–333
Kubo H, Nagata R (1983) J Opt Soc Am 73(12):1719–1724
Kuiper L, Hermsen W, Mendez M (2004) Astrophys J 613:1173–1178
Lai D (2001) Rev Mod Phys 73:629
Lai D, Ho WCG (2002) Astrophys J 566(1):373
Lai D, Salpeter EE (1997) Astrophys J 491:270–285
Lieu R (1983) Mon Not R Astron Soc 205:973–981
Lloyd DA (2003) Astrophysics, ArXiv e-prints, astro-ph/0303561

Lyutikov M, Gavriil FP (2006) Mon Not R Astron Soc 368:690–706
Lyutikov M, Thompson C (2005) Astrophys J 634:1223–1241
Marsden D, White NE (2001) Astrophys J 551:L155–L158
Medin Z, Lai D (2006) Phys Rev A 74:062508
Medin Z, Lai D (2007) Mon Not R Astron Soc 382:1833–1852
Medin Z, Lai D, Potekhin AY (2008) Mon Not R Astron Soc 383:161–172
Mereghetti S, Götz D, Mirabel IF, Hurley K (2005) Astron Astrophys 433:L9–L12
Mészáros P (1992) High-energy radiation from magnetized neutron stars. University of Chicago Press, Chicago
Mignani R (2018) Galaxies 6:36
Mignani RP (2011) Adv Space Res 47:1281–1293
Mignani RP et al (2012) Astron Astrophys 544:9 pp., id. A100. https://doi.org/10.1051/0004-6361/201219177
Mignani RP, Testa V, Marelli M, De Luca A, Salvetti D, Belfiore A, Pierbattista M, Razzano M, Shearer A, Moran P (2016) Astrophys J 825:151–15
Mignani RP, Testa V, González Caniulef D, Taverna R, Turolla R, Zane S, Wu K (2017) Mon Not R Astron Soc 465:492–500
Mignani RP, Shearer A, de Luca A, Marshall FE, Guillemot L, Smith DA, Rudak B, Zampieri L, Barbieri C, Naletto G, Gouiffes C, Kanbach G (2018), arXiv e-prints
Mignani RP et al (2019) Astrophys J 871(2): article id. 246. https://doi.org/10.3847/1538-4357/aafb04
Moran P, Mignani RP, Collins S, de Luca A, Rea N, Shearer A (2013) Mon Not R Astron Soc 436:401–412
Mori K, Ho WCG (2007) Mon Not R Astron Soc 377:905–919
Nagel W, Ventura J (1983) Astron Astrophys 118:66–74
Nobili L, Turolla R, Zane S (2008a) Mon Not R Astron Soc 386:1527–1542
Nobili L, Turolla R, Zane S (2008b) Mon Not R Astron Soc 389:989–1000
Potekhin AY (2014) Phys Usp 57:735-770
Potekhin AY, Lai D (2007) Mon Not R Astron Soc 376:793–808
Potekhin AY, Suleimanov VF, van Adelsberg M, Werner K (2012) Astron Astrophys 546:A121
Potekhin AY, Ho WCG, Chabrier G (2016) Proc Sci, PoS(MPCS2015)016
Pavlov GG, Kaminker AD (1975) Sov Astron Lett 1:181–183
Pavlov GG, Shibanov IA (1979) Zhurnal Eksperimentalnoi i Teoreticheskoi Fiziki 76:1457–1473
Pavlov GG, Shibanov YA, Ventura J, Zavlin VE (1994) Astron Astrophys 289:837–845
Rajagopal M, Romani RW, Miller MC (1997) Astrophys J 479:347–356
Rangelov B, Pavlov GG, Kargaltsev O, Reisenegger A, Guillot S, van Kerkwijk MH, Reyes C (2017) Astrophys J 835:264
Rea N, Turolla R, Zane S, Tramacere A, Stella L, Israel GL, Campana R (2007) Astrophys J 661:L65–L68
Rea N, Zane S, Turolla R, Lyutikov M, Götz D (2008) Astrophys J 686:1245–1260
Ruderman M (1971) Phys Rev Lett 27:1306–1308
Santangelo A, Zane S, Feng H, Xu R, Doroshenko V, Bozzo E, Caiazzo I, Zelati FC, Esposito P, González-Caniulef D, Heyl J, Huppenkothen D, Israel G, Li Z, Lin L, Mignani R, Rea N, Orlandini M, Taverna R, Tong H, Turolla R, Baglio C, Bernardini F, Bucciantini N, Feroci M, Fürst F, Göğüş E, Güngör C, Ji L, Lu F, Manousakis A, Mereghetti S, Mikusincova R, Paul B, Prescod- Weinstein C, Younes G, Tiengo A, Xu Y, Watts A, Zhang S, Zhan SN (2019) Sci China Phys Mech Astron 62:29505
Sartore N, Tiengo A, Mereghetti S, De Luca A, Turolla R, Haberl F (2012) Astron Astrophys 541:A66
Schwinger J (1951) Phys Rev 82:664–679
Shannon RM, Heyl JS (2006) Mon Not R Astron Soc 368(3):1377–1380
Shaviv NJ, Heyl JS, Lithwick Y (1999) Mon Not R Astron Soc 306:333–347

She R, Feng H, Muleri F, Soffitta P, Xu R, Li H, Bellazzini R, Wang Z, Spiga D, Minuti M, Brez A, Spandre G, Pinchera M, Sgrò C, Baldini L, Wen M, Shen Z, Pareschi G, Tagliaferri G, Tayabaly K, Salmaso B, Zhan Y (2015) LAMP: a micro-satellite based soft X-ray polarimeter for astrophysics. In: UV, X-ray, and gamma-ray space instrumentation for astronomy XIX, Proceedings of SPIE, vol 9601, p 96010I
Soffitta P (2017) IXPE the imaging X-ray polarimetry explorer. In: Society of photo-optical instrumentation engineers (SPIE) conference series, vol 10397, p 103970I
Stoneham RJ (1980) Opt Acta 27:545–548
Suleimanov V, Potekhin AY, Werner K (2009) Astron Astrophys 500:891–899
Suleimanov VF, Pavlov GG, Werner K (2012) Astrophys J 751:15
Suleimanov VF, Klochkov D, Pavlov GG, Werner K (2014) Astrophys J Suppl Ser 210:13
Taverna R, Muleri F, Turolla R, Soffitta P, Fabiani S, Nobili L (2014) Mon Not R Astron Soc 438:1686–1697
Taverna R, Turolla R, Gonzalez Caniulef D, Zane S, Muleri F, Soffitta P (2015) Mon Not R Astron Soc 454:3254–3266
Tendulkar SP, Hascöet R, Yang C, Kaspi VM, Beloborodov AM, An H, Bachetti M, Boggs SE, Christensen FE, Craig WW, Guillot S, Hailey CA, Harrison FA, Stern D, Zhang W (2015) Astrophys J 808:32
Thirumalai A, Heyl JS (2014) Adv At Mol Opt Phys 63:323
Thompson C, Lyutikov M, Kulkarni SR (2002) Astrophys J 574:332–355
Tiengo A, Mereghetti S (2007) Astrophys J 657:L101–L104
Turolla R, Zane S, Drake JJ (2004) Astrophys J 603:265–282
van Adelsberg M, Perna R (2009) Mon Not R Astron Soc 399:1523–1533
van Adelsberg M, Lai D, Potekhin AY, Arras P (2005) Astrophys J 628:902–913
Wadiasingh Z, Baring MG, Gonthier PL, Harding AK (2018) Astrophys J 854:98
Wang C, Lai D (2009) Mon Not R Astron Soc 398:515–527
Weisskopf V (1936) K Dan Vidensk Selsk Biol Skr 14:714–732
Zane S, Turolla R (2006) Mon Not R Astron Soc 366:727–738
Zane S, Turolla R, Stella L, Treves A (2001) Astrophys J 560:384–389
Zane S, Rea N, Turolla R, Nobili L (2009) Mon Not R Astron Soc 398:1403–1413
Zane S, Turolla R, Nobili L, Rea Na (2011) Adv Space Res 47:1298–1304
Zhang SN et al (2016) eXTP: enhanced X-ray timing and polarisation mission. In: Space telescopes and instrumentation 2016: ultraviolet to gamma ray, Proceedings of SPIE, vol 9905, p 99051Q
Zhang S et al (2019) Sci China Phys Mech Astron 62:29502

Chapter 13
Magnetic Fields in Gamma-Ray Bursts and Their Polarised Emission

Shiho Kobayashi

Abstract A widely accepted model for producing gamma-ray bursts (GRBs) is based on the dissipation of an ultra-relativistic jet. Synchrotron emission is believed to be the dominant radiation process for the main emission components of GRBs, i.e. prompt gamma-rays, reverse shock emission and afterglow. Although the presence of strong magnetic fields in the emission regions is crucial to produce the bright emission, the origin and its role in the dynamics are still unknown. Understanding the nature of the relativistic jet, especially the energy content, acceleration and collimation, is a major focus of international theoretical and observational effort. Polarisation is one of characteristics of synchrotron emission, and polarimetric observations would allow us to study the magnetic field structure and energy distribution in GRB jets.

13.1 Introduction

Gamma-ray bursts (GRBs) are the brightest photon sources in the universe. GRBs provide direct access to regions of extreme physics (e.g. relativistic shocks, strong gravity) as well as acting as important probes of the high redshift universe (e.g. Piran 1999; Zhang and Mészáros 2004). These events are broadly classified as either long or short depending on burst duration. Long bursts are more common and last for more than 2 s, while short bursts last less than 2 s and tend to have harder spectra. The core-collapses of massive stars and the mergers of binary compact stellar objects such as neutron stars and black holes are possible progenitors for long and short bursts, respectively. Long bursts have been linked to star forming galaxies, and in many cases to core-collapse supernovae. The binary neutron star merger GW 170817 was the first multi-messenger event observed in both gravitational and electromagnetic waves (Abbott et al. 2017a,b,c). The gravitational wave detection

S. Kobayashi (✉)
Astrophysics Research Institute, Liverpool John Moores University, Liverpool, UK
e-mail: S.Kobayashi@ljmu.ac.uk

R. Mignani et al. (eds.), *Astronomical Polarisation from the Infrared to Gamma Rays*, Astrophysics and Space Science Library 460,
https://doi.org/10.1007/978-3-030-19715-5_13

in coincidence with the short burst GRB 170817A confirms that at least a subset of short GRBs is produced by binary neutron star mergers, although the peak isotropic luminosity of GRB 170817A is abnormally low.

A widely accepted model for producing GRBs is based on the dissipation of a relativistic outflow, the internal energy produced by shocks is believed to be radiated via the synchrotron process. Relativistic outflow from a GRB central engine is conventionally assumed to be a baryonic jet, producing synchrotron emission with tangled magnetic fields generated locally by instabilities in shocks (e.g. Medvedev and Loeb 1999). An alternative magnetic model is now attracting more attention from researchers (e.g., Drenkhahn and Spruit 2002; Fan et al. 2004; Zhang and Kobayashi 2005; Lyutikov 2006; Giannios 2008; Mimica et al. 2009, 2010; Zhang and Pe'er 2009; Zhang and Yan 2011; Narayan et al. 2011; Granot 2012). The rotation of a black hole and an accretion disk might cause a helical outgoing magnetohydrodynamic (MHD) wave which accelerates material frozen into the field lines (Tchekhovskoy et al. 2008; McKinney and Blandford 2009; Komissarov et al. 2009). In the magnetic models, the outflow is expected to be threaded with globally ordered magnetic fields.

13.2 The Fireball Model

Ultra-relativistic motion (Lorentz factor $\Gamma \gtrsim 100$) is an essential ingredient in the standard GRB model, and it was originally introduced into GRB models to bypass the compactness problem (how very compact and energetic sources can emit non-thermal radiation, e.g. Piran 1999). Later observations show that GRBs are actually associated with relativistic jets. The detection of GeV photons by *Fermi* requires $\Gamma \gtrsim 10^3$ based on the opacity arguments for some short and long GRBs (Abdo et al. 2009a,b). The onset times of many afterglow indicate relativistic motion with $\Gamma \sim$ a few hundreds. Superluminal motion was also detected for two near-by events, GRB 030329 and the neutron star merger event GW 170817 (Taylor et al. 2004; Mooley et al. 2018b). The standard scenario (the so-called fireball model, see Fig. 13.1)

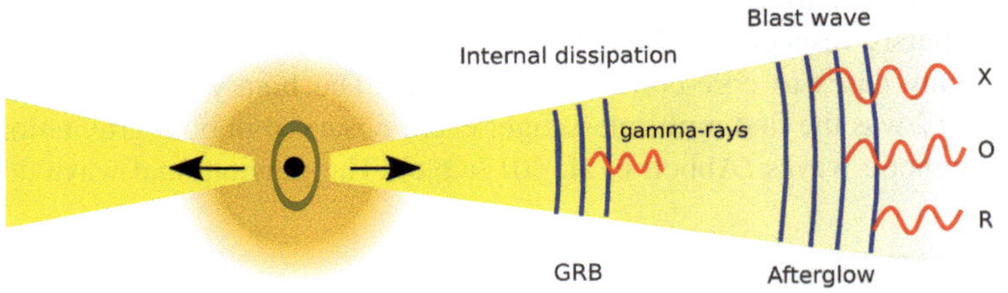

Fig. 13.1 Schematic of the Fireball model, illustrating a relativistic jet from a collapsing star. The prompt gamma-rays (GRB) is due to an internal dissipation process in the original jet, the X-ray (X), optical (O) and radio (R) afterglow originate from the collimated blast wave

has been extensively tested against observations, and it has become the leading paradigm to understand GRB events (e.g. Kumar and Zhang 2015). The scenario consists of several stages,

1. The energy source (or sometimes called the central engine) is believed to be rapid mass accretion onto a newly formed stellar-mass black hole (or highly magnetised neutron star). The energy release drives an ultra-relativistic jet with Lorentz factor $\Gamma \gtrsim 100$. Unlike other astrophysical jets, GRB jets are transient (i.e. explosive events). Since the matter and energy are highly concentrated in a thin layer at the forefront of the "jets", GRB jets are sometimes called "flying pancakes" (Piran 1999).
2. The prompt gamma-ray emission is assumed to be due to internal shocks that arise in an irregular jet (or due to magnetic reconnections in a magnetised jet) (e.g. Rees and Mészáros 1994; Kobayashi et al. 1997; Daigne and Mochkovitch 1998; Zhang and Yan 2011). This dissipation occurs at a large distance from the central engine around 10^{13}–10^{15} cm. Electrons are accelerated to high energy within the collisionless shocks, and produce the observed gamma-rays via the synchrotron process.
3. At further larger distances $\sim 10^{16}$ cm, the jet impacts on the surrounding medium, and the explosion energy is transferred from the original jet (the ejecta from the central engine) to the ambient matter. A short-lived, *optical flash* is emitted from the reverse shocked ejecta at this stage (Mészáros and Rees 1997; Sari and Piran 1999b; Kobayashi 2000). After the energy transfer, the forward shock ambient medium (a collimated blast wave) carries most of the explosion energy.
4. The collimated blast wave further expands and approaches the self-similar Blandford-McKee solution (Blandford and Mckee 1976, strictly speaking, it is described by a conical section of the spherical solution). The blast wave continuously accelerates electrons to high energies. This causes long lasting synchrotron emission which accounts for *afterglow*. As more material accumulates, the blast wave slows down. The typical frequency of the afterglow is gradually going down to lower frequencies (X-ray, optical and radio).
5. Around 10^{18} cm, the blast wave collects sufficient material to slow it down to sub-relativistic velocities, and it approaches the Sedov–Taylor solution. This happens about a few month to a year after the burst. The emission is mostly in the radio band in this final epoch.

Early X-ray afterglow presents complex temporal behaviour (Nousek et al. 2006; Zhang et al. 2006). The light curve initially decays very steeply (known as the steep decay phase), and then shows a very shallow declining phase (the plateau phase) until $10^4 - 10^5$ s after the explosion and before settling at the regular decay expected from the blast wave. About 1/3 of GRBs show *X-ray flares* in the early afterglow phase which are significant re-brightening (up to a factor of 500 in some cases) with short variability timescales of $\Delta t/t \sim 0.2$ on average where t is the time of the flare's onset and Δt is the duration of the flares (Burrows et al. 2005). Although this complex behaviour (especially the plateau phase) is not fully understood yet, the highly variable nature of X-ray flares implies that they are likely to be the results of the same mechanism producing the prompt gamma-ray emission.

While the usual radio afterglow rise on timescale of weeks or months, some events have a *radio flare* peaking at about 1 day, rising quickly before the peak and decaying quickly after the peak. These flares have been interpreted as the emission from the ejecta earlier shocked by the reverse shock. The same particles in the ejecta producing the optical flash have been cooled adiabatically, and their emission shifts quickly toward lower frequencies while weakening (Kulkarni et al. 1999b; Sari and Piran 1999a,b; Kobayashi and Sari 2000).

13.3 Afterglow and Shock Instability

The production of synchrotron radiation requires the presence of a magnetic field in the shocked ejecta and shocked ambient material, but the degree of magnetisation and the configuration of the magnetic fields still remain a matter of debate (e.g. Piran 2005; Sironi et al. 2015).

For the afterglow study, it is usually assumed that the energy densities of the relativistic electrons and of the magnetic field can be characterised by equipartition parameters, ϵ_e and ϵ_B that measure the ratios of these energies to the internal energy produced by shocks. The Blandford-McKee self-similar solution combined with the synchrotron radiation model (Sari et al. 1998) provides an excellent description of the afterglow light curves and spectra. Possibly this is the most successful part of the GRB theory. Afterglow modelling has been carried out mainly for long GRBs, because they produce brighter afterglow due to the larger explosion energy and higher ambient density. Fits to the observed afterglow spectra suggest $\epsilon_e = 10^{-2} - 10^{-1}$ and $\epsilon_B = 10^{-4} - 10^{-2}$ for most events (Panaitescu and Kumar 2001). Considering that the internal energy density in the blast wave (the forward shock ambient matter) is $e = 4\Gamma^2 n m_p c^2$ (Blandford and Mckee 1976; Sari and Piran 1995) where $n \sim 1$ proton/cm^3 is the ISM density and m_p is the proton mass, the magnetic field strength in the blast wave is given by

$$B = (32\pi m_p \epsilon_B n)^{1/2} \Gamma c \sim 0.4\, n \left(\frac{\epsilon_B}{10^{-3}}\right)^{1/2} \left(\frac{\Gamma}{30}\right) \text{G}. \qquad (13.1)$$

This is much larger than the value obtained by a simple shock compression of the intergalactic magnetic field of $\sim \mu$G.

Magnetic fields in massive stars might be dragged out with the stellar wind before the explosion, and the blast wave could propagate in magnetised wind medium. However, since the strength of the fields quickly decay at large radii (e.g Usov and Melrose 1992), it is not sufficiently strong in the afterglow region $R \sim 10^{16}$–10^{17} cm. In other words, the magnetisation of the massive stars would not be high enough to produce a highly magnetised wind medium (see Walder et al. 2012, for magnetic fields in massive stars and their winds). Afterglow modelling has been carried out by assuming the wind density distribution $n = \dot{M}/4\pi m_p v r^2 \propto r^{-2}$ or the ISM distribution n = constant, where $\dot{M} \sim 10^{-5} M_{\odot}\text{year}^{-1}$ is the mass loss rate of the massive star and $v \sim 10^3\,\text{km}\,\text{s}^{-1}$ is the stellar wind velocity (Li and Chevalier 2001). Interestingly, such modelling often favours the ISM ambient medium (Panaitescu and Kumar 2002), although massive stars are supposed to be the progenitors of long GRBs.

The central engine (e.g. a black hole accretion disk system) might produce highly magnetised ejecta at the explosion. However, the afterglow originates from a different emission region (i.e. the forward shock ambient medium rather than the reverse shock ejecta). The strong magnetic field of $\sim$ 1 G, which the afterglow modelling requires, is believed to be generated locally in the shock. In the context of GRB studies, the most discussed instability is the Weibel instability which develops through a charge separation in the background plasma, triggered by magnetic fluctuations. It has been shown that such an instability could lead to growth of the magnetic field and to particle acceleration (e.g. Medvedev and Loeb 1999; Gruzinov and Waxman 1999; Nishikawa et al. 2003; Spitkovsky 2008; Sironi et al. 2015). However, it is not well-understood yet how the magnetised turbulence evolves in the downstream of the shock, and it has a direct implication to afterglow observations.

It has been suggested (Gruzinov and Waxman 1999; Gruzinov 2001) that the coherence length of the fields generated in Weibel instability should be comparable to the plasma skin-depth δ, and that magnetic power on such small scales is expected to decay rapidly. Magnetic fields would maintain its equipartition magnitude only over a skin depth δ scale in the downstream, while modelling of afterglow observations requires that magnetic fields should persist over scales at least several orders of magnitudes larger than δ. To reconcile the problem, a few scenarios have been discussed (Sironi et al. 2015): the small-scale turbulence might evolve to larger scales through inverse cascade effects (Medvedev et al. 2005), it might be dissipated but at a rate which allows to explain the observations (Lemoine 2013), or large-scale fields might be seeded in the downstream by external instabilities (Sironi and Goodman 2007; Levinson 2009).

13.4 Forward and Reverse Shock Modelling

Determination of the magnetic field properties of the original jets (the original ejecta from the central engine) is a key to understanding the driving mechanism of the explosion. Two methods, (1) the forward and reverse shock modelling of early afterglow and (2) early afterglow and prompt gamma-ray polarimetry have been performed in this context. Since the former is sensitive to the strength of magnetic fields, while the latter depends on the coherence length and geometry of the magnetic fields, the two methods are complementary. We discuss the method (1) in this section, and polarimetric studies will be discussed in the next section.

Because of the relativistic beaming effect, the radiation from a jet can be described by a spherical model with an isotropic explosion energy E. The actual energy in the jet with a solid angle Ω is given by $(\Omega/4\pi)E$. The original jet decelerates when a significant fraction of the energy is transferred to the ambient medium. Equalising the energy E with the blast wave energy $(4/3)\pi m_p c^2 n R_d^3 \Gamma^2$, we can estimate the deceleration radius R_d of the ejecta. One factor of Γ in the blast wave energy comes from the bulk expansion of the blast wave, and the other from the shock heating. It takes $R_d/\beta c$ for the jet to reach to the deceleration radius where $\beta = (1 - 1/\Gamma^2)^{1/2}$ is the jet velocity in the unit of the speed of light c. Since the jet propagates toward the observer, the apparent deceleration time t_d for the observer is shorter by a factor of $1 - \beta \sim 1/2\Gamma^2$. Using $E_{52} = E/10^{52}$ erg, we get the deceleration time as

$$t_d \sim \frac{R_d}{2\Gamma^2 c} = \left(\frac{3E}{32\pi\Gamma^8 n m_p c^5}\right)^{1/3} \sim 90\ (E_{52}/n)^{1/3} \left(\frac{\Gamma}{100}\right)^{-8/3} \text{ s.} \tag{13.2}$$

The energy transfer from the ejecta to the ambient medium is due to two shocks: a forward shock propagating into the ambient medium and a reverse shock propagating into the original ejecta (Sari and Piran 1995; Kobayashi et al. 1999; Kobayashi and Zhang 2007). At the deceleration time, the two shocked regions separated by the contact discontinuity have the same bulk Lorentz factor and internal energy density e. The typical frequency of synchrotron emission depends on the Lorentz boost Γ and the magnetic field B and the electron's random Lorentz factor γ_e as $\nu_m \propto \Gamma B \gamma_e^2$ (e.g. Sari et al. 1998). The first two factors are the same for the two shocked regions if ϵ_B is the same for the two regions, while the electron's Lorentz factor is different by a factor of Γ (the two shocked regions contain the same amount of the internal energy, but the ejecta is heavier by a factor of Γ. The internal energy is distributed to a larger number of electrons. Then, the effective temperature is lower by a factor of Γ). Since the typical frequencies of the reverse shock emission is lower by a factor of Γ^2, if the forward shock emits X-rays, the reverse shock would emit optical emission (i.e. optical flash).

Since the ejecta might be endowed with primordial magnetic fields at the central engine, introducing a new parameter $\mathcal{R}_B = \epsilon_{B,r}/\epsilon_{B,f}$ where the subscripts f and r indicate forward shock and reverse shock, respectively, we have a relation for the typical frequencies of the forward and reverse shock emission at the deceleration time (Zhang et al. 2003; Kobayashi and Zhang 2003),

$$\frac{\nu_{m,r}}{\nu_{m,f}} = \Gamma^{-2} \mathcal{R}_B^{1/2}. \tag{13.3}$$

We can give similar discussion on the peak flux of the two shock emission (the peaks in the spectral domain at the deceleration time). The peak flux $F_{\nu,\max}$ is proportional to the number of accelerated electrons N_e, the magnetic field B and the Lorentz boost Γ (e.g. Sari et al. 1998). From the energy conservation, the mass of the ejecta is larger by a factor of Γ at the deceleration time than that of the ambient medium swept by the forward shock. Since the number of electrons is proportional to the mass, we get another relation at the deceleration time t_d,

$$\frac{F_{\nu,\max,r}}{F_{\nu,\max,f}} = \Gamma \mathcal{R}_B^{1/2}. \tag{13.4}$$

After this onset of the afterglow at t_d, the reverse shock optical emission decays as $\sim t^{-2}$, while the forward shock optical emission starts to decay as $\sim t^{-1}$ later when the typical frequency of the forward shock emission $\nu_{m,f}$ crosses the optical band (see the top-left panel of Fig. 13.2; Kobayashi 2000, Zhang et al. 2003). If $\nu_{m,f}$ is already below the optical band at t_d (it means $\nu_{m,r}$ is in the microwave or radio band), the forward shock emission also peaks at t_d, and the reverse shock emission is likely to be masked by the forward shock emission (the bottom-left panel Harrison and Kobayashi 2013). If the deceleration time t_d (the onset of afterglow) is detected, the Lorentz factor of the original jet can be estimated by using Eq. (13.2) (e.g. Melandri et al. 2010; Liang et al. 2010; Cui et al. 2018).

The strength of the reverse shock emission depends on the magnetisation of the ejecta. If the ejecta is endowed with primordial magnetic fields, the optical flash could be more prominent as illustrated in the right panel of Fig. 13.2. Using the relative strength of the two emission components and Eqs. (13.3) and (13.4), we can estimate the magnetisation parameter $\mathcal{R}_B$ (see Zhang et al. 2003; Gomboc et al. 2008; Harrison and Kobayashi 2013; Gao et al. 2015, for the details). The magnetisation parameter has been evaluated for the events which show a clear

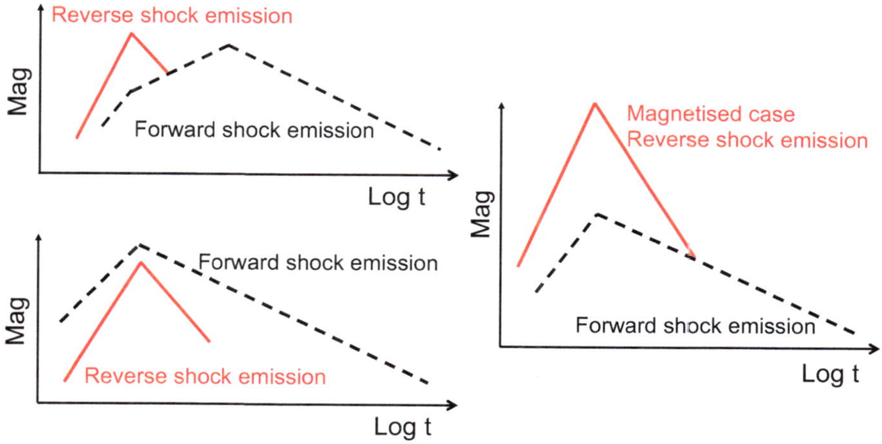

Fig. 13.2 Early optical afterglow light curves with different forward reverse shock contributions. Forward shock emission (black dashed lines) and reverse shock emission (red solid lines). The two left panels show non-magnetised cases $\mathcal{R}_B \sim 1$. If $\nu_{m,f}$ is above the optical band at t_d, the forward shock emission peaks when $\nu_{m,f}$ crosses the optical band (top-left panel). If $\nu_{m,f}$ is below the optical band at t_d, the forward shock emission also peaks at t_d (bottom-left panel). The right panel shows a magnetised case $\mathcal{R}_B \gg 1$

reverse shock signature (e.g. an early steep decay t^{-2} in the optical afterglow light curve) including GRB 990123, GRB 021004, GRB 021211, GRB 041219A, GRB 061126, GRB 090102, GRB 090510, GRB 110205A, GRB 120308A, GRB 130427A, GRB 140512A, GRB 161219B (e.g. Fan et al. 2002, 2005; Kumar and Panaitescu 2003; Gendre et al. 2012; Zheng et al. 2012; Laskar et al. 2013, 2018; Japelj et al. 2014; Zhang et al. 2015; Fraija et al. 2016; Huang et al. 2016). Although $\mathcal{R}_B \lesssim 10$ have been obtained for most of them, very high values $\mathcal{R}_B = 200 - 8000$ are reported for GRB 990123, GRB 090102 and GRB 140512A (Zhang et al. 2003; Harrison and Kobayashi 2013; Huang et al. 2016). This implies that at least some fraction of GRB ejecta carry significant primordial magnetic fields (Fan et al. 2002; Zhang et al. 2003). Nevertheless, since $\epsilon_{B,f} \sim 10^{-4} - 10^{-2}$ is usually inferred from afterglow modelling (Panaitescu and Kumar 2001), and a further lower value of $\epsilon_{B,f} \sim 10^{-6}$ is obtained for GRB 990123 (Panaitescu and Kumar 2004), it follows that $\epsilon_{B,r} \ll 1$. Therefore, all above cases correspond to mildly magnetised outflow and not a Poynting flux dominated outflow at the deceleration radii. The ratio σ between the Poynting flux energy and the kinetic energy is estimated as $\sigma \lesssim 10^{-3}$ for these events (Harrison and Kobayashi 2013).

The detection of bright optical flash suggests that a moderately magnetised jet in which the magnetic field is strong enough to enhance the reverse shock emission but not strong enough to suppress the formation of reverse shock or the Fermi particle acceleration (Zhang and Kobayashi 2005; Giannios et al. 2008; Mimica et al. 2009; Sironi and Spitkovsky 2010; Sironi et al. 2015). If future events indicate high values $\sigma \gtrsim 0.1$, an interesting possibility to reconcile the problem is that optical flash (and the prompt gamma-rays) would be produced through a dissipative MHD process rather than shocks (Giannios and Spruit 2006; Lyutikov 2006; Giannios 2008; Zhang and Yan 2011).

13.5 Polarisation Signals

Since synchrotron emission is believed to be the dominant radiation process for the main emission components of GRBs, i.e. prompt gamma-rays, reverse shock emission and afterglow, these components could be polarised. The observed degree of linear polarisation ($\mathrm{PD_L}$) and polarisation position angle (PA) depend on the properties and geometry of the emitting region. Polarimetric observations allow us to study the magnetic field structure and energy distribution in GRB jets and their dynamics (Lazzati 2006; Toma 2013; Covino and Götz 2016; Gill et al. 2018). Since circular polarisation is expected to be zero for optically thin synchrotron emission, we mainly discuss linear polarisation.

13.5.1 Visible Region

A unique feature of GRB jets is the small angular extent of the *visible region*, which is much smaller than the actual jet opening angle θ_j (see Fig. 13.3). The visible region is an important concept when we discuss polarised emission from a relativistic outflow. As the blast wave is described by a conical section of the spherical Blandford-McKee solution, the jet head has a curvature, and each fluid element at the jet head propagates radially with a high Lorentz factor. Because of the relativistic beaming effect, it radiates practically only in the radial direction (i.e. their propagation direction). Since the radiation is beamed in a forward cone of opening angle of $\sim 1/\Gamma \ll 1$, it is not visible for the observer if the propagation direction of the fluid element makes an angle large than $1/\Gamma$ with the line of the sight. In Fig. 13.3, the emission from element A is visible, but the one from element B is not visible. The observer cannot see the entire surface of the jet head. The visible region is the small area with an angular size of $1/\Gamma$, located around the point at which the line of sight intersects the jet head. In the figure, the edge of the visible region is indicated by the red dotted lines.

13.5.2 Afterglow Emission

Synchrotron emission is highly polarised if magnetic field is coherent in the visible region. For a power-law distribution of electron energies with an index $p \sim 2$, the degree of linear polarisation is $\mathrm{PD}_{\mathrm{L},0} = (p+1)/(p+7/3) \sim 70\%$ (Rybicki and Lightman 1979). However, if magnetic field is fully tangled and isotropic at any point (or in any point-like region), the emission is unpolarised, regardless of the geometry of the visible region. To produce polarised emission, there must be asymmetry in the configuration of magnetic field.

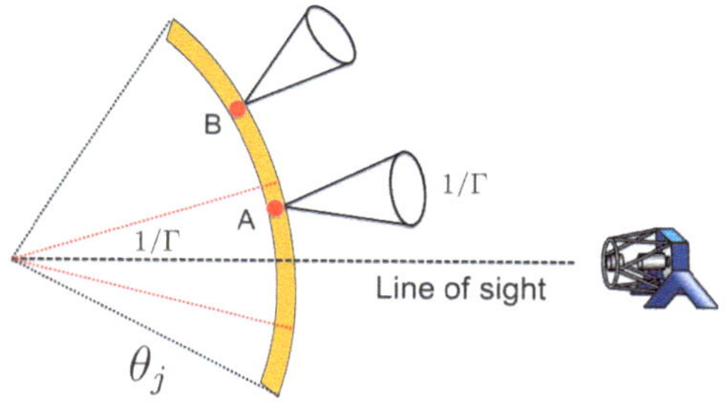

Fig. 13.3 The visible region is bound by the two red dotted lines in this figure. For the observer (the telescope in the figure), the visible region is a small circle, and the line of sight goes through the centre. The emission from each fluid element is beamed in a forward cone of opening angle $\sim 1/\Gamma$. The relativistic jet is characterised by the opening angle θ_j and Lorentz factor Γ

In the afterglow phase, magnetic fields in the emission region (i.e. the blast wave) are produced by shock instabilities, and expected to be tangled. However, the magnetic field generation in collisionless shocks is not fully understood yet, the ultimate fate of the field many skin depths behind the shock front is not clear. If the coherence length of the magnetic field grows at about the speed of light in the local fluid frame, independent, coherent patches of magnetic fields might be created in the visible region (Loeb and Perna 1998; Gruzinov and Waxman 1999). The observed degree of linear polarisation is

$$PD_{\rm L} = \frac{PD_{\rm L,0}}{\sqrt{N}} \sim 10\%, \tag{13.5}$$

where $N \sim 50$ is the number of coherent patches within the visible region of the blast wave (Gruzinov and Waxman 1999). The degree and direction of polarisation should depend on time as the process is stochastic, and the polarisation coherence time is $\Delta t \sim t$ (the time since the explosion).

Another asymmetric configuration of magnetic field also has been discussed (Gruzinov 1999; Ghisellini and Lazzati 1999; Sari 1999). Since at any point in the shock front, the shock normal (the radial direction) is a special direction, the magnetic fields parallel and perpendicular to the shock normal could have significantly different averaged strengths (e.g. Medvedev and Loeb 1999). To discuss how anisotropic field leads to polarised emission, we first consider a small region around the shock front (the dashed circle in the right panel of Fig. 13.4). If the region is sufficiently small, the curvature of the shock front would be neglected, the shocked ambient behind the shock front can be described by a slab of plasma (see the left panel of Fig. 13.4). We first discuss the emission in the slab's rest frame.

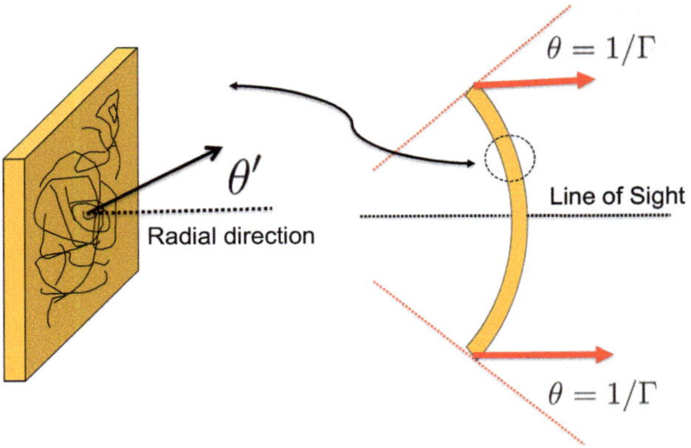

Fig. 13.4 Anisotropic magnetic field and blast wave. Left panel: a small region around the shock front in the local fluid frame. Right panel: the visible region of the shock front in the lab frame (the rest frame of the progenitor which caused an explosion). The polar angles θ' (the local fluid frame) and θ (the lab frame) are measured from the radial direction (the shock-normal direction)

If the slab is observed face-on (the radial direction $\theta' = 0$), magnetic field would be completely tangled and isotropic. The emission should be unpolarised. However, if the strength of the magnetic field parallel to the shock normal $B_{\parallel}$ is different (let us say it is significantly weaker than that of the perpendicular one $B_{\perp}$), the magnetic field has some degree of alignment along the shock front. If the slab is observed edge-on ($\theta' = \pi/2$), the emission should be polarised and the polarisation vector (the electric field vector of the radiation) is in the shock normal direction. The linear polarisation degree at a viewing angle θ' is given (Gruzinov 1999; Sari and Piran 1999a,b) by

$$PD_{\rm L}(\theta') = PD_{\rm L,0} \sin^2\theta' \frac{2\langle B_{\parallel}^2\rangle - \langle B_{\perp}^2\rangle}{2\sin^2\theta'\langle B_{\parallel}^2\rangle + (1+\cos^2\theta')\langle B_{\perp}^2\rangle}. \tag{13.6}$$

If we choose the z-direction of the local fluid frame coordinate to be in the shock normal direction, we have $\langle B_{\perp}^2\rangle = \langle B_x^2\rangle + \langle B_y^2\rangle$ and $\langle B_{\parallel}^2\rangle = \langle B_z^2\rangle$. We obtain $\mathrm{PD_L} = 0$ for three-dimensional isotropic field. If the field is in the shock plane $B_{\perp} \gg B_{\parallel}$, the estimate reduces to $PD_{\rm L}/PD_{\rm L,0} = -\sin^2\theta'/(1+\cos^2\theta')$.

Now we consider the emission from the slab in the lab frame in which the slab is moving in the radial direction with Lorentz factor Γ. Because of the relativistic aberration of photons, the photons emitted at $\theta' = \pi/2$ (the edge-on polarised emission) in the slab's co-moving frame are observed at $\theta \sim 1/\Gamma$ in the lab frame. Considering that the viewing angle of a fluid element is $\theta = 1/\Gamma$ if it is located at the edge of the visible region (the red dotted lines in the right panel of Fig. 13.4), the visible region would be highly polarised at the edge. The polarisation vector is also rotated by the Lorentz transformation (e.g. Lyutikov et al. 2003). The polarisation vector is perpendicular (tangential) to the edge if $\mathrm{PD_L} < 0$ ($\mathrm{PD_L} > 0$). The linear polarisation degree inside the visible region can be evaluated by combining Eq. (13.6) and the aberration formula $\cos\theta' = (\cos\theta - \beta)/(1 - \beta\cos\theta)$.

The visible region appears as a ring on the sky especially at frequencies above the typical synchrotron frequency ν_m due to a relativistic limb-brightening effect (Waxman 1997; Panaitescu and Mészáros 1998; Sari 1998; Granot et al. 1999a,b). Although each segment of the ring is polarised, the net polarisation is zero because of the circular geometry (see the left panel of Fig. 13.5). However, if the surface brightness of the jet (and consequently that of the ring) is not uniform or/and the entire ring is not inside the jet (Gruzinov 1999; Waxman 2003), the net polarisation would be expected. As shown in the right panel of Fig. 13.5, if the line of sight runs slightly inside the jet edge (the black solid line), the bottom part of the ring is located outside the jet, and there is actually no emission from this part. Considering that the polarisation vector in the missing part is vertical, the net polarisation is in the horizontal direction. On the other hand, if the line of sight runs slightly outside the jet edge (the black dashed line), the emission comes only from the top part of the ring. The net emission is dimmer, but more polarised. The polarisation direction is vertical this time.

Afterglow observations suggest that the median opening angle of long GRB jets is about $\theta_j \sim 0.2\,\mathrm{rad}$. It is not well constrained for short GRBs, but they also

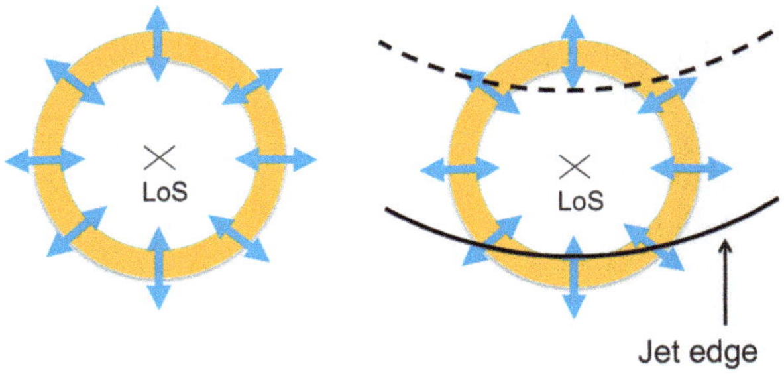

Fig. 13.5 The visible region. Left panel: The blue arrows indicate the polarisation vectors for $PD_L < 0$. The cross shows the point at which the line-of-sight intersects with the visible region. Right panel: a part or most part of the ring is missing if the line of sight runs almost along the jet edge (the black solid line or the black dashed line)

might have similar or slightly wider opening angles (Fong et al. 2015). Since the angular extent of the visible region is very small $1/\Gamma \lesssim 0.01$ in the prompt gamma-ray or early afterglow phase, this off-axis mechanism requires fine tuning of the viewing angle to produce high polarisation in the early phases. Such a viewing angle happens by chance possibly only for narrow jets $\theta_j \sim 1/\Gamma$ (Gruzinov 1999; Waxman 2003; Nakar et al. 2003b). However, the blast wave decelerates as it expands and the angular extent of the visible region grows in the afterglow phase. A favourable viewing angle could be naturally achieved in the afterglow phase (Sari 1999; Ghisellini and Lazzati 1999).

When the Lorentz factor $\Gamma(t) \sim 6(E_{52}/n)^{1/8} t_{\rm day}^{-3/8}$ of the blast wave is sufficiently large, the afterglow emission is well described by the spherical model where $t_{\rm day}$ is the observer time in unit of day. However, once the growing visible region runs off the edge of the jet (the left panel of Fig. 13.6), the emission becomes dimmer than what is expected in the spherical model. This induces a steepening (jet break) in the afterglow light curve (Rhoads 1999; Sari et al. 1999; Panaitescu and Mészáros 1999). Since the typical viewing angle is $\sim 2\theta_j/3$ for the on-axis observer, equalising the viewing angle with the angular extent of the visible region $\sim 1/\Gamma(t)$, we obtain the jet break time as

$$t_j \sim 1.6(E_{52}/n)^{1/3}(\theta_j/0.2)^{8/3}\text{days}. \quad (13.7)$$

If the jet break time is detected, we can estimate the jet opening angle $\theta_j \propto (E/n)^{-1/8} t_j^{3/8}$. If the collimated blast wave expands sideways at the speed of light in its own rest frame, the jet opening angle is given by $\theta_j \sim \theta_{j,0} + ct_p/R \sim \theta_{j,0} + 1/\Gamma$ where t_p is the proper time in the rest frame and R is the blast wave radius in the lab frame. The sideways expansion could become significant at the jet break time. However, for a modest jet opening angle, the sideways expansion is known to be rather slow (e.g. Zhang and MacFadyen 2009; Meliani and Keppens 2010; van Eerten et al. 2010; Granot and Piran 2012). Here we discuss the non-spreading case.

Since it is unlikely that the observer is directed exactly at the centre of the jet, the afterglow emission is expected to be polarised around the jet break time. The light curve of the degree of polarisation would have two maxima, with the polarisation position angle changing by 90° between the first and the second maximum (Ghisellini and Lazzati 1999; Sari 1999). This can be easily understood as follows, even if the entire ring is initially inside the jet (zero polarisation), eventually a part of the visible region runs off the edge of the jet (see the left panel of Fig. 13.6), and the net polarisation is in the horizontal direction in the figure. Since the visible region further expands, at a later time only the top part which is polarised in the vertical direction is inside the jet (the right panel). During the ring expands from the left configuration to the right one in Fig. 13.6, there should be a moment of zero polarisation at which a half of the ring is inside the jet. Although the maximal polarisation degree around the jet break depends on the viewing angle measured from the jet axis θ_{obs}/θ_j, it would be about $\mathrm{PD_L} = 10 - 20\%$ if $B_\perp \gg B_\parallel$. If the surface brightness of the jet is not uniform but structured (e.g. explosion energy per unit solid angle is a function of the angle from the jet axis. Rossi et al. 2002; Zhang and Mészáros 2002), as the dominant emission always comes from the jet-centre side of the (distorted) ring, the polarisation position angle remains constant throughout the whole evolution. The polarisation temporal behaviour is expected to be sensitive to the luminosity distribution of the jet (Rossi et al. 2004).

Optical linear polarisation measurements have been carried out for many GRB afterglows typically several hours to a few days after the prompt gamma-ray emission (Covino and Götz 2016 and the references therein). This is the period in which a jet break is expected to occur. As we will discuss below, the detection or upper limits of the linear polarisation degree are generally low $\lesssim$ a few percent, although at least two events show a 90° rotation of the polarisation position angle around a possible jet break. The low polarisation results imply that $B_\perp$ and $B_\parallel$ have a similar strength in general. In the discussion below, time t indicates the time since the prompt gamma-ray emission.

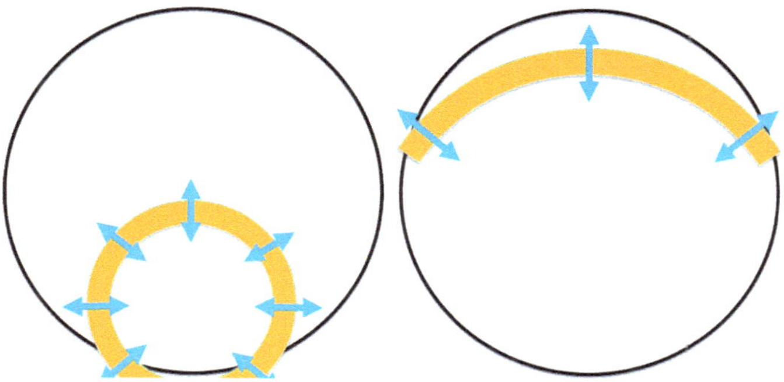

Fig. 13.6 Jet break and polarisation position angle. The black circles show the jet seen face-on. The yellow rings show the visible regions at an early time (left panel) and a late time (right panel)

- **GRB 990123** was a remarkable burst with very high prompt gamma-ray fluence and with very bright optical flash (Akerlof et al. 1999; Sari and Piran 1999a). The reverse shock also produced the radio flare at $t \sim 1$ day (Kulkarni et al. 1999b; Sari and Piran 1999a; Kobayashi and Sari 2000). A jet break was detected at $t_j \sim 2$ days (Kulkarni et al. 1999a). Polarisation of the optical afterglow was measured at $t \sim 18$ h, and an upper limit of 2.3% on the linear polarisation was obtained by using the Nordic Optical Telescope (NOT) (Hjorth et al. 1999).
- **GRB 990510:** The well-sampled optical light curves and simultaneous early time radio observations can be well fit by a simple model for the jet evolution with a jet break time $t_j \sim 1$ day Harrison (1999). The linear polarisation of the optical afterglow was measured by two independent teams at a level of $1.7\% \pm 0.2\%$ at $t = 18.5$ h (Covino et al. 1999), and $1.6\% \pm 0.2\%$ at $t = 20.6$ h (Wijers et al. 1999) with the Very Large Telescope (VLT, Antu). The polarisation degree and angle show no evidence of change.
- **GRB 990712:** Rol et al. (1999) report the detection of polarisation in the optical afterglow with VLT on three instances $t = 0.44$, 0.70 and 1.45 days. The linear polarisation degrees are $2.9\% \pm 0.4\%$, $1.2\% \pm 0.4\%$ and $2.2\% \pm 0.7\%$, respectively, and the polarisation position angle does not vary significantly during these observations. Although optical follow-up observations were performed for more than 30 days, there is no evidence of the jet break in the light curve.
- **GRB 991216:** A gradual optical steepening around $t_j \sim 2$ days has been discussed in the context of the jet break. The steep X-ray afterglow at earlier times might imply that the higher energy emission is concentrated in a narrower jet core that began to spread earlier (Halpern et al. 2000). VLT observations provide upper limits of linear polarisation degree $< 2.7\%$ at $t = 35$ h and $< 5\%$ at $t = 60$ h (Covino et al. 2004).
- **GRB 010222:** The light curve of the optical afterglow shows the decay of the brightness with a broken power law, of which the indices are -0.7 before and -1.3 after the break. The break point of the light curve is $t \sim 0.5$ day (Masetti et al. 2001; Watanabe et al. 2001). Polarimetric observations were obtained at NOT at $t \sim 22.65$ h, the linear polarisation degree is $1.36\% \pm 0.64\%$ (Björnsson et al. 2002).
- **GRB 011211:** The optical afterglow decays as a power law with a slope of -0.83 for the first ~ 2 days after the burst, at which time there is evidence of a break (Holland et al. 2002). Polarimetric observations performed with the VLT on the afterglow at $t \sim 35$ h, the observations yielded an upper limit of 2.7% (Covino et al. 2002).
- **GRB 020405:** The optical light curves of the afterglow are consistent with a single power-law decay of temporal index -1.54 between 1 and 10 days after the prompt event (Bersier et al. (2003a) reports a much steeper decay -1.72 between $t = 1.24$ days and 4.3 days). The red bump detected at $t \sim 20$ days has been modelled with an emerging SN. The light curve at $t \gtrsim 20$ days is consistent with a power-law decay with index -1.85 (Masetti et al. 2003). Linear polarisation measurements were carried out with VLT at $t = 29.5$ h ($1.5\% \pm 0.4\%$) by Masetti et al. (2003) and at $t = 52.0$ h ($1.96\% \pm 0.33\%$) and $t = 76.2$ h ($1.47\% \pm 0.43\%$)

by Covino et al. (2003). Interestingly, Bersier et al. (2003a) measured a high polarisation 9.9% $\pm$ 1.3% at $t = 31.7$ h by using the Multiple Mirror Telescope (MMT). The patchy emission region model (e.g. Gruzinov and Waxman 1999) is not consistent with the rapid change in the polarisation degree (1.5–9.9% in a short period of $\Delta t/t \sim 0.1$) nor the almost constant polarisation position angle. Although early radio data indicate that the ejecta underwent a jet break at $t \sim 1$ day (Berger et al. 2003), there is no signature of a jet break in the optical light curve. The jet-edge models also do not work.

- **GRB 020813:** The optical light curves of the afterglow indicate a jet break $0.33 < t_j < 0.88$ days (Gorosabel et al. 2004 and the references therein). The linear polarisation degree is measured at a level of 1.8–2.4% during $t = 4.7 - 7.9$ h by Barth et al. (2003) with the Keck telescope. VLT polarimetric monitoring campaign of the optical afterglow carried out in three nights, from $t = 0.88$–4.20 days (Gorosabel et al. 2004, see also Lazzati et al. 2004). The linear polarisation is detected at $\sim$ 1% level with an almost constant polarisation position angle, ruling out the 90° rotation expected in a (homogeneous) jet-break model.
- **GRB 021004:** The optical light curves of the afterglow can be interpreted as a power-law decay with 10–20% fluctuations on several timescales, ranging from a few hours down to 20–30 min (Bersier et al. 2003b). The fluctuations can be explained in several models including fluctuations in the ambient matter density, energy injections (refreshed shocks) and angular fluctuations in the jet energy (e.g. Lazzati et al. 2003; Nakar et al. 2003a; Heyl and Perna 2003). A late time steepening can be modelled as a jet break with $t_j = 9$ days (Mirabal 2003). Rol et al. (2003) performed polarimetric observations of the optical afterglow, obtained with NOT and VLT between $t = 8$ and 17 h. The linear polarisation degree is almost constant at a level of 1–2%, but the polarisation position angle rotated by about 45° between the NOT and VLT observations. Late VLT observations even show a 90° change between t =9 and 89 h (Lazzati et al. 2003). The main features in both the light curve and the polarisation fluctuations could be reproduced by the patchy-shell model (Nakar and Oren 2004).
- **GRB 030226:** The optical light curve shows a jet break around $t = 0.8$ days. VLT polarimetry provides an upper limit of linear polarisation degree < 1.1% at $t \sim 1$ day (Klose et al. 2004).
- **GRB 030328:** The light curve of the optical afterglow shows the decay of the brightness with a broken power law, of which the indices are -0.76 before and -1.50 after the break time 0.48 days. VLT polarimetry shows a liner polarisation degree of 2.4% $\pm$ 0.6% at $t \sim 18.5$ h (Maiorano et al. 2006).
- **GRB 030329:** This is a nearby event ($z \sim 0.17$). The very bright optical afterglow makes it the most well-studied afterglow to date. The optical emission steepens from a decay index of -0.873 to -1.97 at $t_j \sim 0.48$ days. The light curve also shows a very large variability a few days after the burst possibly due to refreshed shocks (Granot et al. 2003). VLT polarimetry shows linear polarisation degree at a level of 0.3–2.5% throughout a 38-day period, with

significant variability in polarisation degree and angle on timescales down to hours Greiner et al. (2003).

- **GRB 071010A:** The later evolution of the afterglow requires a break, consistent with being fully achromatic in the optical/NIR and X-ray bands, at $t_j \sim 1$ day, with a post-break decay index of -2.07. VLT polarimetry provides an upper limit of linear polarisation degree $< 1.3\%$ at $t \sim 0.9$ days (Covino et al. 2008).
- **GRB 091018:** The X-ray light curve of the afterglow shows at least two breaks, as well as flaring activity. Although the densely sampled optical light curve also shows several flares, an achromatic break at $t \sim 9$ h was identified (Wiersema et al. 2012). VLT polarimetric monitoring took place over three observing nights. Initially the linear polarisation degree gradually increase up to $\sim 2\%$. After the break, the polarisation degree and polarisation position angle are more variable possibly due to the additional flaring components in the afterglow. Although the scatter is large after the break, the polarisation position angle seems to show a 90° rotation at the break (Wiersema et al. 2012, 2014).
- **GRB 121024A:** Both the X-ray and optical/NIR light curves of the afterglow are best fitted by a broken power law with similar break times $t \sim 10$ h (Wiersema et al. 2014; Varela et al. 2016). The occurrence of a light curve break simultaneously in X-rays and optical wavelengths is suggestive of a jet break origin. However, the post-break decay indices are rather shallow -1.67 (X-ray) and -1.25 (optical) even for non-spreading jets. VLT polarimetry shows linear polarisation degrees at a level of 3–5% and a constant polarisation position angle during $t \sim 3 - 6$ h. Further observations carried out the night after (and after the jet break) show a lower polarisation level with a clear 90° rotation of the polarisation position angle (Wiersema et al. 2014). This would be another identification of the polarisation position angle rotation expected in the jet break model. Surprisingly, circular polarisation $PD_{\mathrm{C}} = 0.61\% \pm 0.13\%$ is detected at $t \sim 4$ h (Wiersema et al. 2014). The linear polarisation degree during the circular polarimetry interval is about 4%, the afterglow shows a ratio $PD_{\mathrm{C}}/PD_{\mathrm{L}} \sim 0.15$. If the electron pitch angle distribution is isotropic, the circular polarisation contributions of electrons with pitch angles $\theta + \epsilon$ and $\theta - \epsilon$ nearly cancel out, and the remaining $\mathrm{PD_C}$ scales with the angular size of the beaming cone of the synchrotron emission $1/\gamma_e$ where γ_e is the random Lorentz factor of electrons. Then, $PD_{\mathrm{C}}/PD_{\mathrm{L}} \sim 1/\gamma_e \sim 10^{-4}$ is expected. The origin of the circular polarisation in the optically thin synchrotron emission is puzzling and unexpected. Even if the electron pitch angle distribution is extremely anisotropic, cancellation would arise from the integration over the emitting region (Nava et al. 2016). Plasma propagation effects (e.g. Faraday conversion) within the source could be strong at low frequencies (close to the synchrotron self-absorption frequency), but negligible at optical frequencies (Matsumiya and Ioka 2003; Sagiv et al. 2004; Toma et al. 2008).

The binary neutron star merger GW 170817 was the first multi-messenger event observed in both gravitational and electromagnetic waves (Abbott et al. 2017a,b). The electromagnetic signal began with a weak, short burst GRB 170817A (Abbott et al. 2017c), which was followed over the next hours and days by the ultraviolet, optical and near-infrared emission from a radio actively powered kilonova (e.g. Arcavi et al. 2017; Coulter et al. 2017; Evans et al. 2017; Pian et al. 2017; Smartt et al. 2017; Tanvir et al. 2017). The late onset of the X-ray, optical and radio afterglow emission suggests emission from a narrow relativistic jet viewed off-axis, initially dominated by material along the line of sight and gradually overtaken by the more energetic core parts of the jet (e.g. Alexander et al. 2017; Granot et al. 2017; Haggard et al. 2017; Hallinan et al. 2017; Kasliwal et al. 2017; Kim et al. 2017; Margutti et al. 2017; Troja et al. 2017a,b; D'Avanzo et al. 2018; Lazzati et al. 2018; Lamb and Kobayashi 2018; Lynman et al. 2018; Mooley et al. 2018a; Nakar et al. 2018; Ruan et al. 2018).

VLT polarimetry shows a linear polarisation degree of $0.50\% \pm 0.07\%$ at 1.46 days after the detection of the gravitational waves. The polarisation measurement was made when the kilonova was still in its blue phase. The low degree of polarisation is consistent with intrinsically unpolarised emission scattered by Galactic dust, suggesting a symmetric geometry of the emitting region and low inclination of the merger system (Covino et al. 2017). VLA radio polarimetry provides an upper limit of 12% at 244 day since the gravitational waves (Corsi et al. 2018). Within the structured jet scenario, the upper-limit on the radio linear polarisation degree constrains the magnetic field configuration in the blast wave (Corsi et al. 2018; Gill and Granot 2018), and it excludes $B_{\parallel} = 0$ (magnetic field fully contained in the plane of the shock), and requires $2\langle B_{\parallel}^2\rangle/\langle B_{\perp}^2\rangle > 0.5$ (semi-isotropic magnetic field). The magnetic configuration seems to be similar to that in the regular GRB blast wave.

13.5.3 Early Afterglow: Reverse Shock Emission

Polarimetric observations of early afterglow have been carried out in recent years by using fully automated "robotic" telescopes. Once the reverse shock crosses the thin ejecta, no new electrons are accelerated anymore, the reverse shock emission rapidly decays in time. Optical flash lasts only for 10–20 mins after the prompt gamma-ray emission. To detect optical flash from GRB events which happen in random directions and at random times, we need robotic telescopes that promptly respond to satellite triggers (X-ray/gamma-ray satellites keep watching the sky with a large field of view) and make observations without the intervention of a human.

The significance of reverse shock emission observations in the GRB study can be understood if we consider that the late (>20 mins) afterglow originates from the blast wave (the forward shocked ambient medium). Since the dynamics of the blast wave is determined only by the ambient density and the explosion energy (or energy per solid angle for structured jets), the late afterglow is insensitive to the

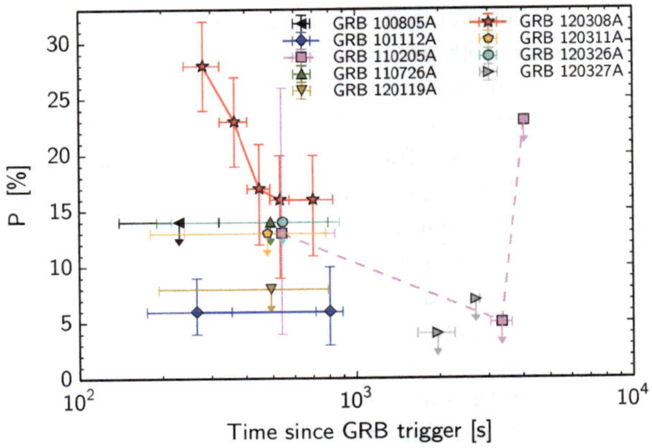

Fig. 13.7 Linear polarisation degree as a function of time after the burst for nine GRBs from the Liverpool Telescope RINGO2 sample. The temporal error bars show the duration of the exposure. From Steele et al. (2017)

properties of the original ejecta from the central engine. The initial Lorentz factor and/or magnetisation of the ejecta can be examined only through the observations of the prompt gamma-rays, reverse shock emission (i.e. optical flashes and radio flares) or possibly X-ray flares. These emission components are believed to originate from the original ejecta from the central engine.

Figure 13.7 shows the optical polarisation measurements (detections or upper limits) of nine events as a function of the observing time since the GRB trigger. We note that all polarisation detection cases in the figure (GRB101112A, GRB110205A and GRB120308A) were achieved at relatively early times $t < 10^3$ s. This reinforces the point that prompt measurements are essential to characterise the polarimetric properties of GRB afterglow; the polarisation degree decays very rapidly as the tight upper limits at late times show (Steele et al. 2017). We list notable polarisation measurements of early afterglow below.

- **GRB 090102:** The steep-shallow decay of optical emission is characteristic of an afterglow whose early time light is dominated by fading radiation generated in the reverse shock. Linear polarisation degree at a 10% level was measured with the Liverpool Telescope (LT) at $t \sim 3$ mins (Steele et al. 2009). The detection of high polarisation provides the first direct evidence that large-scale magnetic fields are present when significant reverse shock emission is produced.
- **GRB 091208B:** Optical linear polarisation in the afterglow is measured with the 1.5 m Kanata telescope at $t = 149 - 706$ s and the polarisation degree is $10.4\% \pm 2.5\%$ (Uehara et al. 2012). The early time light curve with a decay index -0.75 is consistent with the forward shock emission. Uehara et al. (2012) suggest that the magnetic fields are amplified by the MHD instabilities. Another possibility could be that the forward shock propagated in the magnetised wind from the progenitor star (Granot and Königl 2003), although the afterglow light curve implies that the density profile of the external medium is constant rather than wind type.
- **GRB 120308A:** Early afterglow polarimetry was performed by using LT (see Fig. 13.8; Mundell et al. 2013). Linear polarisation degree 28% was detected

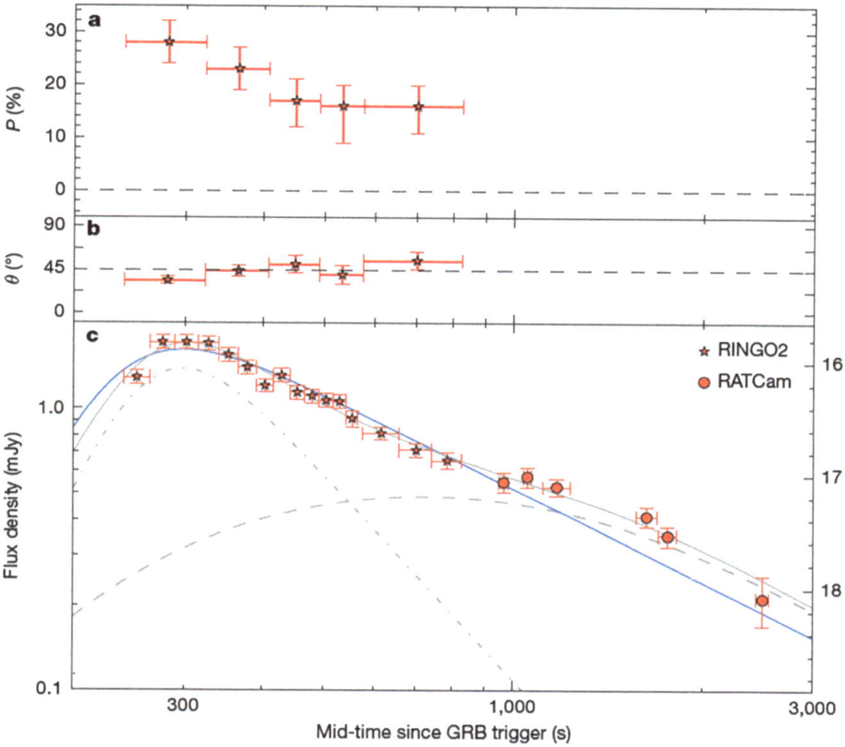

Fig. 13.8 Evolution of optical polarisation and brightness in GRB120308A. Evolution of (**a**) polarisation degree P and (**b**) polarisation position angle. (**c**) Optical light curve. Model fits using one peak (blue solid line) or two peaks (broken grey line for each component). Figure from Mundell et al. (2013)

around the peak at $t \sim 300$ s, and it declined to 16% by $t \sim 800$ s. The polarisation position angle was almost constant over this period. The high polarisation degree and the constancy of the polarisation position angle rule out local shock instability models. Instead, the polarisation properties imply a magnetised baryonic jet with large-scale uniform fields that can survive long after the initial explosion. The amplification of magnetic fields by the rotation of a black hole accretion disk system is often discussed in magnetic jet models. For an axially symmetric field configuration, the poloidal component of the magnetic field decays faster with radius. Therefore, the large-scale magnetic field is likely to be dominated by a toroidal component. Although internal dissipation processes preceding the onset of afterglow can distort the toroidal field (Zhang and Yan 2011; Troja et al. 2017a,b), the visible region with angular scale $\sim 1/\Gamma$ might have a rather uniform magnetic field. In such a case, the polarisation vector points toward the jet axis, and the polarisation position angle is expected to be constant.

Granot and Taylor (2005) used VLA data to constrain the polarisation of radio flares (the late time reverse shock emission from the original ejecta). They find only upper limits for both linear and circular polarisation. Their best limits are for GRB 991216, for which upper limits on the linear and circular polarisation are 7% and 9%, respectively. The upper limits challenge a structure jet model with a large-scale ordered toroidal fields.

13.5.4 Prompt Gamma-Rays

The claim by Coburn and Boggs (2003) that the prompt gamma-ray emission of GRB 021206 was polarised at 80% level has stimulated the discussion on polarisation models, even though the claim was later contradicted by Rutledge and Fox (2004) and Wigger et al. (2004). A large number of mechanisms potentially can produce highly polarised prompt emission (e.g. Lazzati 2009; Toma et al. 2009; Gill et al. 2018 and the references therein), which include a large-scale magnetic fields in the original ejecta, coherent magnetic patches within the visible region, specific viewing angle with anisotropic magnetic fields, Compton drag and photospheric emission.

- **GRB 041219A:** Using multiple-detector coincidence events in the 100–350 keV energy band, Kalemci et al. (2007) searched for polarisation in the prompt gamma-ray emission with the SPI instrument on *INTEGRAL*. Although they reported a linear polarisation degree 98% ± 33%, they could not constrain the systematics of their measurements. The multiple event data was analysed and compared with the predicted instrument response obtained from Monte-Carlo simulations using the GEANT 4 INTEGRAL mass model, McGlynn et al. (2007) measured the linear polarisation in the brightest pulse of the GRB to be about 60%, but at a low level of significance $\sim 2\sigma$. Götz et al. (2009) used the imager on board the *INTEGRAL* (IBIS) to measure the polarisation of the prompt gamma-ray emission. However, no polarisation signal was found integrating over the same pulse, and the upper limit is 4%.
- **GRB 100826A:** The average linear polarisation degree of 27% ± 11% was detected with the GAP on board the *IKAROS* (Yonetoku et al. 2011). By dividing the GRB into two 50 s time intervals, the change of the polarisation position angle was detected.
- **GRB 110301A, GRB 110721A:** With GAP on board the *IKAROS*, linear polarisation degrees of 70%±22% and 84^{+16}_{-28}% were detected for GRB 110301A and GRB 110721A, respectively. They did not detect any significant change of polarisation position angle for these events, which have shorter durations and dimmer brightness compared with GRB 100826A.
- **151006A, 160106A, 160131A, 160325A, 160509A, 160607A, 160623A, 160703A, 160802A, 160821A, 160910A:** Chattopadhyay et al. (2017) presented the polarisation data for the brightest 11 GRBs detected by CZTI on board the

AstroSat satellite during the 1st year of operation. Most of the GRBs show clear polarisation signatures with $\geq 3\sigma$ detection significance for 4 GRBs and $\sim 2.5\sigma$ significance for another 3 GRBs.

13.6 Conclusions

Polarimetric study has been carried out at various stages of the GRB phenomena (almost at the moment of the explosion to a few hundred days after the explosion) and at various frequencies (radio, optical, X-ray/gamma-ray). Although polarimetry at a few percent level can be still demanding for rapidly fading sources, there are already a large number of optical polarisation measurements about several hours to a few days after the prompt gamma-ray emission. The predicted rotation of polarisation position angle has been detected in events which show a possible jet break. The typically low linear polarisation degree of the afterglow emission provides constraints on the plasma/shock physics. The detection of circular polarisation in the optical afterglow of GRB 121024A is totally unexpected, further circular polarisation observations would be needed to understand the origin. High magnetisation parameters obtained by the early afterglow modelling and high polarisation degree of the early optical afterglow imply that ordered magnetic field is advected by the ejecta from the central engine. Although gamma-ray polarimetry is still technically challenging, considering that the high polarisation of the ejecta emission (reverse shock emission), the prompt emission of GRBs should be good targets for gamma-ray polarimeters. High energy polarimetry will enable us to test various GRB models. In coming years, a large number of neutron star–neutron star merges (and possibly neutron star–black hole mergers) will be detected by gravitational wave observatories such as LIGO, Virgo and KAGRA. Optical and radio polarimetry of the electromagnetic counterparts is likely to constrain the geometry of the outflow from the mergers.

References

Abbott BP et al (2017a) Phys Rev Lett 119:161101
Abbott BP et al (2017b) Astrophys J 848:L12
Abbott BP et al (2017c) Astrophys J 848:L13
Abdo AA et al (2009a) Nature 462:331
Abdo AA et al (2009b) Astrophys J 706:L138
Akerlof CW et al (1999) Nature 398:400
Alexander K et al (2017) Astrophys J 848:L21
Arcavi I et al (2017) Nature 551:64
Barth, AJ et al. (2003), Astrophys J 584:L47
Berger E et al (2003) Astrophys J 587:L5
Bersier D et al (2003a) Astrophys J 583:L63
Bersier D et al (2003b) Astrophys J 584:L43

Björnsson et al (2002) Astrophys J 579:L59
Blandford RD, Mckee CF (1976) Phys Fluids 19:1130
Burrows DN et al (2005) Science 309:1833
Chattopadhyay T et al (2017) ArXiv eprints. arXiv:1707.06595
Coburn W, Boggs SE (2003) Nature 423:415
Corsi A et al (2018) Astrophys J 861:L10
Coulter DA et al (2017) Science 358:1556
Covino S, Götz D (2016) A&AT 29:205
Covino S et al (1999) Astron Astrophys 348:L1
Covino S et al (2002) Astron Astrophys 392:865
Covino S et al (2003) Astron Astrophys 400:L9
Covino S et al (2004) In: Astronomical society of the Pacific conference series. Gamma-ray bursts in the afterglow era, vol. 312. p 169
Covino S et al (2008) Mon Not R Astron Soc 388:347
Covino S et al (2017) Nat Astron 1:791
Cui X-H, Zou Y-C, Wei J-J (2018) Mon Not R Astron Soc 473:5142
D'Avanzo P et al (2018) A&A 613:L1
Daigne F, Mochkovitch R (1998) Mon Not R Astron Soc 296:275
Drenkhahn G, Spruit HC (2002) A&A 391:1141
Evans PA et al (2017) Science 358:1565
Fan Y-Z, Dai Z-G, Huang Y-F, Lu T (2002) Chin J Astron Astrophys 2:449
Fan Y-Z, Wei DM, Wang CF (2004) A&A 424:477
Fan YZ, Zhang B, Wei DM (2005) Astrophys J 628:L25
Fong W, Berger E, Margutti R, Zauderer BA (2015) Astrophys J 815:102
Fraija N, Lee WH, Veres P et al (2016) Astrophys J 831:22
Gao H, Wang X-G, Mészáros P et al (2015) Astrophys J 810:160
Gendre B, Atteia JL, Boer M et al (2012) Astrophys J 748:59
Ghisellini G, Lazzati D (1999) Mon Not R Astron Soc 309:L7
Giannios D (2008) A&A 480:305
Giannios D, Spruit HC (2006) A&A 450:887
Giannios D, Mimica P, Aloy MA (2008) A&A 478:747
Gill R, Granot J (2018) Mon Not R Astron Soc 478:4128
Gill R, Granot J, Kumar P (2018) ArXiv eprints. arXiv:1811.11555
Gomboc A, Kobayashi S, Guidorzi C et al (2008) Astrophys J 687:443
Gorosabel J et al (2004) A&A 422:113
Götz D et al (2009) Astrophys J 695:L208
Granot J (2012) Mon Not R Astron Soc 421:2442
Granot J, Königl A (2003) Astrophys J 594:L83
Granot J, Piran T (2012) Mon Not R Astron Soc 421:570
Granot J, Taylor GB (2005) Astrophys J 625:263
Granot J, Piran T, Sari R (1999a) Astrophys J 513:679
Granot J, Piran T, Sari R (1999b) Astrophys J 527:236
Granot J, Nakar E, Piran T (2003) Nature 426:138
Granot J et al (2017) Astrophys J 850:L24
Greiner J et al (2003) Nature 426:2
Gruzinov A (1999) Astrophys J 525:L29
Gruzinov A (2001) Astrophys J 563:L15
Gruzinov A, Waxman E (1999) Astrophys J 511:852
Haggard D et al (2017) Astrophys J 848:L25
Hallinan G et al (2017) Science 358:1579
Halpern JP et al (2000) Astrophys J 543:697
Harrison FA (1999) Astrophys J 523:L121
Harrison R, Kobayashi S (2013) Astrophys J 772:101
Heyl J, Perna R (2003) Astrophys J 586:L13

Hjorth J et al (1999) Science 283:2073
Holland ST et al (2002) Astrophys J 124:639
Huang X-L, Xin L-P, Yi S-X (2016) Astrophys J 833:100
Japelj J, Kopac D, Kobayashi S et al (2014) Astrophys J 785:84
Kalemci E et al (2007) Astrophys J Suppl S 169:75
Kasliwal MM et al (2017) Science 358:1559
Kim S, Schulze S, Resmi L et al (2017) Astrophys J 850:L21
Klose S et al (2004) Astrophys J 128:1942
Kobayashi S (2000) Astrophys J 545:807
Kobayashi S, Sari R (2000) Astrophys J 542:819
Kobayashi S, Zhang B (2003) Astrophys J 582:L75
Kobayashi S, Zhang B (2007) Astrophys J 655:973
Kobayashi S, Piran T, Sari R (1997) Astrophys J 490:92
Kobayashi S, Piran T, Sari R (1999) Astrophys J 513:669
Komissarov SS, Vlahakis N, Königl A, Barkov MV (2009) Mon Not R Astron Soc 394:1182
Kulkarni SR et al (1999a) Nature 398:389
Kulkarni SR et al (1999b) Astrophys J 522:L97
Kumar P, Panaitescu A (2003) Mon Not R Astron Soc 346:905
Kumar P, Zhang B (2015) Phys Rep 561:1
Lamb G, Kobayashi S (2018) Mon Not R Astron Soc 478:733
Laskar T, Berger E, Zauderer BA et al (2013) Astrophys J 776:119
Laskar T, Alexander KD, Berger E et al (2018) Astrophys J 862:94
Lazzati D (2006) New J Phys 8:131
Lazzati D (2009) ArXiv eprints. arXiv:0906.4346
Lazzati D et al (2003) A&A 410:823
Lazzati D et al (2004) A&A 422:121
Lazzati D et al (2018) Phys Rev Lett 120:241103
Li Z, Chevalier R (2001) Astrophys J 551:940
Liang E-W, Yi S-X, Zhang J (2010) Astrophys J 725:2209
Lemoine M (2013) Mon Not R Astron Soc 428:845
Levinson A (2009) Astrophys J 705:L213
Loeb A, Perna R (1998) Astrophys J 495:597
Lynman JD et al (2018) Nat Astron 2:751
Lyutikov M (2006) New J Phys 8:119
Lyutikov M, Pariev VI, Blandford R (2003) Astrophys J 597:998
Maiorano E et al (2006) A&A 455:423
Margutti R et al (2017) Astrophys J 848:L20
Masetti N et al (2001) A&A 374:382
Masetti N et al (2003) A&A 404:465
Matsumiya M, Ioka K (2003) Astrophys J 595:L25
McGlynn S et al (2007) A&A 466:895
McKinney JC, Blandford RD (2009) Mon Not R Astron Soc 394:L126
Medvedev MV, Loeb A (1999) Astrophys J 526:697
Medvedev MV, Fiore M, Fonseca RA et al (2005) Astrophys J 618:L75
Melandri A, Kobayashi S, Mundell C et al (2010) Astrophys J 723:1331
Meliani Z, Keppens R (2010) A&A 520:L3
Mészáros P, Rees M (1997) 476:232
Mimica P, Giannios D, Aloy MA (2009) A&A 494:879
Mimica P, Giannios D, Aloy MA (2010) Mon Not R Astron Soc 407:2501
Mirabal N (2003) Astrophys J 595:935
Mooley KP et al (2018a) Nature 554:207
Mooley KP et al (2018b) Nature 561:355
Mundell CG et al (2013) Nature 504:119
Nakar E, Oren Y (2004) Astrophys J 602:L97

Nakar E, Piran T, Granot J (2003a) New Astron 8:495
Nakar E, Piran T, Waxman E (2003b) J Cosmol Astropart Phys 10:005
Nakar E et al (2018) Astrophys J 867:18
Narayan R, Kumar P, Tchekhovskoy A (2011) Mon Not R Astron Soc 416:2193
Nava L, Nakar E, Piran T (2016) Mon Not R Astron Soc 455:1594
Nishikawa K, Hardee P, Richardson G et al (2003) Astrophys J 595:555
Nousek JA, Kouveliotou C, Grupe D et al (2006) Astrophys J 642:389
Panaitescu A, Kumar P (2001) Astrophys J 560:L49
Panaitescu A, Kumar P (2002) Astrophys J 571:779
Panaitescu A, Kumar P (2004) Mon Not R Astron Soc 353:511
Panaitescu A, Mészáros P (1998) Astrophys J 493:L31
Panaitescu A, Mészáros P (1999) Astrophys J 526:707
Pian E et al (2017) Nature 551:67
Piran T (1999) Phys Rep 314:575
Piran T (2005) AIP Conf Proc 784:164. arXiv:astro-ph/0503060
Rees MJ, Mészáros P (1994) Astrophys J 430:L93
Rhoads JE (1999) Astrophys J 525:737
Rol E et al (1999) Astrophys J 544:707
Rol E et al (2003) A&A 405:L23
Rossi E, Lazzati D, Rees JM (2002) Mon Not R Astron Soc 332:945
Rossi EM, Lazzati D, Salmonson JD, Ghisellini G (2004) Mon Not R Astron Soc 354:86
Ruan JJ et al (2018) Astrophys J 853:L4
Rutledge RE, Fox DB (2004) Mon Not R Astron Soc 350:1288
Rybicki GB, Lightman AP (1979) Radiative processes in astrophysics. Wiley-Interscience, New York
Sagiv A, Waxman E, Loeb A (2004) Astrophys J 615:366
Smartt SJ et al (2017) Nature 551:75
Sironi J, Goodman J (2007) Astrophys J 671:1858
Sironi J, Spitkovsky A (2010) Astrophys J 726:75
Sironi L, Keshet U, Lemoine M (2015) Space Sci Rev 191:519
Sari R (1998) Astrophys J 494:L49
Sari R (1999) Astrophys J 524:L43
Sari R, Piran T (1999a) Astrophys J 517:L109
Sari R, Piran T (1999b) Astrophys J 520:641
Sari R, Piran T (1995) Astrophys J 455:L143
Sari R, Piran T (1999) Astrophys J 520:641
Sari R, Piran T, Narayan R (1998) Astrophys J 497:L17
Sari R, Piran T, Halpern JP (1999) Astrophys J 519:L17
Spitkovsky A (2008) Astrophys J 682:L5
Steele IA, Mundell CG, Smith RJ et al (2009) Nature 462:767
Steele IA, Kopac D, Arnold DM et al (2017) Astrophys J 843:143
Tanvir NR et al (2017) Astrophys J 848:L27
Taylor GB et al (2004) Astrophys J 609:L1
Tchekhovskoy A, McKinney JC, Narayan R (2008) Mon Not R Astron Soc 388:551
Toma K (2013) ArXiv eprints. arXiv:1308.5733
Toma K, Ioka K, Nakamura T (2008) Astrophys J 673:L123
Toma K et al (2009) Astrophys J 698:1042
Troja E et al (2017a) Nature 547:425
Troja E et al (2017b) Nature 551:71
Uehara T et al (2012) Astrophys J 752:L6
Usov VV, Melrose DB (1992) Astrophys J 395:575
van Eerten HJ, Zhang W, MacFadyen A (2010) Astrophys J 722:235
Varela K et al (2016) A&A 589:A37
Walder R, Folini D, Meynet G (2012) Space Sci Rev 166:145

Watanabe J et al (2001) Publ Astron Soc Jpn 53:L27
Waxman E (1997) Astrophys J 491:L19
Waxman E (2003) Nature 423:388
Wiersema K et al (2012) Mon Not R Astron Soc 426:2
Wiersema K et al (2014) Nature 509:201
Wigger C, Hajdas W, Arzner K, Güdel M, Zehnder A (2004) Astrophys J 613:1088
Wijers RAMJ et al (1999) Astrophys J 523:L33
Yonetoku D et al (2011) Astrophys J 743:L30
Zhang B, Kobayashi S (2005) Astrophys J 628:315
Zhang W, MacFadyen AI (2009) Astrophys J 698:1261
Zhang B, Mészáros P (2002) Astrophys J 571:876
Zhang B, Mészáros P (2004) Int J Mod Phys A 19:2385
Zhang B, Pe'er A (2009) Astrophys J Lett 700:L65
Zhang B, Yan H (2011) Astrophys J 726:90
Zhang B, Kobayashi S, Mészáros P (2003) Astrophys J 595:950
Zhang B, Fan YZ, Dyks J et al (2006) Astrophys J 642:354
Zhang S, Jin Z-P, Wei D-M (2015) Astrophys J 798:3
Zheng W, Shen RF, Sakamoto T et al (2012) Astrophys J 751:90

Chapter 14
The Panchromatic Polarisation Signatures of Active Galactic Nuclei

Frédéric Marin

Abstract Among all the astronomical sources investigated through the prism of polarimetry, active galactic nuclei (AGN) have proven to be the richest in terms of complex yet fundamental signatures that helped to understand their true nature. Indeed, AGN exhibit a wide range of wavelength-dependent polarimetric features that are intrinsically related to their multi-scale emission mechanisms. Each waveband is characterised by a different set of polarimetric signatures that can be related to various physical mechanisms that, from the radio band to the soft-γ rays, probe increasingly smaller AGN regions. In fact, panchromatic polarisation measurements are the key to understand how and when AGN form, accrete, and impact the host galaxy they reside in. In this chapter, I will first introduce AGN without focusing on a particular observational technique. I will then review the discoveries and constraints that spectroscopy, imaging, and broadband polarimetry have achieved. Finally, I will highlight the important questions that remain unanswered and how they can be solved with future large millimetre and radio antennas, 30-m class optical and infrared telescopes, and high energy satellites equipped with state-of-the-art polarimeters.

14.1 A Brief Introduction to AGN

"Active galactic nuclei" (AGN) is the most general term used to describe a compact region at the centre of a galaxy that often outshines the starlight contribution from the host, with characteristics indicating that the radiation is not produced by stars. With luminosities ranging from $\sim 10^{40}$ to $10^{47}\,\mathrm{erg\,s^{-1}}$ for the most distant ones (Mortlock et al. 2011), AGN are considered as the most powerful, long-lived objects in the Universe. To explain the production of such tremendous amounts

F. Marin (✉)
Université de Strasbourg, CNRS, Observatoire Astronomique de Strasbourg, UMR 7550, Strasbourg, France
e-mail: frederic.marin@astro.unistra.fr

R. Mignani et al. (eds.), *Astronomical Polarisation from the Infrared to Gamma Rays*, Astrophysics and Space Science Library 460,
https://doi.org/10.1007/978-3-030-19715-5_14

of radiation, accretion of gas and dust onto a supermassive black hole (SMBH) is invoked (Pringle and Rees 1972; Shakura and Sunyaev 1973). The dissipative processes in the viscous accretion disc that forms around the SMBH transport matter inwards and angular momentum outwards, causing the accretion disc to heat up. The resulting thermal multi-temperature black body emission (see Fig. 14.1) peaks in the ultraviolet and shapes the spectral energy distribution (SED) of AGN in the form of the *Big Blue Bump* (Malkan et al. 1982; Sanders et al. 1989). The structure of the disc is still debated because of its spatially unresolvable sizes (smaller than a milliparsec for a 10^8 solar masses black hole). Depending on the dimensionless mass accretion rate $\dot{m}$, normalised to the Eddington rate of the central object, the accretion flow could sustain various geometries and natures (Frank et al. 2002). For $\dot{m} \leq 1$, the disc is likely geometrically thin, optically thick, and radiatively efficient (Shakura and Sunyaev 1973). For $\dot{m} > 1$, the disc becomes geometrically slim (not thin), stays optically thick, and is radiatively inefficient (Abramowicz et al. 1998). However, if $\dot{m} \leq 0.01$, the accreted gas density is low so the gas may be unable to

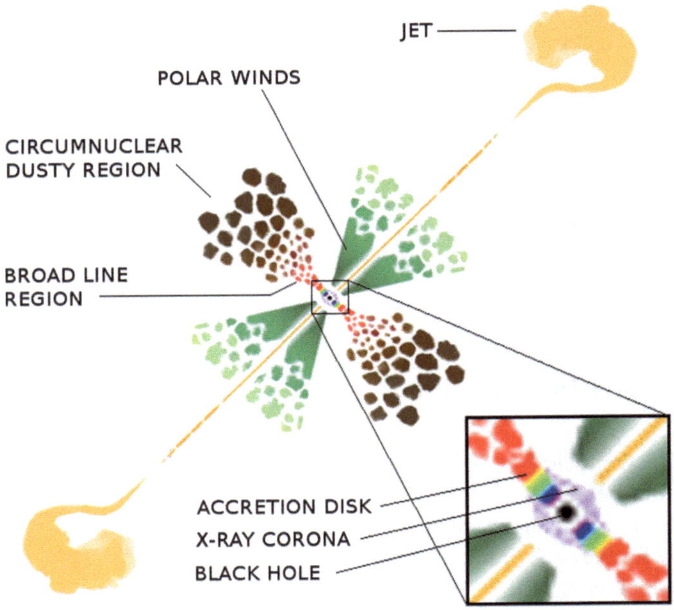

Fig. 14.1 Unscaled sketch of an AGN. At the centre lies a supermassive black hole around which a multi-temperature accretion disc spirals (shown with the colour pattern of a rainbow). The accreted matter is illuminated by an X-ray corona (shown in violet) of unknown size. The disc extends from $\sim 10^{-5}$ pc to $\sim 10^{-3}$ pc for a 10^8 solar masses black hole. The region responsible for the emission of broad lines (BLR) is in red and light brown. It extends up to $\sim 10^{-1}$ pc, where the circumnuclear dusty region (shown in dark brown) onsets. The collimated polar ionised winds (in green) are created in sub-parsec scale regions and their final extension interacts with the interstellar medium (shown in yellow-green at a few hundreds of parsecs). A double-sided, kilo-parsec jet (in yellow) is added to account for radio-loud AGNs (Marin 2016)

radiate energy at a rate that balances viscous heating. The flow becomes optically thin and radiatively inefficient (Narayan and Yi 1994). In all cases, it is commonly admitted that the inner edge of the accretion flow is set by the innermost stable circular orbit (ISCO) radius (Misner et al. 1973). Outside the ISCO, particles can orbit indefinitely in stable circular orbits while inside the ISCO they spiral rapidly past the event horizon into the SMBH. The location of the ISCO depends on the angular momentum (spin) of the central object. For a non-spinning massive object, where the gravitational field can be expressed with the Schwarzschild metric, the ISCO is located at $3R_S$ (Schwarzschild radius $R_S = \frac{2GM}{c^2}$). In the case of rotating black holes, the ISCO in the Kerr metric depends on whether the orbit is prograde or retrograde, resulting in a sign difference for the spin parameter. For massive particles around a maximally spinning black hole, the ISCO is at $0.5R_S$ (prograde), and at $4.5R_S$ for retrograde orbits.

Nearby the inner edge of the accretion disc should lie the region responsible for the emission of X-ray photons. Once again, its nature and geometry are poorly constrained, but it is believed that a plasma of hot electrons is situated on top of the disc, in an atmosphere-like region (Haardt et al. 1993; Walter and Fink 1993; Magdziarz et al. 1998) or in a compact or patchy corona situated at an unknown distance above the disc (Martocchia et al. (1996); Miniutti and Fabian (2004), see the zoomed region in Fig. 14.1). The corona may be generated by buoyancy instabilities in the disc and heated by magnetic reconnection, i.e. shocks at reconnection sites where strong impulsive heating occurs when magnetic field lines are brought together (Di Matteo et al. 1998). The temperature of the plasma is such that ultraviolet photons, thermally emitted by the disc, are boosted up to the X-ray energies by multiple inverse-Compton scattering. This results in a source of X-rays that can be variable in emission, such as observed with X-ray satellites (Lawrence et al. 1985; McHardy and Czerny 1987). The power-density spectrum shows a power-law form with a turnover at low frequencies and a high frequency break, similar to galactic black-hole candidates (see the chapter by Tariq Shabaz). Part of the X-ray photons that are emitted by the corona are bent back to the disc by general relativistic effects. Soft X-rays are mainly absorbed by photoelectric absorption followed by fluorescent line emission, or photoelectric absorption followed by Auger de-excitation. Hard X-ray photons mainly scatter on the disc due to Compton scattering by free or bound electrons (George and Fabian 1991). The most prominent signature of these physical effects is the presence of a strong iron fluorescence line at approximately 6.4 keV (Nandra and Pounds 1994). The emission line profile is often broad, skewed towards the soft X-rays by Doppler shifts and relativistic boosting due to the motion of the disc and the gravitational redshifting of the black hole (Fabian et al. 2000). Iron line fitting then allows to determine the radius of the ISCO, hence the dimensionless spin parameter and the inclination of the accretion disc.

The maximal extension of the outer rim of the accretion disc is, however, unconstrained. It is thought to coincide with the innermost parts of another fundamental constituents of AGN: the broad line region (BLR). Broad emission lines in the optical and ultraviolet bands are a key signature of AGN. They have been detected more than 50 years ago (Sandage 1965) and exhibit a variety of Doppler widths, from 1000 km s^{-1} to velocities larger than 15,000 km s^{-1} (Strateva et al. 2003). Line emission results from photoionisation of the BLR gas by the thermal emission from accretion disc, with the BLR reaching a photoionisation equilibrium temperature of 10,000–20,000 K. Different gas densities, input ionising spectra, gas dynamics, and elemental abundances impact the broad line profiles. Temperatures of the order of 10^4 K correspond to thermal line widths around 10 km s^{-1}, while different elemental abundances play a minor role in line broadening. It is then clear that the gas in the BLR must be dense ($n_e > 10^9$ cm^{-3}) and move supersonically in order to broaden the lines up to several thousands of kilometres per second (Peterson 2006). But what is the geometry of this region? An uniformly filled, rotating BLR would not be able to produce the diversity of line emissions, such as the HeII, NV, and CIV high-ionisation lines or the MgII, CaII, and FeII low-ionisation lines. A distribution of discrete clouds, with different densities and distances from the centre, is able to reproduce the total spectra of AGN (Baldwin et al. 1995), but its geometrical arrangement cannot be spherical otherwise we would observe Lyman continuum absorption by the BLR clouds. This, in fact, was never convincingly observed (Antonucci 1989; Gaskell 2009), suggesting that the BLR cloud distribution is likely flattened along the equatorial plane (often considered as coplanar to the accretion disc plane, see the light brown region in Fig. 14.1). The resulting radial ionisation stratification of the BLR, due to cloud self-shielding, allows the emission of both low- and high-ionisation lines (Gaskell 2009). The same idea was used by different models to explain the physical cause of the formation BLR, either from accretion disc gravitational instabilities (Collins and Hurè 2001), a failed radiatively accelerated dusty wind (Czerny et al. 2017) or a fully formed, radiation-pressure-driven disc outflow that is self-shielded from the ionising continuum (Elvis 2017). The question of where does the disc ends and where does the BLR onset is yet to be answered but what is sure is (1) there is a tight observational relation between the BLR radius and the AGN bolometric luminosity (Kaspi et al. 2000) and (2) the BLR most probably ends where dust grains can start to survive the intense ultraviolet radiation field (Netzer and Laor 1993).

From the dust sublimation radius and beyond (Barvainis 1987; Kishimoto et al. 2007), dust grains are thought to form an optically thick, equatorial region historically coined with the term "torus" (Antonucci 1993). This is a necessary AGN constituent to explain the disappearance of the broad emission lines from the spectra of certain objects (Osterbrock 1984). An asymmetric dusty obscurer in the form of a donut prevents the detection of the BLR and accretion disc if the observer's viewing angle is crossing the compact dusty layers. Viewed from the top, the hole in the pastry allows the observer to have a direct view of the central engine. This lead to the still used denomination of type-1 (dust-free, polar view) and type-2 (dust-obscured, equatorial view) AGN terminology. The donut picture evolved, thanks to observational and numerical constraints (Bianchi et al. 2012), and our actual representation of the torus is a complex, clumpy medium that is in dynamical motion (Garcìa-Burillo et al. 2016) (see Fig. 14.1). From a type-1 view, once could see in principle that the inner part of this circumnuclear region is heated up to 1500 K by the accretion disc emission and cools mainly by thermal dust infrared emission. This phenomenon naturally produces the ubiquitous near-infrared bump/excess observed in the spectral energy distribution of AGN (Neugebauer et al. 1979; Robson et al. 1986; Sanders et al. 1989). The inner radius of this equatorial region is set by the luminosity of the photoionising radiation originating in the central accretion disc (Lawrence 1991). For ultraviolet luminosities between 10^{43} and 10^{44} erg s^{-1}, the inner torus radius is typically comprised between 0.01 and 0.1 pc (Marin et al. 2016). The maximal extension of this opaque dusty region was first though to expand as large as 100 pc (Jaffe et al. 1993), which is challenged by dynamical stability arguments, but the compact core of the torus is now observationally constrained to be only a dozen parsecs wide (Jaffe et al. 2004). A clumpy distribution of graphite grains embedded in a non-spherical geometry can easily reproduce the 2–4 μm excess for temperatures close to the dust sublimation limit (Barvainis 1987). With increasing distances from the torus inner edges, the outer shells present colder dusty grains, only heated by reprocessed infrared radiation with characteristic temperatures lower than 100 K. The resulting infrared emission peaks at near- to mid-infrared wavelengths (3–40 μm, Fritz et al. 2006). Silicate grains are responsible for the observed absorption feature near 9.7 μm we observe in dust-obscured AGN, while graphite grains are responsible for the rapid decline of the emission at wavelength shortwards of a few microns. Infrared emission from the torus should then decrease in the far-infrared/millimetre domain but observations often indicate a peak of emission between 50 and 100 μm (Rowan-Robinson et al. 1984). This is due to the contribution of starburst regions, in which ultraviolet radiation produced by the obscured star-formation is absorbed by the enveloping dust layers and re-radiated around 100 μm. Strong correlations have been found between star formation activity and presence of polycyclic aromatic hydrocarbons in the torus (Woo et al. 2012), demonstrating the importance of the host galaxy in the energy budget of the AGN SED. If large-scale magnetic field is indeed present within the torus (Lopez-Rodriguez et al. 2013), magnetically aligned dust grains could produce polarised infrared emission and rotating grains could result in polarised radio continuum emission. This makes the circumnuclear dusty region of AGN an important part of the global puzzle.

An interesting question is: what constrains the opening angle of the torus? AGN are very often associated with extended polar outflows that carry mass and energy from the sub-parsec central engine to the interstellar/intergalactic medium (Wilson 1996). The most observed geometrical form of such outflows is a double cone that lies along the direction of the AGN radio axis (see the dark green region in Fig. 14.1). The base of this bi-conical, extended region is photo-ionised by oppositely directed beams of radiation emitted from the accretion disc (Macchetto et al. 1994). The polar outflows have sharp, straight edges that could indicate prior collimation by the torus itself. Hence, there might be a direct correlation between the opening angle of the torus and of the winds below 100 pc (Storchi 2015). As the radial distance from the central engine increases, dust from the interstellar medium starts to appears and mix with the electrons. This slow-moving ($<$1000 km s^{-1}), low-density ($10^3 \leq n_e \leq 10^6$ cm^{-3}) wind produces multiple narrow emission lines such as H I, He I, He II, [O III]λ4959, [O III]λ5007, [N II]λ6548, or [N II]λ6583. This narrow line region (NLR) is not uniform; it shows evidence for gas clouds, emission filaments, arcs-like structures, and a large gradient of the line-of-sight gas velocities, with absorption features uniformly blue-shifted relative to the systemic velocity (Afanasiev et al. 2007a,b). This is indicative of a variety of processes happening in the NLR that are ultimately responsible for the presence of warm absorber-emitters (Porquet and Dubau 2000) or the detection of ultra-fast outflows close to the wind launching site (Tombesi et al. 2010). To explain the presence of ultra-fast outflows and the various NLR components we observe, several scenarios have been investigated and it was found that essentially three mechanisms could participate in launching a wind: thermal driving, radiation pressure, or magnetic forces. Accretion discs in hydrostatic equilibrium can produce a thermal wind if the cooler, outer part of the disc is irradiated by its hotter, inner regions. This would result in a high temperature at the disc surface that would puff up the upper gas layers in the forms of a static corona or an outflow (Begelman et al. 1983a,b). Another formation mechanism relies on radiation pressure by electron scattering (if the wind is fully ionised) or by line scattering. Spectral lines increase the scattering coefficient, giving rise to powerful line-driven winds (Proga et al. 1998, 1999; Higginbottom et al. 2014). According to the authors, an accretion disc around a 10^8 solar masses SMBH accreting at the rate of 1.8 solar masses per year can launch a wind at 0.003 pc from the central engine, reaching velocities up to $0.05c$ (at 0.03 pc from the potential well). Finally, massive turbulent strongly magnetised thick gas disc can give rise to magnetocentrifugal winds (Vollmer et al. 2018). In fact, the magnetorotational instability is a universal mechanism to produce turbulence and transport angular momentum in discs at all radii (Balbus et al. 1998).

Strongly magnetised accretion discs are not only responsible for bulk outflows; they can produce highly collimated, powerful jet that can reach to the intergalactic medium. The most impressive, largest jets can reach up to a few mega-parsecs in projected linear sizes (Riley et al. 1989; Bhatnagar et al. 1998; Schoenmakers et al. 2001), while more regular jets easily reach dozens of kilo-parsecs. Unscaled jets are presented in yellow in Fig. 14.1, with narrow bases and with two lobes of radio emission more or less symmetrically located on either side of the AGN. There are two main methods to extract enough power from the central AGN engine to form such jets: the Blandford-Znajek (Blandford et al. 1977) and the Blandford-Payne (Blandford et al. 1982) mechanisms. The Blandford-Znajek process requires an accretion disc with a strong poloidal magnetic field around a spinning black hole. Open magnetic field lines are transferring energy and angular momentum from the disc to the polar direction, resulting in Poynting flux dominated jets. On the other hand, the Blandford-Payne process does not require a spinning black hole. The magnetic field threading the disc extracts energy from the rotating gas to power a jet within the co-rotating large-scale magnetic fields. Which mechanism dominates is not yet determined (Chai et al. 2012; Cao 2018). What is certain, however, is that jets are detected only in a small fraction of AGN. Depending on their ratio of 5 GHz to B-band flux density and radio power, about 15–20% of all AGN can be considered as radio-loud (Kellerman et al. 1989; Urry and Padovani

1995). The difference between radio-loud and radio-quiet objects (in fact radio-faint, because there is some radio emission in radio-quiet AGN) echoes over the whole electromagnetic spectrum (Padovani 2017). Radio-loud AGN are dominated by non-thermal emission from their jet from the radio to the γ-ray band, while radio-quiet AGN are dominated by thermal emission. In addition, the radio-loud fraction of AGN is a function of redshift and optical and X-ray luminosities (della Ceca et al. 1994; La Franca et al. 1994; Jiang et al. 2007). It means that AGN, with their tremendous intrinsic brightness, should act as flashlights to illuminate the intergalactic medium at different cosmological periods, allowing us to study the conditions and transitions of the Universe early in cosmic history.

As they propagate from the AGN up to galactic of intergalactic medium, jets deposit radiation and kinetic energy into the host galaxy. This feedback effect is responsible for quenching starburst activities, creating red extended and dispersion-dominated galaxies. AGN activity in the form or jets, winds, or intense radiation can heat up the hydrogen gas in the galaxy or blow it out completely, thus preventing the gas from cooling and contracting to form stars (Bower et al. 2006; Dubois et al. 2013). Thus, there is a direct action of the AGN on the galaxy it resides in, but the opposite may be also true. Both recent observations and simulations indicate that there is a delay of 50–250 Myr between the onset of starburst and AGN activity (Di Matteo et al. 2005; Davies et al. 2007). This delay is due to a viscous time-lag as the gas from the host takes time to flow down to the AGN central engine (Blank et al. 2016). There are increasing evidences for this mass accretion duty cycle in AGN, showing that the AGN-host system is tightly coupled. This coupling also impact any AGN observation: since the central engine is spatially unresolved, starlight and dust emission from the host often contaminates the AGN signal. This is particularly true in the optical, near-infrared, and far-infrared domains, where parasitic light often dominates in the case of lesser luminous AGN (low-luminosity AGN and Seyfert-galaxies). It results that any complete AGN picture cannot be dissociated from its host galaxy, a statement that will reach its highest significance when we will discuss polarimetry.

14.2 Successes of Polarimetric Observations in Constraining the AGN Physics

The global picture of AGNs seems rather well understood. In fact, details about the accretion physics, formation mechanisms, magnetic fields, AGN constituent morphologies, and kinetic and radiative interactions between the various AGN components are very elusive. The main problem we have to cope with is the cosmological distance of AGNs. Among the nearest type-1 AGNs is NGC 3227 (10h23m30.5790s, +19d51m54.180s), situated at a heliocentric redshift of 0.00386 $\pm$ 0.00001 (H_0 = 73 km s^{-1} Mpc^{-1}, $\Omega_{\rm matter}$ = 0.27, $\Omega_{\rm vacuum}$ = 0.73). This redshift-distance corresponds to a Hubble distance of 20.28 $\pm$ 1.45 Mpc. If we aim at resolving the ISCO of the accretion disc in NGC 3227, considering a central SMBH of 3.6×10^7 solar masses (Onken et al. 2003) and its associated 3.45×10^{-6} pc Schwarzschild radius, about 35 nano-arcsec spatial resolution would be required. This is beyond any prospects, even for X-ray interferometry (Skinner and Krizmanic 2009). Since regular observational techniques such as photometry, spectroscopy, or timing analyses have proven to be not sufficient to spatially resolve the AGN inner regions, a work-around has been found: polarimetry. Polarisation has the advantage of being wavelength-dependent and its main strength is that it is extremely sensitive to the geometry and magnetic fields of the emitting/scattering region. Unlike spectroscopy that is limited by the physical size of the emitting region, polarisation allows us to probe spatially unresolved volumes and still determine their geometry. Polarisation is fundamentally linked to the internal properties of the sources of radiation: the strength and orientation of magnetic fields, the distribution and orientation of scattering particles like dust grains, the microscopic structure of reflecting surfaces, or intrinsic anisotropies of the host galaxy surrounding AGNs. Hence, by measuring the polarisation of AGNs in different wavebands of the electromagnetic spectrum it is possible to constrain the geometry of the innermost, spatially unresolvable, components. In the following I will review the greatest successes of polarimetric observations in constraining the AGN physics, complementing the work of previous authors (Antonucci 2002). The list of discoveries is roughly in the chronological order. I will not review in detail the mechanisms of polarisation production as they are covered in a dedicated chapter of this book.

14.2.1 The Origin of the "Featureless" Continuum

The origin of the optical and ultraviolet continuum emission in AGNs has been long discussed in the past (Schmidt et al. 1980; Serote-Roos et al. 1996). The general form of the white light SED is a power-law $F_\lambda \sim \lambda^\alpha$, with α the spectral photon index of the order of 1–3 (Page et al. 2005; Corral et al. 2011). The origin of this power-law emission was for long puzzling as both synchrotron emission and thermal emission from a multiple-temperature black body (the accretion disc) could explain the spectral shape. In addition, continuum photons and absorption lines from the stellar atmospheres of old stellar populations within the host galaxy tend to complicate the problem by diluting the signal (Kruszewski 1971). This is especially true in type-2 AGNs where the host dominates over nuclear emission.

Spectroscopy and imaging methods were, at the time (1970–1980), not able to solve the origin of the "featureless" continuum (equivalent width of emission lines < 5 Å). Polarisation, on the other hand, was a key to elucidate this mystery. Indeed, if the measured continuum polarisation comes from non-thermal emission by electrons in an anisotropic magnetic field, the polarisation is intrinsic. The polarisation should then vary together with the total flux as the magnetic configuration is evolving, resulting in a polarised flux that should be proportional to the total flux at each wavelength (Angel et al. 1980; Webb et al. 1993). Synchrotron polarisation is then expected to be strongly variable and different to that of the polarisation from the emission lines. On the other hand, if the polarisation of the featureless continuum originates from scattering by electrons, dust grains, or atoms, the polarisation should be insensitive to flux variations as the *geometry* of the scattered cannot change over rapid (hours–days) timescales. The polarisation of the continuum should be similar to that of the polarisation in the emission lines if they originate from the same source. The scattering-induced polarisation is also carrying wavelength-dependent signatures that can help to identify the nature of the scatterer. If Thomson scattering prevails, the polarisation fraction is wavelength-independent. If scattering and transmission of light by a non-spherical distribution of dust grains is responsible for the polarisation of the featureless continuum, the polarisation fraction will vary as λ^{-w} (with $1 \leq w \leq 4$ depending on the dust prescription (Mathis et al. 1977)).

Using polarisation measurements, it was then possible to conclude that the featureless continuum of 3C 84 (NGC 1275) is clearly related to synchrotron emission due to its month-to-month polarisation variations, strengthening up by 3% and rotating from 100° to 150° in polarisation angle (Walker 1968; Babadzhanyants et al. 1975). Later studies showed that the polarisation of 3C 110 and 3C 246 is also attributable to non-thermal emission due to the rise of their polarised fluxes at longer wavelengths with respect to their total fluxes (Webb et al. 1993). In the same paper, the low levels of wavelength-independent, non-variable polarisations in NGC 4151, Mrk 509, NGC 5548, and Mrk 290 are better explained by thermal photons originating from and scattering in an accretion disc. Polarimetry was also used to separate the nuclear and galaxy components, the polarisation of the former being diluted by the unpolarised starlight emission of the latter (Visvanathan and Oke 1968). In fact, it was shown that in several sources such as NGC 7130, NGC 5135, and IC 3639 the nuclear starburst light dominates in the ultraviolet band, being directly responsible for the featureless continuum (Gonzàlez Delgado et al. 1998). Polarimetry was thus of invaluable help to better understand the complex and numerous emission processes that are shaping the SED of AGNs.

14.2.2 The Correlation Between the AGN Radio Axis and Optical Polarisation Angle

While acquiring optical polarimetric measurements of a large sample of radio-loud AGNs with broad emission lines, it was discovered that their polarisation position angle was strongly correlated with the direction of their jets (Stockman et al. 1979). Over the 24 initial targets, 20 had their optical polarisation angle and radio position angle aligned within 30°, and all presented low, $\leq$1%, polarisation degrees. This correlation could not be of random nature but the physical mechanism of such alignment remained unknown at that time. Latter studies of radio-loud AGNs with and without broad emission lines showed that a given population (AGN with broad emission lines) displayed an alignment between the position angle of optical polarisation and the large-scale radio structure, while a second population (AGN without broad emission lines) had a perpendicular relationship (Antonucci 1982). The same bi-modality was found in radio-quiet AGNs (Antonucci 1983; Martin et al. 1983), thanks to deep radio observations that discovered small and weak radio structures despite the presence of jets. All sources exhibited weak polarisation degrees, preventing a clear determination of the physical causes of such alignment. Scattering, transmission through dust grains, or synchrotron radiation were suggested as possible answers but another key was necessary to unlock the secret. Spectropolarimetry brought it a few years later.

14.2.3 Unifying Type-1 and Type-2 AGNs

The puzzling question of the correlation between the AGN radio axis and optical polarisation angle, and its bi-modality between AGN with and without optical broad emission lines, was investigated in further details using high signal-to-noise, high resolution spectropolarimetry. Among a small sample of radio-loud AGNs, 3C 234 stood out due to its high polarisation degree (about 14%) and its polarisation position angle that is perpendicular to the radio axis (Antonucci 1984). More importantly its continuum polarisation, that is almost wavelength-independent, shares the same polarisation properties than the broad Hα line that is observed in polarised flux. The authors concluded that the polarisation could be produced by a geometrically thick, optically thick electron scattering ring (Antonucci 1984). The key was there but the lock squeaked. The final click was brought by Miller and Antonucci (1983) and Antonucci (1985) who discovered a highly polarised, $\sim$16%, wavelength-independent non-stellar continuum in NGC 1068. The polarised flux spectrum of this radio-quiet type-2 AGN revealed the presence of very broad (about 7500 km s^{-1}) symmetric Balmer and permitted FeII lines, that are very similar to what is observed in type-1 AGNs. The same polarisation was measured in the polarised broad line and in the continuum, indicating a common origin. The perpendicular polarisation angle and the high polarisation degree of NGC 1068 became the most egregious evidence for the presence of an obscuring circumnuclear medium that is hiding the source of emission. Photons are scattering inside an electron-filled, polar region situated above and below the obscuring disc, carrying broad line photons into the line-of-sight (Antonucci 1985). This was the first compelling evidence for the presence of a type-1 nucleus inside a type-2 object. Searches for additional type-1 signatures in the polarised flux of other types-2 AGNs were successful (Miller and Goodrich 1990; Tran et al. 1992; Wills et al. 1992; Tran 1995; Ogle et al. 1997), see also Fig. 14.2, proving that *nuclear orientation* is one of the key parameters to understand the AGN zoology (Antonucci 1993).

Spectropolarimetry has shown that, for type-2 objects, broad lines are hidden behind an obscuring dusty layer which is now also observed in the radio and infrared, thanks to adaptive optics (Jaffe et al. 2004) and interferometry (Kishimoto et al. 2009). However, it is impossible to prove that all type-2 AGNs have a broad

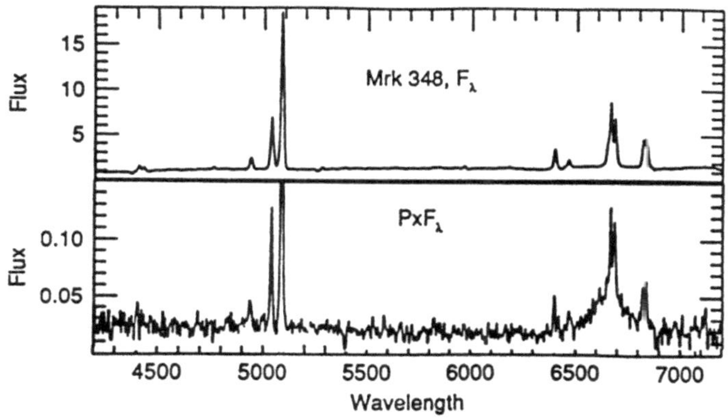

Fig. 14.2 Total flux (top) and polarised flux (bottom) spectra of the radio-quiet, type-2 AGN Mrk 348. In polarised flux a very broad Hα line can be seen. The relative flatness of the polarised continuum suggests that electron (Thomson) scattering is the prevalent mechanism in Mrk 348 (Miller and Goodrich 1990). Reproduced from Miller and Goodrich (1990) with permission

emission line region. High-resolution spectra and broadband observations (covering several Balmer lines) have proven to be necessary to observe faint broad lines in polarised flux (Ramos Almeida et al. 2016), decreasing the estimated number of true type-2s, i.e. AGN that are genuinely lacking a BLR. No strong constraints have been found between the absence of broad emission lines observed in polarised fluxes and a given physical parameter such as bolometric luminosity, torus sizes, or galaxy inclination (Ramos Almeida et al. 2016). However, the discovery of changing-look AGNs in which strong flux variations (up to 2 orders of magnitude) and the appearance/disappearance of broad emission lines may be a step to solve this issue (Matt et al. 2003; LaMassa et al. 2015; Mathur et al. 2018). The disappearance of BLR signatures and the flux dimming could be explained by a variety of processes, from variable obscuration of the source to rapid mass accretion rate drop (Noda and Done 2018). Polarimetry offers a natural way to investigate this question since any change in the geometry of the scatterer directly impact the polarisation degree and angle (Marin 2017). In the case of the changing-look AGN J1011+5442 that switched between type-1 and type-2 classification between 2003 and 2015, the almost null polarisation recorded in the type-2 phase indicates that the observed change of look is not due to a change of obscuration in the torus hiding the BLR and the central engine (Hutsemekers et al. 2017). It is rather due to a rapid decrease of the SMBH accretion rate that is responsible for the vanishing of the BLR region (Hutsemekers et al. 2017; Marin 2017). Additional polarimetric observations are necessary to determine the physical reasons behind the change of look of those AGNs, but this can only be done case-by-case.

14.2.4 Revealing the Hidden Nuclear Location

Closely following the foundation of the Unified Model of AGN (Antonucci 1993), polarisation mapping of the extended structures around the hidden core of type-2 AGNs allowed to pinpoint the source of emission. Back in 1990–1995, the exact location of the central engine in dust-obscured sources was not easily determined (see Fig. 14.3). Infrared (Braatz et al. 1993) and radio (Ulvestad et al. 1987) images were not in agreement regarding the question of the hidden nuclear location. For the closest objects, they diverged by more or less one arcsecond, which represents several tens of parsecs in the case of NGC 1068 (which is among the most observed type-2s ever). Considering that the central engine has a sub-parsec scale, such offset

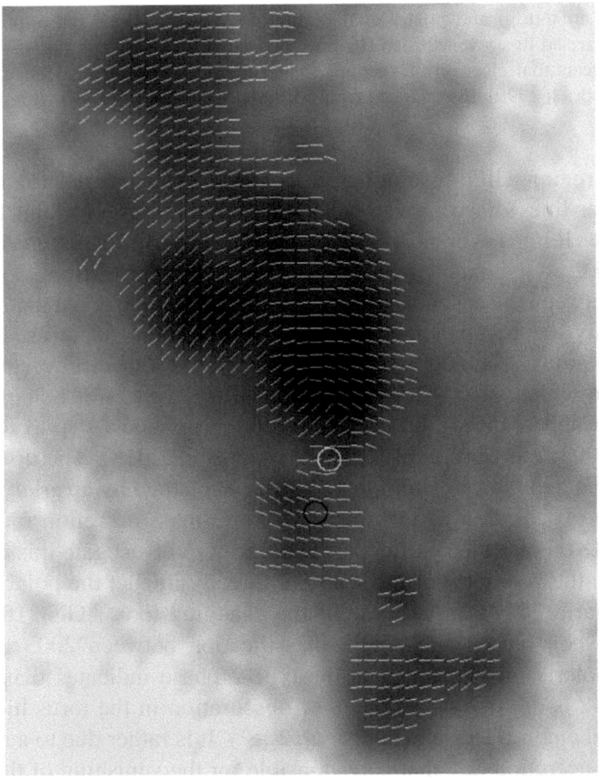

Fig. 14.3 *Hubble Space Telescope* (*HST*) imaging polarimetry of the inner region of the radio-quiet type-2 AGN NGC 1068 taken with the COSTAR-corrected Faint Object Camera in the 2400–2700 Å (Capetti et al. 1995a). The polarisation vectors are superimposed to the total intensity map. The length of the vectors are not representative of the polarisation degree. The position of the source of scattered radiation is indicated using a white circle. The black circle marks previous estimates. Reproduced from Capetti et al. (1995a) with permission

can be dramatic for the understanding of the internal physics of type-2 objects. Polarimetric attempts to determine the true location of the nucleus of NGC 1068, using scanning polarimeters, were done back in the 1970s (Elvius 1978) but the polarisation imaging capabilities brought by the *Hubble Space Telescope* (*HST*) allowed to settle the debate. Using the Faint Object Camera (ultraviolet) and the Wide Field Planetary Camera (visual), with spatial resolutions of 0.06 and 0.08 arcseconds, respectively, it was possible to map the extended NLR of nearby type-2 AGNs (Capetti et al. 1995b,c). The spatial resolution of the maps associated with the polarisation information allowed to reveal a double-conical shape with sharp edges, standing out with much more contrast than in total flux (Macchetto et al. 1994). The centro-symmetric pattern of the polarisation vectors are pointing towards the true, hidden location of the central engine, such as illustrated in Fig. 14.3. The uncertainties in the source location are of the order of 0.02 arcseconds. A deeper look at the polarisation structure in the NLR was achieved with repeated *HST* polarimetric mapping and allowed to isolate 30 different clouds in the upper and lower NLR bicone (Kishimoto 1999). The author used the clouds polarisation degree as an indicator of the scattering angles to reconstruct the three-dimensional structure of the nuclear region. This lead to the first and, 20 years later, still unique three-dimensional view of an AGN that only polarimetry can achieve.

14.2.5 The Laing-Garrington Effect

Another very important piece of information about the physics of AGN has been brought by imaging polarisation, this time in the radio band. Between the crucial observation of NGC 1068 and the establishment of the Unified Model of AGNs, Very Large Array (VLA) observations shed light on the complex question of the one-sidedness in the jets of otherwise symmetrical extragalactic radio sources. If today we are almost certain that special relativistic effects (Doppler boosting) strongly favours relativistic jets and components approaching us, the absence of (detected) counter-jets was quite puzzling at that time (1970–1990). Polarimetric mapping of the jets and lobes in a small sample of radio-loud AGNs have shown that, for double jetted AGN with one jet being brighter than the other, the lobe containing the brighter jet is almost always less depolarised with increasing wavelength than the lobe on the other counter-jet side (Laing 1988; Garrington et al. 1988). This depolarisation asymmetry, now known as the "Laing-Garrington effect" and visible in Fig. 14.4, is of particular importance as it can be explained either by internal differences between the two lobes if the jet one-sidedness is intrinsic (Garrington et al. 1988), or by a differential Faraday rotation through irregularities in a magneto-ionic medium surrounding the radio source if the one-sidedness is due to Doppler beaming (Laing 1988). In the second scenario, the side with the stronger jet closer

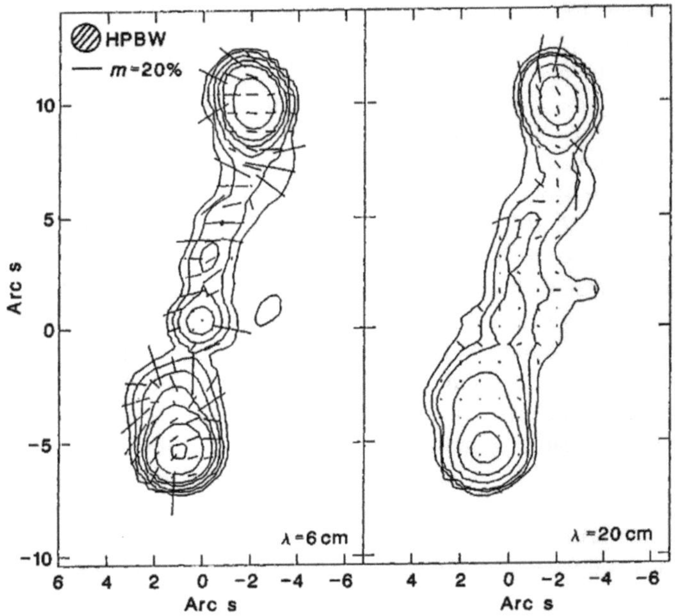

Fig. 14.4 Radio maps of 4C16.49, a type-2 radio-loud AGN that shows prominent twin jets. The left map is taken at 6 cm while the right map is taken at 20 cm. The solid contours are proportional to the total intensity, while the vectors indicate the polarisation angles. The length of the vectors are proportional to the fractional polarisation. The lobe containing the brighter jet depolarises less rapidly with increasing wavelength than the lobe on the counter-jet side (Laing 1988; Garrington et al. 1988). Reproduced from Garrington et al. (1988) with permission

to us is seen through a smaller amount of material and therefore shows less depolarisation. This is intrinsically related to the orientation of the AGN, another major step that lead to the Unified Model a few years later.

14.2.6 The Highly Non-spherical Structure of Outflows

Outflows, together with jets, play a major role in the enrichment of the AGN host galaxy (see Sect. 14.2.6). Polar winds are known to exhibit a wide range of emission and absorption lines, among which broad absorption lines (BAL, only seen for the most luminous AGNs so far) have several fascinating properties. First, the BAL correspond to resonance lines of highly ionised species such as C IV, S IV, or N V. Second, the absorption features are blue-shifted with respect to the corresponding emission lines, with ejection velocity of the order of a fraction of the speed of light. Only 10–20% of AGNs show such features (Lamy and Hutsemèkers 2004) but the close similarity between the emission-line and continuum properties of BAL and non-BAL AGNs rules out a different nature. It is more likely that orientation plays a role in the detection of BAL (Weymann et al. 1991). To decipher the origin of the outflows and the mechanisms responsible for their production, continuum, absorption and emission-lines polarisation was scrupulously observed (Hines and Wills 1995; Brotherton et al. 1997; Hutsemekers 1998b; Ogle et al. 1999; Lamy and Hutsemèkers 2004). BAL AGNs appear to be more highly polarised than non-BAL objects, with polarisation degrees neighbouring 4% on average, which was first thought to be consistent with a more equatorial viewing direction (Ogle et al. 1999). However, their broad emission lines are often unpolarised (Hines and Wills 1995) and the polarisation degree and position angle are wavelength-depend both in the continuum and absorption lines (Brotherton et al. 1997). This suggests multiple scattering mechanisms, resulting in a complex system geometry (Lamy and Hutsemèkers 2004). The increase of continuum polarisation in the ultraviolet, associated with a regular rotation of the polarisation position angle observed in [HB89] 0059-274 tend to point to two different polarisation mechanisms (Lamy and Hutsemèkers 2000). This is supported by polarisation microlensing observations, where the microlensed polarised continuum comes from a compact region coplanar to the accretion disc while the non-microlensed continuum arises from an extended region located along the polar axis (Hutsemekers et al. 2015). The origin of the BAL wind might very well be related to an outflow originating from the disc or from the torus, and extending towards the polar axis. The continuum radiation would then be scattered inside the accretion disc and wind base, then inside the polar outflows, producing roughly perpendicular polarisations (Hutsemekers 1998b; Marin and Goosmann 2013). This non-spherical structure is supported by the strongly polarised residual light within the broad absorption lines, implying that the outflows must have a small opening angle (Ogle et al. 1999; Marin and Goosmann 2013). Spectropolarimetry has proven to be particularly insightful in this regard: the structures observed in polarised light across the broad Hα emission line in PG 1700+518 indicate that the outflows (showing $\sim$4000 km s^{-1} rotational motions) must originate close to the accretion disc and rise nearly vertically (Young et al. 2007). The geometry and acceleration mechanisms of BAL winds are now actively discussed and computed as they appear to be a universal signature of massive and accreting objects (Lucy and Solomon 1970; Proga et al. 2000; Elvis 2000, 2017).

14.2.7 Alignment of AGN Polarisation with Large-Scale Structures

The orientation of AGNs is apparently not strongly correlated to the orientation of their host galaxy. At best, the extended radio structures of type-1 AGN are found to avoid alignment with their host galaxy major axis while type-2 AGN are more randomly distributed. In addition, both AGN types apparently avoid close alignment between their radio axis and their host galaxy plane axis (Schmidt et al. 1997). Apart from those specific zones of avoidance, there is no observational proofs nor reasons for a preferred direction of AGN polarisation position angles. Yet, compiling the polarisation position angle of 170 moderate-to-high redshift ($z \sim$1–2) AGNs, a concentration of polarisation vectors with a preferential direction was found in a $\sim$1000 h^{-1} Mp region of the sky (Hutsemekers 1998a). Subsequent measurements confirmed that the polarisation vectors are coherently oriented in several groups of 20–30 AGNs, which are roughly parallel to the plane of the Local Supercluster (Hutsemekers and Lamy 2001). However this result only holds at $z \geq 1$, which rules out large-scale ($\sim$50 Mpc) magnetic fields that are either (1) converting photons into massless or extremely light pseudo-scalars (Harari and Sikivie 1992) or (2) being responsible for dichroic extinction and scattering (Wood 1997). This means that the causes for AGN polarisation orientation might very well be cosmological. Early large-scale primordial magnetic fields could have played a role on galaxy formation and orientation during the epoch of inflation, explaining the $z \geq 1$ constraints (Battaner et al. 1997; Hutsemekers and Lamy 2001). More recent observations of radio-loud AGN at various redshifts have confirmed the alignment of AGN polarisation with large-scale structures, such as shown in Fig. 14.5 and detailed in Taylor and Jagannathan (2016), Contigiani et al. (2017). Comparing the

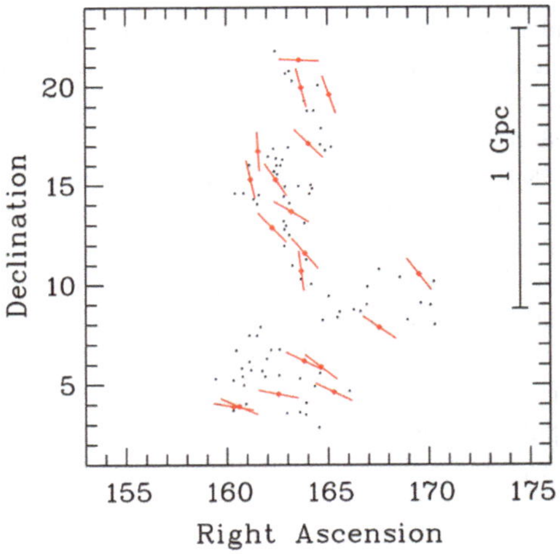

Fig. 14.5 Polarisation vectors (in red) of 19 radio-quiet AGNs with a polarisation degree larger than interstellar contaminant polarisation. The polarisation vectors are superimposed on the large-scale structure belonging to Gpc scale AGN groups at redshift z $\approx$1.3 (Hutsemekers 1998a; Hutsemekers and Lamy 2001; Hutsemekers et al. 2014). Despite a clear correlation, the interpretation of this orientation effect remains puzzling. A solution might be that the spin axes of AGNs are possibly parallel to their host large-scale structures. Reproduced from Hutsemekers et al. (2014), with permission, copyright ESO

AGN optical polarisation position angle to their host large-scale structures, it was additionally deduced that type-1 AGNs are preferentially perpendicular to the host structure (Hutsemekers et al. 2014). This suggests that the spin axis of AGNs is parallel to their host large-scale structures.

14.2.8 Accretion Signatures in Polarised Fluxes

One of the most recent discoveries achieved with polarimetric measurements concerns the spatially non-resolvable inner regions of AGNs. The accretion disc that powers the SMBH is thought to radiate from the far-ultraviolet to the near-infrared (Shakura and Sunyaev 1973), but the disc near-infrared emission is diluted by quantitatively much larger dust emission from the bulk circumnuclear region (the torus). The spectral shape of the disc emission decreases rapidly at longer wavelength, up to 1–2 μm, such as $F_\nu \sim \nu^{1/3}$. The addition of the re-emitted infrared light from the torus causes the spectral slope to be much redder (Zheng et al. 1997; Vanden Berk et al. 2001). As a consequence, it is impossible to observe the total flux from the outer rim of the disc, and thus confirm the disc paradigm (Hubeny et al. 2000). A fascinating aspect of polarimetry is that it produces a blueprint of the original spectrum and keeps memory of it as long as no additional scattering or absorption events happen. Looking at type-1 inclinations, the central engine is directly visible through the optically thin polar outflows. Light does not suffer from additional scattering other than in the equatorial plane, where the accretion disc is located. So, it is theoretically possible to observe the near-infrared component of the accretion disc in polarised light. Such piece of work was achieved in six different, local (z ~0.2–0.6) AGNs (Kishimoto et al. 2009). It was found that, as expected, the total flux spectra are dominated by hot dust re-emission around 1 μm, while the polarised flux spectra consistently and systematically decreasing towards the infrared, see Fig. 14.6. The weighted mean of the spectral slope is 0.44 ± 0.11, which is very close to the theoretical limit of 0.33. Interestingly, the authors do not find any correlation between the spectral slopes and the AGN black hole masses or Eddington luminosities. The marginal, bluer spectral slope found could, however, be interpreted as a hint for a truncated (or a gravitationally unstable) disc. This opens the door for future investigations of the continuum polarimetric signal in nearby AGNs.

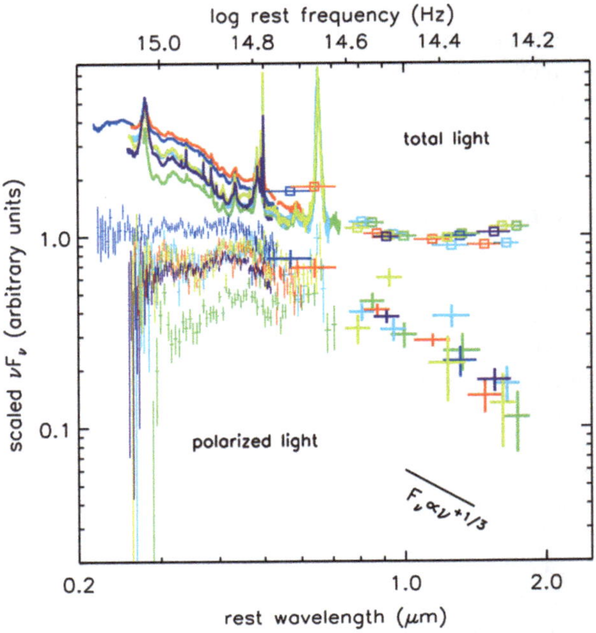

Fig. 14.6 Overlay of the total and polarised flux spectra of six different radio-loud, type-1 AGN (Kishimoto et al. 2008). The polarised flux allows to shave off both the emission lines and the unpolarised re-emission by dust. All the objects behave in a similar and systematic way, revealing the expected blue polarised-light spectra of accretion discs. This is a perfect example of how polarimetry can probe spatially unresolved regions, up to ~800 R_S in this case. Reproduced from Kishimoto et al. (2008), with permission

14.3 Current Status on the Panchromatic Polarisation of AGN

We have seen that polarimetric measurements, from the radio to the ultraviolet band, have led to considerable discoveries. The most significant one is the establishment of the Unified Model of AGNs, but polarimetry has also proven to be able to achieve great discoveries from the scale of the accretion disc to the kilo-parsec jets. Each waveband is characterised by a different set of polarimetric signatures that can be related to various physical mechanisms that, from the radio band to the soft-γ rays, probe increasingly smaller AGN regions. When we compile all the polarimetric data ever recorded for NGC 1068, by far the most observed radio-quiet type-2 AGN in polarisation (Marin 2018a), we can take stock of our knowledge and the work that remains to be done.

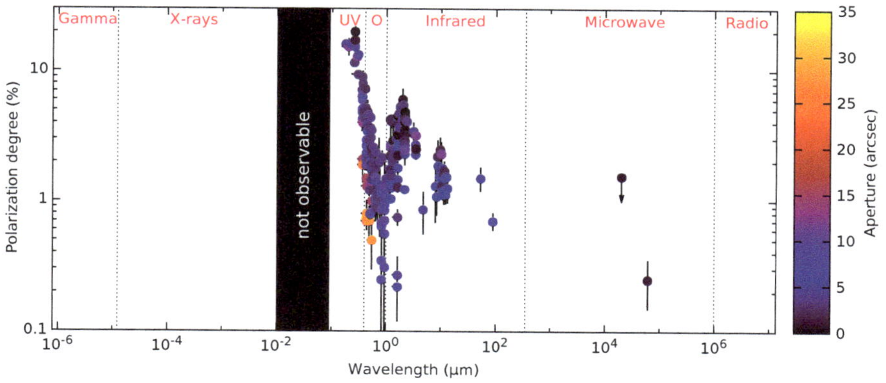

Fig. 14.7 Summary of all the polarimetric measurements ever published for NGC 1068, by far the most observed radio-quiet type-2 AGN in polarisation (Marin 2018a). Instrument apertures are colour-coded (in arc-seconds) and the black area shows where no measurements are achievable due to the hydrogen Galactic absorption downward 91.2 nm

- As it can be seen from Fig. 14.7, the γ-ray and X-ray bands are completely uncharted territories. The reason is simple: in the past, there was never a high energy polarimeter sensitive enough to detect any polarisation signature from AGNs. Note that the polarisation degree is expected to be too low, but this was simply never tried for purely technical and economic reasons. This statement is true for all kind of AGNs, independently of the presence of jets. The high energy band is expected, in fact, to show detectable polarisation levels (McNamara et al. 2009; Marin et al. 2018a,b). A list of mechanisms that could produce γ and X-ray polarisation includes Compton and inverse-Compton scattering, synchrotron emission, Faraday rotation, bremsstrahlung, and collisionally excited line radiation from low energy cosmic protons, electrons, and atoms (Dolan et al. 1967).
- The extreme and far-ultraviolet bands are also lacking existing polarisation measurements for similar reasons. In the optical band, polarisation is observed using birefringent crystals. However, in the case of far-ultraviolet polarimetry, there is no birefringent material that can transmit light as crystals become opaques. A work around must be found using reflection polarimetry but the technique is yet to be optimised and tested (Bolcar et al. 2017). Scattering by electrons, atoms, and dust grains, together with non-thermal emission, is expected to provide strong polarisation degrees since the diluting, unpolarised starlight flux from the host galaxy strongly diminishes in the ultraviolet.
- The mid- and near-ultraviolet, the optical, and the near-infrared polarisation of AGNs has been recorded the most. White light polarisation measurements are achievable from ground telescopes, and the technology is mature. Multiple apertures and detectors have been used, leading to a wealth of polarimetric data in those wavebands. The polarisation is, however, strongly diluted by the host galaxy starlight, starburst light, and interstellar polarisation (see the strong

polarisation deep in the optical band in Fig. 14.7). Removing the parasitic light is a complicated process that implies to observe both stars close to the AGN line-of-sight (to evaluate the importance of interstellar polarisation) and the host galaxy. By minimising the stellar absorption features in the residuals of the host/AGN flux ratio spectra, it is possible to estimate the starlight fractions in the observed AGN continuum flux. The situation can be even more complicated if synchrotron radiation from a jet superimposes its intrinsic polarisation to the observation. So, even if white light polarimetry is common, it is not an easy task to interpret data.

- The infrared band, loosely defined as 1–300 μm here, has been less observed in polarimetric mode. Technologically speaking, it is more complex to observe infrared polarisation since cryogenic cooling of the detectors in necessary in order to minimise the presence of thermal emission from warm optics. Below 2.5 μm, thermal emission from external hardwares is not too strong, so polarimeters can be constructed using half-wave plate retarders and birefringent crystals (Jones et al. 1988), explaining the presence of many measurements at 1–2.2 μm in Fig. 14.7. In the mid- and far-infrared, there is a clear lack of polarimeters. Only a couple of points have been measured in the 19–100 μm waveband using the SOFIA High-resolution Airborne Wideband Camera-plus (Lopez-Rodriguez et al. 2018; Marin 2018a). Scattering, extinction, and re-emission by aligned dust grains, together with magnetic effects, are responsible for most of the polarisation that one can detect in the infrared band of AGNs.
- Finally, the millimetre and centimetre (alternatively called microwave and radio) polarisation have been very little explored in the case of radio-quiet AGNs. When strong, kilo-parsec, collimated jets are present, they radiate a strong (polarised) flux. Non-thermal total and polarised synchrotron emission, differential Faraday rotation (with a λ^2 dependence), Faraday interconversion (from linear polarisation to circular polarisation, and vice versa), and spectral depolarisation (the polarisation of a component changes sharply with frequency as the component becomes opaque) are responsible for the high millimetre and centimetre polarisation in radio-loud AGNs (Leppanen et al. 1995). In the case of radio-quiet AGNs, the source of radio emission and its associated polarisation are much less known (Laor and Behar 2008; Ishibashi et al. 2011). As an example, the degree of polarisation at 4.9 GHz and 15 GHz measured for the radio-quiet AGN NGC 1068 (see Fig. 14.7) was at the sensitivity limits of the VLA observations. New polarimetric studies in the millimetre and centimetre bands are thus necessary.

14.4 Open Questions that Polarimetry Could Solve

Our knowledge of the panchromatic polarisation of AGNs is fragmentary. In fact, looking at Fig. 14.7, only a small fraction of the whole electromagnetic spectrum has been observed for radio-quiet AGNs. In the case of radio-loud objects, this fraction is much greater, but the high energy part of the spectrum is missing for both AGN types. New observations, both in the uncharted high energy band and in the other parts of the electromagnetic spectrum, are severely necessary. Numerous constraints on various physical scenarios and AGN hypotheses can be achieved with current and future polarimeters. I will review a few of them, roughly by increasing wavelength, with their associated future instruments.

14.4.1 Probing the Origin of the X-Ray Fluxes

The morphology and physics of the region responsible for the emission of X-rays in AGNs is still highly debated. Almost all models rely on a compact corona situated above the accretion disc that is heated up by magnetic reconnection. The corona is believed to reprocess thermal ultraviolet photons emitted by the disc up to the X-rays by Inverse-Compton scattering, resulting in a corona that illuminates back the disc due to gravitational effects. Distorted line emission around 6.4 keV indicates that the X-ray corona should be close to the disc so that strong gravity effects are dominant (Merloni and Fabian 2003). However, recent X-ray polarimetric observations by the PoGO+ pathfinder have shaken our certitudes about this scenario for the case of X-ray binaries in the hard state, i.e. when thermal emission from the disc dominates (Chauvin et al. 2018). Future X-ray polarimetric measurements with the *Imaging X-ray Polarimetry Explorer* (*IXPE*, Weisskopf et al. 2016) or with the *enhanced X-ray Timing and Polarimetry Mission* (*eXTP*, Zhang et al. 2016) will shed light on this question. The geometry of the corona can be probed, thanks to the sensitivity of polarimetry to the geometrical shape of the emission source. A slab corona will produce a more asymmetric radiation pattern than a spherical one, and strong gravity effects will change the polarisation in a predictable way. The spin and the mass of the SMBH, together with the inclination of the accretion disc, will be within the reach of future X-ray polarimeters (Schnittman and Krolig 2010; Dočiak et al. 2011).

14.4.2 Testing the Accretion Disc Paradigm

Some key signatures of accretion discs can be revealed only in polarised light (see Sect. 14.2.8), and with higher contrast at ultraviolet than at longer wavelengths. Specifically, models of disc atmospheres usually assume Compton scattering in an electron-filled plasma, resulting in inclination-dependent polarisation signatures. We would expect a polarisation degree in the range 0–11.7%, and always lying along the (projected) disc plane (Chandrasekhar et al. 1960). Yet optical polarisation in type-1 objects is almost always detected at less than a percent and parallel to the radio jets if any (so perpendicular to our expectations) (Stockman et al. 1979; Antonucci 2002). Whether these low levels can be attributed to dominant absorption opacity (Laor and Netzer 1989) or complete Faraday depolarisation (Agol et al. 1996) is unclear. This degeneracy may be broken by looking at the numerous ultraviolet lines that are formed in the innermost AGN regions (e.g. Lyα λ1216, CII λ1335, CIV λ1549, MgII λ2800). These lines are the key to test the accretion disc paradigm and only ultraviolet polarimetric observations with high signal-to-noise ratio and high spectral resolution can distinguish between the two effects. In this context, a high resolution spectropolarimeter such as *POLLUX* can provide the most exquisite results. *POLLUX* is an instrument envisioned for the 15-m primary mirror option of *LUVOIR* (a multi-wavelength space observatory concept being developed by the Goddard Space Flight Center and proposed for the 2020 Decadal Survey Concept Study (Bouret et al. 2018)). The ultraviolet coverage and large resolution of *POLLUX* would also allow to test models of accretion disc atmospheres, where the ionisation of hydrogen should give rise to spectral features at the Lyman limit (912 Å). Yet, the Lyman edge feature has never been convincingly observed in total flux (Kriss et al. 1999; Shull et al. 2012). Polarised flux observations could remove the parasitic light from the other AGN components and allow to better detect the presence of a sharp discontinuity in the continuum at the Lyman limit (Kishimoto et al. 2005).

14.4.3 Revealing the Location, Composition, and Geometric Arrangement of Dust

Decomposition of the SED of AGNs suggests that the mid-infrared component corresponds to equatorial emission, approximately aligned with the plane of the inner accretion disc, while the weaker near-infrared peak might be associated with hot dust in the inner polar region. However, recent studies revealed that this picture is probably erroneous (Hönig et al. 2013; Asmus et al. 2016). The bulk of the infrared emission seems to originate from the polar region above the circumnuclear dust, where only little dust should be present. Using new high angular resolution polarimetric observations with adaptive optics systems (Gratadour et al. 2015; Lopez-Rodriguez et al. 2016), it will be possible to investigate the true location of dust in AGNs and derive its polarisation-sensitive composition. Moreover, as polarisation is enhanced by paramagnetic dust grains alignment, it is possible to go beyond the capabilities of interferometry and probe the parsec-scale magnetic fields. The key argument here is that the magnetic fields align the dust grains according to their intensity and direction. By fitting the infrared polarisation data with different numerical models, the topology of parsec scales magnetic fields can be constrained (Peest et al. 2017; Grosset et al. 2018). Understanding the observed gradual rotation of the polarisation angle towards the far-infrared in NGC 1068 or 3C 273 can allow a characterisation of the global coherent magnetic field structure impacting dust scattering, absorption, and emission over the full infrared band emission.

14.4.4 The Physics and the Internal Structure of the Innermost Jet Regions

AGNs routinely accelerate significant fractions of a solar mass to near light speeds in the form of ballistic jets. Due to the ordered magnetic fields and directionality of emission, the emitted synchrotron radiation is naturally highly polarised (Zheleznyakov and Koragin 2002). Blazars, a sub-class of AGNs that have the orientation of their jets close to the line of sight, are particularly interesting on this case since their orientation makes their non-thermal radiation highly relativistically beamed, and thus very bright. The combination of brightness and high polarisation allows new polarimetric imaging in the (sub)millimetre and radio bands by the Atacama Large Millimeter/submillimeter Array (ALMA) to identify magnetic field configurations at unprecedented scales (Liuzzo et al. 2015). Past NRAO telescopes (VLA, GBT, VLBA) have proven to be particularly useful by, e.g., revealing a rotation of the position angle of linear polarisation in the jet of BL Lacertae, following a compressive feature propagating down the helical jet field (Marscher et al. 2008). However, the case is more complex for radio-quiet AGN where the expected polarised radio flux is too low compared to the sensitivity of old radio telescopes. There is essentially no exploitable data and only future observations of radio-quiet AGNs, using the most recent radio observatories, will provide polarimetric information at millimetre and centimetre wavelengths. Detecting the radio polarisation signal from radio-quiet AGNs is fundamental to shed light on the physical mechanisms producing the faint radio emission. Looking at high angular resolution polarisation maps of jet's bases will also allow to better constrain magneto-hydrodynamic and plasma models (Agudoet al. 2018), especially through the prism of the enigmatic variability processes.

14.5 Summary

Panchromatic polarisation measurements of AGNs are necessary to better understand how such objects can form, accrete matter, and re-emit copious amounts of particles and radiation, ultimately impacting the host galaxy they reside in. In the past, polarimetry has brought the most important constraints on the true nature of AGNs, and it keeps revealing more and more complex features associated with accretion, magnetic, emission, absorption, and re-emission processes. Polarisation, at least in the case of quasi-stellar radio sources, is complex to decipher due to the presence of a contaminating flux from host galaxy. Yet, one can access fundamental signatures of spatially unresolvable AGN components, thanks to polarisation. Many questions linked with AGN formation and evolution processes remain open, and only future polarimetric measurements using 30-m class telescopes or satellites equipped with state-of-the-art polarimeters can resolve them. In particular, accretion disc theories, disc atmosphere models, and jet production mechanisms will be within our reach in the next decade, thanks to the renewed enthusiasm of the community about high energy polarimetry (Marin 2018b). The construction of large radio and millimetre telescopes will also fully benefit the field by probing subarcsecond scales. The visible and infrared bands maybe lagging behind in new instrument projects, yet those wavebands are equally important as the other parts of the electromagnetic spectrum for a complete understanding of AGNs.

Acknowledgements I would like to deeply thank Damien Hustemékers, Delphine Porquet, and Robert "Ski" Antonucci for their insightful comments that helped to improve the quality of this chapter.

References

Abramowicz MA, Czerny B, Lasota JP, Szuszkiewicz E (1988) Astrophys J 332:646
Afanasiev VL, Dodonov SN, Khrapov SS, Mustsevoi VV, Moiseev AV (2007a) Astrophys Bull 62:1
Afanasiev VL, Dodonov SN, Khrapov SS, Mustsevoi VV, Moiseev AV (2007b) Astrophys Bull 62:15
Agol E, Blaes O (1996) Mon Not R Astron Soc 282:965
Agudo I, Thum C, Ramakrishnan V et al (2018) Mon Not R Astron Soc 473:1850
Angel JRP, Stockman HS (1980) Annu Rev Astron Astrophys 18:321
Antonucci RRJ (1982) Nature 299:605
Antonucci RRJ (1983) Nature 303:158
Antonucci RRJ (1984) Astrophys J 278:499
Antonucci R (1993) Annu Rev Astron Astrophys 31:473
Antonucci R (2002) Astrophysical spectropolarimetry. Proceedings of the XII Canary Islands Winter School of Astrophysics, p 151. https://ui.adsabs.harvard.edu/abs/2002apsp.conf..151A/abstract
Antonucci RRJ, Miller JS (1985) Astrophys J 297:621
Antonucci RRJ, Kinney AL, Ford HC (1989) Astrophys J 342:64
Asmus D, Hönig SF, Gandhi P (2016) Astrophys J 822:109
Babadzhanyants MK, Hagen-Thorn VA (1975) Astrophysics 11:259
Balbus SA, Hawley JF (1998) Rev Mod Phys 70:1
Baldwin J, Ferland G, Korista K, Verner D (1995) Astrophys J Lett 455:L119
Barvainis R (1987) Astrophys J 320:537
Battaner E, Florido E, Jimenez-Vicente J (1997) Astron Astrophys 326:13
Begelman MC, McKee CF (1983) Astrophys J 271:89
Begelman MC, McKee CF, Shields GA (1983) Astrophys J 271:70
Bhatnagar S, Krishna G, Wisotzki L (1998) Mon Not R Astron Soc 299:L25
Bianchi S, Maiolino R, Risaliti G (2012) Adv Astron 2012:782030
Blandford RD, Payne DG (1982) Mon Not R Astron Soc 199:883
Blandford RD, Znajek RL (1977) Mon Not R Astron Soc 179:433
Blank M, Duschl WJ (2016) Mon Not R Astron Soc 462:2246
Bolcar MR, Aloezos S, Bly VT et al (2017) Society of Photo-Optical Instrumentation Engineers (SPIE) Conference Series 10398:1039809
Bouret J-C, Neiner C, Gómez de Castro AI et al (2018) Society of Photo-Optical Instrumentation Engineers (SPIE) Conference Series 10699:106993B
Bower RG, Benson AJ, Malbon R et al (2006) Mon Not R Astron Soc 370:645
Braatz JA, Wilson AS, Gezari DY, Varosi F, Beichman CA (1993) Astrophys J Lett 409:L5
Brotherton MS, Tran HD, van Breugel W, Dey A, Antonucci R (1997) Astrophys J Lett 487:L113
Cao X (2018) Mon Not R Astron Soc 473:4268
Capetti A, Macchetto F, Axon DJ, Sparks WB, Boksenberg A (1995a) Astrophys J Lett 452:L87

Capetti A, Axon DJ, Macchetto F, Sparks WB, Boksenberg A (1995b) Astrophys J 446:155
Capetti A, Macchetto F, Axon DJ, Sparks WB, Boksenberg A (1995c) Astrophys J 448:600
Chai B, Cao X, Gu M (2012) Astrophys J 759:114
Chandrasekhar S (1960) Radiative transfer, LCCN: 60-3117. Dover, New York. https://ui.adsabs.harvard.edu/abs/1960ratr.book.....C/abstract
Chauvin M, Florén H-G, Friis M et al (2018) Nat Astron 2:652
Collin S, Huré J-M (2001) Astron Astrophys 372:50
Contigiani O, de Gasperin F, Miley GK et al (2017) Mon Not R Astron Soc 472:636
Corral A, Della Ceca R, Caccianiga A et al (2011) Astron Astrophys 530:A42
Czerny B, Li Y-R, Hryniewicz K et al (2017) Astrophys J 84:154
Davies RI, Müller Sánchez F, Genzel R et al (2007) Astrophys J 671:1388
della Ceca R, Lamorani G, Maccacaro T et al (1994) Astrophys J 430:533
Di Matteo T (1998) Mon Not R Astron Soc 299:L15
Di Matteo T, Springel V, Hernquist L (2005) Nature 433:604
Dolan JF (1967) Space Sci Rev 6, 579
Dovčiak M, Muleri F, Goosmann RW, Karas V, Matt G (2011) Astrophys J 731:75
Dubois Y, Gavazzi R, Peirani S, Silk J (2013) Mon Not R Astron Soc 433:3297
Elvis M (2000) Astrophys J 545:63
Elvis M (2017) Astrophys J 847:56
Elvius A (1978) Astron Astrophys 65:233
Fabian AC, Iwasawa K, Reynolds CS, Young AJ (2000) Publ Astron Soc Pac 112:1145
Frank J, King A, Raine DJ (2002) Accretion power in astrophysics. Cambridge University Press, Cambridge
Fritz J, Franceschini A, Hatziminaoglou E (2006) Mon Not R Astron Soc 366:767
García-Burillo S, Combes F, Ramos Almeida C et al (2016) Astrophys J Lett 823:L12
Garrington ST, Leahy JP, Conway RG, Laing RA (1988) Nature 331:147
Gaskell CM (2009) New Astron Rev 53:140
George IM, Fabian AC (1991) Mon Not R Astron Soc 249:352
González Delgado RM, Heckman T, Leitherer C et al (1998) Astrophys J 505:174
Gratadour D, Rouan D, Grosset L, Boccaletti A, Clénet Y (2015) Astron Astrophys 581:L8
Grosset L, Rouan D, Gratadour D et al (2018) Astron Astrophys 612:A69
Haardt F, Maraschi L (1993) Astrophys J 413:507
Harari D, Sikivie P (1992) Phys Lett B 289:67
Higginbottom N, Proga D, Knigge C et al (2014) Astrophys J 789:19
Hines DC, Wills BJ (1995) Astrophys J Lett 448:L69
Hönig SF, Kishimoto M, Tristram KRW et al (2013) Astrophys J 771:87
Hubeny I, Agol E, Blaes O, Krolik JH (2000) Astrophys J 533:710
Hutsemekers D (1998) Astron Astrophys 332:410
Hutsemékers D, Lamy H (2001) Astron Astrophys 367:381
Hutsemekers D, Lamy H, Remy M (1998) Astron Astrophys 340:371
Hutsemékers D, Braibant L, Pelgrims V, Sluse D (2014) Astron Astrophys 572:A18
Hutsemékers D, Sluse D, Braibant L, Anguita T (2015) Astron Astrophys 584:A61
Hutsemékers D, Agís González B, Sluse D, Ramos Almeida C, Acosta Pulido J-A (2017) Astron Astrophys 604:L3
Ishibashi W, Courvoisier TJ-L (2011) Astron Astrophys 525:A118
Jaffe W, Ford HC, Ferrarese L, van den Bosch F, O'Connell RW (1993) Nature 364:213
Jaffe W, Meisenheimer K, Röttgering HJA et al (2004) Nature 429:47
Jiang L, Fan X, Ivezić Ž et al (2007) Astrophys J 656:680
Jones TJ, Klebe D (1988) Publ Astron Soc Pac 100:1158
Kaspi S, Smith PS, Netzer H et al (2000) Astrophys J 533:631
Kellermann KI, Sramek R, Schmidt M, Shaffer DB, Green R (1989) Astron J 98:1195
Kishimoto M (1999) Astrophys J 518:676
Kishimoto M, Antonucci R, Boisson C, Blaes O (2005) Astron Polarim Curr Status Futur Dir 343:435

Kishimoto M, Hönig SF, Beckert T, Weigelt G (2007) Astron Astrophys 476:713
Kishimoto M, Antonucci R, Blaes O et al (2008) Nature 454:492
Kishimoto M, Hönig SF, Antonucci R et al (2009) Astron Astrophys 507:L57
Kriss GA, Davidsen AF, Zheng W, Lee G (1999) Astrophys J 527:683
Kruszewski A (1971) Acta Astronomica 21:311
La Franca F, Gregorini L, Cristiani S, de Ruiter H, Owen F (1994) Astron J 108:1548
Laing RA (1988) Nature 331:149
LaMassa SM, Cales S, Moran EC et al (2015) Astrophys J 800:144
Lamy H, Hutsemékers D (2000) Astron Astrophys 356:L9
Lamy H, Hutsemékers D (2004) Astron Astrophys 427:107
Laor A, Behar E (2008) Mon Not R Astron Soc 390:847
Laor A, Netzer H (1989) Mon Not R Astron Soc 238:897
Lawrence A (1991) Mon Not R Astron Soc 252:586
Lawrence A, Watson MG, Pounds KA, Elvis M (1985) Mon Not R Astron Soc 217:685
Leppanen KJ, Zensus JA, Diamond PJ (1995) Astron J 110:2479
Liuzzo E, Nagai H, Giovannini G, Mignano A (2015) Revolution in astronomy with ALMA: the third year, vol 499. Astronomical Society of the Pacific, San Francisco, p 129
Lopez-Rodriguez E, Packham C, Young S et al (2013) Mon Not R Astron Soc 431:2723
Lopez-Rodriguez E, Packham C, Roche PF et al (2016) Mon Not R Astron Soc 458:3851
Lopez-Rodriguez E, Fuller L, Alonso-Herrero A et al (2018) Astrophys J 859:99
Lucy LB, Solomon PM (1970) Astrophys J 159:879
Macchetto F, Capetti A, Sparks WB, Axon DJ, Boksenberg A (1994) Astrophys J Lett 435:L15
Magdziarz P, Blaes OM, Zdziarski AA, Johnson WN, Smith DA (1998) Mon Not R Astron Soc 301:179
Malkan MA, Sargent WLW (1982) Astrophys J 254:22
Marin F (2016) Mon Not R Astron Soc 460:3679
Marin F (2017) Astron Astrophys 607:A40
Marin F (2018a) Mon Not R Astron Soc 479:3142
Marin F (2018b) Galaxies 6:38
Marin F, Goosmann RW (2013) Mon Not R Astron Soc 436:2522
Marin F, Goosmann RW, Petrucci P-O (2016) Astron Astrophys 591:A23
Marin F, Dovčiak M, Kammoun ES (2018a) Mon Not R Astron Soc 478:950
Marin F, Dovčiak M, Muleri F, Kislat FF, Krawczynski HS (2018b) Mon Not R Astron Soc 473:1286
Marscher AP, Jorstad SG, D'Arcangelo FD et al (2008) Nature 452:966
Martin PG, Thompson IB, Maza J, Angel JRP (1983) Astrophys J 266:470
Martocchia A, Matt G (1996) Mon Not R Astron Soc 282:L53
Mathis JS, Rumpl W, Nordsieck KH (1977) Astrophys J 217:425
Mathur S, Denney KD, Gupta A et al (2018) Astrophys J arXiv:1810.06616
Matt G, Guainazzi M, Maiolino R (2003) Mon Not R Astron Soc 342:422
McHardy I, Czerny B (1987) Nature 325:696
McNamara AL, Kuncic Z, Wu K (2009) Mon Not R Astron Soc 395:1507
Merloni A, Fabian AC (2003) Mon Not R Astron Soc 342:951
Miller JS, Antonucci RRJ (1983) Astrophys J Lett 271:L7
Miller JS, Goodrich RW (1990) Astrophys J 355:456
Miniutti G, Fabian AC (2004) Mon Not R Astron Soc 349:1435
Misner C, Thorne KS, Wheeler J (1973) Gravitation. W.H.Freeman & Co Ltd, San Francisco
Mortlock DJ, Warren SJ, Venemans BP et al (2011) Nature 474:616
Nandra K, Pounds KA (1994) Mon Not R Astron Soc 268:405
Narayan R, Yi I (1994) Astrophys J Lett 428:L13
Neugebauer G, Oke JB, Becklin EE, Matthews K (1979) Astrophys J 230:79
Netzer H, Laor A (1993) Astrophys J Lett 404:L51
Noda H, Done C (2018) Mon Not R Astron Soc 480:3898
Ogle PM, Cohen MH, Miller JS et al (1997) Astrophys J Lett 482:L37

Ogle PM, Cohen MH, Miller JS et al (1999) Astrophys J Suppl S 125:1
Onken CA, Peterson BM, Dietrich M, Robinson A, Salamanca IM (2003) Astrophys J 585:121
Osterbrock DE (1984) Q J R Astron Soc 25:1
Padovani P (2017) Nat Astron 1:0194
Page KL, Reeves JN, O'Brien PT, Turner MJL (2005) Mon Not R Astron Soc 364:195
Peest C, Camps P, Stalevski M, Baes M, Siebenmorgen R (2017) Astron Astrophys 601:A92
Peterson BM (2006) Physics of active galactic nuclei at all scales, vol 693. Springer, Berlin, p 77
Porquet D, Dubau J (2000) Astron Astrophys Suppl 143:495
Pringle JE, Rees MJ (1972) Astron Astrophys 21:1
Proga D, Stone JM, Drew JE (1998) Mon Not R Astron Soc 295:595
Proga D, Stone JM, Drew JE (1999) Mon Not R Astron Soc 310:476
Proga D, Stone JM, Kallman TR (2000) Astrophys J 543:686
Ramos Almeida C, Martínez González MJ, Asensio Ramos A et al (2016) Mon Not R Astron Soc 461:1387
Riley JM, Warner PJ, Rawlings S et al (1989) Mon Not R Astron Soc 236:13P
Robson EI, Gear WK, Brown LMJ, Courvoisier TJ-L, Smith MG (1986) Nature 323:134
Rowan-Robinson M, Clegg PE, Beichman CA et al (1984) Astrophys J Lett 278:L7
Sandage A (1965) Astrophys J 141:1560
Sanders DB, Phinney ES, Neugebauer G, Soifer BT, Matthews K (1989) Astrophys J 347:29
Schmidt GD, Miller JS (1980) Astrophys J 240:759
Schmitt HR, Kinney AL, Storchi-Bergmann T, Antonucci R (1997) Astrophys J 477:623
Schnittman JD, Krolik JH (2010) Astrophys J 712:908
Schoenmakers AP, de Bruyn AG, Röttgering HJA, van der Laan H (2001) Astron Astrophys 374:861
Serote-Roos M, Boisson C, Joly M, Ward MJ (1996) Mon Not R Astron Soc 278:897
Shakura NI, Sunyaev RA (1973) Astron Astrophys 24:337
Shull JM, Stevans M, Danforth CW (2012) Astrophys J 752:162
Skinner GK, Krizmanic JF (2009) Exp Astron 27:61
Stockman HS, Angel JRP, Miley GK (1979) Astrophys J Lett 227:L55
Storchi Bergmann T (2015) Galaxies in 3D across the universe, vol 309. p 190. https://doi.org/10.1017/S1743921314009648
Strateva IV, Strauss MA, Hao L et al (2003) Astron J 126:1720
Taylor AR, Jagannathan P (2016) Mon Not R Astron Soc 459:L36
Tombesi F, Cappi M, Reeves JN et al (2010) Astron Astrophys 521:A57
Tran HD (1995) Astrophys J 440:565
Tran HD, Miller JS, Kay LE (1992) Astrophys J 397:452
Ulvestad JS, Neff SG, Wilson AS (1987) Astron J 93:22
Urry CM, Padovani P (1995) Publ Astron Soc Pac 107:803
Vanden Berk DE, Richards GT, Bauer A et al (2001) Astron J 122:549
Visvanathan N, Oke JB (1968) Astrophys J Lett 152:L165
Vollmer B, Schartmann M, Burtscher L et al (2018) Astron Astrophys 615:A164
Walker MF (1968) Astrophys J 151:71
Walter R, Fink HH (1993) Astron Astrophys 274:105
Webb W, Malkan M, Schmidt G, Impey C (1993) Astrophys J 419:494
Weisskopf MC, Ramsey B, O'Dell S et al (2016) In: Proceedings of the SPIE, vol 9905. p 990517
Weymann RJ, Morris SL, Foltz CB, Hewett PC (1991) Astrophys J 373:23
Wills BJ, Wills D, Evans NJ II et al (1992) Astrophys J 400:96
Wilson AS (1996) Vistas Astron 40:63
Woo J-H, Kim JH, Imanishi M, Park D (2012) Astron J 143:49
Wood K (1997) Astrophys J Lett 477:L25
Young S, Axon DJ, Robinson A, Hough JH, Smith JE (2007) Nature 450:74
Zhang SN, Feroci M, Santangelo A et al (2016) Proceedings of the SPIE, vol 9905. p 99051Q
Zheleznyakov VV, Koryagin SA (2002) Astron Lett 28:727
Zheng W, Kriss GA, Telfer RC, Grimes JP, Davidsen AF (1997) Astrophys J 475:469

Index

R. Mignani et al. (eds.), *Astronomical Polarisation from the Infrared to Gamma Rays*, Astrophysics and Space Science Library 460,
https://doi.org/10.1007/978-3-030-19715-5

MIX
Papier aus verantwortungsvollen Quellen
Paper from responsible sources
FSC® C105338

If you have any concerns about our products,
you can contact us on
ProductSafety@springernature.com

In case Publisher is established outside the EU,
the EU authorized representative is:
Springer Nature Customer Service Center GmbH
Europaplatz 3, 69115 Heidelberg, Germany

Printed by Libri Plureos GmbH
in Hamburg, Germany